Carbohydrates:

The Essential Molecules of Life

The front cover shows a representation of the solution structure of a heparin fragment, determined by NMR spectroscopy (Protein Data Bank code: 1hpn).

The investigations on sugars are proceeding very gradually. It will perhaps interest you that mannose is the geometrical isomer of grape sugar. Unfortunately, the experimental difficulties in this group are so great, that a single experiment takes more time in weeks than other classes of compounds take in hours, so only very rarely a student is found who can be used for this work. Thus, nowadays, I often face difficulties in trying to find themes for the doctoral theses.

Emil Fischer

Carbohydrates:
The Essential Molecules of Life

Second Edition

Robert V. Stick

School of Biomedical, Biomolecular and Chemical Sciences
The University of Western Australia
35 Stirling Hwy
Crawley
Western Australia 6009
Australia

Spencer J. Williams

School of Chemistry and Bio21 Molecular Science and
Biotechnology Institute
University of Melbourne
30 Flemington Rd
Parkville
Victoria 3010
Australia

ELSEVIER

Amsterdam • Boston • Heidelberg • London • New York • Oxford
Paris • San Diego • San Francisco • Singapore • Sydney • Tokyo

Elsevier
Linacre House, Jordan Hill, Oxford OX2 8DP, UK
Radarweg 29, PO Box 211, 1000 AE Amsterdam, The Netherlands

First edition 2001
Second edition 2009

Notice
No responsibility is assumed by the publisher for any injury and/or damage to persons
or property as a matter of products liability, negligence or otherwise, or from any use
or operation of any methods, products, instructions or ideas contained in the material
herein. Because of rapid advances in the medical sciences, in particular, independent
verification of diagnoses and drug dosages should be made

British Library Cataloguing in Publication Data
A catalogue record for this book is available from the British Library

Library of Congress Cataloging-in-Publication Data
A catalog record for this book is available from the Library of Congress

ISBN: 978-0-240-52118-3

For information on all Elsevier publications
visit our website at elsevierdirect.com

Printed and bound in the United Kingdom
Transferred to Digital Printing, 2011

For Rob,
unrealized artist

Also, in memory of Bruce Stone and his beloved 1,3-β-glucans and
wattle-bloom arabinogalactan proteins

Contents

Preface and Acknowledgements

The year 2000 marked a watershed in the sciences with the sequencing of the human genome. Along with other sequencing efforts, we now know the blueprint for life in an ever-increasing number of organisms. Not unexpectedly, whole new areas of science have flourished: genomics, ribonomics, proteomics, metabolomics and, not to be left out, glycomics. Glycomics has been defined as 'the functional study of carbohydrates in living organisms' (de Paz, J. L. and Seeberger, P. H. *QSAR Comb. Sci.*, 2006, **25**, 1027).

Glycomics would not have even been considered a century ago because carbohydrates and, in particular the sugars, were viewed simply as essential molecules for the survival of most organisms. For example, sucrose and glucose provided energy, starch stored energy, and cellulose was responsible for structure and strength. Decades of research then provided novel carbohydrate structures where the function was not always obvious. What were these molecules doing in the world of biology, often being present on the surface of bacteria, viruses and cancer cells, the vanguard of these life forms?

Well, these molecules have a function, and it is now recognized that carbohydrate–protein and even carbohydrate–carbohydrate interactions are of fundamental importance in modulating protein structure and localization, signalling in multicellular systems and cell–cell recognition, including bacterial and viral infection processes, inflammation and aspects of cancer. Some of these carbohydrates have high molecular weights and, not surprisingly, complex chemical structures that challenge the chemists, biochemists and biologists. A pertinent example would be that of the N-glycans, complex molecules in which the carbohydrate is linked, through nitrogen, to a peptide chain (thus forming a glycopeptide or glycoprotein); a small change in the structure of the carbohydrate can lead to all sorts of human diseases.

This book will provide all of the background for a successful study of carbohydrates. Also, it will give a taste for the subject of glycobiology, concentrating especially on the structures and the biosynthesis of carbohydrates and glycoconjugates, and to a lesser extent on their function. A question often asked is 'Why study carbohydrate chemistry?'. The answer is simple: 'It is fundamental to the study of biology'. An organic chemist trained in carbohydrates will move smoothly into the worlds of biochemistry, molecular biology and cell biology; the reverse is much more difficult.

We are indebted, in particular, to David Vocadlo, and to Steve Withers, Harry Brumer III, Adrian Scaffidi, Andrew Watts, Keith Stubbs, Ethan Goddard-Borger, Tanja Wrodnigg, Arnold Stütz and Malcolm McConville for insightful comments into the structure and content of this new book. Also, Keith Stubbs, Adrian Scaffidi, Ethan Goddard-Borger and Nathan McGill spent tireless hours in the proofreading of the manuscript and made many useful suggestions. Frieder Lichtenthaler is again thanked for the photographs of Fischer. RVS acknowledges the hospitality of the Institut für Organische Chemie, Technische Universität Graz and the Institut für Chemie, Karl-Franzens Universität Graz in the writing of part of the manuscript. SJW thanks his wife Jilliarne for her patience and support through the writing of this book.

Robert Stick and Spencer Williams

Abbreviations

Ac	acetyl
AIBN	2,2′-azobis(isobutyronitrile)
All	allyl (prop-2-enyl)
AMP/ADP/ATP	adenosine 5′-mono/di/triphosphate
Ar	aryl
ATIII	antithrombin III
BMS	*tert*-butyldimethylsilyl
Bn	benzyl (phenylmethyl)
Boc	*tert*-butoxycarbonyl
BPS	*tert*-butyldiphenylsilyl
Bz	benzoyl
CAN	cerium(IV) ammonium nitrate
Cbz	benzyloxycarbonyl
C_6H_{11}	cyclohexyl
ClAc	chloroacetyl
CMP/CDP/CTP	cytidine 5′-mono/di/triphosphate
CoA	coenzyme A
CSA	camphor-10-sulfonic acid
DABCO	1,4-diazabicyclo[2.2.2]octane
DAST	(diethylamino)sulfur trifluoride
DBU	1,8-diazabicyclo[5.4.0]undec-7-ene
DCC	*N,N′*-dicyclohexylcarbodiimide
DCE	1,2-dichloroethane
DDQ	2,3-dichloro-5,6-dicyanobenzoquinone
DEAD	diethyl azodicarboxylate
DIAD	diisopropyl azodicarboxylate
DMAP	4-(dimethylamino)pyridine
DMDO	dimethyldioxirane
DME	1,2-dimethoxyethane
DMF	dimethylformamide
DMSO	dimethyl sulfoxide
DMTST	dimethyl(methylthio)sulfonium triflate

DNP	2,4-dinitrophenyl
DTBMP	2,6-di-*tert*-butyl-4-methylpyridine
DTBP	2,6-di-*tert*-butylpyridine
DTPM	(dimethyltrioxopyrimidinylidene)methyl
DTT	1,4-dithiothreitol
ER	endoplasmic reticulum
ERAD	endoplasmic reticulum–associated degradation
FADH	flavin adenine dinucleotide
Fmoc	9-fluorenylmethoxycarbonyl
GAG	glycosaminoglycan
GH	glycoside hydrolase
GMP/GDP/GTP	guanosine 5′-mono/di/triphosphate
GPI	glycosylphosphatidylinositol
GT	glycosyltransferase
HIT	heparin-induced thrombocytopenia
HIV	human immunovirus
HMPA	hexamethylphosphoramide
IDC	iodonium dicollidine
Im	1-imidazolyl
IPTG	isopropyl 1-thio-β-D-galactopyranoside
KLH	keyhole limpet hemocyanin
LDA	lithium diisopropylamide
Lev	levulinyl (4-oxopentanoyl)
LPG	lipophosphoglycan
LPS	lipopolysaccharide
*m*CPBA	3(*meta*)-chloroperbenzoic acid
Ms	mesyl (methanesulfonyl)
NADH	nicotinamide adenine dinucleotide
NADPH	nicotinamide adenine dinucleotide phosphate
NBS	*N*-bromosuccinimide
NIS	*N*-iodosuccinimide
NMO	*N*-methylmorpholine *N*-oxide
Ns	4-nitrobenzenesulfonyl
PAPS	3′-phosphoadenosine-5′-phosphosulfate
PCC	pyridinium chlorochromate
PDC	pyridinium dichromate
PEG	poly(ethylene glycol)
PEP	phosphoenolpyruvate
Ph	phenyl
Phth	phthalyl
PI	phosphatidylinositol
Piv	pivalyl (2,2-dimethylpropanoyl)

PLP	pyridoxal-5′-phosphate
*p*MB	4(*para*)-methoxybenzyl
*p*NP	4(*para*)-nitrophenyl
*p*TSA	4(*para*)-toluenesulfonic acid
py	pyridine
rt	room temperature
SF	selectfluor {1-chloromethyl-4-fluoro-1,4-diazoniabicyclo [2.2.2]octane bis(tetrafluoroborate)}
TBP	2,4,6-tri-*tert*-butylpyridine
TCP	tetrachlorophthalyl
TDS	thexyldimethylsilyl
TEMPO	2,2,6,6-tetramethylpiperidine-1-oxyl
Tf	triflyl (trifluoromethanesulfonyl)
THF	tetrahydrofuran
THP	tetrahydropyran-2-yl
TIPS	triisopropylsilyl
TMP	2,2,6,6-tetramethylpiperidide
Tol	tolyl (4-methylphenyl)
TPAP	tetrapropylammonium perruthenate
Tr	trityl (triphenylmethyl)
Ts	tosyl (4-toluenesulfonyl)
TTBP	2,4,6-tri-*tert*-butylpyrimidine
UMP/UDP/UTP	uridine 5′-mono/di/triphosphate

Chapter 1

The 'Nuts and Bolts' of Carbohydrates

The Early Years

A Bunsen (1811–1899) burner, a Claisen (1851–1930) flask, a Liebig (1803–1873) condenser, an Erlenmeyer (1825–1909) flask, a Büchner (1860–1917) funnel and flask, all common tools for the practising chemist and also a reflection of the origins of much of the chemistry of the nineteenth century – Europe and, in particular, Germany. Although the name of Emil Fischer[a] never graced a piece of apparatus, it became deeply embedded in the same period, so much so that Fischer is considered by many to be the pioneer of organic chemistry and biochemistry and, undoubtedly, the father of carbohydrate chemistry.[1,2]

What exactly is a carbohydrate? As the name implies, an empirical formula $C \cdot H_2O$ (or CH_2O) was often encountered, with molecular formulae of $C_5H_{10}O_5$ and $C_6H_{12}O_6$ being the most common. The appreciable solubility of these molecules in water was commensurate with the presence of hydroxyl groups, and there was often evidence for the carbonyl group of an aldehyde or ketone. These polyhydroxylated aldehydes and ketones were termed *aldose*s and *ketoses*, respectively, with the more common members referred to as *aldopentoses/aldohexoses* and *ketopentoses/ketohexoses*. Very early on, it became apparent that larger molecules existed that could be converted, by hydrolysis, into smaller and more common units – *monosaccharides* from *polysaccharides*. Nowadays, the definition of what is a carbohydrate has been much expanded to include oxidized or reduced molecules and those that contain other types of atoms (often nitrogen). The term 'sugar' is used to describe monosaccharides and the somewhat higher molecular weight di- and trisaccharides.

To try to appreciate the genius and elegance of Fischer's work with sugars, let us consider the conditions and resources available in a typical German laboratory of the

---

[a] Emil Hermann Fischer (1852–1919), Ph.D. (1874) under von Baeyer at the University of Strassburg, professorships at Munich, Erlangen (1882), Würzburg (1885) and Berlin (1892). Nobel Prize in Chemistry (1902).

References start on page 32

Figure 1 Photograph of the Baeyer group in 1878 at the laboratory of the University of Munich (room for combustion analysis), with inscriptions from Fischer's hand; in the centre is Adolf Baeyer; seated to the right is the 25-year-old Emil Fischer, in a peaked cap and strikingly self-confident 3 years after his doctorate; standing to the left of Baeyer is Wilhelm Koenigs.[1] This, and the photograph on page 16, are reproduced with permission from the 'Collection of Emil Fischer Papers' (Bancroft Library, University of California, Berkeley) and the kind assistance of Professor Dr. Frieder W. Lichtenthaler (Darmstadt, Germany).

time. The photograph (Figure 1) of von Baeyer's[b] research group in Munich speaks volumes.

Fischer is surrounded by formally attired, austere men, some wearing hats (for warmth?) and many sporting a beard or a moustache. The large hood in the background carries an assortment of apparatus, presumably for the purpose of microanalysis.

Microanalysis, performed meticulously by hand, was the cornerstone of Fischer's work on sugars. Melting point and optical rotation were essential adjuncts in the determination of chemical structure and equivalence. All of these required *pure* chemical compounds, necessitating crystallinity at every possible opportunity as sugar 'syrups' often decomposed on distillation, and the concept of chromatography was

[b] Johann Friedrich Wilhelm Adolf von Baeyer (1835–1917), Ph.D. under Kekulé and Hofmann at the Universities of Heidelberg and Berlin, respectively, professorships at Strassburg and Munich. Nobel Prize in Chemistry (1905).

barely embryonic in the brains of Day[c] and Tswett.[d] Fortunately, many of the naturally occurring sugars were found to be crystalline; however, upon chemical modification, their products often were not crystalline. These observations, coupled with the need to investigate the chemical structure of sugars, encouraged Fischer and others to invoke some of the simple reactions of organic chemistry, and to invent new ones.

Oxidation was an operationally simple task for the early German chemists. The aldoses, apart from showing the normal attributes of a *reducing sugar* (forming a beautiful silver mirror when treated with Tollens'[e] reagent or causing the precipitation of brick-red cuprous oxide when subjected to Fehling's[f] solution), were easily oxidized by bromine water to carboxylic acids, termed *aldonic acids*:

$$
\text{CHO} \quad \xrightarrow[\text{H}_2\text{O}]{\text{Br}_2} \quad \text{COOH}
$$

Moreover, heating the newly formed aldonic acid often formed cyclic esters, or lactones:

$$
\begin{array}{c}\text{COOH}\\ | \\ \\ |\\ \text{—OH}\end{array} \quad \xrightarrow{-\text{H}_2\text{O}} \quad \begin{array}{c}\text{CO—}\\ | \quad\;\; \text{O}\\ |\end{array}
$$

Ketoses, not surprisingly, were not oxidized by bromine water and could thus be simply distinguished from aldoses.

Dilute nitric acid was also used for the oxidation of aldoses, this time to dicarboxylic acids, termed *aldaric acids*:

$$
\begin{array}{c}\text{CHO}\\ | \\ | \\ \text{CH}_2\text{OH}\end{array} \quad \xrightarrow{\text{dil. HNO}_3} \quad \begin{array}{c}\text{COOH}\\ | \\ | \\ \text{COOH}\end{array}
$$

[c] David Talbot Day (1859–1925), Ph.D. at the Johns Hopkins University, Baltimore (1884), chemist, geologist and mining engineer.

[d] Mikhail Semenovich Tswett (1872–1919), D.Sc. at the University of Geneva, Switzerland (1896), chemist and botanist.

[e] Bernhard C.G. Tollens (1841–1918), professor at the University of Göttingen.

[f] Hermann von Fehling (1812–1885), professor at the University of Stuttgart.

References start on page 32

Lactone formation from these diacids was still observed, with the formation of more than one lactone not being uncommon:

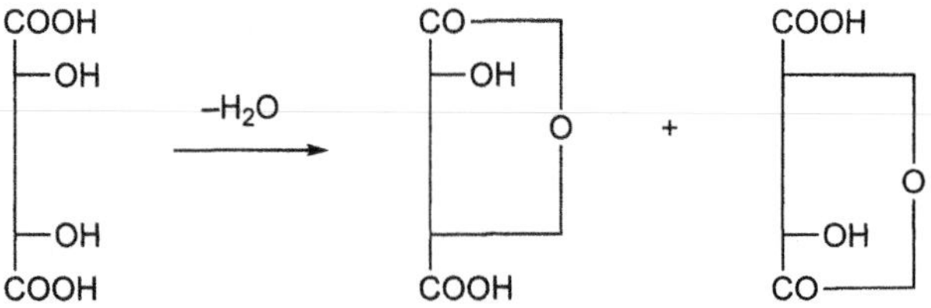

Reduction of sugars was most conveniently performed with sodium amalgam (NaHg) in ethanol. Aldoses yielded one unique *alditol* whereas ketoses, for reasons that may already be apparent, gave a mixture of two alditols:

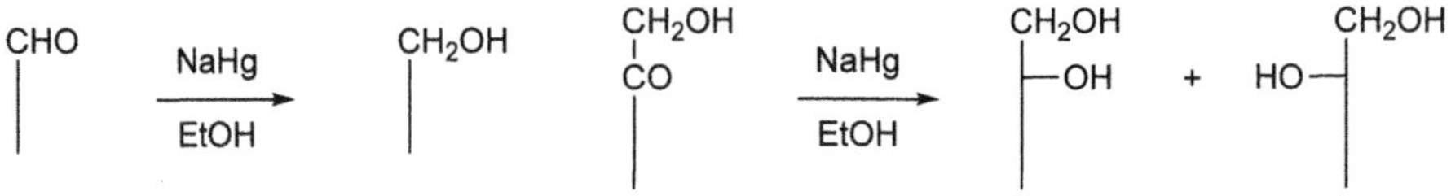

Fischer, with interests in chemicals other than carbohydrates, treated a solution of benzenediazonium ion (the cornerstone of the German dye-stuffs industry) with potassium hydrogen sulfite and, in doing so, discovered phenylhydrazine by chance:

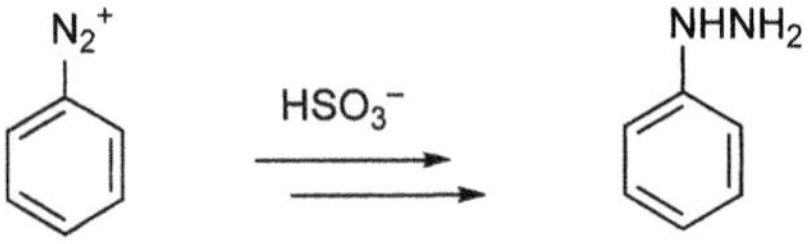

Fischer soon found that phenylhydrazine was useful for the characterization of the somewhat unreliable sugar acids by converting them into their very crystalline phenylhydrazinium salts:

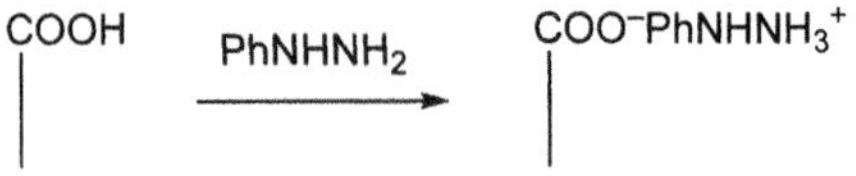

Phenylhydrazine also transformed aldehydes and ketones into phenylhydrazones and, not remarkably, similar transformations were possible with aldoses and ketoses:

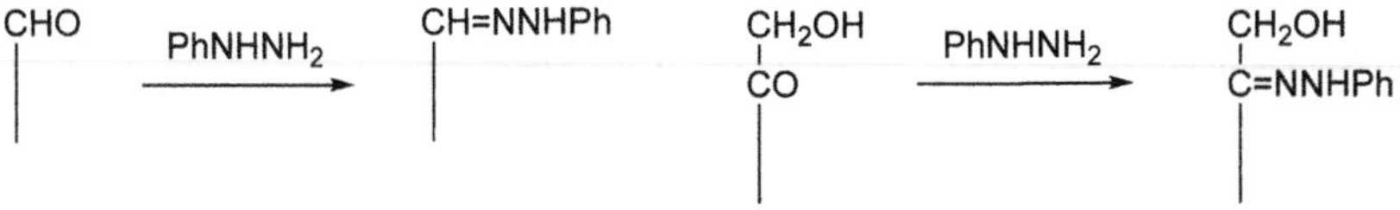

The remarkable aspect of this work was that both aldoses and ketoses, when treated more vigorously with an *excess* of phenylhydrazine, were converted into unique derivatives, *phenylosazones*:

$$
\begin{array}{c}
\text{CHO} \\
| \\
\text{CHOH} \\
|
\end{array}
\quad \text{or} \quad
\begin{array}{c}
\text{CH}_2\text{OH} \\
| \\
\text{CO} \\
|
\end{array}
\quad \xrightarrow{\text{PhNHNH}_2} \quad
\begin{array}{c}
\text{CH=NNHPh} \\
| \\
\text{C=NNHPh} \\
|
\end{array}
$$

The different phenylosazones had distinctive crystalline forms and, also, were formed at different rates from the various parent sugars.

Another carbohydrate chemist of the time, Kiliani,[g] amply acknowledged by Fischer but generally underrated by his peers, had applied some well-known chemistry to aldoses and ketoses, namely the addition of hydrogen cyanide. The products, after acid hydrolysis, were aldonic acids. Fischer took the lactones derived from these acids and showed that they could be reduced to aldoses, containing an extra carbon atom:

$$
\begin{array}{c}
\text{CHO} \\
| \\
| \\
|\!-\!\text{OH} \\
|
\end{array}
\xrightarrow[\text{pH 9}]{\text{NaCN, H}_2\text{O}}
\begin{array}{c}
\text{CN} \\
| \\
\text{CHOH} \\
| \\
|\!-\!\text{OH} \\
|
\end{array}
\xrightarrow{\text{H}_3\text{O}^+}
\begin{array}{c}
\text{COOH} \\
| \\
\text{CHOH} \\
| \\
|\!-\!\text{OH} \\
|
\end{array}
\xrightarrow{-\text{H}_2\text{O}}
\begin{array}{c}
\text{CO}\!-\! \\
| \quad\ | \\
\text{CHOH} \quad \text{O} \\
| \\
|\!-\! \\
|
\end{array}
\xrightarrow[\text{EtOH}]{\text{NaHg}}
\begin{array}{c}
\text{CHO} \\
| \\
\text{CHOH} \\
| \\
|\!-\!\text{OH} \\
|
\end{array}
$$

$$
\begin{array}{c}
\text{CH}_2\text{OH} \\
| \\
\text{CO} \\
| \\
|\!-\!\text{OH} \\
|
\end{array}
\xrightarrow[\text{pH 9}]{\text{NaCN, H}_2\text{O}}
\begin{array}{c}
\text{CH}_2\text{OH} \\
| \\
\text{HOCCN} \\
| \\
|\!-\!\text{OH} \\
|
\end{array}
\xrightarrow{\text{H}_3\text{O}^+}
\begin{array}{c}
\text{CH}_2\text{OH} \\
| \\
\text{C(OH)COOH} \\
| \\
|\!-\!\text{OH} \\
|
\end{array}
\xrightarrow{-\text{H}_2\text{O}}
\begin{array}{c}
\text{CH}_2\text{OH} \\
| \\
\text{HOCCO}\!-\! \\
| \qquad | \\
| \qquad \text{O} \\
|\!-\! \\
|
\end{array}
$$

$$
\xrightarrow[\text{EtOH}]{\text{NaHg}}
\begin{array}{c}
\text{CH}_2\text{OH} \\
| \\
\text{C(OH)CHO} \\
| \\
|\!-\!\text{OH} \\
|
\end{array}
\equiv
\begin{array}{c}
\text{CHO} \\
| \\
\text{C(OH)CH}_2\text{OH} \\
| \\
|\!-\!\text{OH} \\
|
\end{array}
$$

---

[g] Heinrich Kiliani (1855–1945), Ph.D. under Erlenmeyer and von Baeyer, professor at the University of Freiburg.

References start on page 32

Not so obviously, this synthesis converts an aldose or a ketose into *two* new aldoses (an early example of a stereoselective synthesis). Fischer used and developed this *ascent* (adding one carbon) of the homologous aldose series so well that it is known as the Kiliani–Fischer synthesis.

It was logical that if one could ascend the aldose series, then one should also be able to descend it, and so were developed various methods for this *descent*. Perhaps, the most well known is that devised by Ruff;[h] the aldose is first oxidized to the aldonic acid, and subsequent treatment of the calcium salt of the acid with hydrogen peroxide gives the aldose:

CHO
|
CHOH $\xrightarrow[\text{then Ca(OH)}_2]{\text{Br}_2,\ \text{H}_2\text{O}}$ $(\text{COO}^-)_2\text{Ca}^{2+}$
| |
                                                                           CHOH $\xrightarrow[\text{Fe}^{3+}]{\text{H}_2\text{O}_2}$ CHO   + CO_2
| |

It is an interesting complement to the ascent of a series that the (Ruff) descent converts two aldoses into a single new aldose.

The final transformation that was available to Fischer, albeit somewhat late in the piece, was of an informative, rather than a preparative, nature. *Lobry de Bruyn* and *Alberda van Ekenstein*[3,4] announced the *rearrangement* of aldoses and ketoses upon treatment with dilute alkali:

CHO   CHO$^-$ CHO
| $\xrightarrow{\text{HO}^-}$ ‖ $\rightleftharpoons$ |
H–C–OH   C–OH   HO–C–H
| | |

                        ⇅

                     CH_2OH
                       |
                       CO
                       |

This simple, enolate-driven sequence allowed the isomerization of one aldose into its C2 epimer, together with the formation of the structurally related ketose. It also explained the observation that ketoses, although not oxidizable by bromine

---

[h] Otto Ruff (1871–1939), professorships at Danzig and Breslau.

water (at a pH below 7), gave positive Tollens' and Fehling's tests (conducted with each reagent under alkaline conditions).

Fischer now had the necessary chemical tools (and intellect!) to launch an assault on the structure determination of sugars.

The Constitution of Glucose and Other Sugars

(+)-Glucose from a variety of sources (fruits and honey), (+)-galactose from the hydrolysis of 'milk sugar' (lactose), (−)-fructose from honey, (+)-mannitol from various plants and algae, and (+)-xylose and (+)-arabinose from the acid treatment of wood and beet pulp were the sugars available to Fischer when he started his seminal structural studies in 1884 in Munich.

What were the established facts about (+)-glucose at that time? (+)-Glucose was a reducing sugar that could be oxidized to gluconic acid with bromine water and to glucaric acid with dilute nitric acid. That the six carbon atoms were in a contiguous chain had been shown by Kiliani: the conversion of (+)-glucose into a mixture of heptonic acids (by conventional Kiliani extension), followed by the treatment of this mixture with red phosphorus and hydrogen iodide (strongly reducing conditions), gave heptanoic acid:

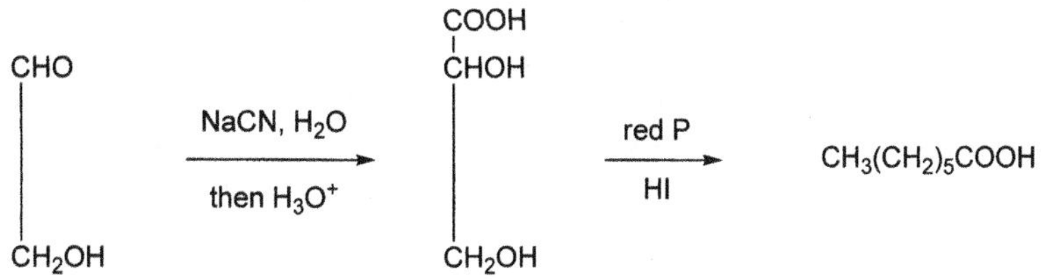
CHO
CH2OH
NaCN, H2O
then H3O+
COOH
CHOH
CH2OH
red P
HI
CH3(CH2)5COOH

Thus, the structure of (+)-glucose was established as a straight-chain, polyhydroxylated aldehyde:[i]

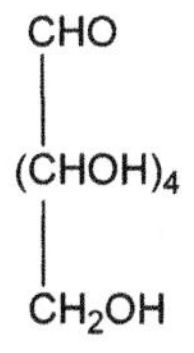
CHO
(CHOH)4
CH2OH

[i] A similar sequence on (−)-fructose produced 2-methylhexanoic acid, establishing the fact that fructose was a 2-keto sugar:

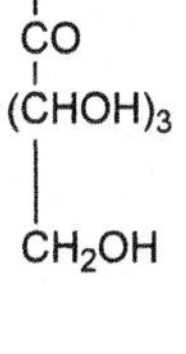
CH2OH
CO
(CHOH)3
CH2OH

References start on page 32

The theories of Le Bel and van't Hoff, around 1874, decreed that a carbon atom substituted by four different groups (as we commonly have for sugars) should be tetrahedral in shape and be able to exist as two separate forms, non-superimposable mirror images and thus isomers. These revolutionary ideas were seized upon and endorsed by Fischer and formed the cornerstone for his arguments on the structure of (+)-glucose.

Let us digress to consider the simplest aldose, the aldotriose, glyceraldehyde (formaldehyde and glycolaldehyde, while formally sugars, are not regarded as such):

$$
\begin{array}{l}
\text{CHO} \\
| \\
\text{CHOH} \\
| \\
\text{CH}_2\text{OH}
\end{array}
$$

The two isomers, in fact enantiomers, may be represented using Fischer projection formulae:[j]

$$
\begin{array}{c}
\text{CHO} \\
\text{H}\!-\!\!\!-\!\!\!-\!\text{OH} \\
\text{CH}_2\text{OH}
\end{array}
\qquad
\begin{array}{c}
\text{CHO} \\
\text{HO}\!-\!\!\!-\!\!\!-\!\text{H} \\
\text{CH}_2\text{OH}
\end{array}
$$

Rosanoff, an American chemist of the time, decreed, quite arbitrarily, that (+)-glyceraldehyde would be represented by the first of the two enantiomers, and its unique *absolute configuration* was described a little later by the use of the small capital letter, D:[5,6,k]

$$
\begin{array}{c}
\text{CHO} \\
\text{H}\!-\!\!\!-\!\!\!-\!\text{OH} \\
\text{CH}_2\text{OH}
\end{array}
\qquad
\begin{array}{c}
\text{CHO} \\
\text{HO}\!-\!\!\!-\!\!\!-\!\text{H} \\
\text{CH}_2\text{OH}
\end{array}
$$

D-(+)-glyceraldehyde   L-(−)-glyceraldehyde

Fischer, in an effort to thread together the jumble of experimental results on sugars, had earlier decided that (+)-glucose would be drawn with the hydroxyl group to the right at its bottommost (highest numbered) 'substituted' carbon, thus sharing the same configuration as (+)-glyceraldehyde:

---

[j] Such formulae were first announced by Fischer in 1891 and, besides simplifying the depiction of the sugars, were universally accepted. Being planar projections, the actual stereochemical information is available only if you know the 'rules' – horizontal lines represent bonds above the plane, vertical lines represent bonds below the plane. Only one 'operation' is hence allowed with Fischer projection formulae – a rotation of 180° in the plane.

[k] Accepted practice is to depict D in font that is (two points) smaller than the regular text.

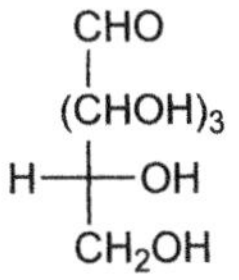

CHO
|
(CHOH)$_3$
H——OH
CH$_2$OH

D-(+)-glucose

The challenge that remained was to elucidate the *relative configuration* of the other three centres (eight possibilities)!

What follows is an account of Fischer's elucidation of the structure of (+)-glucose, interspersed with anecdotal information gleaned from a wonderful article by Professor Frieder Lichtenthaler (Darmstadt, Germany)[7] to celebrate the centenary of the announcement of the structure of (+)-glucose in 1891.[8,9] It is a remarkable fact that these two publications contain no new experimental details – all of the necessary information was already present in the chemical literature! To begin, a passage from a letter by Fischer to von Baeyer:

> *The investigations on sugars are proceeding very gradually. It will perhaps interest you that mannose is the geometrical isomer of grape sugar. Unfortunately, the experimental difficulties in this group are so great, that a single experiment takes more time in weeks than other classes of compounds take in hours, so only very rarely a student is found who can be used for this work. Thus, nowadays, I often face difficulties in trying to find themes for the doctoral theses.*

On top of this 'soul searching' by Fischer, consider the following experimental results:

D-xylose $\xrightarrow[\text{EtOH}]{\text{NaHg}}$ xylitol $[\alpha]_D$ 0°

L-arabinose $\xrightarrow[\text{EtOH}]{\text{NaHg}}$ arabinitol $[\alpha]_D$ 0°

Both alditols would appear to be achiral (meso) compounds, but what about the following experimental result?

D-xylose $\xrightarrow{\text{HNO}_3}$ xylaric acid $[\alpha]_D$ 0°

L-arabinose $\xrightarrow{\text{HNO}_3}$ arabinaric acid $[\alpha]_D$ –22.7°

References start on page 32

In the two sets of experiments, the termini of the chains were identical (both 'CH$_2$OH' or both 'COOH'). Xylitol and xylaric acid are most likely meso compounds, but arabinaric acid is not! This meant that arabinitol *had* to be chiral; only in the presence of borax (which forms 'complexes' with polyols) was Fischer able to obtain a very small, negative rotation for arabinitol.

Bearing in mind these experimental difficulties, let us return to the proof of the structure of (+)-glucose:

1. Because Fischer had arbitrarily placed the hydroxyl group at C5 on the right for (+)-glucose, all interrelated sugars must have the same (D) absolute configuration.
2. Arabinose, on Kiliani–Fischer ascent, gave a mixture of glucose and mannose.[1]

CHO   CHO   CHO
CHOH   H—OH   HO—H
CHOH   CHOH   CHOH
H—OH   H—⁵OH   H—OH
CH₂OH   CH₂OH   CH₂OH

D-arabinose   D-glucose/D-mannose

3. Arabinaric acid was *not* a meso compound and, therefore, the hydroxyl group at C2 of D-arabinose *must* be to the left.

CHO   CHO   CHO
HO—²H   H—OH   HO—H
CHOH   HO—H   HO—H
H—OH   H—OH   H—OH
CH₂OH   CH₂OH   CH₂OH

D-arabinose   D-glucose/D-mannose

---

[1] Mannose was first prepared (1887) in very low yield by the *careful* (HNO$_3$) oxidation of mannitol and later obtained from the acid hydrolysis of 'mannan' (a polysaccharide) present in tagua palm seeds (ivory nut). That glucose and mannose were epimers at C2 was shown by the following transformations:

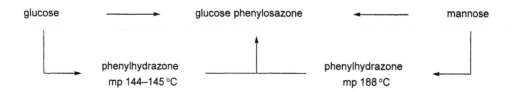

glucose ——→ glucose phenylosazone ←—— mannose

phenylhydrazone   phenylhydrazone
mp 144–145 °C   mp 188 °C

4. *Both* glucaric and mannaric acids are optically active; this places the hydroxyl group at C4 of the two hexoses on the right.[m]

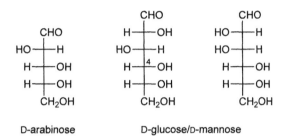

D-arabinose   D-glucose/D-mannose

5. D-Glucaric acid comes from the oxidation of D-glucose, but L-glucaric acid can be obtained from L-glucose *or* D-gulose.[n] This is only possible if D-gulose is related to L-glucose by a 'head to tail' swap:

$$
\begin{array}{cccc}
\text{CHO} & \text{CHO} & \text{CH}_2\text{OH} & \text{CHO} \\
\text{H---OH} & \text{HO---H} & \text{HO---H} & \text{H---OH} \\
\text{HO---H} & \text{H---OH} & \text{H---OH} & \text{H---OH} \\
\text{H---OH} & \text{HO---H} & \text{HO---H} \;\equiv\; & \text{HO---H} \\
\text{H---OH} & \text{HO---H} & \text{HO---H} & \text{H---OH} \\
\text{CH}_2\text{OH} & \text{CH}_2\text{OH} & \text{CHO} & \text{CH}_2\text{OH}
\end{array}
$$

D-glucose   L-glucose   'head to tail' swap   D-gulose

---

[m] The relative configuration of D-arabinose is now established.

[n] L-Glucose, together with L-mannose, had been prepared earlier by Kiliani–Fischer extension of (+)-arabinose (actually L-arabinose) from sugar beet:

$$
\begin{array}{ccc}
 & \text{CHO} & \text{CHO} \\
\text{CHO} & \text{HO---H} & \text{H---OH} \\
\text{H---OH} & \text{H---OH} & \text{H---OH} \\
\text{HO---H} \rightarrow & \text{HO---H} \quad + & \text{HO---H} \\
\text{HO---H} & \text{HO---H} & \text{HO---H} \\
\text{CH}_2\text{OH} & \text{CH}_2\text{OH} & \text{CH}_2\text{OH}
\end{array}
$$

D-Gulose, together with D-idose, arose when (−)-xylose (actually D-xylose) from cherry gum was subjected to a Kiliani–Fischer synthesis:

$$
\begin{array}{ccc}
 & \text{CHO} & \text{CHO} \\
\text{CHO} & \text{H---OH} & \text{HO---H} \\
\text{H---OH} & \text{H---OH} & \text{H---OH} \\
\text{HO---H} \rightarrow & \text{HO---H} \quad + & \text{HO---H} \\
\text{H---OH} & \text{H---OH} & \text{H---OH} \\
\text{CH}_2\text{OH} & \text{CH}_2\text{OH} & \text{CH}_2\text{OH}
\end{array}
$$

References start on page 32

This wonderful piece of analysis thus provided unequivocal structures for three (of the possible eight) D-aldohexoses and one 2-keto-D-hexose:[o]

| D-glucose | D-mannose | D-gulose | D-fructose |

After the elucidation of the structure of D-arabinose and the four D-hexoses above, similar chemical transformations and logic were employed to unravel the structure of D-galactose; Kiliani, in 1888, had secured the structure of D-xylose.

The six aldoses and one ketose are members of the sugar 'family trees', with glyceraldehyde at the base for aldoses and dihydroxyacetone for 2-ketoses (Figures 2 and 3).

There are various interesting aspects of these family trees:

- The trees are constructed systematically, i.e. hydroxyl groups are placed to the 'right' (R) or the 'left' (L) according to the designation in the left-side margin.
- As applied to this system, the various mnemonics enable one to write the structure of any named sugar or, in the reverse, to name any sugar structure.[p]
- As Fischer encountered unnatural sugars through synthesis, additional names had to be found: 'lyxose' is an anagram of 'xylose', and 'gulose' is an abbreviation/ rearrangement of 'glucose'.
- It is well worthwhile to consider the simple name D-glucose; it describes a unique molecule with four stereogenic centres and *must* be superior to the systematic name of (2R,3S,4R,5R)-2,3,4,5,6-pentahydroxyhexanal![q]

---

[o] Glucose and fructose (and for that matter mannose) gave the same phenylosazone and were inter-related products of the Lobry de Bruyn–Alberda van Ekenstein rearrangement.

[p] Figure 2:   the tetroses – 'ET' (the film!)

the pentoses – 'raxl' is perhaps less flowery!

the hexoses – designed by Louis and Mary Fieser (Harvard University)

Figure 3:   dihydroxyacetone – an achiral molecule

the term 'ulose' is formal nomenclature for a ketose

[q] The only other bastion of the D/L system is that of amino acids; for details of the direct chemical correlation of sugars and amino acids, see the elegant work of Wolfrom, Lemieux and Olin (*J. Am. Chem. Soc.*, 1948, **71**, 2870).

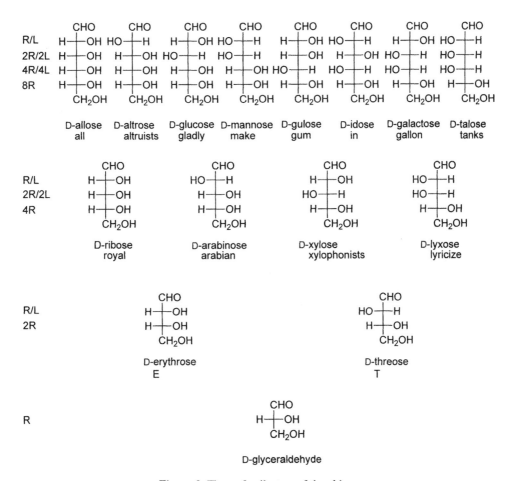

Figure 2 The D- family tree of the aldoses

It was not until 1951 that the D absolute configuration for (+)-glucose, arbitrarily chosen by Fischer some 75 years earlier, was proven to be correct. By a series of chain degradations, (+)-glucose was converted into (−)-arabinose and then (−)-erythrose. Chain extension of (+)-glyceraldehyde also gave (−)-erythrose, together with (−)-threose. Oxidation of (−)-threose gave (−)-tartaric acid, the enantiomer of (+)-tartaric acid.

(+)-Tartaric acid had been converted independently into a beautifully crystalline rubidium/sodium salt; an X-ray structure determination of this salt showed that it has the following absolute configuration:[10]

COORb
H——OH
HO——H
COONa

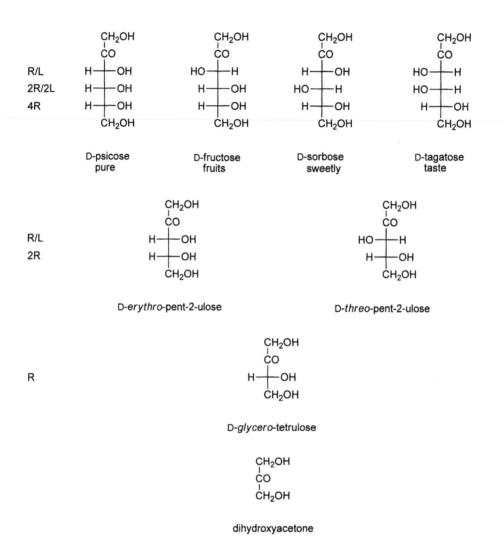

Figure 3 The D- family tree of the 2-ketoses

This defined the structures of (+)-tartaric acid, (−)-tartaric acid and (−)-threose as and allowed the assignment of absolute configuration to (−)-erythrose and to (+)-glyceraldehyde:

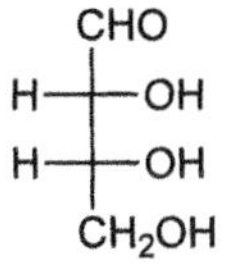

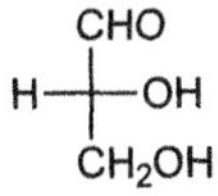

CHO
H——OH
H——OH
CH₂OH

CHO
H——OH
CH₂OH

D-(−)-erythrose   D-(+)-glyceraldehyde

Rosanoff and Fischer had been proven correct.

A photograph of Fischer in his later years at the University of Berlin is exceptional in that it shows 'the master' still actively working at the bench, with a face full of interest and determination (Figure 4).

Finally, the constant exposure to chemicals, particularly phenylhydrazine (osazone formation) and mercury (NaHg reductions), caused chronic poisoning and eczema and, coupled with the loss of his wife in 1895 (due to meningitis) and two of his three sons in events associated with World War I, Fischer took away his own life in 1919, shortly after being diagnosed with cancer. The only remaining eldest son, Hermann O. L. Fischer (1888–1960), went on to become an eminent biochemist at the University of California, Berkeley.

The Cyclic Forms of Sugars, and Mutarotation

Although Fischer had solved the structure of D-glucose, one annoying fact still remained – there were actually two known forms! Crystallization of D-glucose from water at room temperature produced material with melting point 146°C and specific rotation +112° (in water), whereas crystallization at just below the boiling point of water produced material with similar melting point (150°C) but vastly different specific rotation (+19°, in water). How could this be possible?

So far, we have represented the structure of D-(+)-glucose as a Fischer projection, which is a useful convention. However, in real life, as either a solid or in solution, D-(+)-glucose has a molecular structure that may take up an infinite number of shapes, or *conformations*. If one makes a molecular model of D-(+)-glucose, a linear, zig-zag conformation seems attractive:

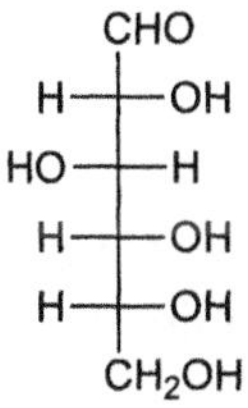

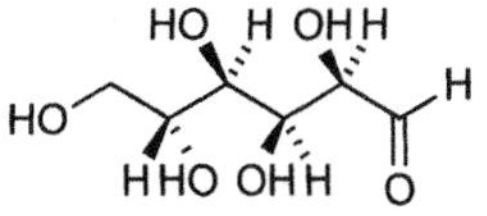

CHO
H——OH
HO——H
H——OH
H——OH
CH₂OH

Figure 4 Emil Fischer around the turn of the century in his 'Privatlaboritorium' at the University of Berlin; the somewhat unusual laboratory stool he inherited from his predecessor, August Wilhelm von Hofmann, who, in 1865, brought it to Berlin on his move from the Royal College of Chemistry, London.[1] This, and the photograph on page 2, are reproduced with permission from the 'Collection of Emil Fischer Papers' (Bancroft Library, University of California, Berkeley) and the kind assistance of Professor Dr. Frieder W. Lichtenthaler (Darmstadt, Germany).

Playing around with this linear conformation, by rotation around the various carbon–carbon bonds, does *nothing* to the configuration of the molecule but leads to an infinite number of other conformations. One of these conformations, on close scrutiny, has the hydroxyl group on C5 adjacent to the aldehyde group (C1). What follows is a chemical reaction, the nucleophilic addition of the C5 hydroxyl group to the aldehyde group, to generate a *hemiacetal*:

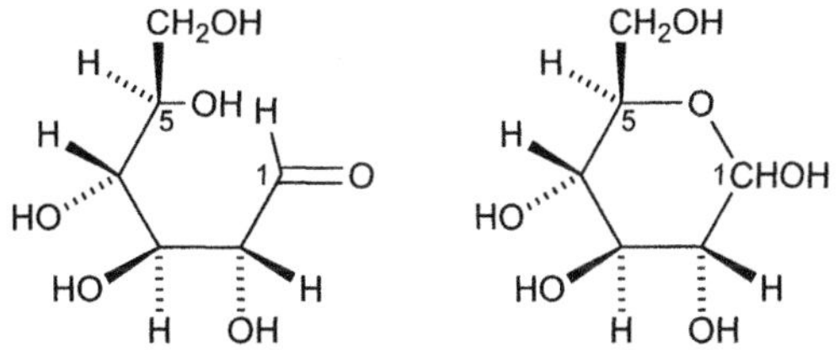

This new chemical structure possesses an extra stereogenic centre (C1), and so the product of the cyclization may exist in two discrete, isomeric forms:[r]

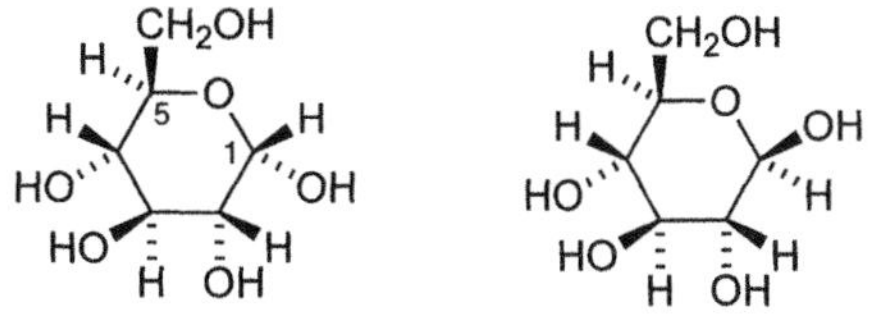

These new cyclic structures for D-glucose explained the existence of two forms of glucose; indeed, such cyclic forms had been suggested by von Baeyer in 1870 and again by Tollens in 1883.[7] Fischer, somewhat surprisingly, never completely accepted these structures. Again, it must be emphasized that the above depictions are each a form of D-glucose; C5 defines the D absolute configuration, and carbons two to four complete the description.

It will be obvious even at this stage that, to the sugar chemists of the 1900s, communicating with structures like the ones discussed above was a tiresome process; a shorthand had to be developed. In 1926, an eminent chemist of the time, W. N. Haworth,[s] made suggestions about the six-membered ring being represented as a hexagon with the front edges emboldened, causing the hexagon to be viewed front-edge-on to the paper:[11]

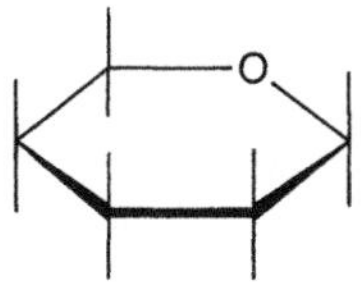

---

[r] Put more formally, the two faces of the aldehyde are diastereotopic (*re* and *si*); addition of the hydroxyl group to the aldehyde thus generates two diastereoisomeric hemiacetals, not necessarily in equal amounts.

[s] Walter Norman Haworth (1883–1950), a student of W.H. Perkin, Ph.D. under Wallach (Göttingen). Nobel Prize in Chemistry (1937).

References start on page 32

The two remaining bonds to each carbon are depicted, one above and one below the plane of the hexagon. Now, the two cyclic forms of D-glucose can be drawn swiftly and accurately:[t]

Some years earlier (in 1913),[13] 'complexation' studies with boric acid had shown that the more highly rotating isomer of D-glucose ($[\alpha]_D$ +112°) possessed a *cis*-relationship between the hydroxyl groups at carbons one and two. Full structural assignments were now possible and, to simplify the matter of communication even further, formal names were given to the two isomers:

mp 146 °C, $[\alpha]_D$ +112° (H_2O)   mp 150 °C, $[\alpha]_D$ +19° (H_2O)

α-D-glucopyranose   β-D-glucopyranose

---

[t]For an excellent discussion on the 'rotational operations' allowed with Haworth formulae, see *Advanced Sugar Chemistry: Principles of Sugar Stereochemistry* by R.S. Shallenberger (AVI Publishing Company Inc., Westport, Connecticut, 1982, p. 110) – 'Haworth structures can be rotated *on* the plane of the paper on which they are drawn if, and only if, the identity of the leading edge of the structure is not lost'. It is also important to recognize that Haworth formulae are indeed that, and not projection formulae.
Another convention, suggested by John A. Mills in 1955, and still in general use,[12] again uses a hexagon but to be viewed in the plane of the paper. Nowadays, hydrogen substituents are not shown but others are, using 'wedge' (above) or 'dash' (below) notation. The style of the 'wedge' and 'dash' bonds is in line with current IUPAC recommendations.

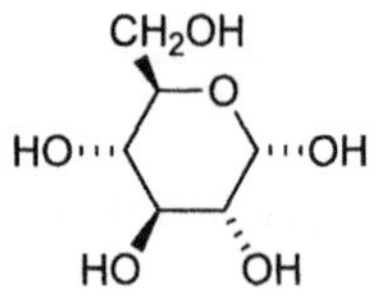

The term '-ose' still indicates a sugar, and 'pyranose' is a sugar having a six-membered cyclic structure.[u] The terms 'α' and 'β' refer to the particular *anomer* (diastereoisomer, epimer) and C1, for aldoses, is the *anomeric carbon*.

Let us return to the question of the shape of the D-(+)-glucose chain and consider what happens when we encounter a different conformation, one where the hydroxyl group on C4 finds itself adjacent to the aldehyde group:

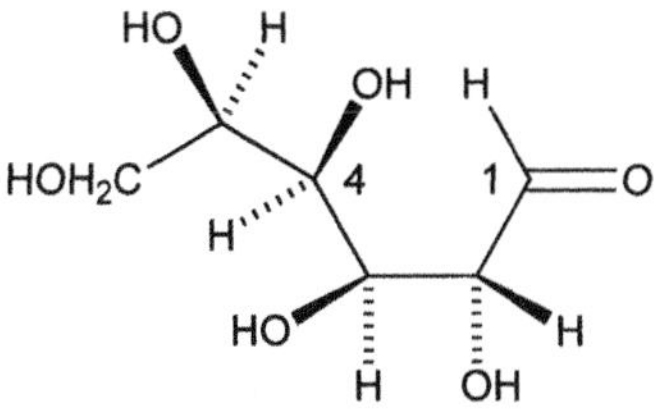

Again, hemiacetal formation is possible, resulting in the formation of two new anomers (drawn according to the Haworth convention):

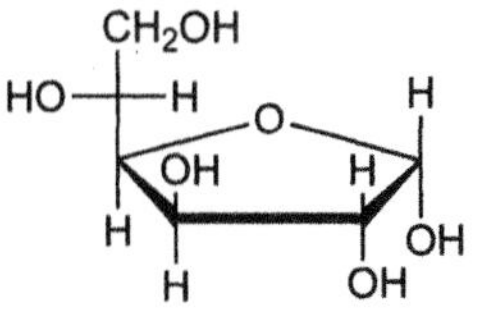

α-D-glucofuranose 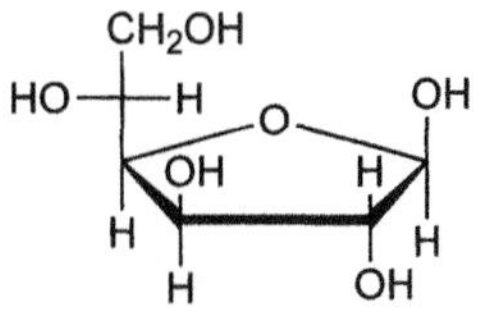   β-D-glucofuranose

There are few data available for these (five-membered ring) 'furanose'[v] forms of D-glucose, simply because they have never been isolated; crystalline D-glucose is either one of the pure pyranose anomers, or a mixture thereof.

---

[u] By analogy with the molecule pyran:

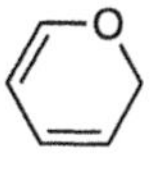

[v] From the molecule furan:

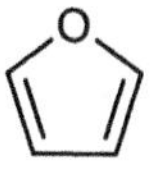

Before proceeding any further, it is worth looking at a few sugars other than D-glucose and considering their cyclic structures:

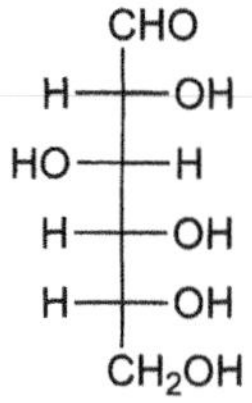

D-glucose

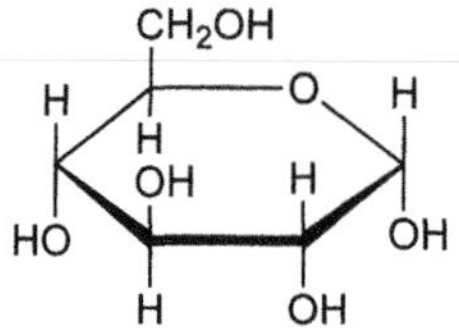

α-D-glucopyranose

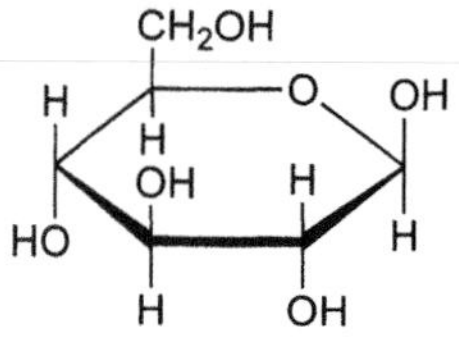

β-D-glucopyranose

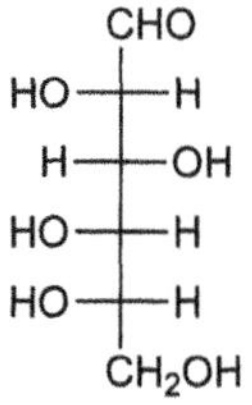

L-glucose

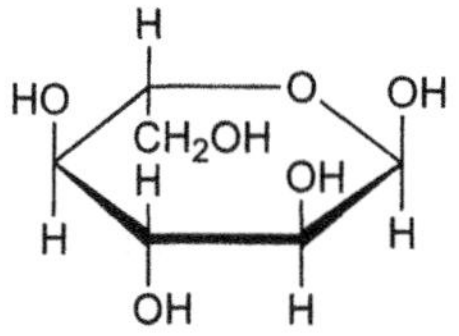

α-L-glucopyranose

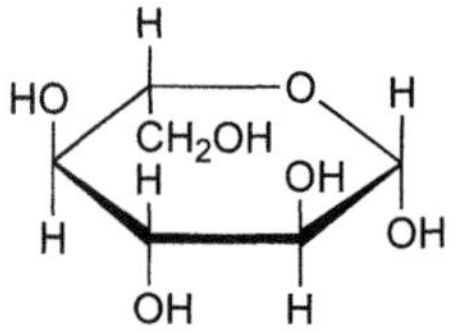

β-L-glucopyranose

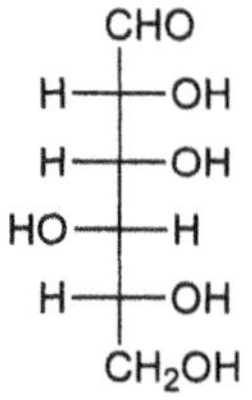

D-gulose

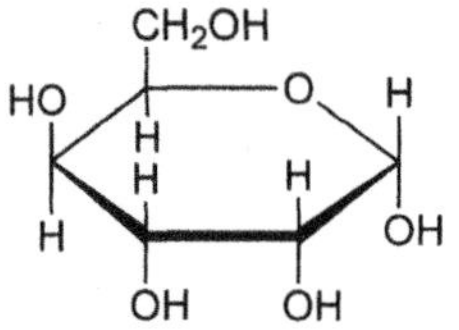

α-D-gulopyranose

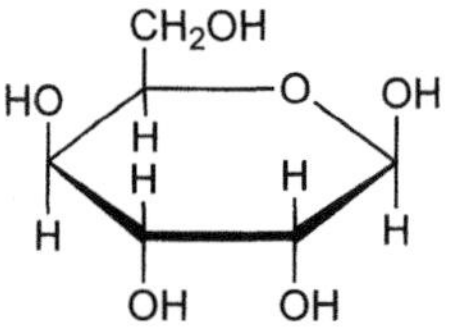

β-D-gulopyranose

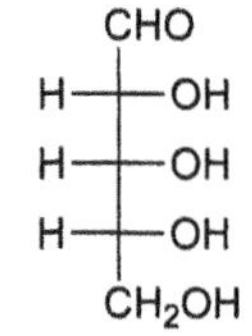

D-ribose

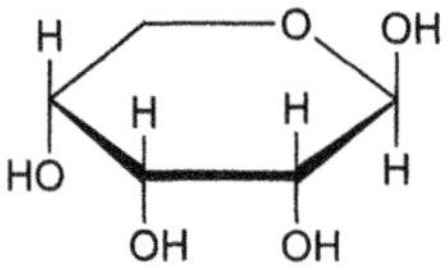

β-D-ribopyranose

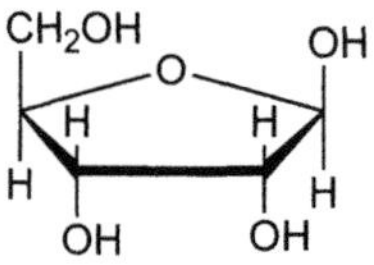

β-D-ribofuranose

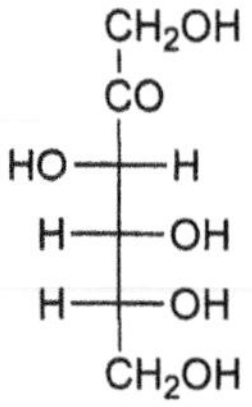

D-fructose

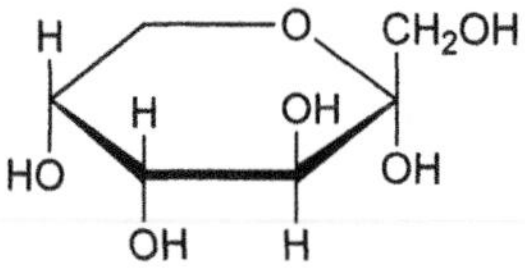

α-D-fructopyranose

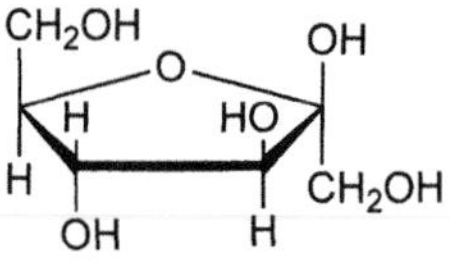

β-D-fructofuranose

Several pertinent points emerge:

- In the Fischer/Haworth 'interconversion', all hydroxyl groups on the 'right' in a Fischer projection are placed 'below' the ring in the Haworth, and all those on the 'left' are placed 'above'.
- In a D-aldohexose, the 'CH$_2$OH' group at C5 is placed 'above' in the Haworth pyranose form; in the L-aldohexose, it is placed 'below'.
- The anomeric descriptions, 'α' and 'β', are obviously related to the absolute configuration; there can be no clearer statement than that enunciated by Collins and Ferrier:[14]

> *For D-glucose and all compounds of the D-series, α-anomers have the hydroxyl group at the anomeric centre projecting downwards in Haworth formulae*; *α-L-compounds have this group projecting upwards. The β-anomers have the opposite configurations at the anomeric centre, i.e., the hydroxyl group projects upwards and downwards for β-D- and β-L-compounds, respectively.*

Thus, the enantiomer of α-D-glucopyranose is α-L-glucopyranose.[w]

- Whereas the pyranose forms dominate in aqueous solutions of most monosaccharides, it is quite common to find the furanose form when the sugar is incorporated into a biomolecule, e.g., β-D-ribofuranose in ribonucleic acid.
- The anomeric carbon atom of the 2-ketoses is naturally C2.

The cyclic structure for sugars now helped to explain several observations that had been made by the German pioneers in the nineteenth century:

- Aldoses did not form addition compounds with sodium bisulfite and failed some of the very sensitive and characteristic colour tests for aldehydes.
- Generally, aldoses tended to react with hydrogen cyanide and with phenylhydrazine more slowly than normal aldehydes.
- With a careful choice of reagents and conditions, D-glucose could be converted into two different penta-acetates:

---

[w] Originally, another famous carbohydrate chemist, Claude S. Hudson (1881–1952, a student of van't Hoff, Ph.D. at Princeton University), proposed a definition for 'α' and 'β' based on the relative magnitude of the specific optical rotation.[15]

References start on page 32

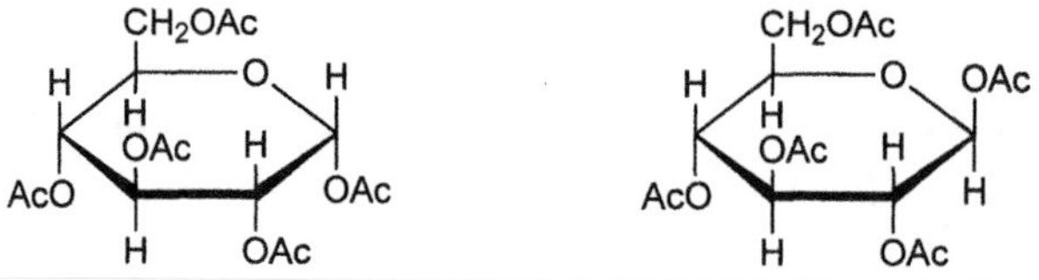

penta-O-acetyl-α-D-glucopyranose   penta-O-acetyl-β-D-glucopyranose

Still, aldoses did eventually show most of the reactions characteristic of an aldehyde; how then to explain this apparent dichotomy? The answer lay in an observation made originally in 1846[16] and corroborated in 1895 by Tanret[17] – the optical rotation of either pure enantiomer of glucose changes with time, the phenomenon of *mutarotation*. For example, a freshly prepared solution of α-D-glucopyranose in water has a specific rotation close to +112°, but this falls with time to a final value of +52°; conversely, the initial value of about +19° for β-D-glucopyranose rises to the same +52°.

For a while, this solution with a specific rotation of +52° was thought to contain a new form of glucose. However, the phenomenon of mutarotation was later easily explained by a consideration of the following equilibrium:[18]

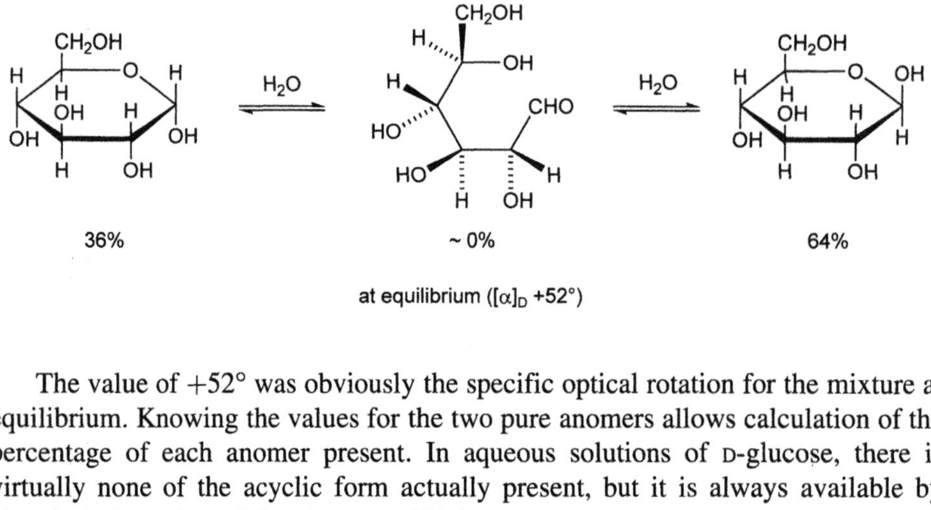

36%   ~ 0%   64%

at equilibrium ([α]$_D$ +52°)

The value of +52° was obviously the specific optical rotation for the mixture at equilibrium. Knowing the values for the two pure anomers allows calculation of the percentage of each anomer present. In aqueous solutions of D-glucose, there is virtually none of the acyclic form actually present, but it is always available by chemical disruption of the above equilibrium.

Finally, a few words on the actual experimental determination of 'ring size' in carbohydrates. Years ago, the classic approaches involved 'methylation analysis'[19–21] and 'periodate cleavage'.[22–25] These methods are still in use, especially where the actual methylation is followed by mass spectrometric analysis; however, it is the power of nuclear magnetic resonance (NMR) spectroscopy, both ^{1}H and ^{13}C, that is nowadays often brought to bear on such problems.[26–30]

The Shape (Conformation) of Cyclic Sugars, and the Anomeric Effect

A century of investigation had unlocked the stereochemical secrets of D-(+)-glucose, depicted as either a Fischer projection or, more accurately as we have seen, a cyclic molecule in a Haworth formula:[x]

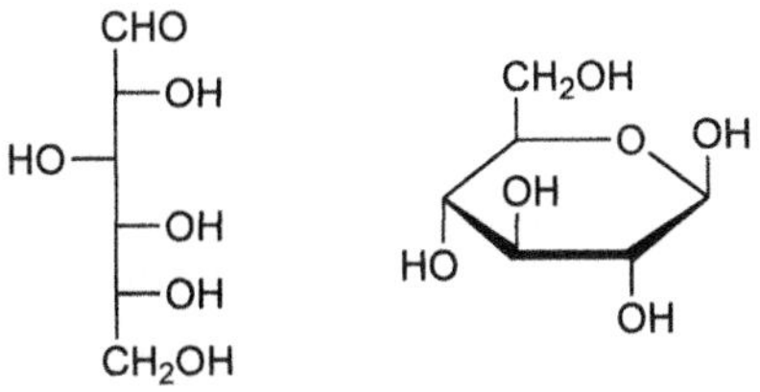

D-(+)-glucose   β-D-glucopyranose

In the early 1900s, most chemists believed that a saturated six-membered ring was non-planar. However, it took the work of Hassel,[y] which employed electron diffraction studies in the gas phase, to put some substance into this notion; the cyclohexane ring was shown to have a non-planar shape (conformation), like that of a *chair*:[31]

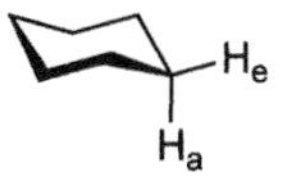

Some years later, Barton recognized the importance of the two different types of bonds present in cyclohexane (*equatorial* and *axial*) and used this revelation to explain the conformation and reactivity in molecules such as the steroids.[32,33] The beauty of these results was that, in the chair conformation for cyclohexane, each carbon was almost exactly tetrahedral in shape – cyclohexane, as predicted and shown, exhibited no Baeyer 'angle strain'.

A further advance by Hassel was to predict that the conformation of the pyranose ring would also be non-planar and, probably, again have the shape of a chair:

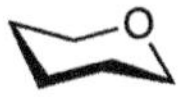

---

[x] From now on, hydrogen atoms bound to carbon will generally not be shown.

[y] Odd Hassel (1897–1981), Ph.D. from the University of Berlin, Norwegian, shared a Nobel Prize in Chemistry (1969) with Derek Harold Richard Barton (1918–1998), British.

References start on page 32

For β-D-glucopyranose, the most common monosaccharide found in the free form, all of the hydroxyl substituents on the pyranose ring are equatorially disposed (otherwise the molecule is no longer β-D-glucose!):

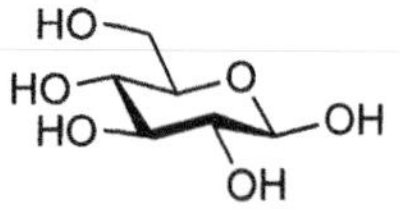

It is well known that cyclohexane, as a neat liquid or as a solution at room temperature, is in rapid equilibrium, via the *boat* conformation, with another, degenerate chair conformation; a result of this equilibrium is that there is a general interchange of equatorial and axial bonds on each carbon atom:

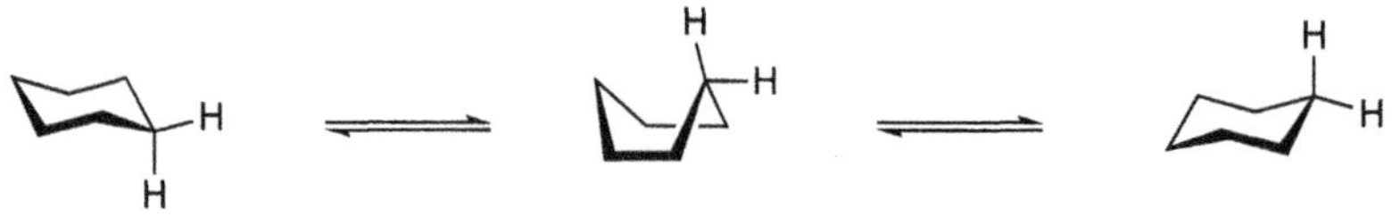

What would be the consequences, if any, of such a process applied to β-D-glucopyranose?

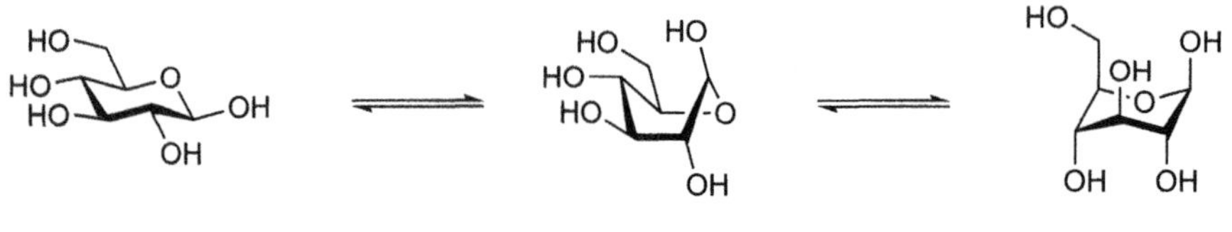

Again an equilibrium is possible, via a boat conformation, but the new chair conformation is obviously different from the original one – with only axial substituents, the energy of the new conformation is significantly higher (some $25 \, \text{kJ mol}^{-1}$).

How, then, do we actually establish the *preferred conformation* for a molecule such as β-D-glucopyranose?

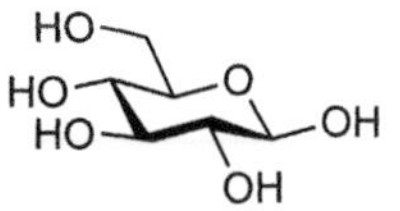

When the molecule in question is crystalline, then a single crystal, X-ray structure determination will yield both the molecular structure and the conformation. When the molecule is a liquid, or in solution, ^1H NMR spectroscopy will often give the answer. For a conformation such as the one discussed above, the value of the coupling constant between, e.g. H2 and H3 ($J_{2,3}$) will normally be 'large' (9–10 Hz) and so will be indicative of a *trans*-diaxial relationship between the coupling protons. The other, higher energy, all-axial conformation will have a 'small' (1–2 Hz) value for $J_{2,3}$, indicative of a diequatorial relationship.

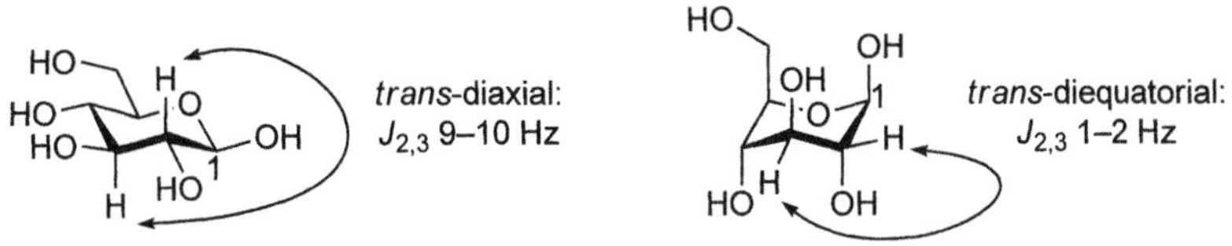

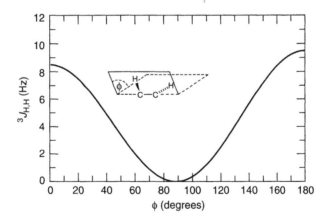

These values in carbohydrates are in general agreement with the early observations by Lemieux[34] and, a little later, with the rule espoused by Karplus,[35] as applied to the relationship between the magnitude of the coupling constant and the size of the torsional angle between vicinal protons.[26–28,z]

A word of caution is necessary here – the conformation of a molecule in the solid state is not *necessarily* the same as those in the liquid state or in solution.

---

z The dihedral angle dependence of vicinal coupling constants is only an approximation. The relationship is sensitive to the local environment within the molecule and can be perturbed by the presence of electron-withdrawing substituents, and changes to bond angles and bond lengths. As Karplus has remarked: 'The person who attempts to estimate dihedral angles to an accuracy of one or two degrees does so at his own peril'.

References start on page 32

As we saw earlier, the D-aldopentoses and D-aldohexoses exist in aqueous solution primarily as a mixture of the α- and β-pyranose forms; occasionally, as with D-ribose, -altrose, -idose and -talose, significant amounts of the furanose forms can also be found.[36] In all of these pyranose forms, it is the 'normal' chair conformation that is almost always preferred; however, α- and β-D-ribose, β-D-arabinose, and α-D-lyxose, -altrose and -idose all show contributions from the 'inverted' chair conformation and, indeed, α-D-arabinose even shows a preference for it.[37]

Apart from these chair conformations for the D-aldopyranoses, there exist other, higher energy conformations, namely the boat and the *skew*. It must be stressed that, although these higher energy forms are not present to any significant extent in aqueous solution, they are discrete conformations encountered in the conversion of one chair into the other. The *half-chair* is a common conformation for some carbohydrate derivatives where chemical modification of the pyranose ring has occurred.

What follows is a summary of the limiting conformations for the pyranose ring, namely the chair (*C*), boat (*B*), half-chair (*H*) and skew (*S*) forms, together with their modern descriptors (it is obviously necessary to avoid such terms as 'normal' and 'inverted').

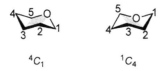

Only two chair forms are possible. The descriptors arise according to the following protocol:[38]

- The lowest-numbered carbon of the ring (C1) is taken as an *exo*-planar atom.
- O, C2, C3 and C5 define the reference plane of the chair.
- Viewed clockwise (O → 2 → 3 → 5), C4 is above (below) this plane and C1 is below (above).
- Atoms that are above (below) the plane are written as superscripts (subscripts), which precede (follow) the letter.
- $^4C_1$ and $^1C_4$ result.

Six boat forms are possible, with only two of these shown (the reference plane in each form is unique and obvious).

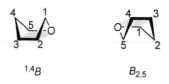

Twelve half-chair forms are possible and, again, only two of these are shown (the reference plane is defined by four contiguous atoms and is again unique).

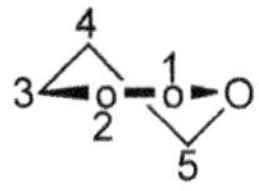

$${}^4H_5 \qquad {}^5H_4$$

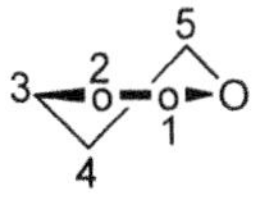

Six skew forms are possible, with only two of these shown (the reference plane is not obvious, being made up of three contiguous atoms and the remaining non-adjacent atom).[38]

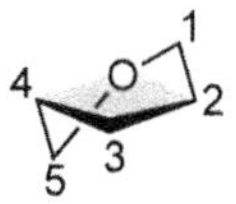

$${}^1S_5 \qquad {}^OS_2$$

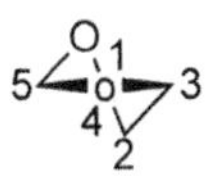

The chair form is more stable than the skew form, which is again more stable than both the boat and half-chair forms. In pyranose rings that contain a double bond, it is the half-chair that is the normal conformation.

The conformations available to the furanose ring are just the envelope (E) and the twist (T); both have 10 possibilities, and the energy differences among all of the conformations are quite small.

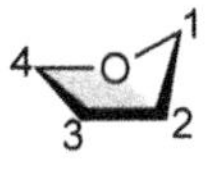

$${}^1E \qquad E_O \qquad {}^2T_3$$

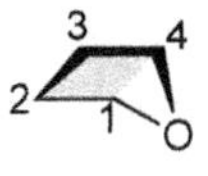

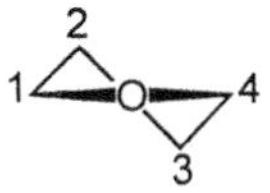

Let us now reflect on the familiar equilibrium that is established when D-(+)-glucose is dissolved in water:

acyclic form

36%   64%

The two main components of the mixture are present in the indicated amounts and each in the preferred ${}^4C_1$ conformation. The free energy difference for such an equilibrium amounts to about 1.5 kJ mol^{-1} in favour of the β-anomer, somewhat short of the accepted value ($3.8 \, \text{kJ mol}^{-1}$) for an equatorial over an axial hydroxyl group, the only difference between the two molecules in question. This propensity for formation of the α-anomer over that which would normally be expected was first noted by Edward,[39] and termed the *anomeric effect* by Lemieux.[40,41] So wide ranging and important is the effect that it virtually ensures the axial configuration of an electronegative substituent at the anomeric carbon in some derivatives:

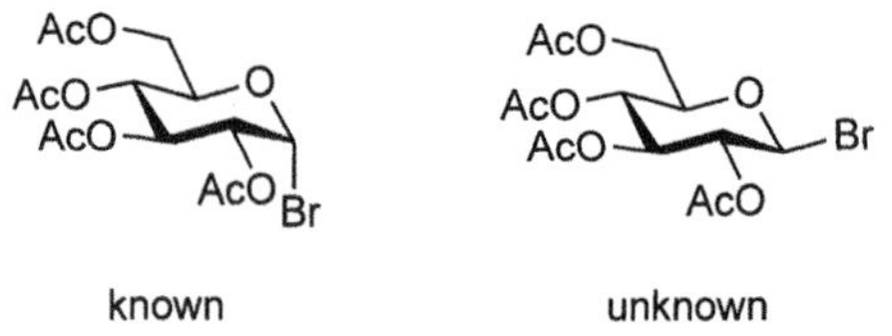

known   unknown

Also, the anomeric effect is responsible for the stabilization of conformations that would otherwise seemingly capitulate to other, unfavourable interactions:

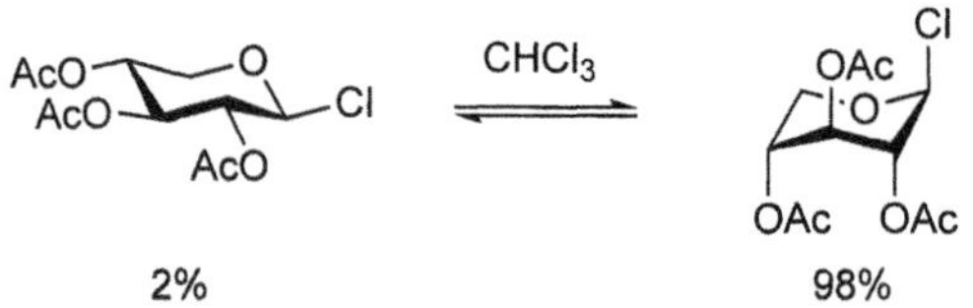

2%   98%

The origin of the anomeric effect, which itself increases with the electronegativity of the substituent and decreases in solvents of high dielectric constant, has been explained in several ways. The first of these, somewhat naively, involves unfavourable lone pair – lone pair interactions (the so-called rabbit ear effect) in the equatorial anomer that are obviously not present in the axial anomer; the second, related to the first, considers unfavourable dipole–dipole interactions in the equatorial anomer that can be minimized in the axial anomer:[42]

However, the third, and generally accepted, explanation involves the interaction between a lone pair of electrons located 'axially' in a molecular orbital (n) on O5 and an (unoccupied) anti-bonding molecular orbital (σ*) of the C1 to X bond (a stabilizing $n_O \rightarrow \sigma^*_{C-X}$ orbital interaction).[43]

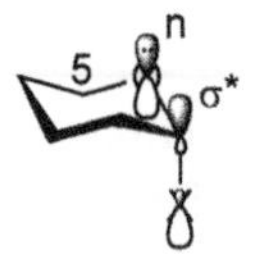

The 'anti-periplanar' arrangement found in the axial anomer favours this 'back-bonding', resulting in a slight shortening of the O5 to C1 bond, a slight lengthening of the C1 to X bond[44] and a general increase in the electron density at X. This explanation is in harmony with the 'bond–no bond' concept that allows two valence-bond structures to be drawn for the axial anomer (and, in so doing, stabilizing the molecule):

The causes of the anomeric effect continue to be discussed,[45–48] and two rigorous treatments have recently appeared.[49,50]

Probably, one of the greatest contributions by Lemieux to the field of carbohydrate chemistry was his delineation of the importance of the *exo-anomeric effect*.[51–54] In a simple acetal derived from a pyranose sugar, the normal anomeric effect operates, which stabilizes the axial anomer over the equatorial anomer:

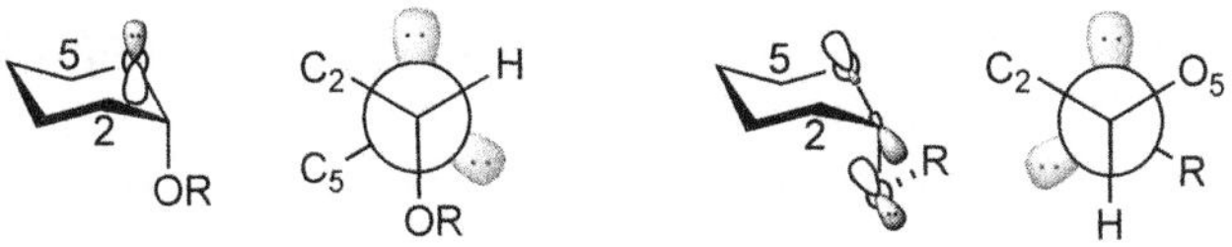

C1-O5   C1-O1

However, in an appropriate conformation of the *exo*-cyclic alkoxy group, there is again an anti-periplanar arrangement of a lone pair on oxygen (of OR) and the C1 to O5 bond, allowing 'back donation' (an $n_O \rightarrow \sigma^*_{C-O}$ orbital interaction) again to stabilize this conformation, the so-called *exo*-anomeric effect.[aa] Because these two anomeric effects (*endo*- and *exo*-) operate in opposite directions, the *exo*-anomeric effect is not considered important with such axial acetals. However, in an equatorial acetal, where there is no contribution from a normal anomeric effect, it is the *exo*-anomeric effect that is dominant and dictates the preferred gauche conformation (of O5 and R) at the anomeric carbon atom:

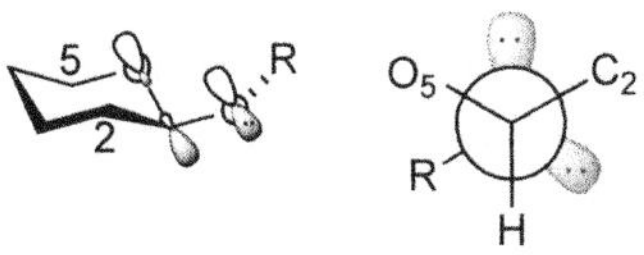

C1-O1

[aa] As a consequence of this new term, the original anomeric effect is often referred to as the *endo-anomeric effect*. Another term, the *kinetic anomeric effect*, is sometimes used in discussions dealing with the transition state of a reaction.[49]

References start on page 32

Taken to its logical conclusion, the *exo*-anomeric effect explains the helical shape of many polysaccharide chains; it is certainly important in determining the shape of many biologically important *oligosaccharides*.[55,ab]

Another aspect of the anomeric effect worth mentioning here is based on observations, first by Lemieux[56] and then by Paulsen,[57] that various *N*-glycosylpyridinium and *N*-glycosylimidazolium salts preferred to exist in conformations where the positively-charged nitrogen atom was not axially oriented, the so-called *reverse anomeric effect*:[49,50]

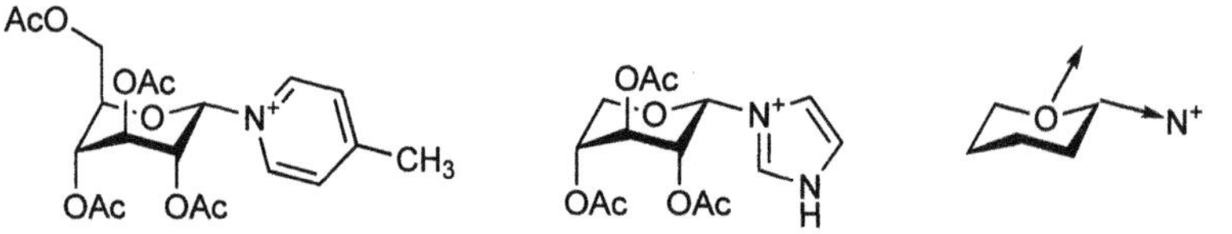

$^1C_4$ (or $B_{2,5}$) $^1C_4$

One explanation for the origin of this 'reverse' effect is a simple and favourable interaction of opposing dipoles; however, steric effects cannot be ignored, and there has been a great deal of discussion recently as to whether the reverse anomeric effect even exists.[58–63]

It would seem appropriate to end this section on another aspect of carbohydrate conformation, namely the one dealing with, for a D-hexopyranose, the substituents at C5 (CH$_2$OH) and at C1 (OR).[50] What are the preferred conformations around the C5–C6 bond of a D-hexopyranose?

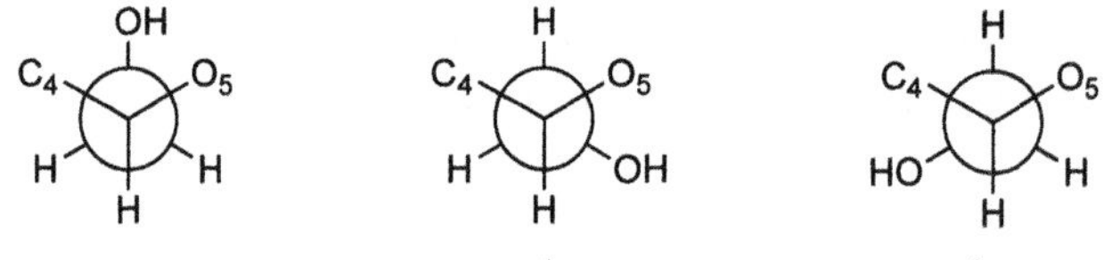

gg   *gt*   *tg*

Not surprisingly, the staggered forms predominate and are given descriptors according to, first, the relationship of the 6-OH to O5 (*gauche* or *trans*), and then the same OH to C4.[64] Typical values for a D-glucopyranose structure are *gg*:*gt*:*tg* = 3:2:0 and, for a D-*galacto* equivalent, they are 1:3:1.[65]

[ab] A general term for small chains of monosaccharides, with up to 10 residues in the chain.

For an OR substituent at the anomeric carbon, the main conformations around the C1 to O1 bond are dominated, as we have already seen, by the *exo*-anomeric effect.

β- (C1-O1)   preferred   α- (C1-O1)   preferred

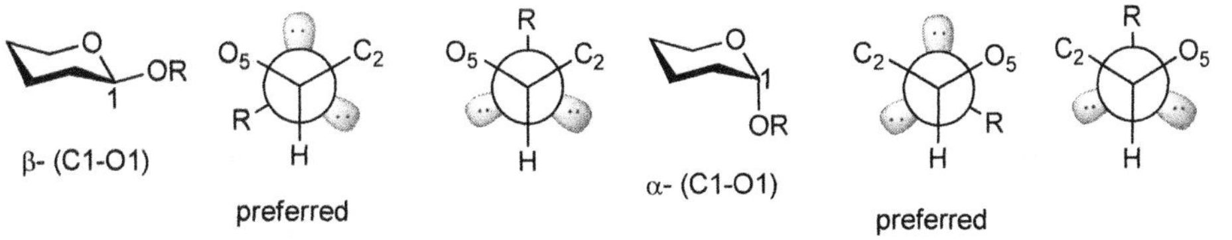

Of the two conformations derived for a β- or α- substituent, it is the one with fewer *gauche* interactions that seems to dominate. When the R group is, in fact, another sugar, as in the disaccharide melibiose (6-*O*-β-D-galactopyranosyl-D-glucopyranose), similar arguments apply, but now various torsional angles must be defined to describe properly the conformation of the molecule. For example, ϕ defines the angle H1′, C1′, O6 and C6; ψ defines C1′, O6, C6 and C5; ω defines O6, C6, C5 and H5. Again, the value of ϕ is generally close to that predicted from the *exo*-anomeric effect.[66]

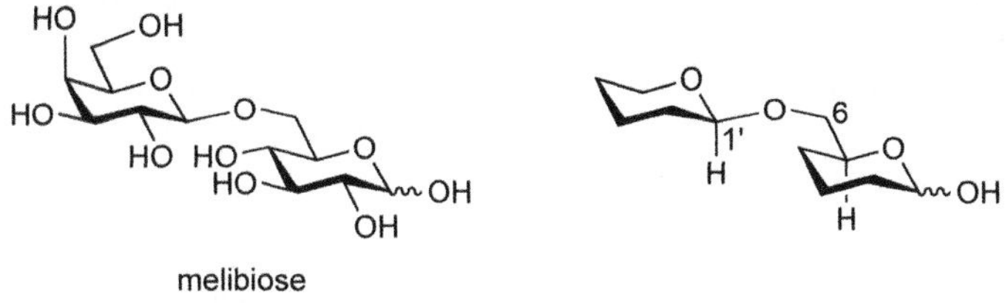

melibiose

We have seen in this introductory chapter that the seminal studies of Fischer were carried along in the early part of the next century by people such as Haworth and Hudson. However, it was Lemieux[67,ac] who dominated carbohydrate chemistry for the major part of the twentieth century, with enormous contributions to NMR spectroscopy, conformational analysis, synthesis and glycobiology; his final words on the factors that govern carbohydrate/protein binding are truly memorable.[68] Lemieux's remarkable synthesis of sucrose in 1953 really set the stage for the material to be discussed in the next few chapters.[69]

[ac] Raymond U. Lemieux (1920–2000), Ph.D. under C.B. Purves (McGill University).

References start on page 32

References

1. Lichtenthaler, F.W. (2002). *Eur. J. Org. Chem.*, 4095.
2. Kunz, H. (2002). *Angew. Chem. Int. Ed.*, **41**, 4439.
3. Lobry de Bruyn, C.A. and Alberda van Ekenstein, W. (1895). *Recl. Trav. Chim. Pays-Bas*, **14**, 203.
4. Angyal, S.J. (2001). *Top. Curr. Chem.*, **215**, 1.
5. Rosanoff, M.A. (1906). *J. Am. Chem. Soc.*, **28**, 114.
6. Hudson, C.S. (1948). *Adv. Carbohydr. Chem.*, **3**, 1.
7. Lichtenthaler, F.W. (1992). *Angew. Chem. Int. Ed. Engl.*, **31**, 1541.
8. Fischer, E. (1891). *Ber. Dtsch. Chem. Ges.*, **24**, 1836.
9. Fischer, E. (1891). *Ber. Dtsch. Chem. Ges.*, **24**, 2683.
10. Bijvoet, J.M., Peerdeman, A.F. and van Bommel, A.J. (1951). *Nature*, **168**, 271.
11. Drew, H.D.K. and Haworth, W.N. (1926). *J. Chem. Soc.*, 2303.
12. Mills, J.A. (1955). *Adv. Carbohydr. Chem.*, **10**, 1.
13. Böeseken, J. (1949). *Adv. Carbohydr. Chem.*, **4**, 189.
14. Collins, P.M. and Ferrier, R.J. (1995). *Monosaccharides: Their Chemistry and Their Roles in Natural Products*, p. 14. John Wiley and Sons.
15. Hudson, C.S. (1909). *J. Am. Chem. Soc.*, **31**, 66.
16. Dubrunfaut, M. (1846). *Compt. Rend.*, **23**, 38.
17. Tanret, M.C. (1895). *Compt. Rend.*, **120**, 1060.
18. Lewis, B.E., Choytun, N., Schramm, V.L. and Bennet, A.J. (2006). *J. Am. Chem. Soc.*, **128**, 5049.
19. Hirst, E.L. and Purves, C.B. (1923). *J. Chem. Soc. (Trans.)*, **123**, 1352.
20. Hirst, E.L. (1926). *J. Chem. Soc.*, 350.
21. Haworth, W.N., Hirst, E.L. and Learner, A. (1927). *J. Chem. Soc.*, 1040, 2432.
22. Jackson, E.L. and Hudson, C.S. (1937). *J. Am. Chem. Soc.*, **59**, 994.
23. Jackson, E.L. and Hudson, C.S. (1939). *J. Am. Chem. Soc.*, **61**, 959.
24. Maclay, W.D., Hann, R.M. and Hudson, C.S. (1939). *J. Am. Chem. Soc.*, **61**, 1660.
25. Bobbitt, J.M. (1956). *Adv. Carbohydr. Chem.*, **11**, 1.
26. Vliegenthart, J.F.G. (2006). In *NMR Spectroscopy and Computer Modeling of Carbohydrates*. ACS Symposium Series 930 (J.F.G. Vliegenthart and R.J. Woods, eds) p. 1. American Chemical Society: Oxford University Press.
27. Widmalm, G. (1998). In *Carbohydrate Chemistry* (G.-J. Boons, ed.) p. 448. Blackie Academic and Professional: London.
28. Duus, J.Ø., Gotfredsen, C.H. and Bock, K. (2000). *Chem. Rev.*, **100**, 4589.
29. Jiménez-Barbero, J. and Peters, T. (2003). *NMR Spectroscopy of Glycoconjugates*. Wiley-VCH: Weinheim.
30. Roslund, M.U., Tähtinen, P., Niemitz, M. and Sjöholm, R. (2008). *Carbohydr. Res.*, **343**, 101.
31. Hassel, O. and Ottar, B. (1947). *Acta Chem. Scand.*, **1**, 929.
32. Barton, D.H.R. (1950). *Experientia*, **6**, 316.
33. Barton, D.H.R. (1953). *J. Chem. Soc.*, 1027.
34. Lemieux, R.U., Kullnig, R.K., Bernstein, H.J. and Schneider, W.G. (1957). *J. Am. Chem. Soc.*, **79**, 1005.
35. Karplus, M. (1963). *J. Am. Chem. Soc.*, **85**, 2870.
36. Angyal, S.J. (1984). *Adv. Carbohydr. Chem. Biochem*, **42**, 15; (1991), **49**, 19.
37. Collins, P.M. and Ferrier, R.J. (1995). *Monosaccharides: Their Chemistry and Their Roles in Natural Products*, p. 33. John Wiley and Sons.
38. Schwarz, J.C.P. (1973). *J. Chem. Soc., Chem. Commun.*, 505.
39. Edward, J.T. (1955). *Chem. Ind.*, 1102.

40. Lemieux, R.U. (1964). In *Molecular Rearrangements* (P. de Mayo, ed.) part 2, p. 709. Interscience Publishers: John Wiley and Sons.
41. Lemieux, R.U. (1971). *Pure Appl. Chem.*, **25**, 527.
42. Wolfe, S., Rauk, A., Tel, L.M. and Csizmadia, I.G. (1971). *J. Chem. Soc.* B, 136.
43. Juaristi, E. and Cuevas, G. (1992). *Tetrahedron*, **48**, 5019.
44. Jones, P.G. and Kirby, A.J. (1984). *J. Am. Chem. Soc.*, **106**, 6207.
45. Ma, B., Schaefer, H.F., III and Allinger, N.L. (1998). *J. Am. Chem. Soc.*, **120**, 3411.
46. Thatcher, G.R.J. ed. (1993). *The Anomeric Effect and Associated Stereoelectronic Effects*. ACS Symposium Series 539. American Chemical Society: Washington DC.
47. Juaristi, E. and Cuevas, G. (1995). *The Anomeric Effect*. CRC Press.
48. Box, V.G.S. (1998). *Heterocycles*, **48**, 2389.
49. Chandrasekhar, S. (2005). *ARKIVOC*, 37.
50. Grindley, T.B. (2001). In *Glycoscience*: *Chemistry and Chemical Biology* (B.O. Fraser-Reid, K. Tatsuta, and J. Thiem, eds) vol. I, p. 3. Springer-Verlag.
51. Lemieux, R.U., Pavia, A.A., Martin, J.C. and Watanabe, K.A. (1969). *Can. J. Chem.*, **47**, 4427.
52. Praly, J.-P. and Lemieux, R.U. (1987). *Can. J. Chem.*, **65**, 213.
53. Tvaroška, I. and Bleha, T. (1989). *Adv. Carbohydr. Chem. Biochem.*, **47**, 45.
54. Tvaroška, I. and Carver, J.P. (1998). *Carbohydr. Res.*, **309**, 1.
55. Meyer, B. (1990). *Top. Curr. Chem.*, **154**, 141.
56. Lemieux, R.U. and Morgan, A.R. (1965). *Can. J. Chem.*, **43**, 2205.
57. Paulsen, H., Györgdeák, Z. and Friedmann, M. (1974). *Chem. Ber.*, **107**, 1590.
58. Perrin, C.L., Fabian, M.A., Brunckova, J. and Ohta, B.K. (1999). *J. Am. Chem. Soc.*, **121**, 6911.
59. Perrin, C.L. (1995). *Tetrahedron*, **51**, 11901.
60. Vaino, A.R., Chan, S.S.C., Szarek, W.A. and Thatcher, G.R.J. (1996). *J. Org. Chem.*, **61**, 4514.
61. Randell, K.D., Johnston, B.D., Green, D.F. and Pinto, B.M. (2000). *J. Org. Chem.*, **65**, 220.
62. Vaino, A.R. and Szarek, W.A. (2001). *J. Org. Chem.*, **66**, 1097.
63. Grundberg, H., Eriksson-Bajtner, J., Bergquist, K.-E., Sundin, A. and Ellervik, U. (2006). *J. Org. Chem.*, **71**, 5892.
64. Bock, K. and Duus, J.Ø. (1994). *J. Carbohydr. Chem.*, **13**, 513.
65. Pan, Q., Klepach, T., Carmichael, I., Reed, M. and Serianni, A.S. (2005). *J. Org. Chem.*, **70**, 7542.
66. Klepach, T.E., Carmichael, I. and Serianni, A.S. (2005). *J. Am. Chem. Soc.*, **127**, 9781.
67. Lemieux, R.U. (1990). *Explorations with Sugars: How Sweet It Was*. American Chemical Society: Washington DC.
68. Lemieux, R.U. (1996). *Acc. Chem. Res.*, **29**, 373.
69. Lemieux, R.U. and Huber, G. (1956). *J. Am. Chem. Soc.*, **78**, 4117.

Chapter 2

Synthesis and Protecting Groups[1–6]

Much of today's chemistry is concerned with synthesis, and carbohydrate chemistry is no exception. The large pharmaceutical companies ('Big Pharma') once employed small armies of chemists to synthesize a myriad of compounds that were necessary for lead development of a potential 'block buster' drug. Nowadays, however, the same companies 'outsource' much of their synthetic work to smaller, private companies, but there is still a (growing) need for synthetic chemists to do the work. If you want to do synthesis, you need to know about protecting groups.

The protecting groups used in carbohydrates are generally the same as those of mainstream organic chemistry; the difference, however, is that even a monosaccharide presents a myriad of hydroxyl groups that need protection, in either an individual (regioselective) or a unique (orthogonal) manner. Also, the introduced protecting groups may affect the reactivity of the resulting molecule or even participate in some of its reactions.

Synthesis with carbohydrates would be a less complicated matter if it were confined to the natural and abundant aldoses, ketoses and oligosaccharides. However, there often arises the need for modified monosaccharides or, perhaps, an unusual or rare oligosaccharide. For example, how would one approach the synthesis of a molecule such as '3-deoxy-D-glucose'[a] starting from D-glucose?

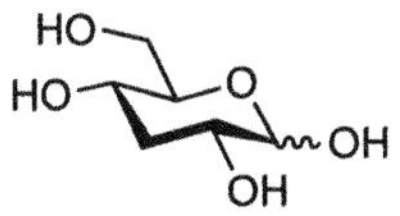

The problems are twofold: first, the need for a chemical reaction that will replace a hydroxyl group by a hydrogen atom; second, the need to carry out this replacement *only* at C3. Also, what about the synthesis of an oligosaccharide, say, a disaccharide?

---

[a] As '3-deoxy-D-allose' is just as good, an unambiguous name should be used: 3-deoxy-D-*ribo*-hexose. The molecule is depicted as an α/β mixture of pyranose forms.

References start on page 67

The problems are not much different from the monosaccharide example: first, a chemical method is needed to join two D-glucose units; second, the two monosaccharides must be manipulated so that the linkage is *specifically* β-1,4.

To set the stage, consider a very early synthesis, performed by Fischer in 1893:

methyl α-D-glucopyranoside

mp 165°C, $[\alpha]_D$ +158° (H_2O)

methyl β-D-glucopyranoside

mp 107°C, $[\alpha]_D$ –33° (H_2O)

By heating D-glucose with methanol containing some hydrogen chloride, two new chemicals, actually anomeric acetals, were formed: a 'synthesis' and, at the same time, a 'protecting group' for the anomeric carbon. More will be discussed about this unique and important reaction later.

The discussion that follows lists the various protecting groups that are in common use, namely, those that actually work! However, pertinent references to newer and emerging protecting groups will also be included. Appendix I presents some of the protecting groups for the hydroxyl and amino groups, together with their possible orthogonal relationships.

Esters

Esters, together with ethers and acetals, constitute the main protecting groups for the hydroxyl groups of sugars. The popularity of esters derives from their ease of formation, generally employing readily available acid anhydrides or acid chlorides, and also the ease of removal. Although the primary role of esters introduced into carbohydrates is to protect the otherwise reactive hydroxyl groups, they can also play a role in precipitating useful chemical reactions at both anomeric and non-anomeric carbon atoms. Esters, like ethers and acetals, reduce the polarity of the carbohydrate and so improve solubility in organic solvents.

Acetates:[b] The acetylation of D-glucose was first performed in the mid-nineteenth century, which helped to confirm the pentahydroxy nature of the molecule. Since then, three sets of conditions are commonly used for the transformation:

---

[b] For purposes of nomenclature, it is important to realize that, while treatment of a hydroxyl group with, for example, acetic anhydride yields an acetate, the actual protecting group is an *acetyl* group.

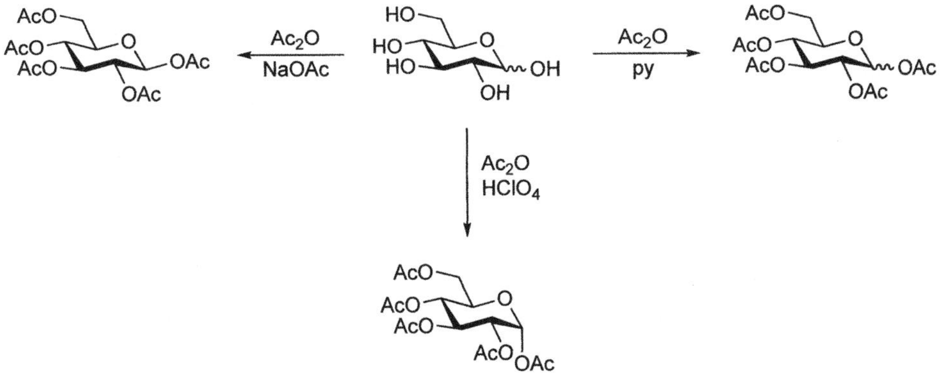

The reaction in pyridine is general and convenient and usually gives the same ratio of anomers of the penta-acetate as found in the parent free sugar.[7,8] With an acid catalyst, the reaction probably operates under thermodynamic control and gives the more stable anomer. Sodium acetate causes a rapid anomerization of the free sugar,[9] and the more reactive anomer is then preferentially acetylated.[c] Iodine has been used for various acetylations, with some interesting and regioselective transformations:[12]

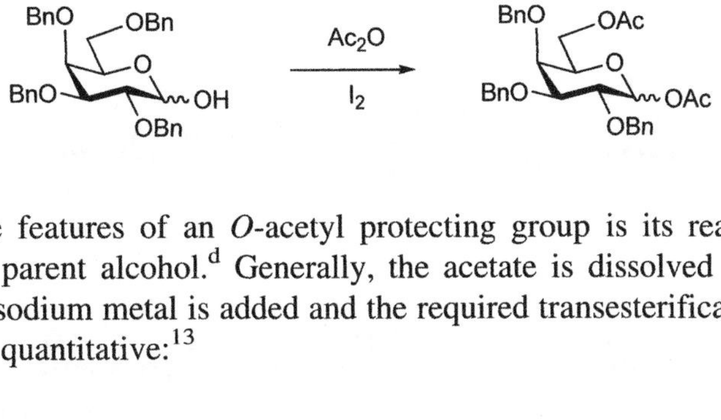

One of the features of an *O*-acetyl protecting group is its ready removal to regenerate the parent alcohol.[d] Generally, the acetate is dissolved in methanol, a small piece of sodium metal is added and the required transesterification reaction is both rapid and quantitative:[13]

$$\text{–OCOCH}_3 + \text{CH}_3\text{OH} \xrightarrow{\text{NaOCH}_3} \text{–OH} + \text{CH}_3\text{COOCH}_3$$

Other systems that carry out this sort of reaction are anion-exchange resin (HO⁻), ammonia or potassium cyanide in methanol,[8,14,15] guanidine (guanidinium

---

[c] Deprotonation of the β-anomer of the free sugar gives a β-oxyanion that interacts unfavourably with the lone pairs of electrons on O5; a rapid acetylation removes this interaction.[10,11]

[d] A high level of crystallinity in simple derivatives is also a much relished feature by the preparative chemist.

References start on page 67

nitrate in methanol)[16] and a mixture of triethylamine, methanol and water.[17] For base-sensitive substrates, hydrogen chloride or tetrafluoroboric acid–ether in methanol is a viable alternative for deacetylation;[18] even dibutyltin oxide is reported to cause the removal of acetyl groups, albeit thus far for a limited range of substrates.[19]

For the selective acetylation of one hydroxyl group over another, one has the choice of lowering the reaction temperature or employing reagents specifically designed for such a purpose.[12,20–22] The selective removal of an acetyl group at the anomeric position can easily be achieved, probably owing to the better leaving-group ability of the anomeric oxygen;[23–25] hydrazinium acetate and benzylamine appear to work well in most instances:[26–28]

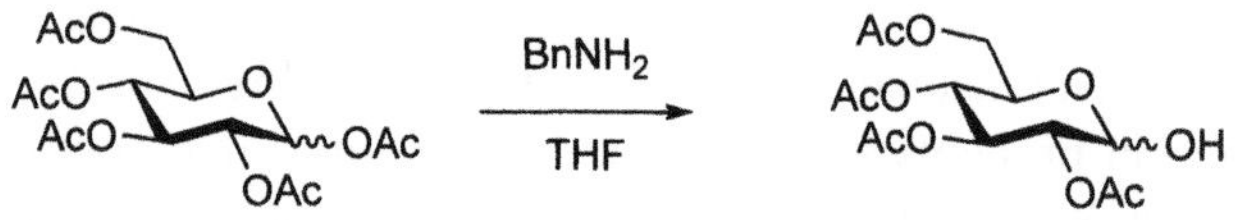

Recently, the use of enzymes, especially lipases, has added another dimension to this concept of selectivity:[29–36]

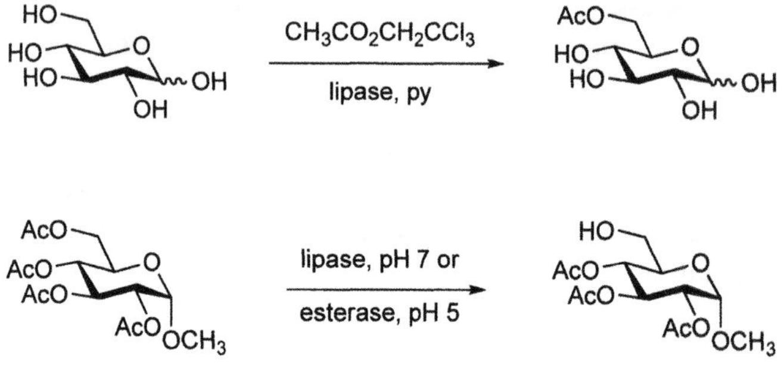

Benzoates: In general, benzoates are more robust groups than acetates and may give rise to derivatives that are useful in X-ray crystallographic determinations (e.g., 4-bromobenzoates). The robustness of benzoates is reflected both in their preparation (benzoyl chloride, pyridine) and in their reversion to the parent alcohol (sodium–methanol for protracted periods). Acetates can be removed in preference to benzoates.[37]

The selective benzoylation of a carbohydrate[38] can be achieved either by careful control of the reaction conditions[39,40] or by the use of a less reactive reagent, such as N-benzoylimidazole[41,42] or 1-benzoyloxybenzotriazole:[43]

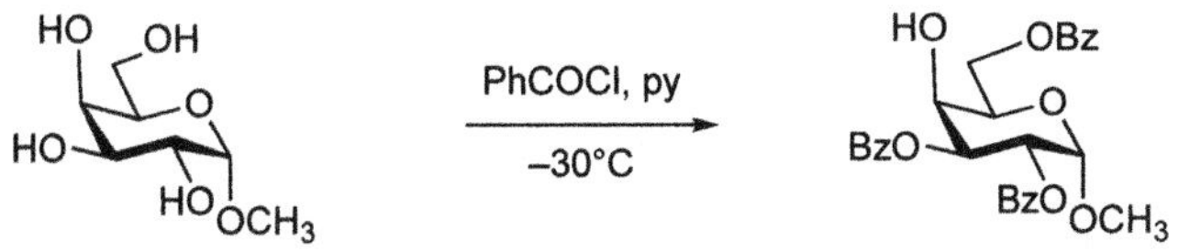

Chloroacetates: Chloroacetates are easily acquired (chloroacetic anhydride in pyridine), are stable enough to survive most synthetic transformations and yet, being more labile than acetates, can be selectively transformed back to the hydroxyl group (thiourea,[44] 'hydrazinedithiocarbonate'[45] or DABCO[46]):

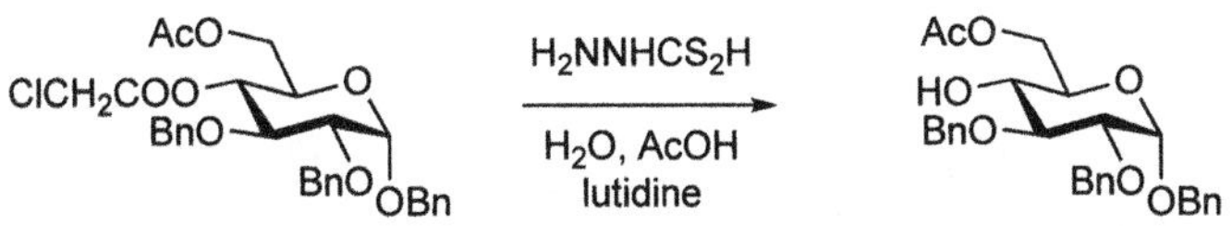

Pivalates: Esters of pivalic acid (2,2-dimethylpropanoic acid), for the reason of steric bulk, can be installed preferentially at the more reactive sites of a sugar but require reasonably vigorous conditions for their removal:[47,48]

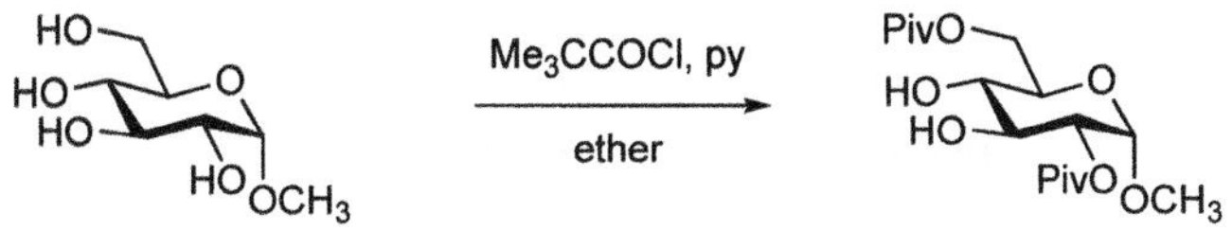

Levulinates: These are esters of levulinic acid (4-oxopentanoic acid) that are easily installed and have the advantage of conversion back to the hydroxyl group under essentially neutral conditions,[49,50] probably by a form of 'assisted cleavage':

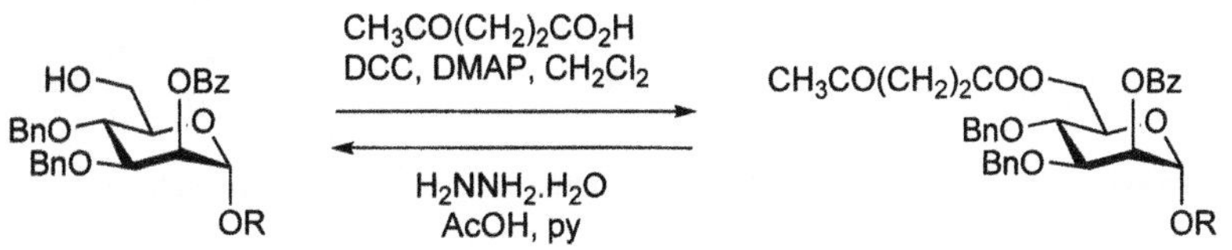

This concept of 'assisted cleavage' has been used in the development of other ester-based protecting groups:[51–54]

References start on page 67

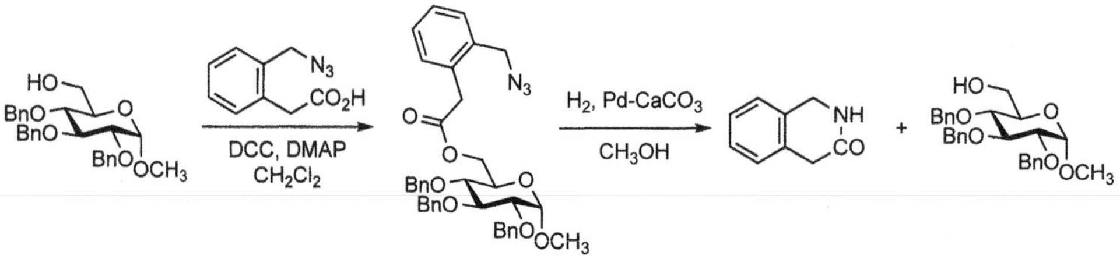

Carbonates, borates, phosphates, sulfates and nitrates: Cyclic carbonates are occasionally used for the protection of vicinal diols, providing the dual advantages of installation under basic (phosgene) or neutral (1,1'-carbonyldiimidazole) conditions, and easy removal.[55,56]

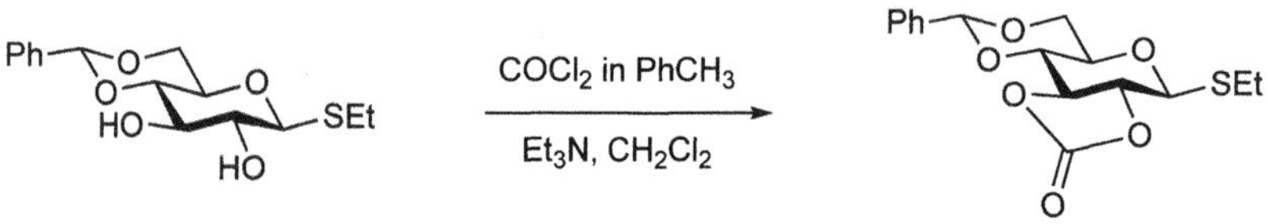

Borates, although rarely used in the protection of the hydroxyl group, are useful in the purification, analysis and structure determination of sugar polyols. Phenylboronates seem to have more potential in synthesis.[57,58]

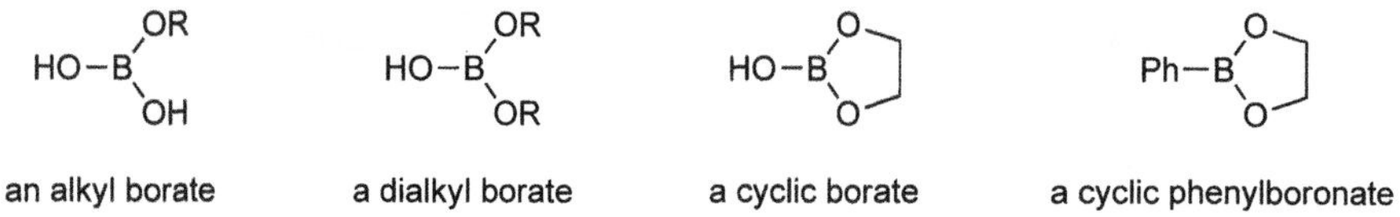

| an alkyl borate | a dialkyl borate | a cyclic borate | a cyclic phenylboronate |

Sugar phosphates, and their oligomers, are found as the keystone of the molecules of life – RNA, DNA and ATP:

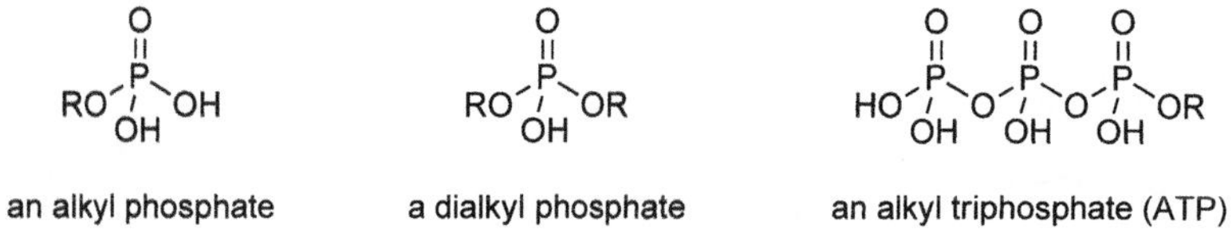

| an alkyl phosphate | a dialkyl phosphate | an alkyl triphosphate (ATP) |

Sulfates are common components of many biologically important molecules (see Chapter 10); sugar nitrates formed the basis of many of the early explosives (see Chapter 9).

an alkyl sulfate   an alkyl nitrate

Cyclic sulfates, which are often accessed via the cyclic sulfite, formally offer protection of a diol but are reactive enough to act as an effective leaving group when treated with a nucleophile:[59]

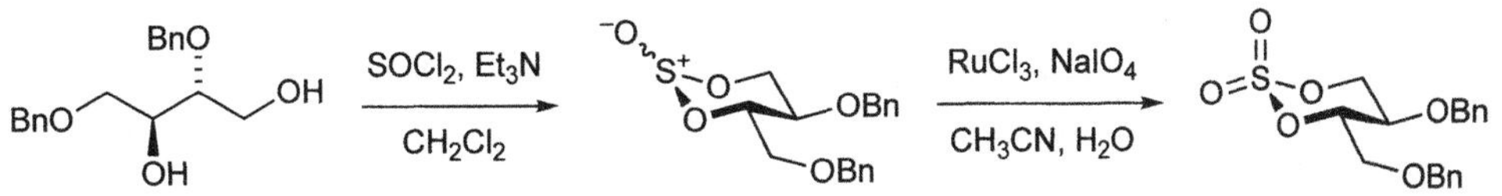

Sulfonates: This last group of esters is not at all characterized by protection of the hydroxyl group but, rather, by its *activation* towards nucleophilic substitution:

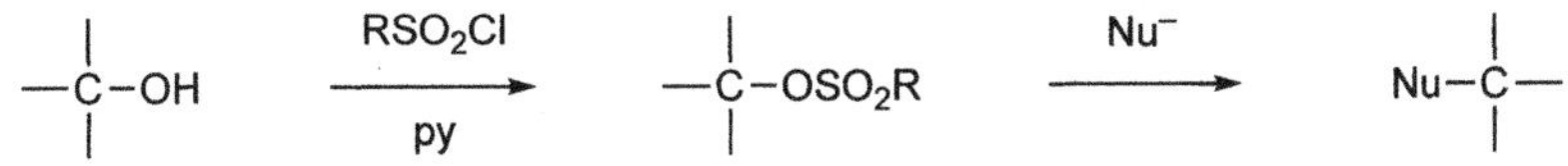

The three sulfonates commonly encountered are tosylate (4-toluenesulfonate), mesylate (methanesulfonate) and triflate (trifluoromethanesulfonate), generally installed in pyridine and using the acid chloride (4-toluenesulfonyl chloride and methanesulfonyl chloride) or trifluoromethanesulfonic anhydride.[60] For alcohols of low reactivity, the combination of methanesulfonyl chloride and triethylamine (which produces the very reactive sulfene, CH_2SO_2) is particularly effective.[61,62] The sulfonates, once installed, show the following order of reactivity towards nucleophilic displacement:

$$CF_3SO_2O- \; >> \; CH_3SO_2O- \; > \; 4\text{-}CH_3C_6H_4SO_2O-$$

An addition to the above trio of sulfonates is the imidazylate (imidazole-1-sulfamate), said to be more stable than the corresponding triflate, but of the same order of reactivity.[63,64]

The selective sulfonylation of a sugar polyol is possible,[65,66] and *N*-tosylimidazole has proven to be of some use in this regard.[67]

References start on page 67

Finally, a few general comments are offered to end this section on esters. 4-(Dimethylamino)pyridine has proven to be an excellent adjunct in the synthesis of carbohydrate esters, especially for less reactive hydroxyl groups.[68] Acyl migration of carbohydrate esters, where possible, can be a problem but can also be put to use:[38,69]

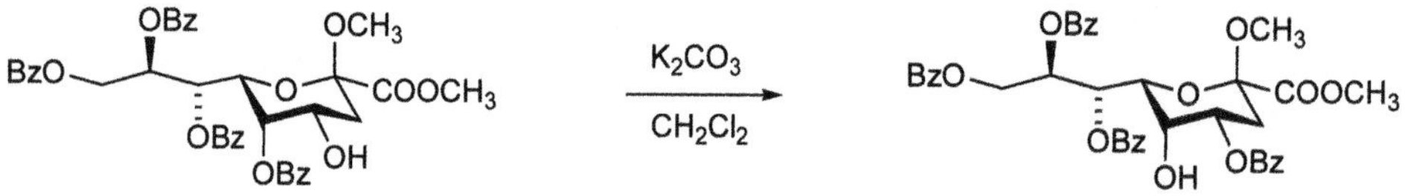

Furanosyl esters, when needed, can often be prepared indirectly from the starting sugar; for example, 1-*O*-acetyl-2,3,5-tri-*O*-benzoyl-β-D-ribose is much used in nucleoside synthesis:[70]

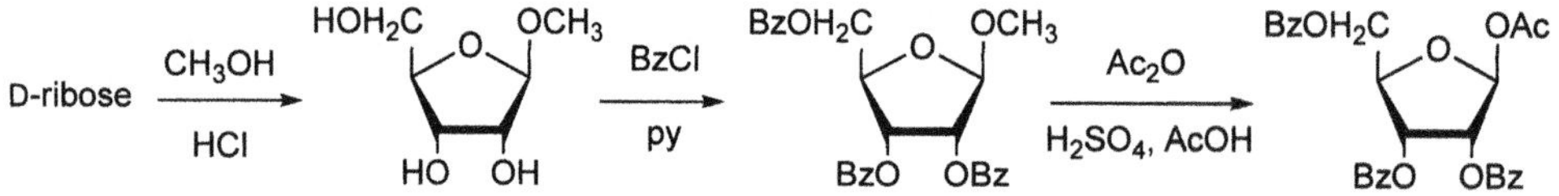

Ethers[71]

We have seen in the above discussion that esters act essentially as 'temporary' protecting groups; they are, after all, generally unstable to basic conditions. Ethers, on the contrary, are much more robust and 'permanent' groups found only at non-anomeric positions (otherwise, they would not be ethers, but the more reactive acetals); they are stable to base and mildly acidic conditions.

Methyl ethers: Methyl ethers are of little value as protecting groups for the hydroxyl group per se, as they are far too stable for easy removal, but they have a place in carbohydrate chemistry in terms of structure elucidation. Since the pioneering work of Purdie (methyl iodide, silver oxide)[72] and Haworth (dimethyl sulfate, aqueous sodium hydroxide),[73] and the improvements offered by Kuhn (methyl iodide, DMF, silver oxide)[74] and Hakomori (methyl iodide, DMSO, sodium hydride),[75] 'methylation analysis' has played a key role in the structure elucidation of oligosaccharides. For example, from enzyme-mediated hydrolysis studies, the naturally occurring reducing disaccharide, gentiobiose, was known to consist of two β-linked D-glucose units. Complete methylation of gentiobiose gave an octamethyl 'ether' that, after acid hydrolysis, yielded 2,3,4,6-tetra-*O*-methyl-D-glucose and 2,3,4-tri-*O*-methyl-D-glucose. Barring the occurrence of an unusual septanose ring form, this result defined gentiobiose as 6-*O*-β-D-glucopyranosyl-D-glucopyranose:

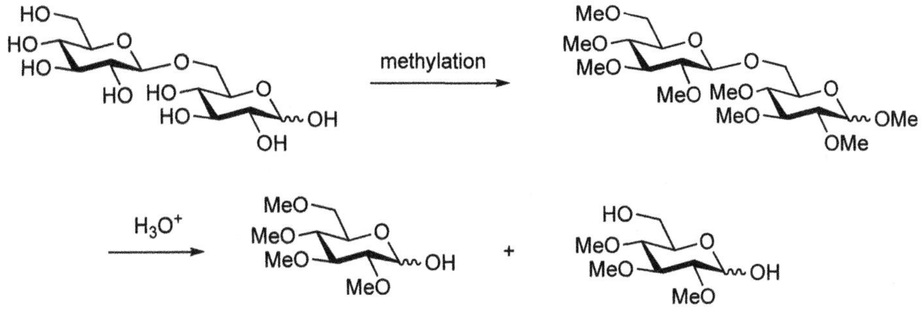

Benzyl ethers: Benzyl ethers offer a versatile means of protection for the hydroxyl group, being installed under basic (benzyl bromide, sodium hydride, DMF; benzyl bromide, sodium hydride, tetrabutylammonium iodide, THF[76,77]), acidic (benzyl trichloroacetimidate, triflic acid;[78,79] phenyldiazomethane, tetrafluoroboric acid[80]) or neutral (benzyl bromide, silver triflate) conditions.[81] Also, many methods exist for the removal of the benzyl protecting group: classical hydrogenolysis (hydrogen, palladium-on-carbon, often in the presence of an acid), catalytic transfer hydrogenolysis (ammonium formate, palladium-on-carbon, methanol),[82,83] reduction under Birch conditions (sodium, liquid ammonia), treatment with anhydrous ferric chloride[84] and even photobromination.[85]

A useful synthesis of tetra-*O*-benzyl-D-glucono-1,5-lactone is shown:[86]

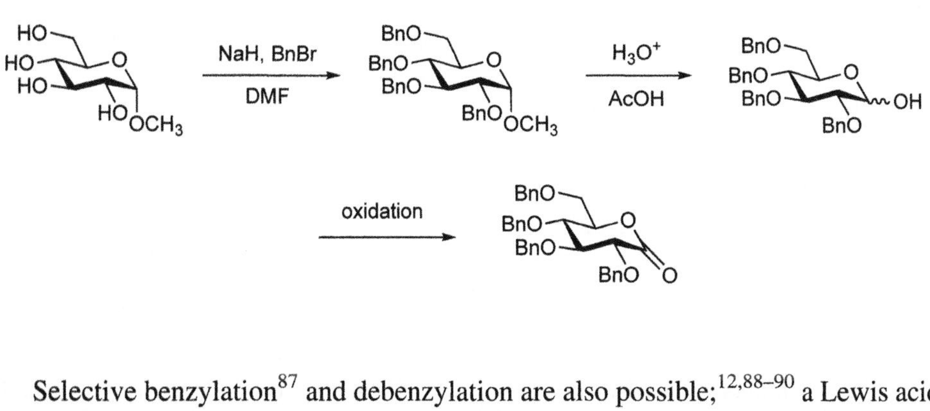

Selective benzylation[87] and debenzylation are also possible;[12,88–90] a Lewis acid in combination with acetic anhydride or benzoyl bromide is a versatile method for the conversion of a benzyl ether into an acetate or a benzoate.[91–94]

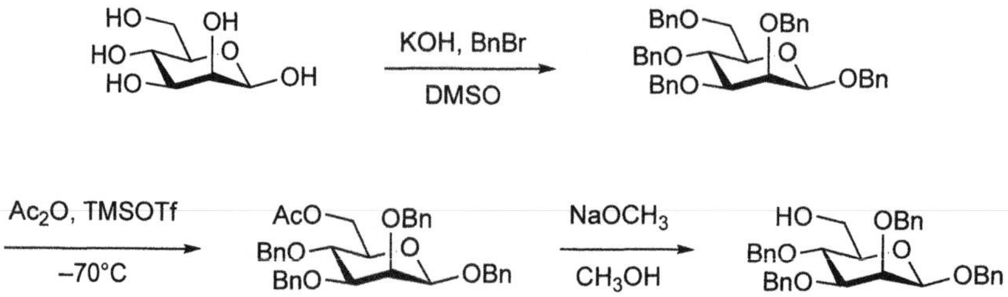

The (2-naphthyl)methyl group is closely related to the benzyl (phenylmethyl) group and may offer some advantages for the protection of the hydroxyl groups of sugars.[95,96] A novel approach installs a cyclic 'dibenzyl ether':[97]

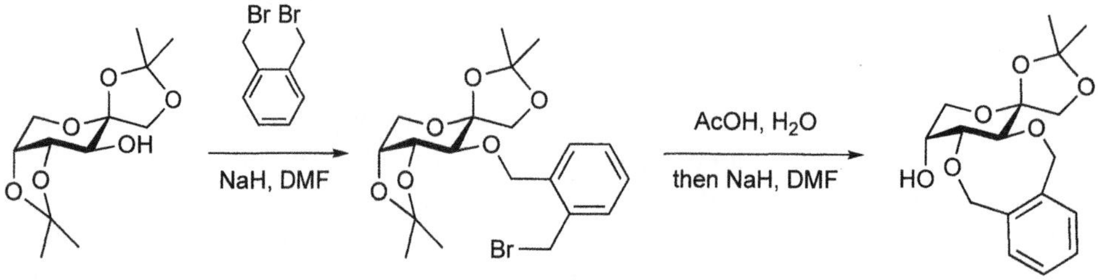

4-Methoxybenzyl ethers: These substituted benzyl ethers have found an increasing use over the past two decades, for reasons of easy installation (4-methoxybenzyl chloride or bromide, sodium hydride, DMF;[98,99] 4-methoxybenzyl trichloroacetimidate[100]) and the availability of an extra, oxidative mode of deprotection:[101]

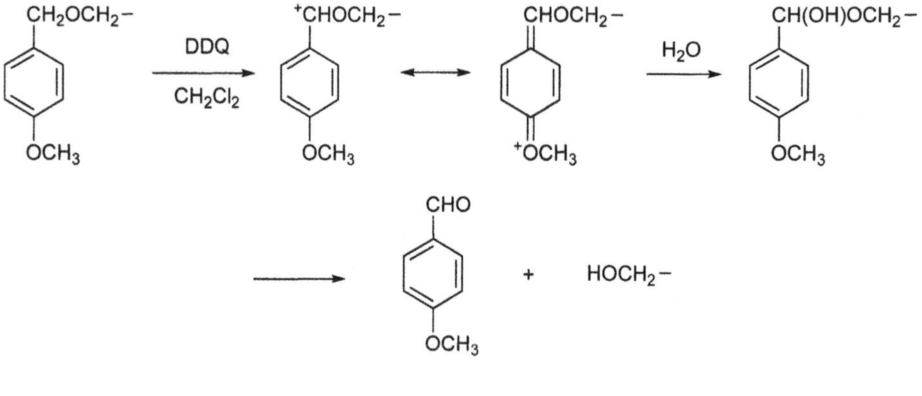

Other oxidants can also be used,[98,102,103] and good selectivity is usually observed.[104] Trifluoroacetic acid and tin(IV) chloride have also been used to remove the 4-methoxybenzyl protecting group.[105,106]

Allyl ethers:[107,108] Roy Gigg, more than anyone else, was responsible for the establishment of the allyl (prop-2-enyl) ether as a useful protecting group in carbohydrate chemistry.[109] Allyl groups may be found at both anomeric and non-anomeric positions, the latter ethers being installed under basic (allyl bromide, sodium hydride, DMF), acidic (allyl trichloroacetimidate, triflic acid)[110] or almost neutral conditions.[111] Many methods exist for the removal of the allyl group,[112] most relying on an initial prop-2-enyl to prop-1-enyl isomerization,[113] and varying from the classical (potassium *tert*-butoxide–dimethyl sulfoxide, followed by mercuric chloride[114] or acid[109]) to palladium-based (palladium-on-carbon, acid)[115,116] and rhodium-based procedures.[117–120] Other variants of the allyl group have found some use in synthesis.[121,122]

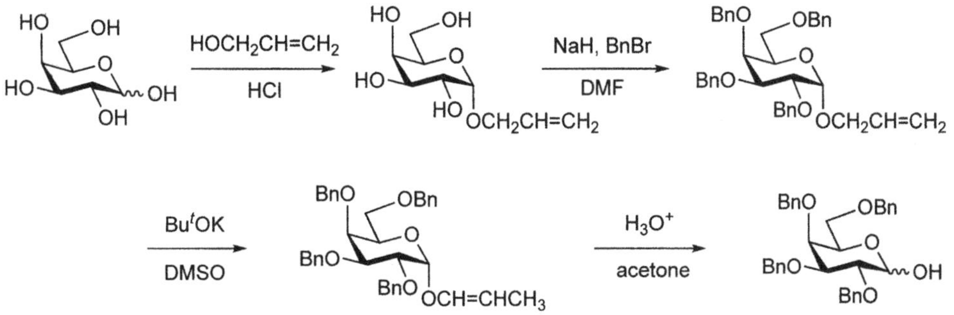

Trityl ethers: The trityl (triphenylmethyl) ether was the earliest group for the selective protection of a primary alcohol. Although the introduction of a trityl group has always been straightforward (trityl chloride, pyridine),[123] various improvements have been made.[124–126] The removal process has been much studied, and the reagents used are generally either Brønsted[127,128] or Lewis acids;[129–131] other methods include either conventional hydrogenolysis or reduction under Birch conditions.[132]

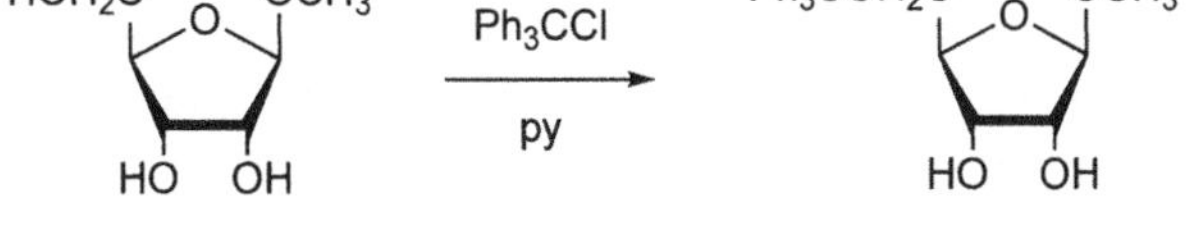

Silyl ethers:[133] The original use of silyl ethers in carbohydrates was not so much for the protection of any hydroxyl group but, rather, for the chemical modification of these normally water-soluble, non-volatile compounds. For example, the *per-O*-silylation of monosaccharides was a necessary preamble to successful analysis by gas–liquid chromatography or mass spectrometry:[134]

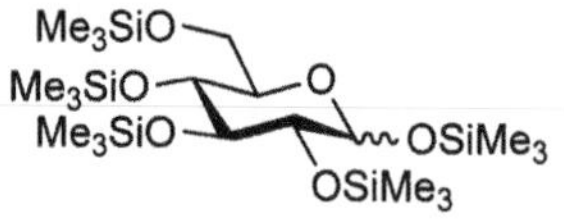

It was not until the pioneering work by Corey that silicon was used in the protection of hydroxyl groups within carbohydrates.[135] Nowadays, trimethylsilyl (TMS), triethylsilyl (TES), *tert*-butyldimethylsilyl (BMS),[e] *tert*-butyldiphenylsilyl (BPS) and triisopropylsilyl (TIPS) ethers are commonly used, with normal installation via the chlorosilane.[136,137] Quite often, the more bulky reagents show preference for a primary alcohol. Diols, especially those found in nucleosides, can be protected as a cyclic derivative:

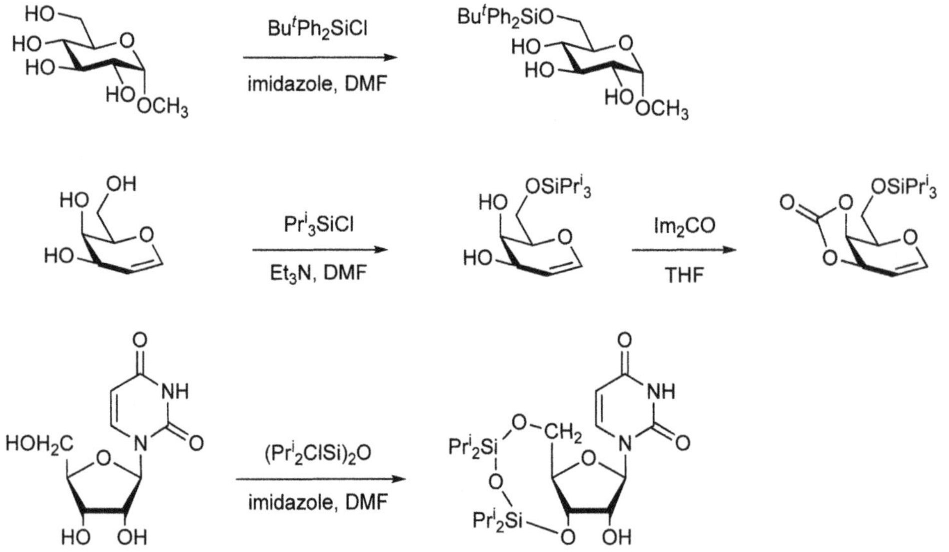

Silyl ethers survive many of the common synthetic transformations of organic chemistry[138] but are readily removed, when required, by treatment with a reagent that

---

[e] One of us (RVS) takes a stand here on the common (but silly) abbreviations used for some of the silicon protecting groups. What is the point in using 'TBDMS' for 'Bu^{t}Me$_2$Si', a saving of a mere three characters, when the more sensible 'BMS' is available? BPS is then a logical and clear abbreviation for Bu^{t}Ph$_2$Si.

supplies the fluoride ion, e.g. tetrabutylammonium fluoride (a basic reagent) or hydrogen fluoride/pyridine (the Si—F bond is extremely strong, 590 kJ mol^{-1}).[133,139] Strongly basic conditions will cleave a silyl ether, and, not surprisingly, migration of the silicon protecting group or other vulnerable residues, e.g. acyl groups, will occur under these conditions.[140,141] Silyl ethers can be cleaved under acidic conditions, and the general ease of acid hydrolysis is $Me_3SiO- > Et_3SiO- \gg Bu^tMe_2SiO- \gg Pr^i{}_3SiO- \gg Bu^tPh_2SiO-$; aqueous hydrofluoric acid is particularly attractive in that it also offers a source of fluoride ions. Some very mild procedures for the removal of silyl ethers have been reported,[142–145] with some selectivity for primary ethers over secondary ones:

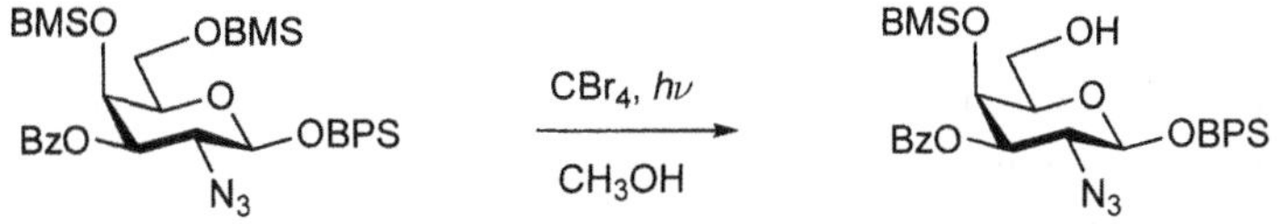

Finally, the humble TMS ether should not be ignored in synthesis:[146]

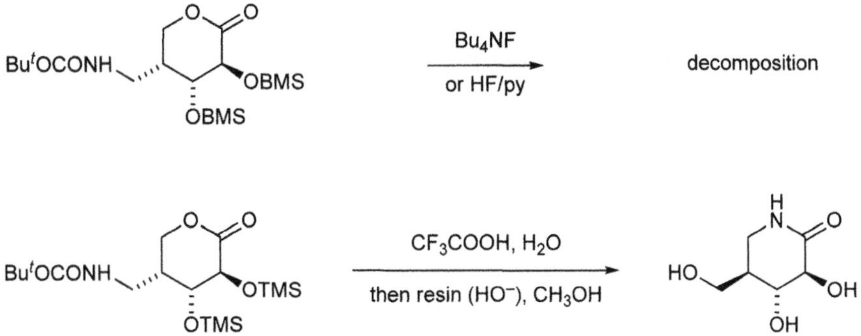

Acetals[1,71,147–150]

Before embarking on a discussion of carbohydrate acetals, it is important to review the nature and reactivity of the various hydroxyl groups within D-glucopyranose:

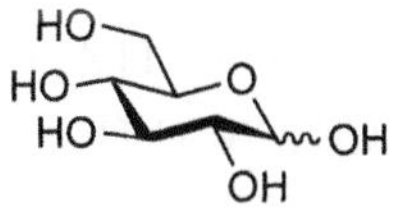

Of the five hydroxyl groups present, the *anomeric* hydroxyl group is unique, being part of a *hemiacetal* structure; all of the other hydroxyl groups show the reactions typical of an alcohol. We have already seen several unique reactions of the anomeric centre, one of which was the formation (by Fischer) of a mixture of *acetals* by the treatment of D-glucose with methanol and hydrogen chloride:

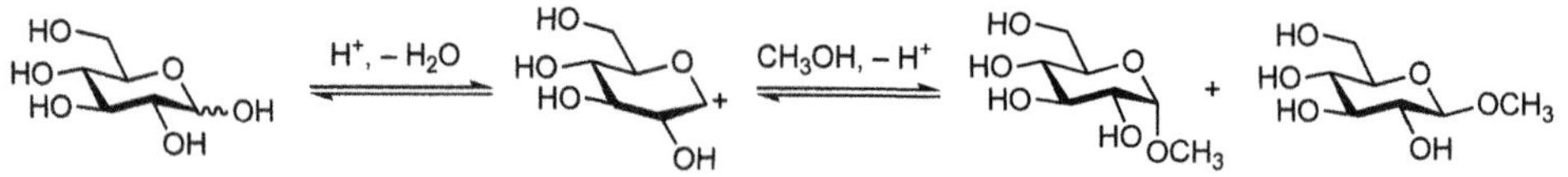

These methyl acetals, methyl α- and β-D-glucopyranoside, offered a form of protection to the anomeric centre and allowed for the useful synthesis of fully protected free sugars:

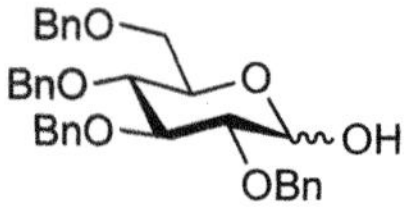

Other acetals have been developed that also offer this unique protection of the anomeric centre but have the added advantage of removal under milder and more selective conditions:

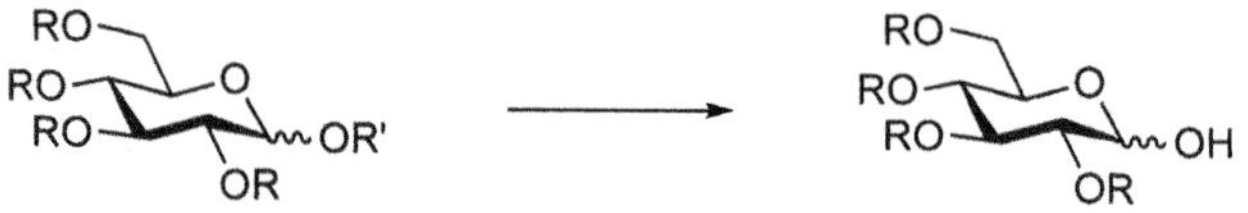

R'	Cleavage conditions	Reference(s)
CH_3	H_3O^+ or Ac_2O, $H_2SO_4/NaOCH_3$, CH_3OH	151
CH_2Ph	H_2, Pd-C or Na, NH_3	
$CH_2CH=CH_2$	Bu^tOK, $DMSO/H_3O^+$ or $(Ph_3P)_3RhCl/H_3O^+$	
CH_2CCl_3	Zn, CH_3CO_2H	152
$CH_2CH_2SiMe_3$	Bu_4NF, THF or TFA, CH_2Cl_2	147, 153
$(CH_2)_3CH=CH_2$	NBS, CH_3CN, H_2O	154

Acetals, apart from being useful in the protection of the anomeric centre, may also be used for the protection of other hydroxyl groups and are easily and selectively removed:

R'	Cleavage conditions	Reference(s)
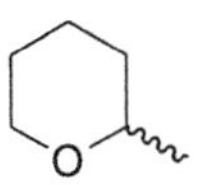	H^+	1, 147, 155–158
CH_3OCH_2	H^+	1, 159
$CH_3OCH_2CH_2OCH_2$	$ZnBr_2$, CH_2Cl_2 or H^+	1, 147, 160, 161

Even though these sorts of acetals find great use in general synthetic chemistry, their use and acceptance has been somewhat limited in carbohydrates; perhaps, the reasons for this can be found in the pages that follow.

Cyclic acetals: Any synthetic endeavours with carbohydrates must recognize the presence, more often than not, of molecules containing more than one hydroxyl group, often in *cis*-1,2 or *cis*-1,3 dispositions. So arose the need to 'protect' such diol systems, and 'cyclic acetals' were the obvious answer. The benzylidene and isopropylidene acetals stand (almost) alone as two prodigious protecting groups of diols, and some general comments are warranted.

In line with the general principles of stereochemistry and conformational analysis,[162] the cyclic acetals of benzaldehyde (benzylidene) and acetone (isopropylidene), when formed under equilibrating conditions, generally result where possible in 1,3-dioxane and 1,3-dioxolane structures, respectively:

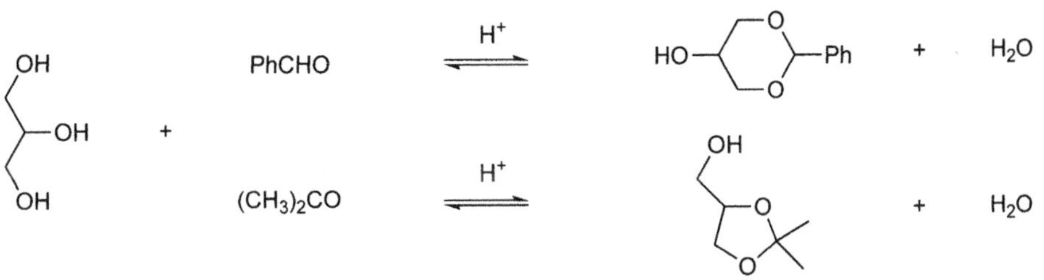

References start on page 67

Also, under these equilibrating conditions, the phenyl group will strive to take up an equatorial orientation:

However, situations sometimes arise where it is necessary to protect a 1,3-diol as an isopropylidene acetal. Then, a reagent must be found that will provide the acetal under non-equilibrating conditions, bearing in mind that the product will suffer from destabilizing 1,3-diaxial interactions involving a methyl group:

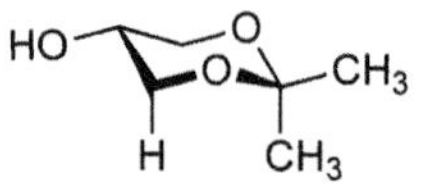

Benzylidene acetals: Treatment of a carbohydrate diol with benzaldehyde under various acidic conditions,[163–165] typically utilizing fused zinc chloride, furnishes the benzylidene acetal(s) in excellent yield:[f]

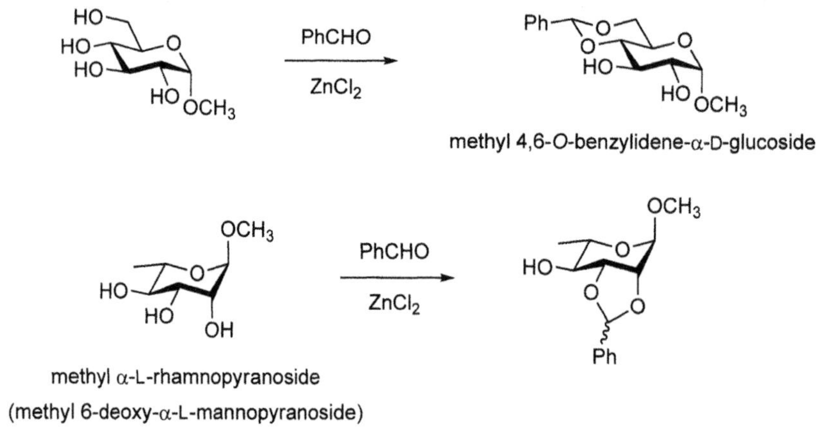

methyl 4,6-O-benzylidene-α-D-glucoside

methyl α-L-rhamnopyranoside
(methyl 6-deoxy-α-L-mannopyranoside)

---

[f] Note that, in the name, the configuration (R) of the new acetal centre is not specified but presumed, and the use of 'glucopyranoside' is an unnecessary tautology.

When this old but reliable method fails, one may resort to a 'transacetalization' process involving treatment of the diol with benzaldehyde dimethyl acetal under acidic conditions:[166–169]

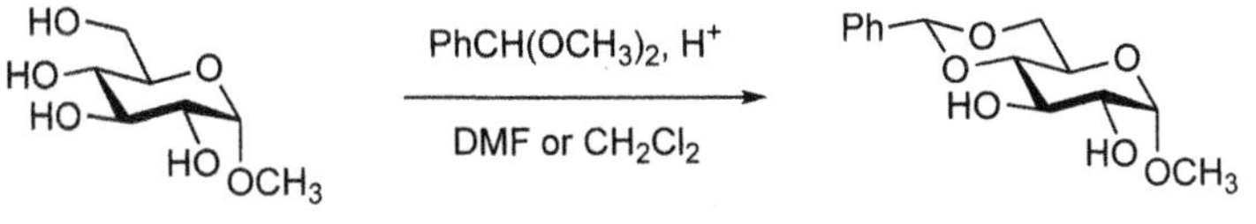

When a benzylidene acetal needs to be installed under non-acidic conditions, α,α-dibromotoluene in pyridine can be used; this is not a common method as, unsurprisingly, a mixture of diastereoisomers often results:[170]

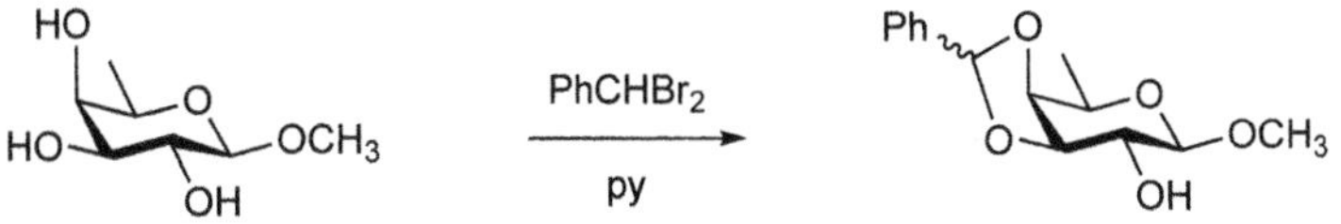

One of the strengths of the benzylidene acetal protecting group is that it may be removed by the normal reagents (acid treatment[165,171–174] or reduction under Birch conditions)[1,147–150] to regenerate the parent diol or, more productively, by methods that involve functional group transformations. Over the past two decades, an array of methods has been devised for the removal of the benzylidene acetal group with concomitant conversion into a benzyl ether, for example:[175,176]

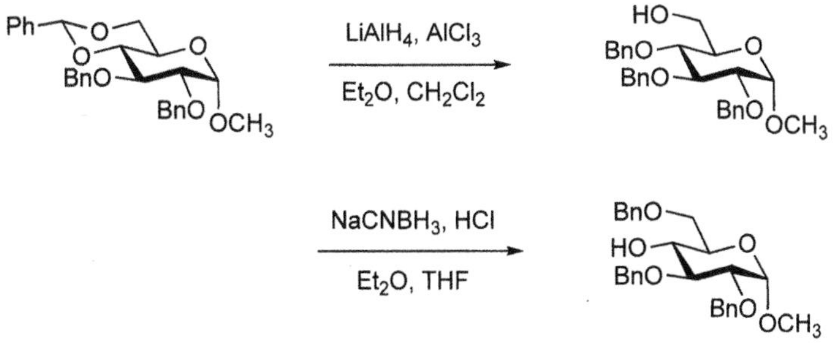

The methods are based on preferential complexation/protonation at O6 or O4, leading to intermediate carbenium ions that are subsequently reduced; the solvent obviously plays an important role in some of these transformations:

Electrophile	Reducing agent	Solvent	Product	Reference
AlCl₃*	LiAlH₄	Et₂O, CH₂Cl₂	6-OH	177
Ph₂BBr	PhSH or THF·BH₃	CH₂Cl₂	6-OH	178
Bu₂BOTf	THF·BH₃	CH₂Cl₂	6-OH	179
VO(OTf)₂ or M(OTf)₃ (M = Sc, Pr, Nd, Sm, Eu or Gd)	THF·BH₃	CH₂Cl₂	6-OH	180
CoCl₂	THF·BH₃	THF	6-OH	181
AlCl₃*	Me₃NBH₃	PhCH₃ or CH₂Cl₂ THF	6-OH 4-OH	182
AlCl₃/H₂O	Me₃NBH₃	THF	4-OH	183
Et₂OBF₃	Me₂NHBH₃	CH₂Cl₂ CH₃CN	6-OH 4-OH	184
PhBCl₂ CF₃SO₃H	Et₃SiH	CH₂Cl₂	6-OH 4-OH	185
Cu(OTf)₂	THF·BH₃ Me₂EtSiH	CH₂Cl₂ CH₃CN	6-OH 4-OH	186
HCl or CH₃SO₃H*	NaCNBH₃	THF	4-OH	187 188
CF₃COOH Et₂OBF₃	Et₃SiH	CH₂Cl₂	4-OH	189 190
CF₃SO₃H	NaCNBH₃	THF	4-OH	191

* These methods are in common use.

No detailed mechanistic studies have been performed on the reductive opening of benzylidene acetals, but it is obvious that the process is governed by a complex interplay among steric, acid–base and solvent effects.[175,182] The transformation is not restricted just to dioxane-type benzylidene acetals; some very interesting observations have been made with dioxolane acetals:[177]

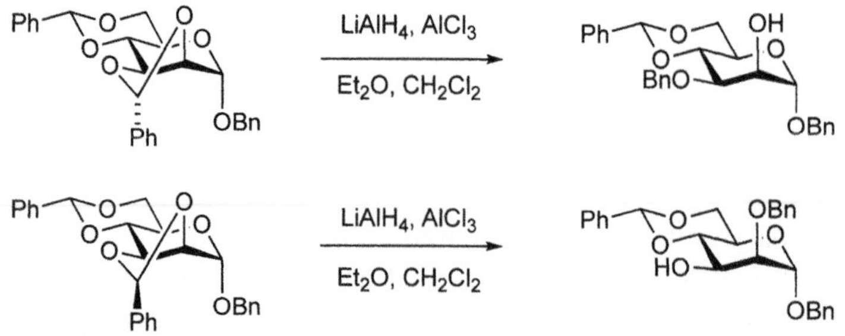

In fact, Yamaura and co-workers have recently reported a novel synthesis of 1,2-*O*-benzylidene acetals; only one diastereoisomer, when treated with diisobutylaluminium hydride, gave the required benzyl α-D-glucopyranoside:[192]

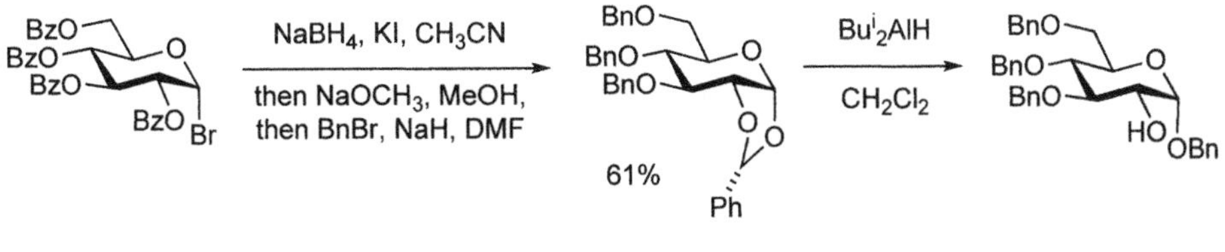

61%

Another useful transformation of benzylidene acetals involves treatment with *N*-bromosuccinimide, to form a bromo benzoate (the Hanessian–Hullar reaction):[193–198]

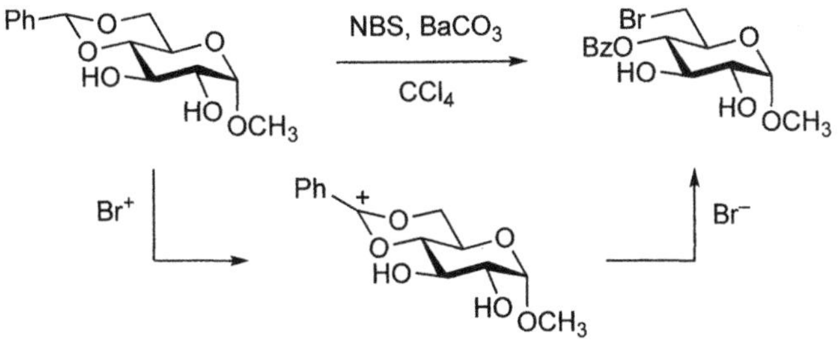

The use of calcium carbonate instead of barium carbonate seems to improve the process,[199] and a related photochemical version employing bromotrichloromethane has been reported.[200]

Finally, another oxidative method employs ozone to convert a benzylidene acetal into a hydroxy benzoate:[201]

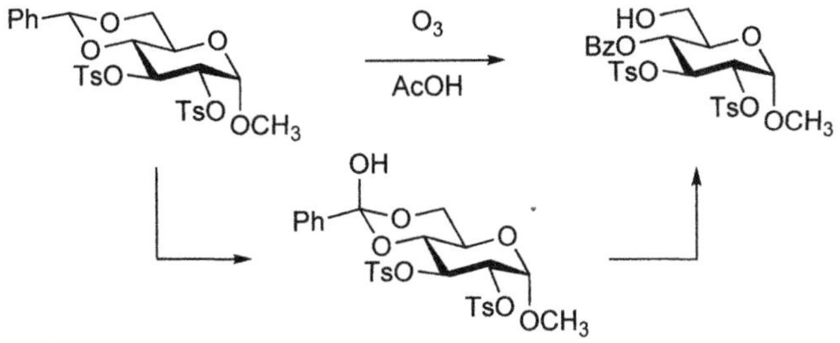

4-Methoxybenzylidene acetals: The 4-Methoxybenzylidene acetals are usually prepared from the carbohydrate diol and 4-methoxybenzaldehyde dimethyl acetal under acidic conditions:[103]

References start on page 67

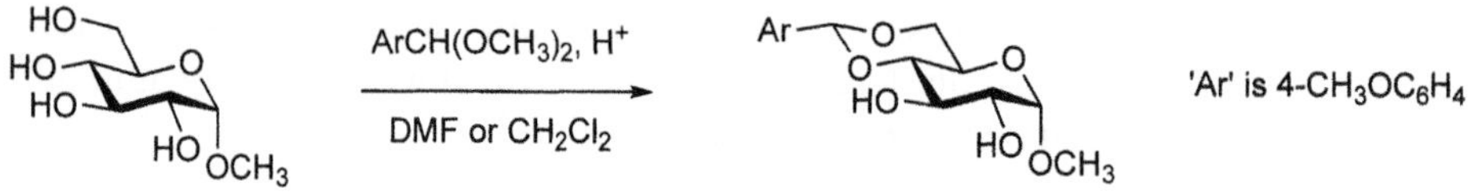

The advantages possessed by this substituted benzylidene acetal, apart from the increased lability to acid, are the somewhat milder conditions for reductive ring-opening:[103,202]

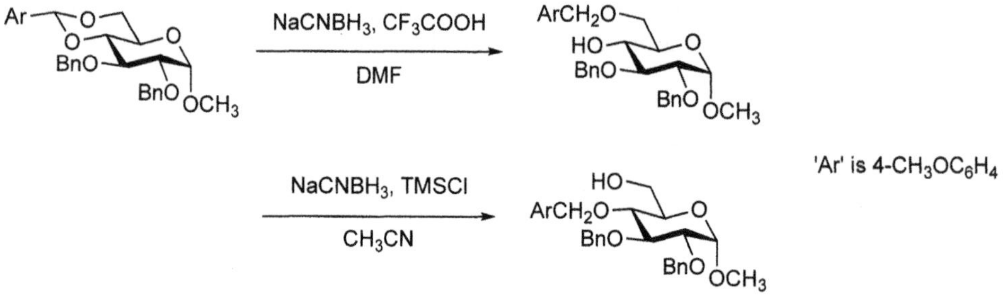

Before leaving this section, mention should be made of (2-naphthyl)methylene acetals (possessing all and more of the features of a benzylidene acetal),[203] 2-(phenylsulfonyl)ethylidene acetals (stable to acid, but removable under reducing conditions)[204,205] and, for pure creativity, an acetal capable of reductive fragmentation (by Crich):[206]

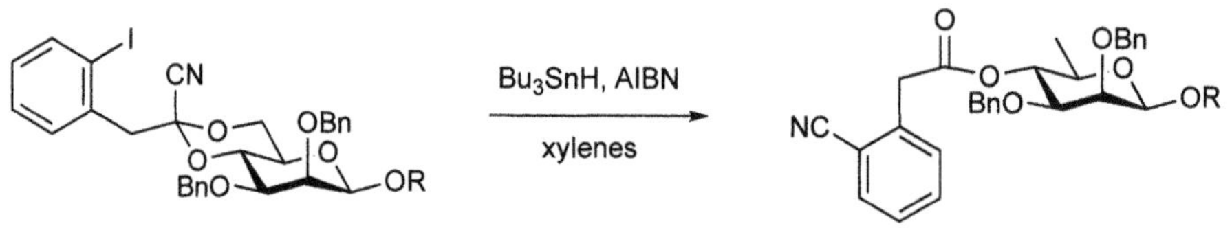

Isopropylidene acetals: Fischer prepared the first isopropylidene acetal of a sugar in 1895,[207] and since then, three main methods have emerged for the installation of this important protecting group under acidic conditions, utilizing acetone, 2,2-dimethoxypropane or 2-methoxypropene.

Nothing warms the heart of a carbohydrate chemist more than the sight of the following three classical transformations:[208]

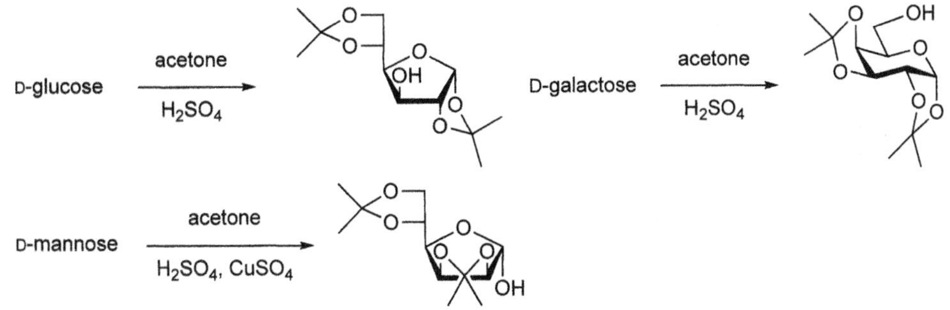

Under the strongly acidic conditions employed, all three products are the thermodynamically favoured ones and, in one step, provide direct access to molecules with just one hydroxyl group available for subsequent transformations. Other protic acids (HBF$_4$–ether,[167] 4-toluenesulfonic acid[209,210]) and some Lewis (FeCl$_3$)[211] acids promote the acetalization process equally well;[212] a polymer-bound reagent has recently been reported.[213]

2,2-Dimethoxypropane generally gives similar results to those with acetone, but useful differences are often observed:[214]

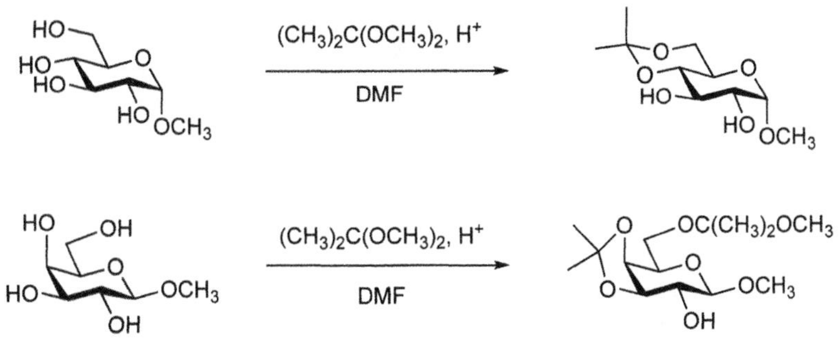

The latter transformation gives a rapid, direct and high-yielding route into a D-galactose unit suitable for further elaboration just at O2.

The last reagent, 2-methoxypropene, was developed in the mid-1970s, largely by the efforts of Gelas and Horton, for the synthesis of isopropylidene acetals. Owing to the high reactivity of the reagent and the trace amounts of acid catalyst used, the products formed were those ascribed to 'kinetic control':[215]

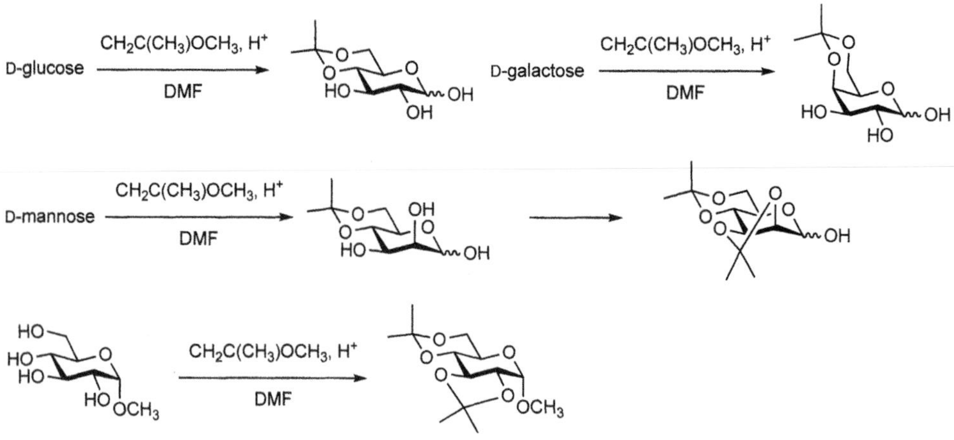

Removal of the isopropylidene protecting group generally offers few problems: trifluoroacetic acid–water (9:1) is particularly effective;[216] other occasions may warrant the use of iodine in methanol,[217] a Lewis acid[218–220] or even cerium(IV) ammonium nitrate.[221] The selective removal of just one acetal from some di-*O*-isopropylidene derivatives is possible:

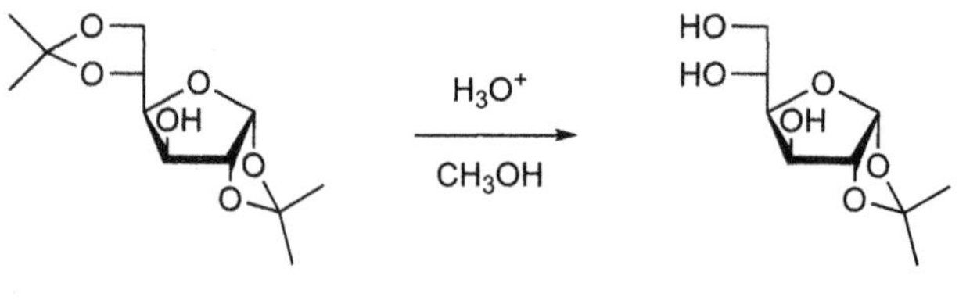

Finally, various acids immobilized on silica have recently been used for the sequential acetalization/acetylation of a range of sugar derivatives:[222]

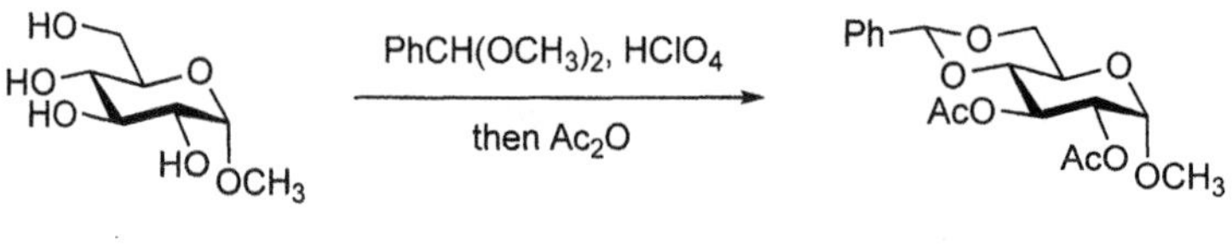

Good selectivity is also observed in the removal of one acetal from a di-*O*-isopropylidene derivative with these immobilized reagents.[223]

Diacetals: One of the triumphs of modern carbohydrate chemistry has been to attract 'into the fold', as it were, outstanding synthetic chemists from mainstream organic chemistry. A major reason for this attraction has been the occurrence of carbohydrates in various natural products and the role that carbohydrates play in many biological processes. These gifted chemists have been able to view carbohydrates in an unbiased light, and so make advances in areas that may have appeared somewhat stagnant.

In the area of acetal protecting groups, Ley has published an elegant sequence of papers, which describe new methods for the protection of diequatorial vicinal diols, as commonly found in carbohydrates.[224] In the early publications, a bisdihydropyran reagent was able to react with just the 2,3-diol of methyl α-D-galactopyranoside, by virtue of forming a dispiroacetal that is uniquely stabilized by *four* individual anomeric effects, a *trans*-decalin-like core, and four equatorial substituents on the central dioxane ring:[225]

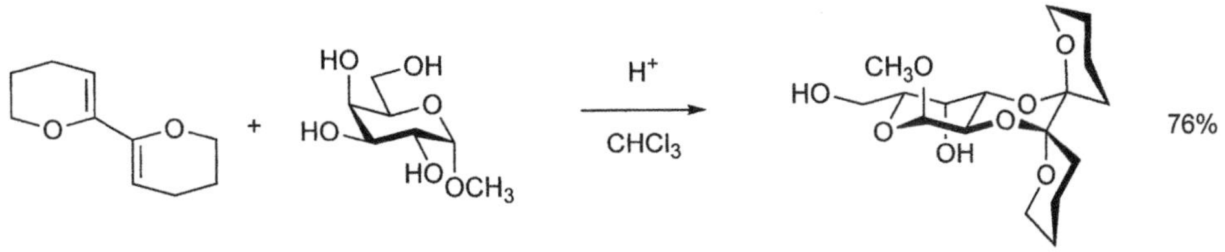

Some limitations were observed with the reaction of various alkyl α-D-mannopyranosides and the bisdihydropyran reagent and, in general, quite acidic conditions were needed to remove the dispiroacetal protecting group.[226]

In an improvement to the whole procedure, it was found that 1,1,2,2-tetramethoxycyclohexane offered the same selectivity for diequatorial vicinal diols, including those of methyl α-D-mannopyranoside:[227]

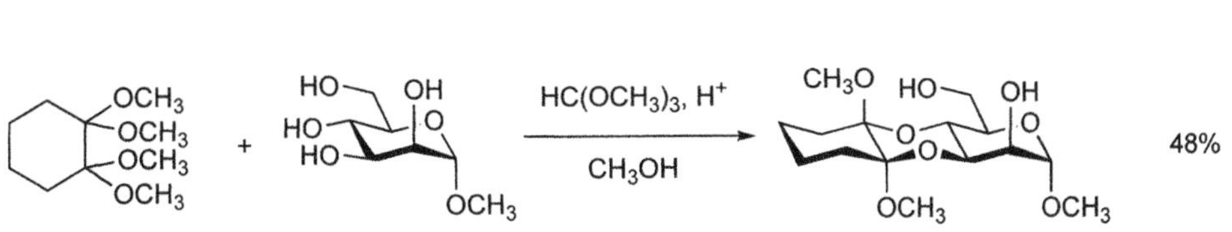

Ultimately, the reagent of choice for the protection of a diequatorial vicinal diol was found to be not a diacetal at all but, rather, a diketone:[228]

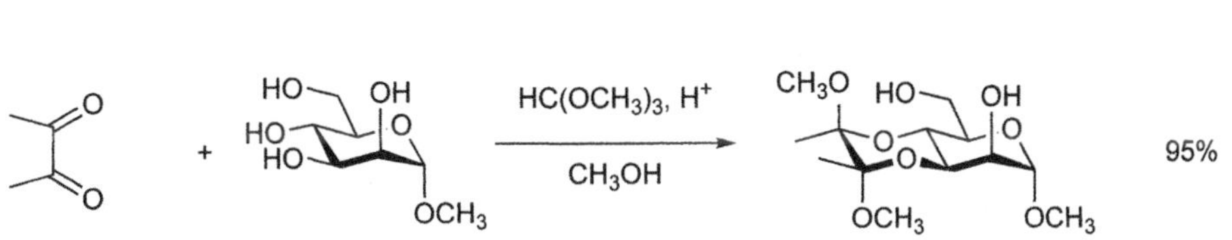

This most remarkable reaction is being increasingly used in synthetic carbohydrate chemistry.[229]

Cyclohexylidene acetals: The cyclohexylidene acetal is a rarely used protecting group (partly because the resulting NMR spectra are quite complex) that offers ease of installation (cyclohexanone, cyclohexanone dimethyl acetal or 1-methoxycyclohexene, all under acidic conditions), a propensity to form 1,3-dioxolanes and a greater stability towards hydrolysis than the corresponding isopropylidene acetal:[230]

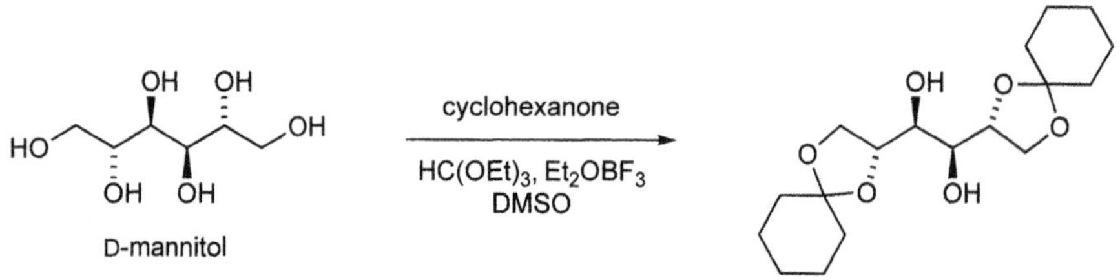

D-mannitol

cyclohexanone

$HC(OEt)_3$, Et_2OBF_3
DMSO

Dithioacetals:[231] Anomeric dithioacetals, since their first preparation by Fischer in 1894,[232] have maintained their importance to synthetic chemists because they offer one of the few ways of locking an aldose in its acyclic form.[233] Subsequent manipulations on the rest of the molecule can offer useful synthetic intermediates:[234]

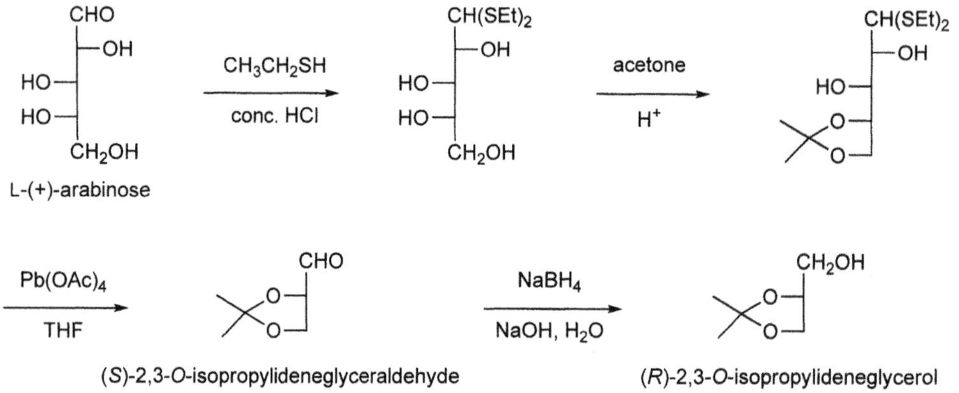

L-(+)-arabinose

CH_3CH_2SH
conc. HCl

acetone
H^+

$Pb(OAc)_4$
THF

$NaBH_4$
NaOH, H_2O

(S)-2,3-O-isopropylideneglyceraldehyde

(R)-2,3-O-isopropylideneglycerol

It is an unfortunate fact that removal of the dithioacetal protecting group, when necessary, often requires the use of environmentally unfriendly heavy metal salts, such as Hg(II). Hence, other methods have been devised.[235]

Thioacetals: Although there has been a renewed interest in acyclic thioacetals,[236] it is the cyclic thioacetals, or 1-thio sugars, that are the most important members of this class. As such, these thioacetals are versatile starting materials for the synthesis of disaccharides and higher oligomers (see Chapters 4 and 5) and owe their popularity to the ease of preparation and handling:[237–240]

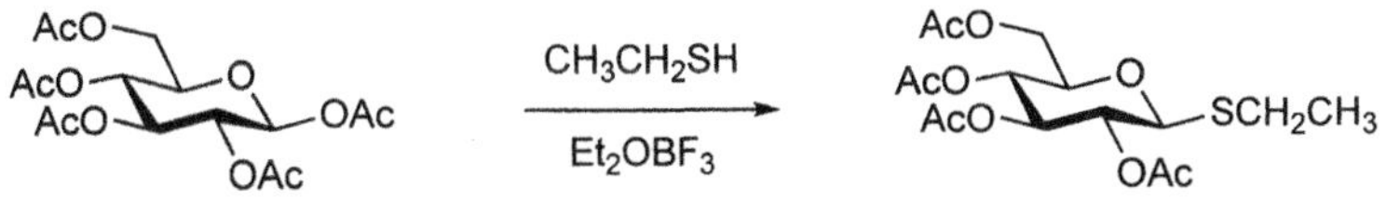

ethyl tetra-O-acetyl-1-thio-β-D-glucopyranoside

Stannylene acetals:[241–243] The treatment of a vicinal diol with dibutyltin oxide gives rise to a cyclic derivative known as a 'stannylene acetal':

Apparently, the size of the tin atom allows such stannylene acetals to form from both *cis* and *trans* vicinal diols; also, the tin atom causes an increase in the reactivity (nucleophilicity) of an attached oxygen atom so that subsequent acylations and alkylations may be performed under very mild conditions:

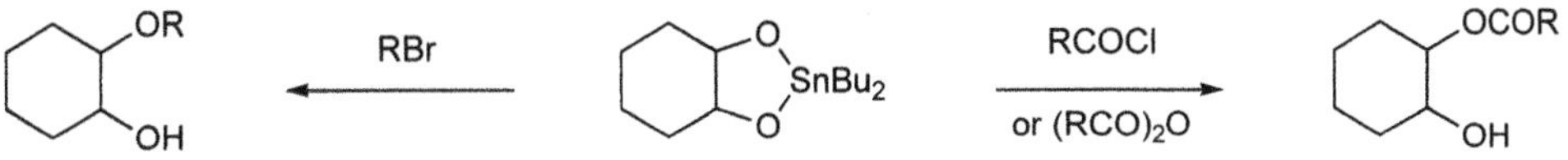

Not surprisingly, this sequence of reactions has found great application in the selective protection of carbohydrate diols and polyols:[244,245]

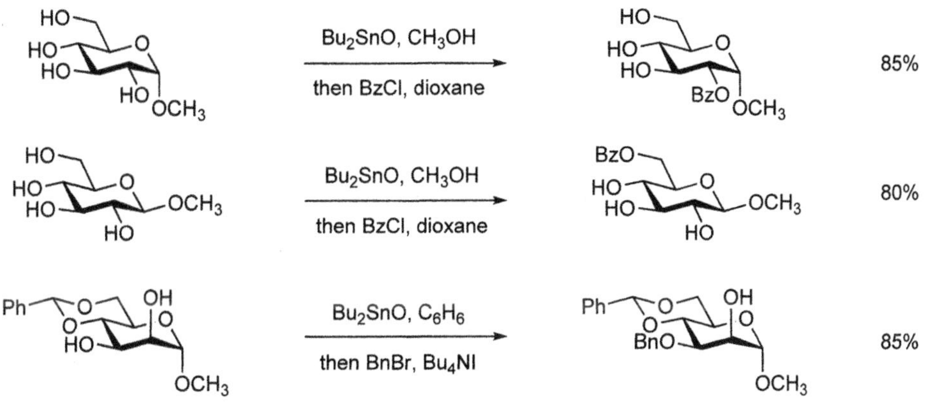

The above transformations show that, even though the acylation/alkylation is regioselective, it is not always possible to predict the outcome of a particular reaction. Acylations with an acid halide reportedly give different products from those conducted with an acid anhydride.[246] In general, an equatorial oxygen is functionalized in preference to one that is axial,[247] and the addition of a tetrabutylammonium halide is necessary to increase the rate of the alkylation reaction.[248,249] Useful building blocks for glycoside synthesis can be prepared from sugar polyols and benzoyl chloride in the presence of an excess of dibutyltin oxide.[40]

Two conflicting publications, both employing dibutyltin dimethoxide as the reagent, have highlighted the care that must be taken in making generalizations about this particularly useful synthetic method.[250,251] A regioselective sulfation of disaccharides that also uses stannylene acetal methodology has been reported.[252,253] Finally, a report on anomeric stannylene acetals allows for the isomerization of 6-O-trityl-D-galactose into the rare sugar, D-talose:[254]

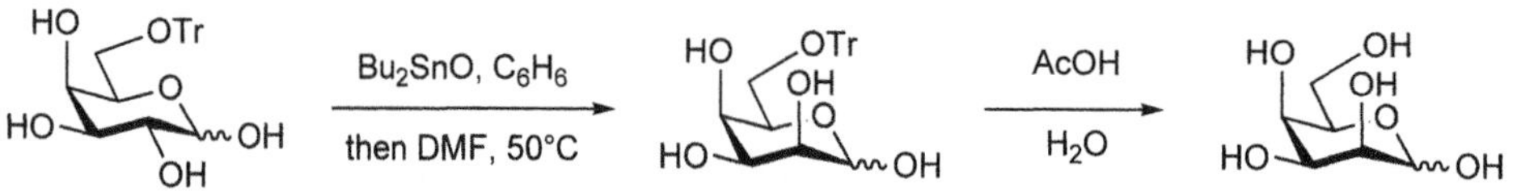

Shortly after the establishment of the stannylene acetal methodology, it was found that the treatment of an alcohol with bis(tributyltin) oxide gave rise to a 'stannyl ether'.[255]

$$(Bu_3Sn)_2O \quad + \quad 2\ HOR \quad \longrightarrow \quad 2\ Bu_3SnOR \quad + \quad H_2O$$

Again, the reactivity of the oxygen in the stannyl ether was greatly enhanced, being able to combine directly with acylating agents, but again needing the presence of a tetrabutylammonium halide for successful alkylation.[256] Some interesting transformations of carbohydrate polyols were observed:

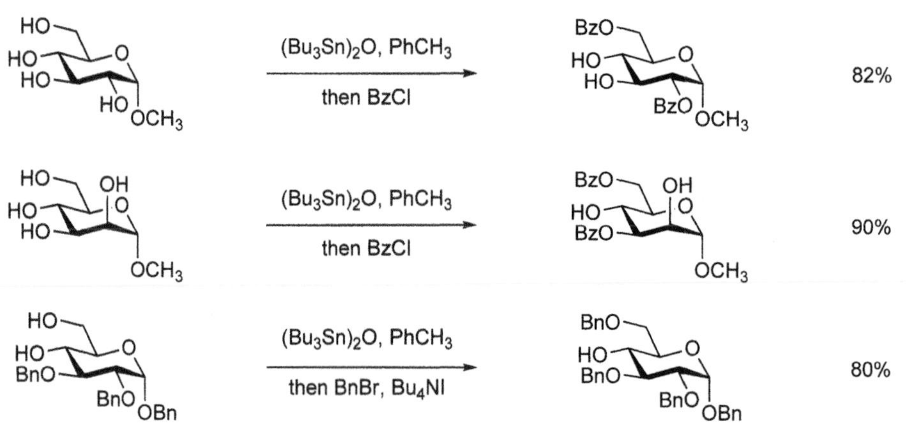

Comments have been made on the variability of the regioselectivity of the process according to the reaction conditions employed.[246,257]

The Protection of Amines

So far, the carbohydrates that we have encountered consist of just carbon, hydrogen and oxygen. However, other heteroatoms, nitrogen and phosphorus in particular, are commonly included in carbohydrate structures, and an important class is that of the 'amino sugars':

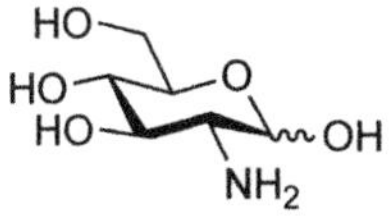

'D-glucosamine'

2-amino-2-deoxy-D-glucopyranose

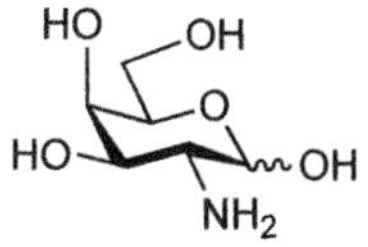

'D-galactosamine'

2-amino-2-deoxy-D-galactopyranose

Traditionally, amino group protection in carbohydrates relied on the chemistry developed earlier in the peptide field (benzyloxycarbonyl, *tert*-butoxycarbonyl). However, removal of such carbamyl groups requires the use of hydrogen, anhydrous acid or strong base, conditions that may adversely affect a carbohydrate, protected or otherwise. So arose the need to develop other protecting groups for the primary amine.

The acetyl group is found in molecules such as *N*-acetyl-D-glucosamine, a common component of many natural oligosaccharides and polysaccharides.

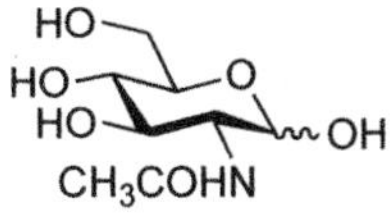

N-acetyl-D-glucosamine

2-acetamido-2-deoxy-D-glucopyranose

However, most attempts to use acetyl as a protecting group for nitrogen in carbohydrate synthesis have failed, mainly owing to the inherent reactivity of what is actually an amide. For example, we will see in Chapter 4 that any attempt to activate the anomeric carbon of a derivative of *N*-acetyl-D-glucosamine

References start on page 67

inevitably leads to an oxazolinium ion, which is often not reactive enough. Even worse, the oxazoline can be formed and isolated as a significant but unwanted by-product:

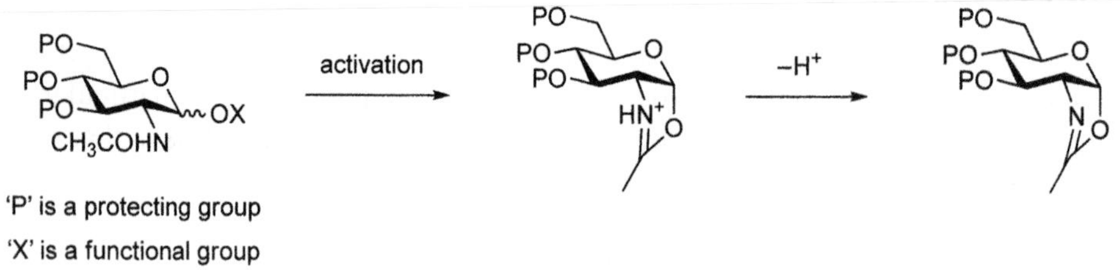

'P' is a protecting group

'X' is a functional group

In order to circumvent such a problem, let us reflect on the functional groups within D-glucosamine: a primary amine, a hemiacetal and three other hydroxyl groups. It is well established that the amine is more nucleophilic than any of the hydroxyl groups and so, not surprisingly, it may be selectively functionalized, bearing in mind that this protecting group for nitrogen must be commensurate with those chosen for the hydroxyl groups, and the task at hand. The protected nitrogen will often now be an amide and the hydroxyl groups often esters, the former more stable to most reagents than the latter, but presenting potential problems in later removal (deprotection). Consequently, there has to be something 'special' about these amides as protecting groups.

Another issue in a simple amide derived from a primary (and for that matter, secondary) amine is the potential for isomerism associated with rotation around the NH—CO bond:

If this rotation is 'slow' on the NMR timescale, then annoying broadening effects can be observed in the resultant spectra; such an issue has plagued the spectra of benzyloxycarbonyl and *tert*-butoxycarbonyl derivatives of amines, but the problem can be minimized at higher instrument temperature. Despite this line broadening in NMR spectra, special amide and carbamate protecting groups have been designed for the primary amine of sugars, and the most useful of these are listed; each product is characterized by possessing a weakly basic nitrogen atom, and being susceptible to a unique deprotection strategy:

Reactant	Reagent	Product	Deprotection	Product	Reference(s)
—NH_2	CCl_3COCl	—$NHCOCCl_3$	Bu_3SnH, AIBN	—NHAc	258
—NH_2	$(CF_3CO)_2O$	—$NHCOCF_3$	K_2CO_3 or NH_3	—NH_2	1
—NH_2	CCl_3CH_2OCOCl	—$NHCOOCH_2CCl_3$	Zn	—NH_2	1, 259–261
—NH_2	CH_2CHCH_2OCOCl	—$NHCOOCH_2CHCH_2$	Pd(0)	—NH_2	1, 262

A logical extension of the above 'monovalent' approach for a primary amine would be a 'divalent' protecting group. In its simplest form, one could imagine, for example, N-acetyl-N-trichloroethoxycarbonyl protection,[260] with the nitrogen now present as an even less basic imide. Although this arrangement could still lead to rotamer problems (in NMR spectra), it has the dual advantages of easy and selective removal of the N-trichloroethoxycarbonyl group, and retention of the (usually desired) acetyl group in the product. An even better arrangement would be to have a *symmetrical* imide, and the simplest is the *N,N*-diacetyl group; again, once deprotected, one acetyl group is retained in the product:

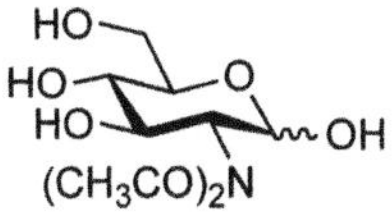

Carbohydrate chemists are always busy, almost preoccupied, with the invention of new protecting groups, and it comes as no surprise to see a whole range of these divalent, symmetrical groups for the primary amine.[263] Once again, these groups are characterized by easy installation and, most importantly, mild and selective removal:

References start on page 67

Reactant	Reagent	Product	Deprotection	Product	Reference(s)
—NHAc	CH₃COCl or CH₂C(CH₃)OAc	—NAc₂	NaOCH₃	—NHAc	264, 265
—NH₂			N₂H₄ or H₂N(CH₂)₂NH₂	—NH₂	23, 266–268
—NH₂			H₂N(CH₂)₂NH₂ or NaBH₄	—NH₂	269–274
—NH₂	(CH₃COCH₂)₂		NH₂OH	—NH₂	275
—NH₂			NaOH, then H₃O⁺	—NH₂	276
—NH₂	(EtOCS)₂S, then ClCOSCl		Zn, THF/H₂O, then Ac₂O	—NHAc	277
—NH₂	then Ac₂O		NaOMe, then Bu₃SnH	—NHAc	278
—NH₂			NH₃	—NH₂	279

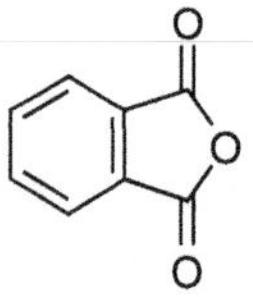

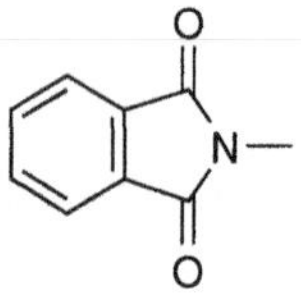

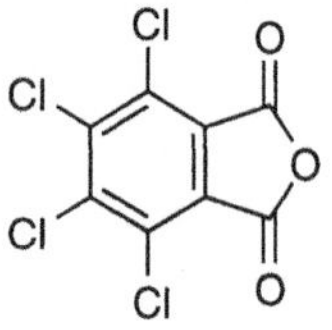

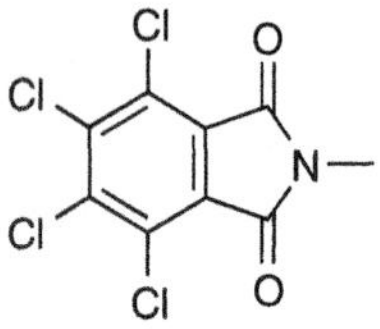

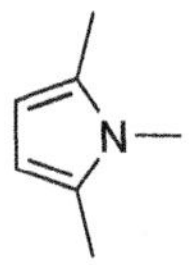

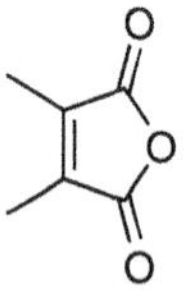

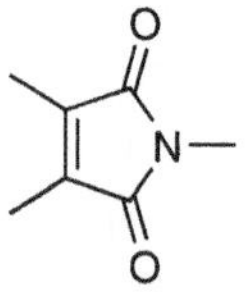

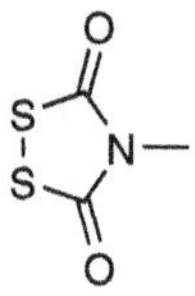

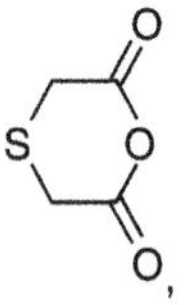

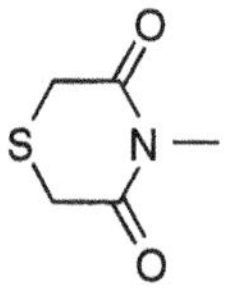

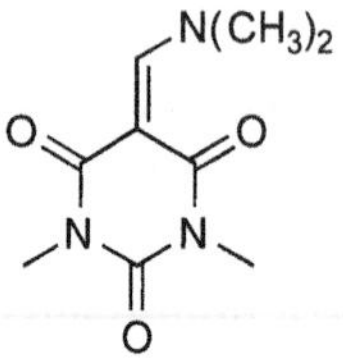

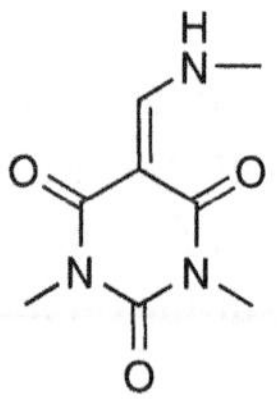

(continued)

Reactant	Reagent	Product	Deprotection	Product	Reference(s)
—NH$_2$	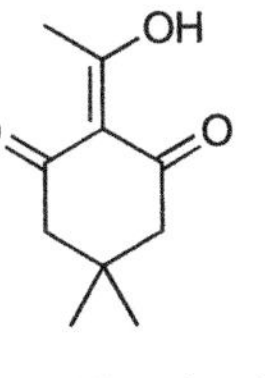 (2-acetyldimedone)	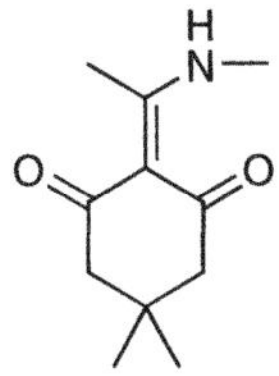	NH$_3$, N$_2$H$_4$ or RNH$_2$	—NH$_2$	280
—NH$_2$	TfN$_3$, Cu(II) or Zn(II)	—N$_3$	H$_2$/Pd-C or PPh$_3$/H$_2$O	—NH$_2$	281–283
—NH$_2$	BnBr	—NBn$_2$	H$_2$/Pd(OH)$_2$-C	—NH$_2$	284

The azide group is a common precursor to the amino group and can be introduced at C2 (usually) of a carbohydrate by a number of means;[285] conversely, a not-so-obvious protection of the amino group is offered by the azide group:

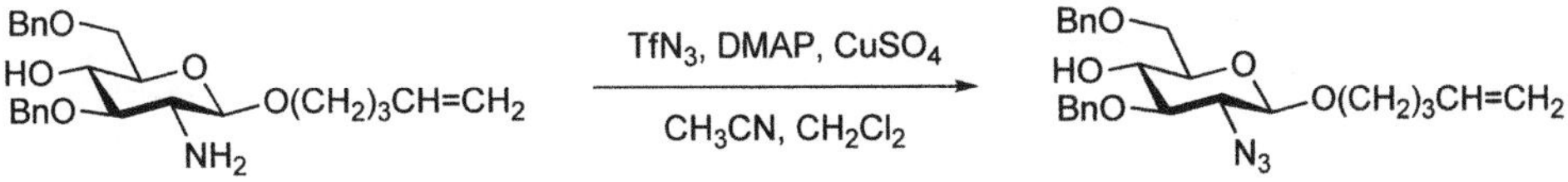

This primary amine to azide conversion is now most conveniently performed with imidazole-1-sulfonyl azide hydrochloride, a shelf-stable, crystalline diazotransfer reagent:[286]

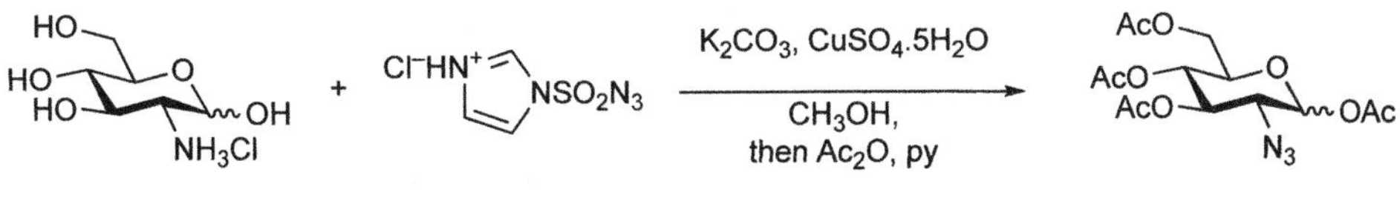

Finally, mention should be made of microwave assistance for the manipulation of protecting groups;[287,288] also, there are those who favour the use of 'fluorous' protecting groups to assist in the matter of purification.[289]

The 'wagon wheel' below utilizes much of the chemistry discussed above to illustrate the flexibility in protection of a molecule such as methyl α-D-galactopyranoside:

References start on page 67

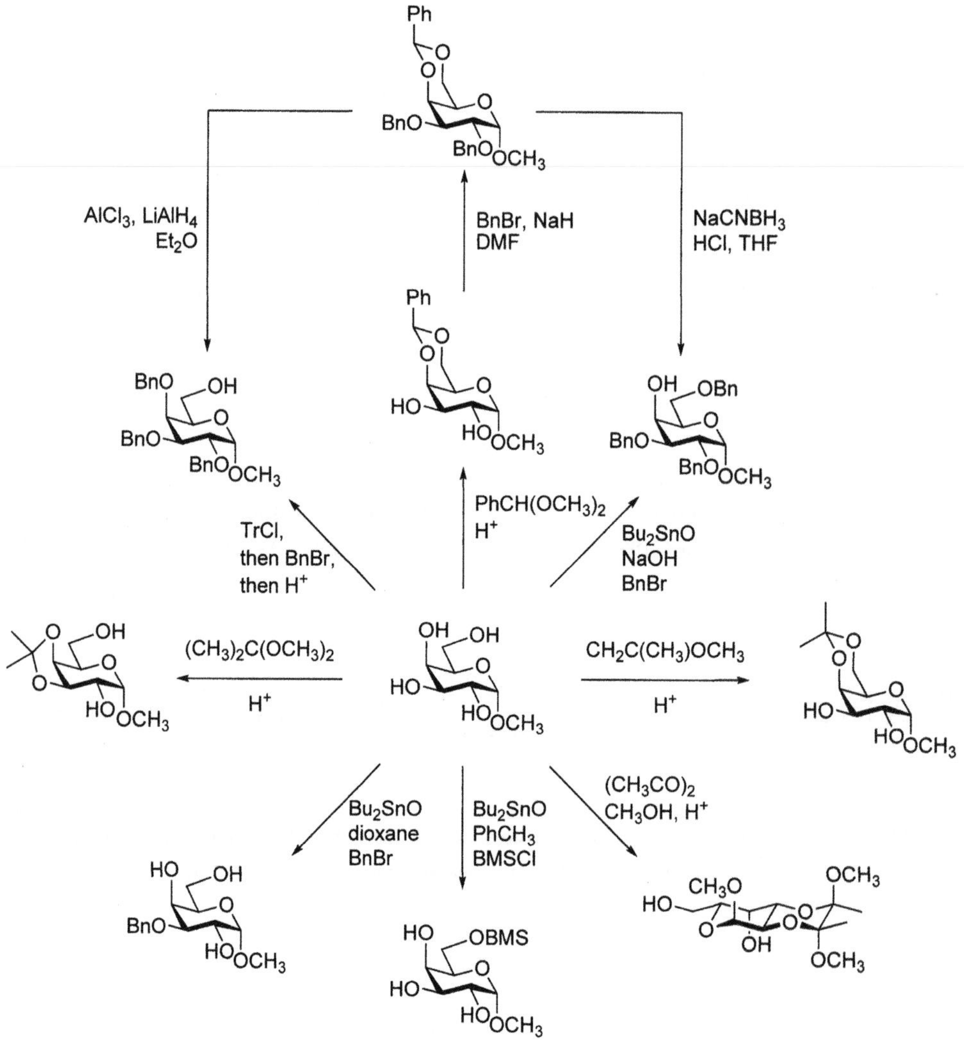

Ph

BnO

BnO OCH₃

AlCl₃, LiAlH₄
Et₂O

BnBr, NaH
DMF

NaCNBH₃
HCl, THF

Ph

HO

HO OCH₃

BnO OH

BnO

BnO OCH₃

OH OBn

BnO

BnO OCH₃

TrCl,
then BnBr,
then H⁺

PhCH(OCH₃)₂
H⁺

Bu₂SnO
NaOH
BnBr

OH

HO OCH₃

(CH₃)₂C(OCH₃)₂
H⁺

OH OH

HO

HO OCH₃

CH₂C(CH₃)OCH₃
H⁺

HO

HO OCH₃

Bu₂SnO
dioxane
BnBr

Bu₂SnO
PhCH₃
BMSCl

(CH₃CO)₂
CH₃OH, H⁺

HO OH

BnO

HO OCH₃

HO OBMS

HO

HO OCH₃

CH₃O   OCH₃

HO   O

OH   OCH₃

Orthogonality

Orthogonal protecting groups are those in a set that require a unique reagent for their individual removal, being unreactive to the conditions required for the removal of the other groups of the set. For example, the acetyl group (of an ester) and the benzyl group (of an ether) are an obvious orthogonal pair of protecting groups for the hydroxyl group. This concept has been much extended over recent years, to include monosaccharides that are protected in a fully orthogonal sense:[290-292]

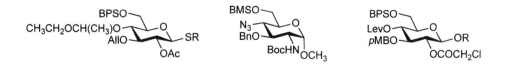

A regioselective one-pot protection of carbohydrates employs many of the principles of orthogonality.[293,294]

An interesting twist, that of 'uni-chemo protection', has been given to the concept of orthogonality; a diol, for example, is protected at one hydroxyl as an ester of a single amino acid and at the other (hydroxyl) again as an ester but of a dipeptide. One cycle of the Edman degradation then releases one hydroxyl group that, after functionalization, is followed by another (Edman) degradation to release the other hydroxyl group:[295]

References

1. Greene, T.W. and Wuts, P.G.M. (1991, 1999). *Protective Groups in Organic Synthesis.* John Wiley and Sons.
2. Kocienski, P.J. (1994). *Protecting Groups.* Thieme.
3. Jarowicki, K. and Kocienski, P. (2001). *J. Chem. Soc., Perkin Trans. 1,* 2109.
4. Hanson, J.R. (1999). *Protecting Groups in Organic Synthesis.* Sheffield Academic Press.
5. Grindley, T.B. (1996). In *Modern Methods in Carbohydrate Synthesis* (S.H. Khan and R.A. O'Neill, eds) p. 225. Harwood Academic.
6. Oscarson, S. (2006). In *The Organic Chemistry of Sugars* (D.E. Levy and P. Fügedi, eds) p. 53. CRC Press.
7. Wolfrom, M.L. and Thompson, A. (1963). *Methods Carbohydr. Chem.,* **2**, 211.
8. Conchie, J., Levvy, G.A. and Marsh, C.A. (1957). *Adv. Carbohydr. Chem.,* **12**, 157.
9. Swain, C.G. and Brown, J.F., Jr. (1952). *J. Am. Chem. Soc.,* **74**, 2538.
10. Schmidt, R.R. (1986). *Angew. Chem. Int. Ed. Engl.,* **25**, 212.
11. Schmidt, R.R. and Michel, J. (1984). *Tetrahedron Lett.,* **25**, 821.
12. Kartha, K.P.R. and Field, R.A. (1997). *Tetrahedron,* **53**, 11753.
13. Zemplén, G. and Pacsu, E. (1929). *Ber. Dtsch. Chem. Ges.,* **62**, 1613.
14. Lemieux, R.U. and Stick, R.V. (1975). *Aust. J. Chem.,* **28**, 1799.

15. Herzig, J., Nudelman, A., Gottlieb, H.E. and Fischer, B. (1986). *J. Org. Chem.*, **51**, 727.
16. Ellervik, U. and Magnusson, G. (1997). *Tetrahedron Lett.*, **38**, 1627.
17. Lemieux, R.U., Hendriks, K.B., Stick, R.V. and James, K. (1975). *J. Am. Chem. Soc.*, **97**, 4056.
18. Vekemans, J.A.J.M., Franken, G.A.M., Chittenden, G.J.F. and Godefroi, E.F. (1987). *Tetrahedron Lett.*, **28**, 2299.
19. Liu, H.-M., Yan, X., Li, W. and Huang, C. (2002). *Carbohydr. Res.*, **337**, 1763.
20. Wang, H., She, J., Zhang, L.-H. and Ye, X.-S. (2004). *J. Org. Chem.*, **69**, 5774.
21. Ishihara, K., Kurihara, H. and Yamamoto, H. (1993). *J. Org. Chem.*, **58**, 3791.
22. Moitessier, N., Englebienne, P. and Chapleur, Y. (2005). *Tetrahedron*, **61**, 6839.
23. Best, W.M., Dunlop, R.W., Stick, R.V. and White, S.T. (1994). *Aust. J. Chem.*, **47**, 433.
24. Mikamo, M. (1989). *Carbohydr. Res.*, **191**, 150.
25. Grynkiewicz, G., Fokt, I., Szeja, W. and Fitak, H. (1989). *J. Chem. Res., Synop.*, 152.
26. Sim, M.M., Kondo, H. and Wong, C.-H. (1993). *J. Am. Chem. Soc.*, **115**, 2260.
27. Johnsson, R. and Ellervik, U. (2005). *Synlett*, 2939.
28. Tiwari, P. and Misra, A.K. (2006). *Tetrahedron Lett.*, **47**, 3573.
29. Gridley, J.J., Hacking, A.J., Osborn, H.M.I. and Spackman, D.G. (1998). *Tetrahedron*, **54**, 14925.
30. Kennedy, J.F., Kumar, H., Panesar, P.S., Marwaha, S.S., Goyal, R., Parmar, A. and Kaur, S. (2006). *J. Chem. Technol. Biotechnol.*, **81**, 866.
31. Horrobin, T., Tran, C.H. and Crout, D. (1998). *J. Chem. Soc., Perkin Trans. 1*, 1069.
32. Bashir, N.B., Phythian, S.J., Reason, A.J. and Roberts, S.M. (1995). *J. Chem. Soc., Perkin Trans. 1*, 2203.
33. Hennen, W.J., Sweers, H.M., Wang, Y.-F. and Wong, C.-H. (1988). *J. Org. Chem.*, **53**, 4939.
34. Rencurosi, A., Poletti, L., Russo, G. and Lay, L. (2003). *Eur. J. Org. Chem.*, 1672.
35. Kadereit, D. and Waldmann, H. (2001). *Chem. Rev.*, **101**, 3367.
36. La Ferla, B. (2002). *Monatsh. Chem.*, **133**, 351.
37. Byramova, N.E., Ovchinnikov, M.V., Backinowsky, L.V. and Kochetkov, N.K. (1983). *Carbohydr. Res.*, **124**, C8.
38. Haines, A.H. (1976). *Adv. Carbohydr. Chem. Biochem.*, **33**, 11.
39. Williams, J.M. and Richardson, A.C. (1967). *Tetrahedron*, **23**, 1369.
40. Zhang, Z. and Wong, C.-H. (2002). *Tetrahedron*, **58**, 6513.
41. Carey, F.A. and Hodgson, K.O. (1970). *Carbohydr. Res.*, **12**, 463.
42. Chittenden, G.J.F. (1971). *Carbohydr. Res.*, **16**, 495.
43. Pelyvás, I.F., Lindhorst, T.K., Streicher, H. and Thiem, J. (1991). *Synthesis*, 1015.
44. Glaudemans, C.P.J. and Bertolini, M.J. (1980). *Methods Carbohydr. Chem.*, **8**, 271.
45. van Boeckel, C.A.A. and Beetz, T. (1983). *Tetrahedron Lett.*, **24**, 3775.
46. Lefeber, D.J., Kamerling, J.P. and Vliegenthart, J.F.G. (2000). *Org. Lett.*, **2**, 701.
47. Jiang, L. and Chan, T.-H. (1998). *J. Org. Chem.*, **63**, 6035.
48. Greene, T.W. and Wuts, P.G.M. (1991). *Protective Groups in Organic Synthesis*, p. 99. John Wiley and Sons.
49. den Hartog, J.A.J., Wille, G. and van Boom, J.H. (1981). *Recl. Trav. Chim. Pays-Bas*, **100**, 320.
50. Liu, X., Stocker, B.L. and Seeberger, P.H. (2006). *J. Am. Chem. Soc.*, **128**, 3638.
51. Arranz, E. and Boons, G.-J. (2001). *Tetrahedron Lett.*, **42**, 6469.
52. Vatèle, J.-M. (2005). *Tetrahedron Lett.*, **46**, 2299.
53. Love, K.R., Andrade, R.B. and Seeberger, P.H. (2001). *J. Org. Chem.*, **66**, 8165.
54. Xu, J. and Guo, Z. (2002). *Carbohydr. Res.*, **337**, 87.
55. Kutney, J.P. and Ratcliffe, A.H. (1975). *Synth. Commun.*, **5**, 47.
56. Zhu, T. and Boons, G.-J. (2001). *Org. Lett.*, **3**, 4201.
57. Cross, G.G. and Whitfield, D.M. (1998). *Synlett*, 487.
58. Duggan, P.J. and Tyndall, E.M. (2002). *J. Chem. Soc., Perkin Trans. 1*, 1325.

59. Liu, H. and Pinto, B.M. (2005). *J. Org. Chem.*, **70**, 753.
60. Binkley, R.W. and Ambrose, M.G. (1984). *J. Carbohydr. Chem.*, **3**, 1.
61. King, J.F. (1975). *Acc. Chem. Res.*, **8**, 10.
62. Crossland, R.K. and Servis, K.L. (1970). *J. Org. Chem.*, **35**, 3195.
63. Hanessian, S. and Vatèle, J.-M. (1981). *Tetrahedron Lett.*, **22**, 3579.
64. Vatèle, J.-M. and Hanessian, S. (1997). In *Preparative Carbohydrate Chemistry* (S. Hanessian, ed.) p. 127. Marcel Dekker.
65. Cramer, F.D. (1963). *Methods Carbohydr. Chem.*, **2**, 244.
66. Yamamura, H., Kawasaki, J., Saito, H., Araki, S. and Kawai, M. (2001). *Chem. Lett.*, 706.
67. Hicks, D.R. and Fraser-Reid, B. (1974). *Synthesis*, 203.
68. Höfle, G., Steglich, W. and Vorbrüggen, H. (1978). *Angew. Chem. Int. Ed. Eng.*, **17**, 569.
69. Danishefsky, S.J., DeNinno, M.P. and Chen, S.-h. (1988). *J. Am. Chem. Soc.*, **110**, 3929.
70. Recondo, E.F. and Rinderknecht, H. (1959). *Helv. Chim. Acta*, **42**, 1171.
71. Stanek, J., Jr (1990). *Top. Curr. Chem.*, **154**, 209.
72. Purdie, T. and Irvine, J.C. (1903). *J. Chem. Soc. (Trans.)*, **83**, 1021.
73. Haworth, W.N. (1915). *J. Chem. Soc. (Trans.)*, **107**, 8.
74. Kuhn, R., Baer, H.H. and Seeliger, A. (1958). *Liebigs Ann. Chem.*, **611**, 236.
75. Hakomori, S. (1964). *J. Biochem. (Tokyo)*, **55**, 205.
76. Czernecki, S., Georgoulis, C., Provelenghiou, C. and Fusey, G. (1976). *Tetrahedron Lett.*, 3535.
77. Rana, S.S., Vig, R.and Matta, K.L. (1982–1983) *J. Carbohydr. Chem.*, **1**, 261.
78. Wessel, H.-P., Iversen, T. and Bundle, D.R. (1985). *J. Chem. Soc., Perkin Trans. 1*, 2247.
79. Jensen, H.S., Limberg, G. and Pedersen, C. (1997). *Carbohydr. Res.*, **302**, 109.
80. Liotta, L.J. and Ganem, B. (1989). *Tetrahedron Lett.*, **30**, 4759.
81. Berry, J.M. and Hall, L.D. (1976). *Carbohydr. Res.*, **47**, 307.
82. Anwer, M.K. and Spatola, A.F. (1980). *Synthesis*, 929.
83. Bieg, T. and Szeja, W. (1985). *Synthesis*, 76.
84. Rodebaugh, R., Debenham, J.S. and Fraser-Reid, B. (1996). *Tetrahedron Lett.*, **37**, 5477.
85. Riley, J.G. and Grindley, T.B. (2001). *J. Carbohydr. Chem.*, **20**, 159.
86. Presser, A., Kunert, O. and Pötschger, I. (2006). *Monatsh. Chem.*, **137**, 365.
87. Fan, Q.-H., Li, Q., Zhang, L.-H. and Ye, X.-S. (2006). *Synlett*, 1217.
88. Yang, G., Ding, X. and Kong, F. (1997). *Tetrahedron Lett.*, **38**, 6725.
89. Lecourt, T., Herault, A., Pearce, A.J., Sollogoub, M. and Sinaÿ, P. (2004). *Chem. Eur. J.*, **10**, 2960.
90. Falck, J.R., Barma, D.K., Venkataraman, S.K., Baati, R. and Mioskowski, C. (2002). *Tetrahedron Lett.*, **43**, 963.
91. Alzeer, J. and Vasella, A. (1995). *Helv. Chim. Acta*, **78**, 177.
92. Lu, W., Navidpour, L. and Taylor, S.D. (2005). *Carbohydr. Res.*, **340**, 1213.
93. Polat, T. and Linhardt, R.J. (2003). *Carbohydr. Res.*, **338**, 447.
94. Brar, A. and Vankar, Y.D. (2006). *Tetrahedron Lett.*, **47**, 5207.
95. Csávás, M., Borbás, A., Szilágyi, L. and Lipták, A. (2002). *Synlett*, 887.
96. Wright, J.A., Yu, J. and Spencer, J.B. (2001). *Tetrahedron Lett.*, **42**, 4033.
97. Balbuena, P., Rubio, E.M., Ortiz Mellet, C. and García Fernández, J.M. (2006). *Chem. Commun.*, 2610.
98. Takaku, H., Kamaike, K. and Tsuchiya, H. (1984). *J. Org. Chem.*, **49**, 51.
99. Kunz, H. and Unverzagt, C. (1992). *J. prakt. Chem.*, **334**, 579.
100. Nakajima, N., Horita, K., Abe, R. and Yonemitsu, O. (1988). *Tetrahedron Lett.*, **29**, 4139.
101. Oikawa, Y., Yoshioka, T. and Yonemitsu, O. (1982). *Tetrahedron Lett.*, **23**, 885.
102. Classon, B., Garegg, P.J. and Samuelsson, B. (1984). *Acta Chem. Scand.*, **B38**, 419.
103. Johansson, R. and Samuelsson, B. (1984). *J. Chem. Soc., Perkin Trans. 1*, 2371.

104. Horita, K., Yoshioka, T., Tanaka, T., Oikawa, Y. and Yonemitsu, O. (1986). *Tetrahedron*, **42**, 3021.
105. Yan, L. and Kahne, D. (1995). *Synlett*, 523.
106. Yu, W., Su, M., Gao, X., Yang, Z. and Jin, Z. (2000). *Tetrahedron Lett.*, **41**, 4015.
107. Guibé, F. (1997). *Tetrahedron*, **53**, 13509.
108. Guibé, F. (1998). *Tetrahedron*, **54**, 2967.
109. Gigg, J. and Gigg, R. (1966). *J. Chem. Soc. C*, 82.
110. Wessel, H.-P. and Bundle, D.R. (1985). *J. Chem. Soc., Perkin Trans. 1*, 2251.
111. Lakhmiri, R., Lhoste, P. and Sinou, D. (1989). *Tetrahedron Lett.*, **30**, 4669.
112. Dahlén, A., Sundgren, A., Lahmann, M., Oscarson, S. and Hilmersson, G. (2003). *Org. Lett.*, **5**, 4085.
113. Gent, P. and Gigg, R. (1974). *J. Chem. Soc., Chem. Commun.*, 277.
114. Gigg, R. and Warren, C.D. (1968). *J. Chem. Soc. C*, 1903.
115. Boss, R. and Scheffold, R. (1976). *Angew. Chem. Int. Ed. Engl.*, **15**, 558.
116. Nukada, T., Kitajima, T., Nakahara, Y. and Ogawa, T. (1992). *Carbohydr. Res.*, **228**, 157.
117. Corey, E.J. and Suggs, J.W. (1973). *J. Org. Chem.*, **38**, 3224.
118. Ziegler, F.E., Brown, E.G. and Sobolov, S.B. (1990). *J. Org. Chem.*, **55**, 3691.
119. Boons, G.-J., Burton, A. and Isles, S. (1996). *Chem. Commun.*, 141.
120. Barbier, M., Grand, E. and Kovensky, J. (2007). *Carbohydr. Res.*, **342**, 2635.
121. Gigg, R. (1980). *J. Chem. Soc., Perkin Trans. 1*, 738.
122. Marković, D. and Vogel, P. (2004). *Org. Lett.*, **6**, 2693.
123. Helferich, B. (1948). *Adv. Carbohydr. Chem.*, **3**, 79.
124. Hanessian, S. and Staub, A.P.A. (1976). *Methods Carbohydr. Chem.*, **7**, 63.
125. Chaudhary, S.K. and Hernandez, O. (1979). *Tetrahedron Lett.*, 95.
126. Murata, S. and Noyori, R. (1981). *Tetrahedron Lett.*, **22**, 2107.
127. Krainer, E., Naider, F. and Becker, J. (1993). *Tetrahedron Lett.*, **34**, 1713.
128. Pathak, A.K., Pathak, V., Seitz, L.E., Tiwari, K.N., Akhtar, M.S. and Reynolds, R.C. (2001). *Tetrahedron Lett.*, **42**, 7755.
129. Kohli, V., Blöcker, H. and Köster, H. (1980). *Tetrahedron Lett.*, **21**, 2683.
130. Randazzo, G., Capasso, R., Cicala, M.R. and Evidente, A. (1980). *Carbohydr. Res.*, **85**, 298.
131. Yadav, J.S. and Subba Reddy, B.V. (2000). *Synlett*, 1275.
132. Kováč, P. and Bauer, S. (1972). *Tetrahedron Lett.*, 2349.
133. Greene, T.W. and Wuts, P.G.M. (1991). *Protective Groups in Organic Synthesis*, p. 68. John Wiley and Sons, Inc.
134. Dutton, G.G.S. (1973). *Adv. Carbohydr. Chem. Biochem.*, **28**, 11.
135. Corey, E.J. and Venkateswarlu, A. (1972). *J. Am. Chem. Soc.*, **94**, 6190.
136. Lalonde, M. and Chan, T.H. (1985). *Synthesis*, 817.
137. Danishefsky, S.J. and Bilodeau, M.T. (1996). *Angew. Chem. Int. Ed. Engl.*, **35**, 1380.
138. Muzart, J. (1993). *Synthesis*, 11.
139. Nelson, T.D. and Crouch, R.D. (1996). *Synthesis*, 1031.
140. Mulzer, J. and Schöllhorn, B. (1990). *Angew. Chem. Int. Ed. Engl.*, **29**, 431.
141. Arias-Pérez, M.S., López, M.S. and Santos, M.J. (2002). *J. Chem. Soc., Perkin Trans. 2*, 1549.
142. Feixas, J., Capdevila, A., Camps, F. and Guerrero, A. (1992). *J. Chem. Soc., Chem. Commun.*, 1451.
143. Farràs, J., Serra, C. and Vilarrasa, J. (1998). *Tetrahedron Lett.*, **39**, 327.
144. Barros, M.T., Maycock, C.D. and Thomassigny, C. (2001). *Synlett*, 1146.
145. Chen, M.-Y., Patkar, L.N., Lu, K.-C., Lee, A.S.-Y. and Lin, C.-C. (2004). *Tetrahedron*, **60**, 11465.
146. Meloncelli, P.J. and Stick, R.V. (2006). *Aust. J. Chem.*, **59**, 827.
147. Kocienski, P.J. (1994). *Protecting Groups*, pp. 68, 96. Thieme.
148. Gelas, J. (1981). *Adv. Carbohydr. Chem. Biochem.*, **39**, 71.

149. Calinaud, P. and Gelas, J. In *Preparative Carbohydrate Chemistry* (S. Hanessian, ed.) p. 3. Marcel Dekker, Inc.
150. Haines, A.H. (1981). *Adv. Carbohydr. Chem. Biochem.*, **39**, 13.
151. Guthrie, R.D. and McCarthy, J.F. (1967). *Adv. Carbohydr. Chem.*, **22**, 11.
152. Lemieux, R.U. and Driguez, H. (1975). *J. Am. Chem. Soc.*, **97**, 4069.
153. Jansson, K., Ahlfors, S., Frejd, T., Kihlberg, J., Magnusson, G., Dahmén, J., Noori, G. and Stenvall, K. (1988). *J. Org. Chem.*, **53**, 5629.
154. Fraser-Reid, B., Udodong, U.E., Wu, Z., Ottosson, H., Merritt, J.R., Rao, C.S., Roberts, C. and Madsen, R. (1992). *Synlett*, 927.
155. Wolfrom, M.L., Beattie, A., Bhattacharjee, S.S. and Parekh, G.G. (1968). *J. Org. Chem.*, **33**, 3990.
156. Miyashita, N., Yoshikoshi, A. and Grieco, P.A. (1977). *J. Org. Chem.*, **42**, 3772.
157. Oriyama, T., Yatabe, K., Sugawara, S., Machiguchi, Y. and Koga, G. (1996). *Synlett*, 523.
158. Zhang, Z.-H., Li, T.-S., Jin, T.-S. and Wang, J.-X. (1998). *J. Chem. Res., Synop.*, 152.
159. Nishino, S. and Ishido, Y. (1986). *J. Carbohydr. Chem.*, **5**, 313.
160. El-Shenawy, H.A. and Schuerch, C. (1984). *Carbohydr. Res.*, **131**, 239.
161. Guindon, Y., Morton, H.E. and Yoakim, C. (1983). *Tetrahedron Lett.*, **24**, 3969.
162. Eliel, E.L., Wilen, S.H. and Mander, L.N. (1994). *Stereochemistry of Organic Compounds.* John Wiley and Sons.
163. Richtmyer, N.K. (1962). *Methods Carbohydr. Chem.*, **1**, 107.
164. Chittenden, G.J.F. (1988). *Recl. Trav. Chim. Pays-Bas*, **107**, 607.
165. Niu, Y., Wang, N., Cao, X. and Ye, X.-S. (2007). *Synlett*, 2116.
166. Evans, M.E. (1980). *Methods Carbohydr. Chem.*, **8**, 313.
167. Albert, R., Dax, K., Pleschko, R. and Stütz, A.E. (1985). *Carbohydr. Res.*, **137**, 282.
168. Ferro, V., Mocerino, M., Stick, R.V. and Tilbrook, D.M.G. (1988). *Aust. J. Chem.*, **41**, 813.
169. Joseph, C.C., Zwanenburg, B. and Chittenden, G.J.F. (2003). *Synth. Commun.*, **33**, 493.
170. Garegg, P.J. and Swahn, C.-G. (1980). *Methods Carbohydr. Chem.*, **8**, 317.
171. Xia, J. and Hui, Y. (1996). *Synth. Commun.*, **26**, 881.
172. Andrews, M.A. and Gould, G.L. (1992). *Carbohydr. Res.*, **229**, 141.
173. Procopio, A., Dalpozzo, R., De Nino, A., Maiuolo, L., Nardi, M. and Romeo, G. (2005). *Org. Biomol. Chem.*, **3**, 4129.
174. Agnihotri, G. and Misra, A.K. (2006). *Tetrahedron Lett.*, **47**, 3653.
175. Garegg, P.J. In *Preparative Carbohydrate Chemistry* (S. Hanessian, ed.) p. 53. Marcel Dekker, Inc.
176. Garegg, P.J. (1984). *Pure Appl. Chem.*, **56**, 845.
177. Lipták, A., Imre, J., Harangi, J., Nánási, P. and Neszmélyi, A. (1982). *Tetrahedron*, **38**, 3721.
178. Guindon, Y., Girard, Y., Berthiaume, S., Gorys, V., Lemieux, R. and Yoakim, C. (1990). *Can. J. Chem.*, **68**, 897.
179. Jiang, L. and Chan, T.-H. (1998). *Tetrahedron Lett.*, **39**, 355.
180. Wang, C.-C., Luo, S.-Y., Shie, C.-R. and Hung, S.-C. (2002). *Org. Lett.*, **4**, 847.
181. Tani, S., Sawadi, S., Kojima, M., Akai, S. and Sato, K.-I. (2007). *Tetrahedron Lett.*, **48**, 3103.
182. Ek, M., Garegg, P.J., Hultberg, H. and Oscarson, S. (1983). *J. Carbohydr. Chem.*, **2**, 305.
183. Sherman, A.A., Mironov, Y.V., Yudina, O.N. and Nifantiev, N.E. (2003). *Carbohydr. Res.*, **338**, 697.
184. Oikawa, M., Liu, W.-C., Nakai, Y., Koshida, S., Fukase, K. and Kusumoto, S. (1996). *Synlett*, 1179.
185. Sakagami, M. and Hamana, H. (2000). *Tetrahedron Lett.*, **41**, 5547.
186. Shie, C.-R., Tzeng, Z.-H., Kulkarni, S.S., Uang, B.-J., Hsu, C.-Y. and Hung, S.-C. (2005). *Angew. Chem. Int. Ed.*, **44**, 1665.
187. Garegg, P.J., Hultberg, H. and Wallin, S. (1982). *Carbohydr. Res.*, **108**, 97.
188. Zinin, A.I., Malysheva, N.N., Shpirt, A.M., Torgov, V.I. and Kononov, L.O. (2007). *Carbohydr. Res.*, **342**, 627.

189. DeNinno, M.P., Etienne, J.B. and Duplantier, K.C. (1995). *Tetrahedron Lett.*, **36**, 669.
190. Debenham, S.D. and Toone, E.J. (2000). *Tetrahedron*: Asymm., **11**, 385.
191. Pohl, N.L. and Kiessling, L.L. (1997). *Tetrahedron Lett.*, **38**, 6985.
192. Suzuki, K., Nonaka, H. and Yamaura, M. (2004). *J. Carbohydr. Chem.*, **23**, 253.
193. Hanessian, S. and Plessas, N.R. (1969). *J. Org. Chem.*, **34**, 1053.
194. Hullar, T.L. and Siskin, S.B. (1970). *J. Org. Chem.*, **35**, 225.
195. Hanessian, S. (1972). *Methods Carbohydr. Chem.*, **6**, 183.
196. Hanessian, S. (1987). *Org. Synth.*, **65**, 243.
197. McNulty, J., Wilson, J. and Rochon, A.C. (2004). *J. Org. Chem.*, **69**, 563.
198. Crich, D., Yao, Q. and Bowers, A.A. (2006). *Carbohydr. Res.*, **341**, 1748.
199. Chrétien, F., Khaldi, M. and Chapleur, Y. (1990). *Synth. Commun.*, **20**, 1589.
200. Chana, J.S., Collins, P.M., Farnia, F. and Peacock, D.J. (1988). *J. Chem. Soc., Chem. Commun.*, 94.
201. Deslongchamps, P., Moreau, C., Fréhel, D. and Chênevert, R. (1975). *Can. J. Chem.*, **53**, 1204.
202. Hernández-Torres, J.M., Achkar, J. and Wei, A. (2004). *J. Org. Chem.*, **69**, 7206.
203. Borbás, A., Szabó, Z.B., Szilágyi, L., Bényei, A. and Lipták, A. (2002). *Tetrahedron*, **58**, 5723.
204. Chéry, F., Rollin, P., De Lucchi, O. and Cossu, S. (2001). *Synthesis*, 286.
205. Roizel, B.C.-du, Cabianca, E., Rollin, P. and Sinaÿ, P. (2002). *Tetrahedron*, **58**, 9579.
206. Crich, D. and Bowers, A.A. (2006). *J. Org. Chem.*, **71**, 3452.
207. Fischer, E. (1895). *Ber. Dtsch. Chem. Ges.*, **28**, 1145.
208. Schmidt, O.Th. (1963). *Methods Carbohydr. Chem.*, **2**, 318.
209. He, D.Y., Li, Z.J., Li, Z.J., Liu, Y.Q., Qiu, D.X. and Cai, M.S. (1992). *Synth. Commun.*, **22**, 2653.
210. Zhao, G., Zhang, Y. and Wang, J. (2007). *Collect. Czech. Chem. Commun.*, **72**, 1214.
211. Singh, P.P., Gharia, M.M., Dasgupta, F. and Srivastava, H.C. (1977). *Tetrahedron Lett.*, 439.
212. Lin, C.-C., Jan, M.-D., Weng, S.-S., Lin, C.-C. and Chen, C.-T. (2006). *Carbohydr. Res.*, **341**, 1948.
213. Pedatella, S., Guaragna, A., D'Alonzo, D., De Nisco, M. and Palumbo, G. (2006). *Synthesis*, 305.
214. Catelani, G., Colonna, F. and Marra, A. (1988). *Carbohydr. Res.*, **182**, 297.
215. Gelas, J. and Horton, D. (1981). *Heterocycles*, **16**, 1587.
216. Christensen, J.E. and Goodman, L. (1968). *Carbohydr. Res.*, **7**, 510.
217. Szarek, W.A., Zamojski, A., Tiwari, K.N. and Ison, E.R. (1986). *Tetrahedron Lett.*, **27**, 3827.
218. Saravanan, P., Chandrasekhar, M., Anand, R.V. and Singh, V.K. (1998). *Tetrahedron Lett.*, **39**, 3091.
219. Raghavendra Swamy, N. and Venkateswarlu, Y. (2002). *Tetrahedron Lett.*, **43**, 7549.
220. He, H., Yang, F. and Du, Y. (2002). *Carbohydr. Res.*, **337**, 1673.
221. Maulide, N., Vanherck, J.-C., Gautier, A. and Markó, I.E. (2007). *Acc. Chem. Res.*, **40**, 381.
222. Mukhopadhyay, B., Russell, D.A. and Field, R.A. (2005). *Carbohydr. Res.*, **340**, 1075.
223. Rajput, V.K., Roy, B. and Mukhopadhyay, B. (2006). *Tetrahedron Lett.*, **47**, 6987.
224. Ley, S.V., Baeschlin, D.K., Dixon, D.J., Foster, A.C., Ince, S.J., Priepke, H.W.M. and Reynolds, D.J. (2001). *Chem. Rev.*, **101**, 53.
225. Ley, S.V., Leslie, R., Tiffin, P.D. and Woods, M. (1992). *Tetrahedron Lett.*, **33**, 4767.
226. Ley, S.V., Boons, G.-J., Leslie, R., Woods, M. and Hollinshead, D.M. (1993). *Synthesis*, 689.
227. Grice, P., Ley, S.V., Pietruszka, J., Priepke, H.W.M. and Warriner, S.L. (1997). *J. Chem. Soc., Perkin Trans. 1*, 351.
228. Hense, A., Ley, S.V., Osborn, H.M.I., Owen, D.R., Poisson, J.-F., Warriner, S.L. and Wesson, K.E. (1997). *J. Chem. Soc., Perkin Trans. 1*, 2023.
229. Ley, S.V. and Polara, A. (2007). *J. Org. Chem.*, **72**, 5943.
230. Yin, H., Franck, R.W., Chen, S.-L., Quigley, G.J. and Todaro, L. (1992). *J. Org. Chem.*, **57**, 644.
231. Wander, J.D. and Horton, D. (1976). *Adv. Carbohydr. Chem. Biochem.*, **32**, 15.
232. Fischer, E. (1894). *Ber. Dtsch. Chem. Ges.*, **27**, 673.

233. Funabashi, M., Arai, S. and Shinohara, M. (1999). *J. Carbohydr. Chem.*, **18**, 333.
234. Rodriguez, E.B., Scally, G.D. and Stick, R.V. (1990). *Aust. J. Chem.*, **43**, 1391.
235. Miljković, M., Dropkin, D., Hagel, P. and Habash-Marino, M. (1984). *Carbohydr. Res.*, **128**, 11.
236. McAuliffe, J.C. and Hindsgaul, O. (1998). *Synlett*, 307.
237. Ferrier, R.J. and Furneaux, R.H. (1976). *Carbohydr. Res.*, **52**, 63.
238. Kartha, K.P.R. and Field, R.A. (1998). *J. Carbohydr. Chem.*, **17**, 693.
239. Tiwari, P., Agnihotri, G. and Misra, A.K. (2005). *J. Carbohydr. Chem.*, **24**, 723.
240. Stubbs, K.A., Macauley, M.S. and Vocadlo, D.J. (2006). *Carbohydr. Res.*, **341**, 1764.
241. David, S. and Hanessian, S. (1985). *Tetrahedron*, **41**, 643.
242. David, S. In *Preparative Carbohydrate Chemistry* (S. Hanessian, ed.) p. 69. Marcel Dekker, Inc.
243. Grindley, T.B. (1998). *Adv. Carbohydr. Chem. Biochem.*, **53**, 17.
244. Munavu, R.M. and Szmant, H.H. (1976). *J. Org. Chem.*, **41**, 1832.
245. Simas, A.B.C., Pais, K.C. and da Silva, A.A.T. (2003). *J. Org. Chem.*, **68**, 5426.
246. Dong, H., Pei, Z., Byström, S. and Ramström, O. (2007). *J. Org. Chem.*, **72**, 1499.
247. Nashed, M.A. and Anderson, L. (1976). *Tetrahedron Lett.*, 3503.
248. David, S., Thieffry, A. and Veyrières, A. (1981). *J. Chem. Soc., Perkin Trans. 1,* 1796.
249. Nikrad, P.V., Beierbeck, H. and Lemieux, R.U. (1992). *Can. J. Chem.*, **70**, 241.
250. Jenkins, D.J. and Potter, B.V.L. (1994). *Carbohydr. Res.*, **265**, 145.
251. Boons, G.-J., Castle, G.H., Clase, J.A., Grice, P., Ley, S.V. and Pinel, C. (1993). *Synlett*, 913.
252. Guilbert, B., Davis, N.J. and Flitsch, S.L. (1994). *Tetrahedron Lett.*, **35**, 6563.
253. Lubineau, A. and Lemoine, R. (1994). *Tetrahedron Lett.*, **35**, 8795.
254. Hodosi, G. and Kováč, P. (1998). *J. Carbohydr. Chem.*, **17**, 557.
255. Ogawa, T. and Matsui, M. (1981). *Tetrahedron*, **37**, 2363.
256. Alais, J. and Veyrières, A (1981). *J. Chem. Soc., Perkin Trans. 1,* 377.
257. Dasgupta, F. and Garegg, P.J. (1994). *Synthesis*, 1121.
258. Blatter, G., Beau, J.-M. and Jacquinet, J.-C. (1994). *Carbohydr. Res.*, **260**, 189.
259. Ellervik, U. and Magnusson, G. (1996). *Carbohydr. Res.*, **280**, 251.
260. Sun, B., Pukin, A.V., Visser, G.M. and Zuilhof, H. (2006). *Tetrahedron Lett.*, **47**, 7371.
261. Tokimoto, H. and Fukase, K. (2005). *Tetrahedron Lett.*, **46**, 6831.
262. Boullanger, P., Banoub, J. and Descotes, G. (1987). *Can. J. Chem.*, **65**, 1343.
263. Aly, M.R.E. and Schmidt, R.R. (2005). *Eur. J. Org. Chem.*, 4382.
264. Castro-Palomino, J.C. and Schmidt, R.R. (1995). *Tetrahedron Lett.*, **36**, 6871.
265. Demchenko, A.V. and Boons, G.-J. (1998). *Tetrahedron Lett.*, **39**, 3065.
266. Banoub, J., Boullanger, P. and Lafont, D. (1992). *Chem. Rev.*, **92**, 1167.
267. Bundle, D.R. and Josephson, S. (1979). *Can. J. Chem.*, **57**, 662.
268. Kanie, O., Crawley, S.C., Palcic, M.M. and Hindsgaul, O. (1993). *Carbohydr. Res.*, **243**, 139.
269. Debenham, J.S., Madsen, R., Roberts, C. and Fraser-Reid, B. (1995). *J. Am. Chem. Soc.*, **117**, 3302.
270. Debenham, J.S., Debenham, S.D. and Fraser-Reid, B. (1996). *Bioorg. Med. Chem.*, **4**, 1909.
271. Castro-Palomino, J.C. and Schmidt, R.R. (1995). *Tetrahedron Lett.*, **36**, 5343.
272. Rodebaugh, R., Debenham, J.S. and Fraser-Reid, B. (1997). *J. Carbohydr. Chem.*, **16**, 1407.
273. Lergenmüller, M., Ito, Y. and Ogawa, T. (1998). *Tetrahedron*, **54**, 1381.
274. Olsson, L., Kelberlau, S., Jia, Z.J. and Fraser-Reid, B. (1998). *Carbohydr. Res.*, **314**, 273.
275. Bowers, S.G., Coe, D.M. and Boons, G.-J. (1998). *J. Org. Chem.*, **63**, 4570.
276. Aly, M.R.E., Castro-Palomino, J.C., Ibrahim, E.-S.I., El-Ashry, E.-S.H. and Schmidt, R.R. (1998). *Eur. J. Org. Chem.*, 2305.
277. Jensen, K.J., Hansen, P.R., Venugopal, D. and Barany, G. (1996). *J. Am. Chem. Soc.*, **118**, 3148.
278. Castro-Palomino, J.C. and Schmidt, R.R. (2000). *Tetrahedron Lett.*, **41**, 629.

279. Dekany, G., Bornaghi, L., Papageorgiou, J. and Taylor, S. (2001). *Tetrahedron Lett.*, **42**, 3129.
280. Bornaghi, L., Dekany, G., Drinnan, N.B., Taylor, S., Toth, I. and West, M.L. (2006). *J. Carbohydr. Chem.*, **25**, 1.
281. Olsson, L., Jia, Z.J. and Fraser-Reid, B. (1998). *J. Org. Chem.*, **63**, 3790.
282. Yan, R.-B., Yang, F., Wu, Y., Zhang, L.-H. and Ye, X.-S. (2005). *Tetrahedron Lett.*, **46**, 8993.
283. Nyffeler, P.T., Liang, C-H., Koeller, K.M. and Wong, C.-H. (2002). *J. Am. Chem. Soc.*, **124**, 10773.
284. Jiao, H. and Hindsgaul, O. (1999). *Angew. Chem. Int. Ed.*, **38**, 346.
285. Teodorović, P., Slättegård, R. and Oscarson, S. (2005). *Carbohydr. Res.*, **340**, 2675.
286. Goddard-Borger, E.D. and Stick, R.V. (2007). *Org. Lett.*, **9**, 3797.
287. Söderberg, E., Westman, J. and Oscarson, S. (2001). *J. Carbohydr. Chem.*, **20**, 397.
288. Couri, M.R.C., Evangelista, E.A., Alves, R.B., Prado, M.A.F., Gil, R.P.F., De Almeida, M.V. and Raslan, D.S. (2005). *Synth. Commun.*, **35**, 2025.
289. Kojima, M., Nakamura, Y. and Takeuchi, S. (2007). *Tetrahedron Lett.*, **48**, 4431.
290. Wong, C.-H., Ye, X.-S. and Zhang, Z. (1998). *J. Am. Chem. Soc.*, **120**, 7137.
291. Opatz, T., Kallus, C., Wunberg, T., Schmidt, W., Henke, S. and Kunz, H. (2003). *Eur. J. Org. Chem.*, 1527.
292. Moitessier, N., Henry, C., Aubert, N. and Chapleur, Y. (2005). *Tetrahedron Lett.*, **46**, 6191.
293. Wang, C.-C., Kulkarni, S.S., Lee, J.-C., Luo, S.-Y. and Hung, S.-C., (2008). *Nat. Protoc.*, **3**, 97.
294. Français, A., Urban, D. and Beau, J.-M. (2007). *Angew. Chem. Int. Ed.*, **46**, 8662.
295. Komba, S., Kitaoka, M. and Kasumi, T. (2005). *Eur. J. Org. Chem.*, 5313.

Chapter 3

The Reactions of Monosaccharides[1]

So far, most of the chemical transformations of a monosaccharide discussed in this book have been concerned with the concept of protecting groups. In general, the order of reactivity of the various hydroxyl groups of D-glucopyranose was observed to be O1 (hemiacetal) > O6 (primary) > O2 (adjacent to C1 and therefore more acidic) > O3 > O4; this order, of course, is dictated very much by the nature of the substrate and the reagent, as well as whether an O–H or a C–O bond is being broken. Now, a host of chemical reactions will be discussed that cause functional group transformations within monosaccharides, either for the purpose of elaboration into oligosaccharides or for the synthesis of a derivative of the monosaccharide. Such derivatives include the deoxy and amino deoxy sugars.

By the nature of the subject, this discussion will be neither exhaustive nor all-inclusive. It will, however, highlight the methods that work well in synthetic carbohydrate chemistry, particularly those that have evolved over the last few decades.

Oxidation[2–6]

We have already seen the importance of oxidation to the German chemists of the nineteenth century. Aldoses were oxidized to aldonic acids using bromine water and to aldaric acids using dilute nitric acid; these reagents are still used today.

The major advances have been concerned with the discovery or 'invention', as some have called it,[a] of chemoselective reagents for oxidation of the various hydroxyl groups within a carbohydrate. The well-established Jones' (CrO_3, H_2SO_4)[8] and Collins' (CrO_3, pyridine)[9] reagents soon gave way to milder and more easily handled oxidizing agents, namely pyridinium chlorochromate[10] and pyridinium

---

[a] Sir Derek Barton co-authored a series of papers that highlighted the 'invention' of new chemical reactions, particularly those concerned with free radical pathways.[7]

References start on page 124

dichromate.[11,12] These reagents oxidize both primary and secondary alcohols and can produce either the aldehyde or the acid from a primary alcohol:

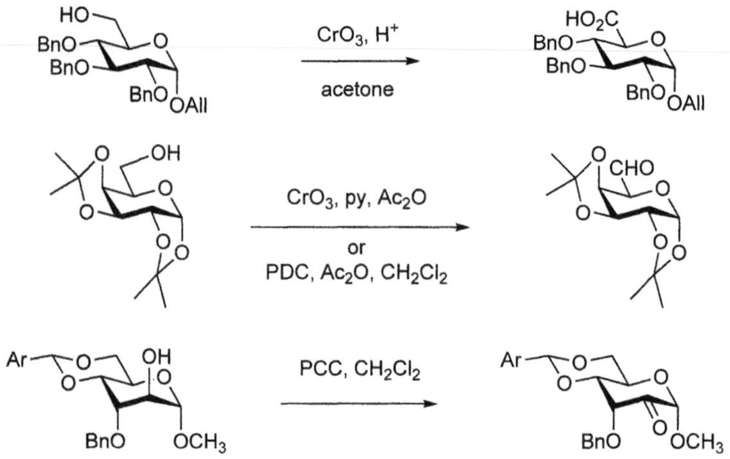

Another major class of oxidizing agents is based on dimethyl sulfoxide.[13] Since the initial report by Pfitzner and Moffatt (dimethyl sulfoxide, dicyclohexylcarbodiimide), a string of reagents has been developed that activate the dimethyl sulfoxide for oxidation of, in particular, secondary alcohols:

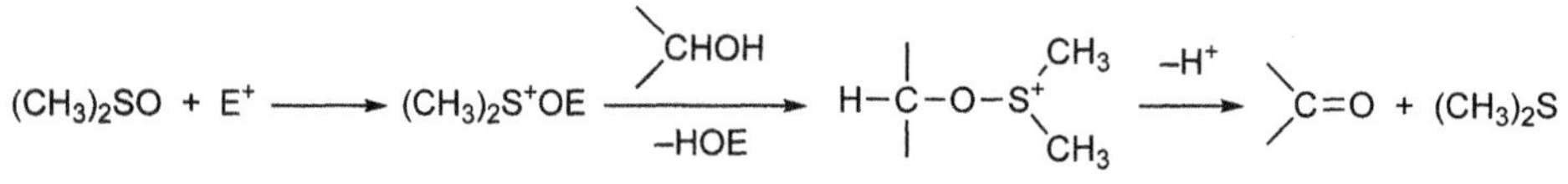

$$(CH_3)_2SO + E^+ \longrightarrow (CH_3)_2S^+OE \xrightarrow[-HOE]{>CHOH} H-\overset{|}{\underset{|}{C}}-O-\overset{+}{\underset{CH_3}{S}}{}^{CH_3} \xrightarrow{-H^+} >C=O + (CH_3)_2S$$

With primary alcohols, the formation of '(methylthio)methyl ethers' (actually thioacetals) can be troublesome:[14]

$$(CH_3)_2S^+OE \xrightarrow{-HOE} CH_3S^+=CH_2 \xrightarrow[-H^+]{RCH_2OH} RCH_2OCH_2SCH_3$$

Some of the most popular activating agents for dimethyl sulfoxide are acetic anhydride, trifluoroacetic anhydride, phosphorus pentaoxide, pyridine/sulfur trioxide and oxalyl chloride:[15,16]

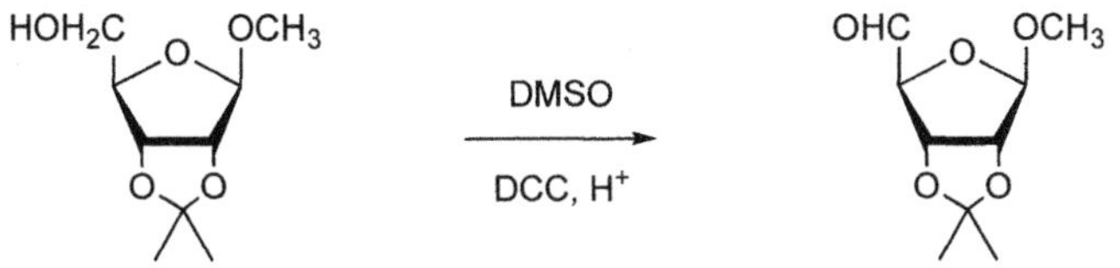

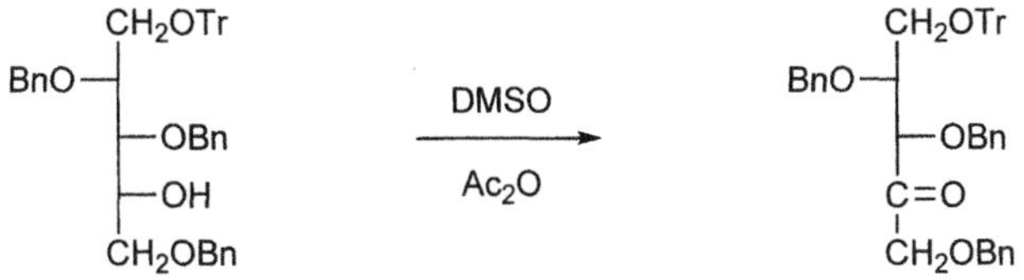

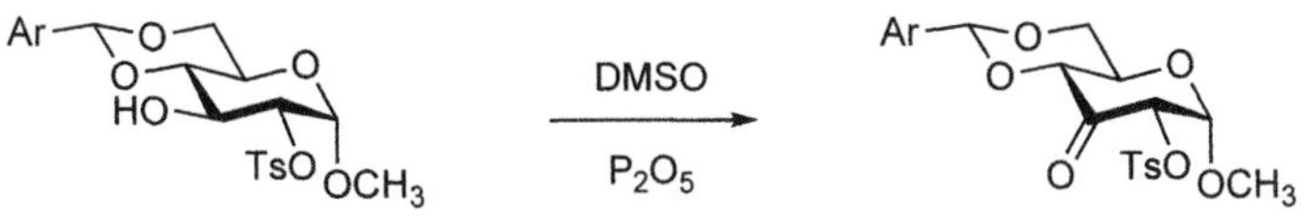

In fact, the activation of dimethyl sulfoxide by oxalyl chloride is such a general, mild and selective method of oxidation that it has taken on a mantle of its own; it is known as the Swern oxidation, after its founder:[17]

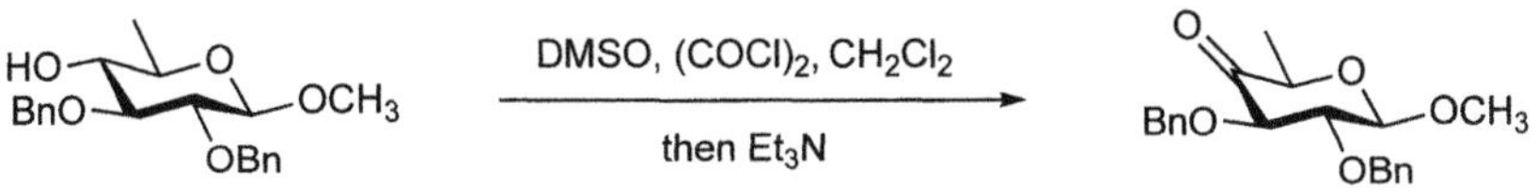

There have been suggestions for the use of recyclable sulfoxides or polymer-bound sulfoxides, the latter to avoid the malodorous dimethyl sulfide that is liberated from oxidations conducted with dimethyl sulfoxide.[18–20]

A named reagent, the Dess–Martin periodinane, has proven its value in synthesis and is being used more regularly in the carbohydrate area:[21–24]

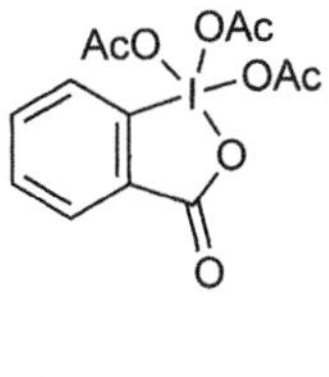

Ruthenium tetraoxide is a powerful oxidizing agent, the action of which can be somewhat moderated by employing small amounts of ruthenium(III) or ruthenium(IV) compounds and stoichiometric amounts of another oxidizing agent:[25–27]

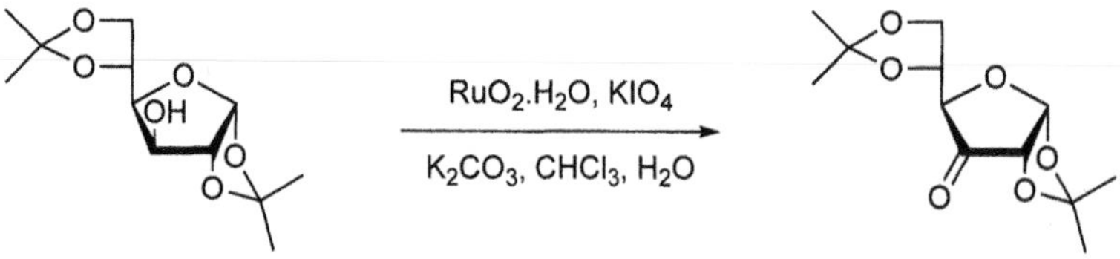

Ley offered a novel variant of ruthenium-based oxidants by showing that tetrapropylammonium perruthenate could be used in catalytic amounts, together with another oxidizing agent:[28]

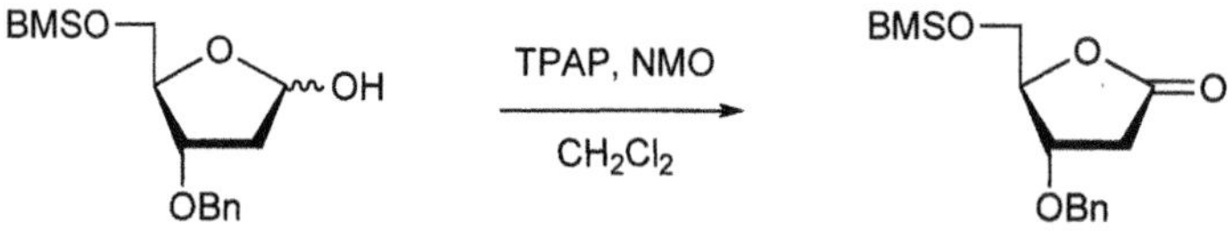

Two significant improvements to the method have been reported: a polymer-supported perruthenate and the use of molecular oxygen as the co-oxidant.[29]

2,2,6,6-Tetramethylpiperidine-1-oxyl, used in catalytic amounts with a co-oxidant, is capable of oxidizing primary alcohols to either the aldehyde or the carboxylic acid:[30–33]

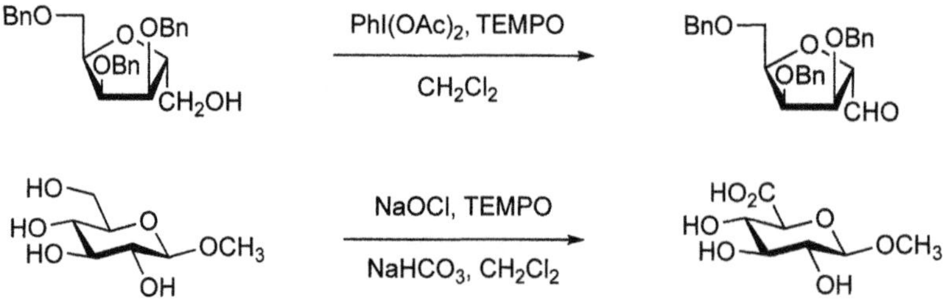

The latter example illustrates the utility of the method, which seems to compare favourably with the more traditional oxygen-over-platinum dehydrogenation[34] for producing uronic acids. The mechanism of action of this interesting oxidant is thought to involve an oxoammonium ion:[35]

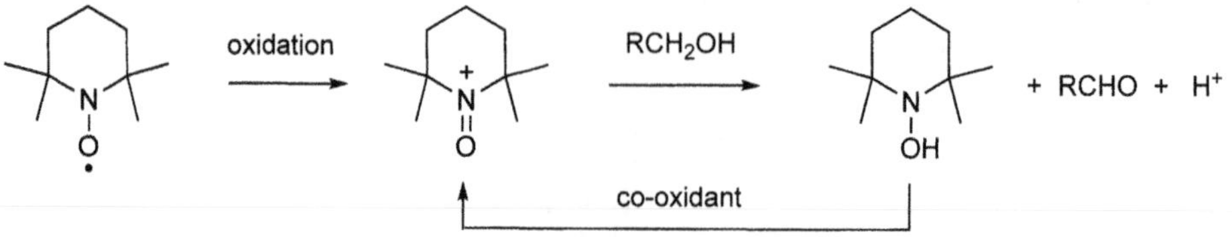

In connection with the work on the reaction of vicinal diols with bis(tributyltin) oxide and with dibutyltin oxide (see Chapter 2), it was found that the resulting tin-containing derivatives could be oxidized rapidly and selectively by molecular bromine or by *N*-bromosuccinimide:[36–39]

3,4-di-*O*-benzyl-D-mannitol

3,4-di-*O*-benzyl-D-fructofuranose

3,4-di-*O*-benzyl-β-D-fructopyranose

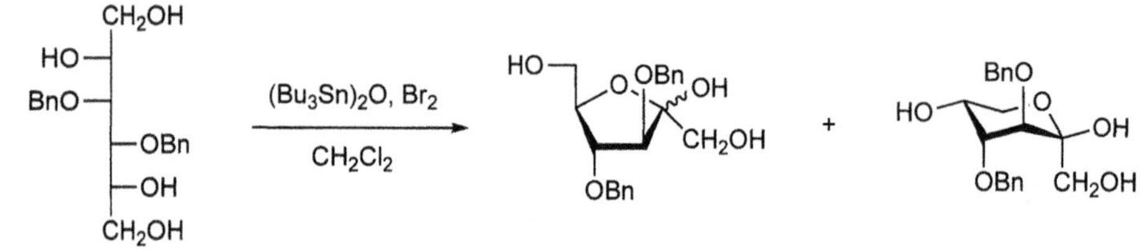

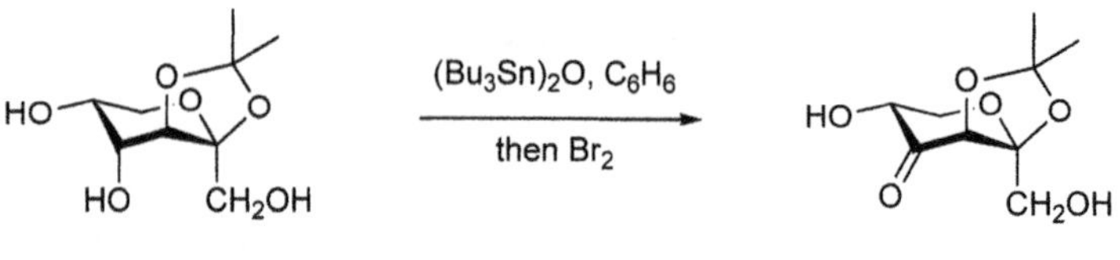

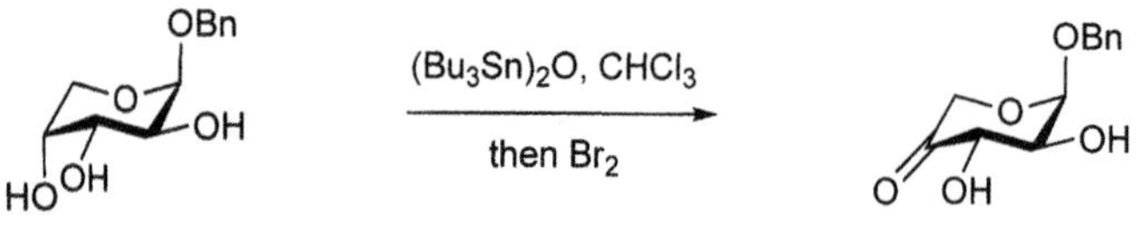

This oxidation is truly remarkable, not only in its activation of both the substrate (diol) and the reagent (bromine) but also in the regiochemistry displayed.

It is worth noting at this stage that many of the ketones produced by the oxidation of carbohydrate secondary alcohols are often found as stable, well-defined hydrates; presumably, the numerous electronegative oxygen atoms present in the molecules raise the reactivity of the carbonyl group. Hydrogen bonding in these hydrates may also be a stabilizing factor.[26] In solution, these hydrates are in

equilibrium with the ketone and exhibit the reactivity expected (as well as releasing an equivalent of water!).

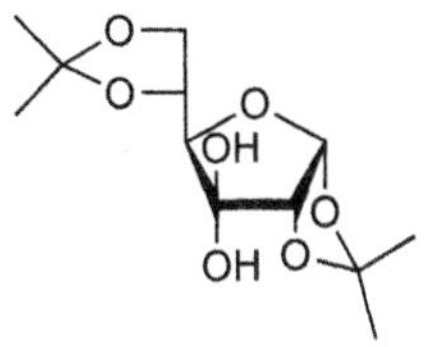

mp 112–114°C
$[\alpha]_D$ +44.5° (EtOH)

This section would not be complete without a reference to the oxidative cleavage of carbohydrate vicinal diols by reagents such as periodic acid and lead(IV) acetate. These sorts of oxidations not only provide valuable information on relative configuration and ring size[40] but also are of synthetic utility:[41]

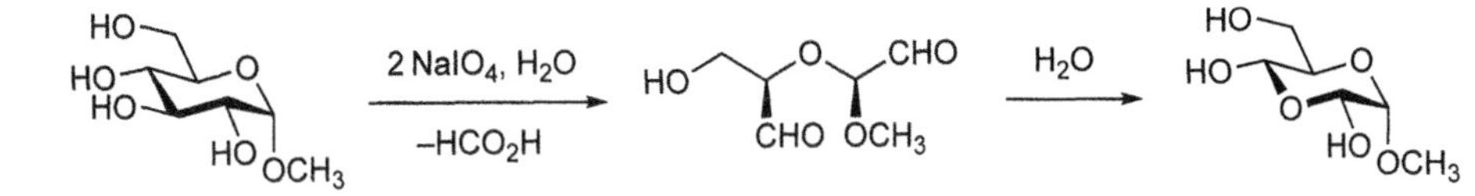

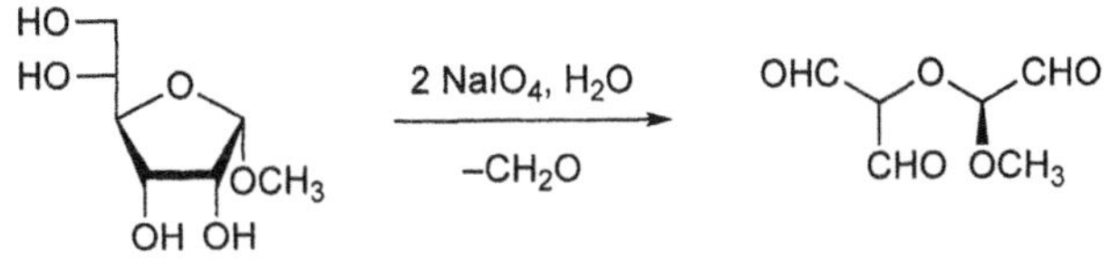

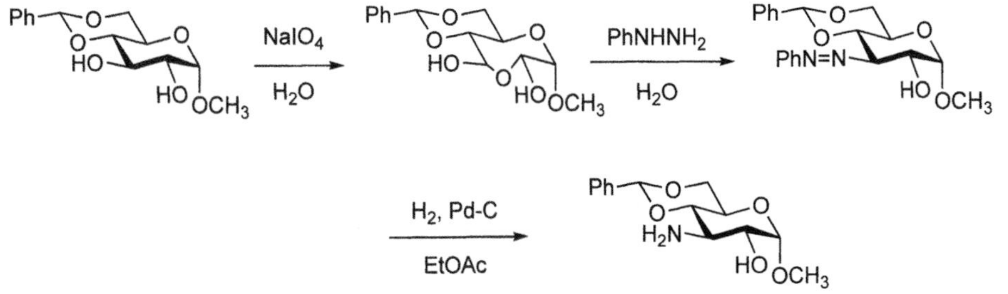

These days, of course, the problems of ring size and relative configuration are much more easily solved by NMR spectroscopy.[42–45]

Reduction

The reduction of an aldose to an alditol is now more safely done with sodium borohydride than with the more toxic sodium amalgam. For many years,

however, chemists were restricted to such 'dissolving metals' and 'catalytic hydrogenation' for the reduction of carbohydrates. Still, many interesting reactions were reported:[46–48]

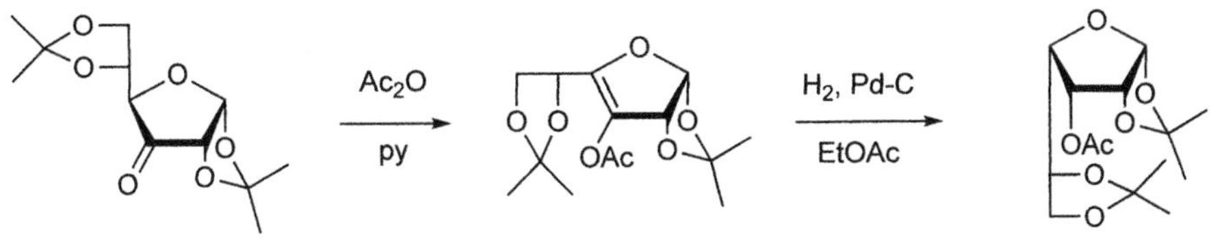

3-O-acetyl-1,2:5,6-di-
O-isopropylidene-
α-D-gulose

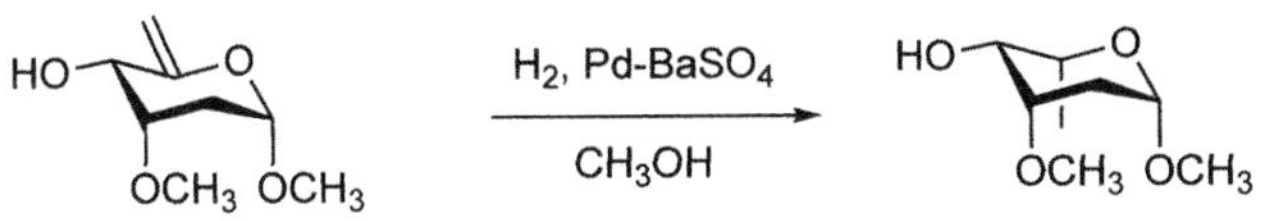

methyl 2,6-dideoxy-3-O-methyl-β-L-*lyxo*-hexopyranoside

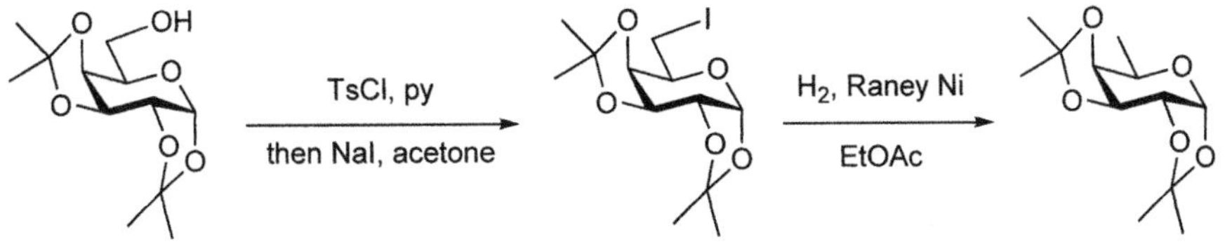

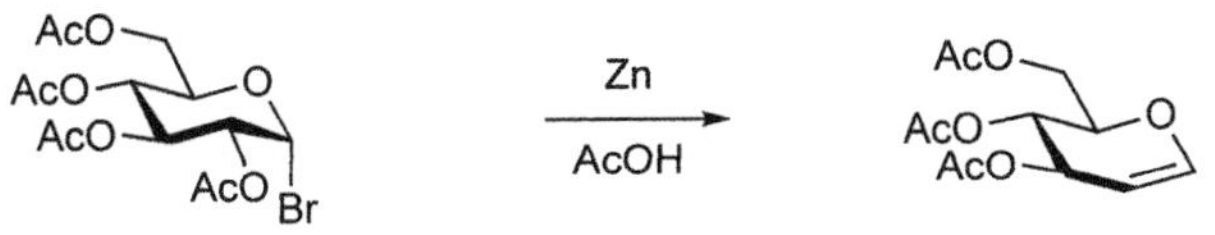

The concept of 'catalytic transfer hydrogenation' has obviated the need for using molecular hydrogen in transformations such as the above.[49,50]

With the advent of sodium borohydride and lithium aluminium hydride, many of the reductions performed on carbohydrates became easier and, with sodium borohydride, safer. A modern Kiliani–Fischer synthesis now uses sodium borohydride for reduction of the lactone to the aldose and, if necessary, further reduction to the alditol:

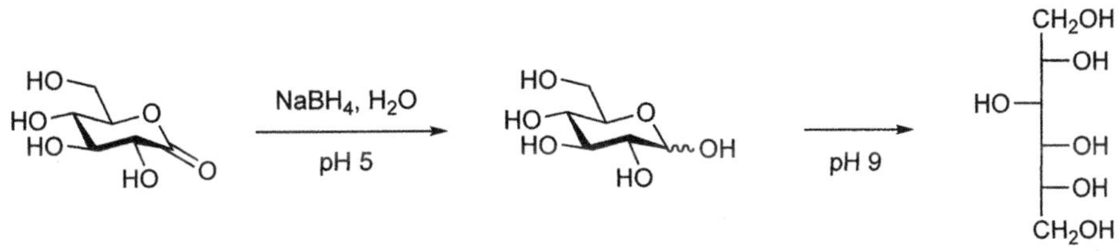

Other interesting transformations followed, some involving modified boro- and aluminium hydrides:[26,51–54]

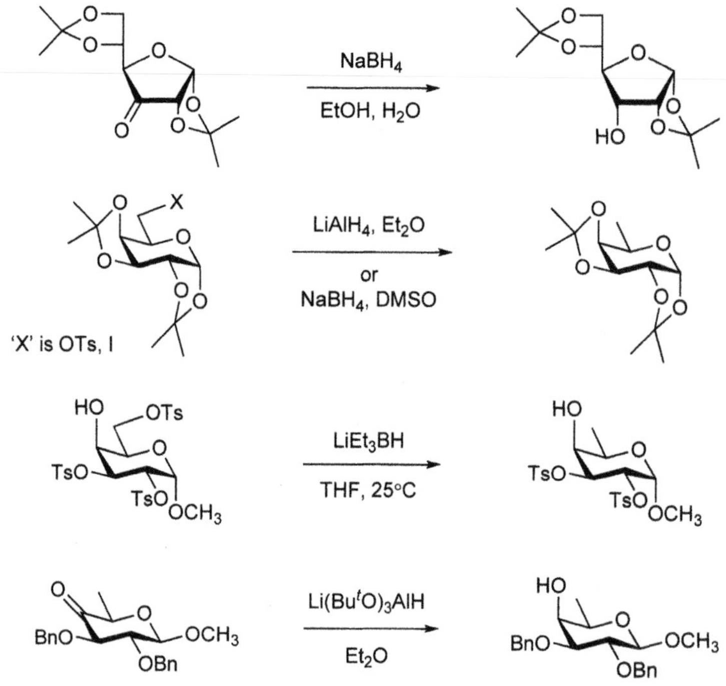

Tributyltin hydride has been used routinely for the removal of halogen from carbohydrate derivatives:[55,56]

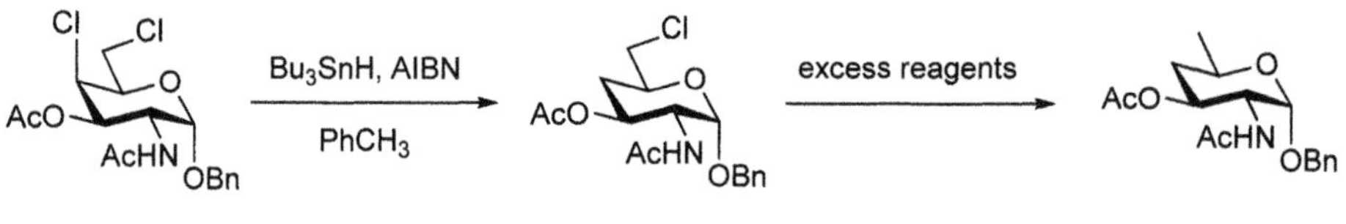

However, this reducing agent received a great boost in 1975 when Barton and McCombie reported the deoxygenation of secondary alcohols via, among other functionalities, the derived xanthates:[57,58]

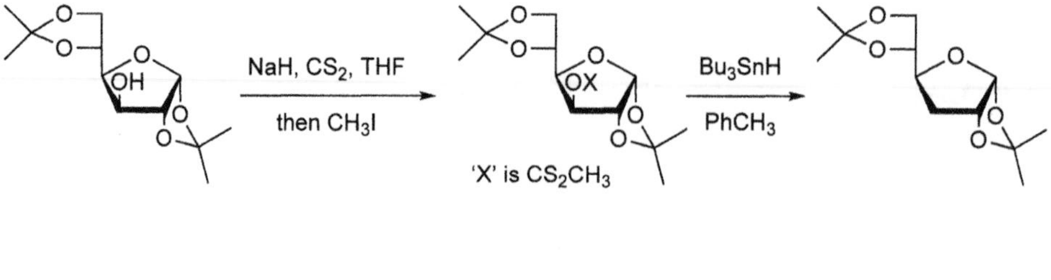

This 'invention' was much touted by Barton because of the free radical nature of the reaction:

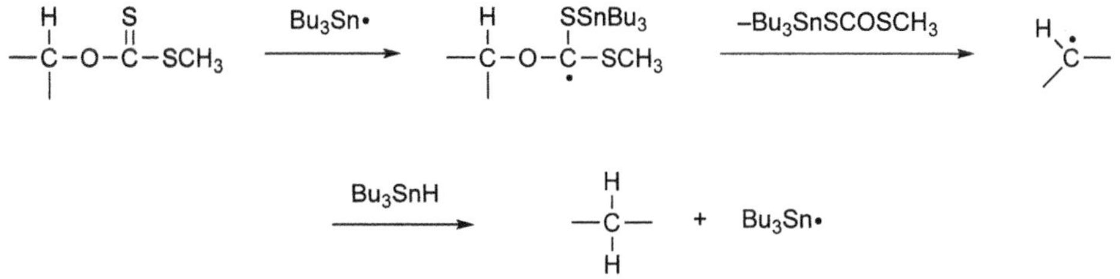

Over the years, the process has been refined and improved and is really the method of choice for the deoxygenation of a secondary alcohol.[59–61] Apart from the original xanthate, equally attractive substrates are a thiocarbonate and a thiocarbamate:[62–64]

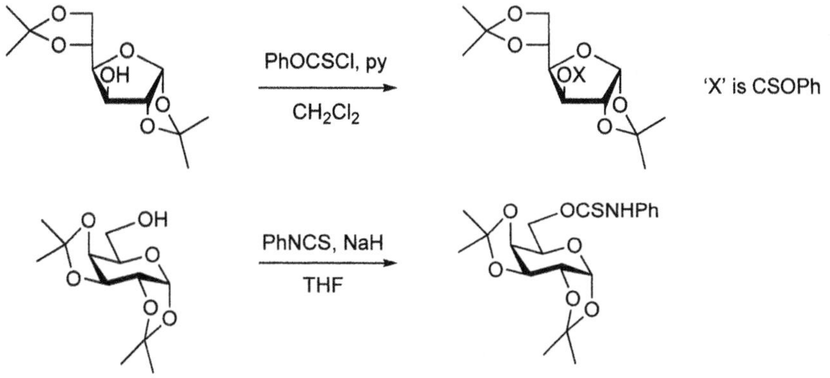

The Barton–McCombie deoxygenation method has been suitably modified for primary and tertiary alcohols.[61,65,66]

Although tin hydrides are still used routinely for the reduction, the waste products are highly toxic and therefore pose a problem in disposal. A useful procedure has been reported for the removal of these toxic tin residues,[67] and there exist methods that use a catalytic amount of the tin reagent and a stoichiometric amount of another reducing agent.[56] Also, a polystyrene-supported organotin hydride has been developed.[68,69] However, it is felt that reagents based on a Si–H or, more likely in terms of cost, a P–H bond, will prove safer for industrial application.[70–72]

References start on page 124

An extension of the above thiocarbonyl-based methodology gives rise to the selective deoxygenation of a diol:[73,74]

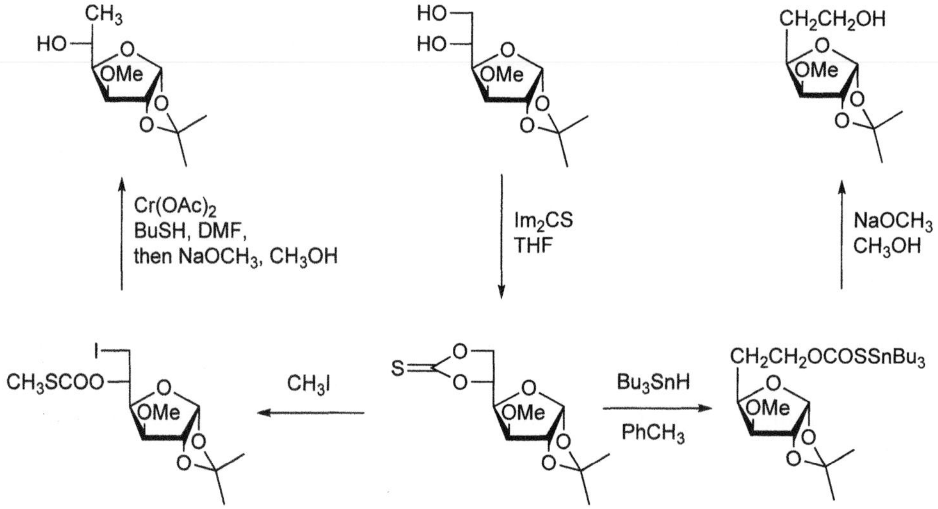

In summary, the best method for the deoxygenation of a primary alcohol probably still involves a conventional sequence:

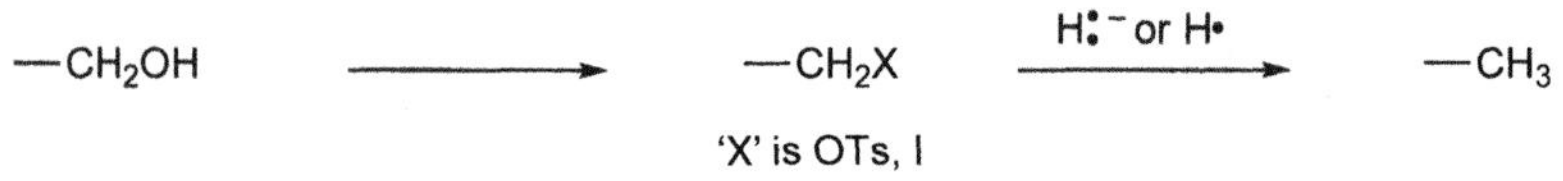

'X' is OTs, I

However, it is the Barton–McCombie free radical procedure that is best for secondary alcohols:

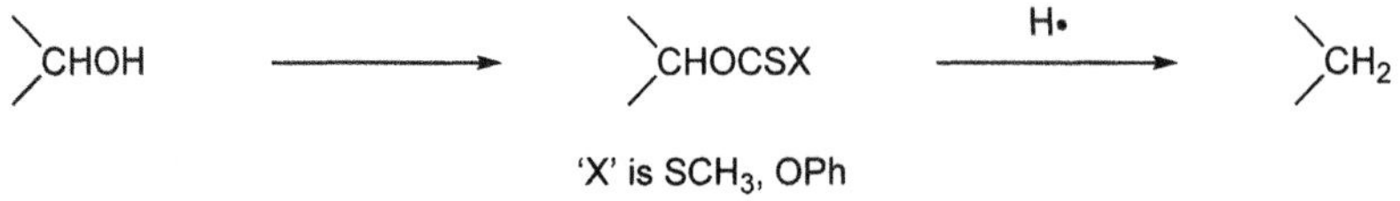

'X' is SCH₃, OPh

Halogenation

We have just seen that deoxyhalogeno sugars are of some importance in the synthesis of deoxy sugars. Although all of the halogens could be introduced into a monosaccharide using the older methods and reagents, there were some limitations as to what could be achieved:[75]

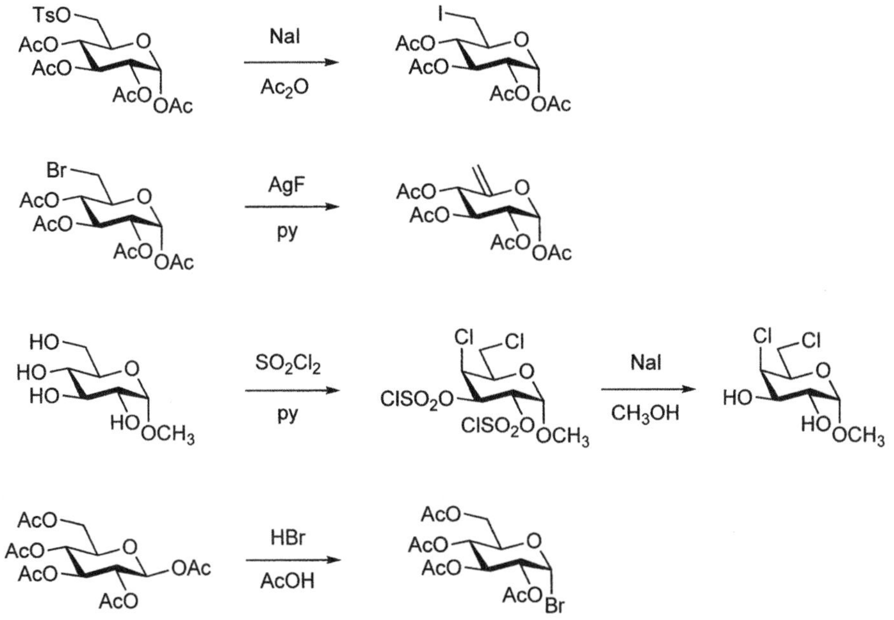

TsO— AcO— AcO— AcO OAc   →(NaI, Ac₂O)   I— AcO— AcO— AcO OAc

Br— AcO— AcO— AcO OAc   →(AgF, py)   AcO— AcO— AcO OAc

HO— HO— HO— HO OCH₃   →(SO₂Cl₂, py)   Cl Cl ClSO₂O— ClSO₂O OCH₃   →(NaI, CH₃OH)   Cl Cl HO— HO OCH₃

AcO— AcO— AcO— AcO —OAc   →(HBr, AcOH)   AcO— AcO— AcO— AcO Br

It soon became obvious that additional processes were needed for the introduction of halogen at the lone primary carbon, in preference to any of the three secondary carbon atoms; the aforementioned Hanessian–Hullar reaction (Chapter 2) was one of these:

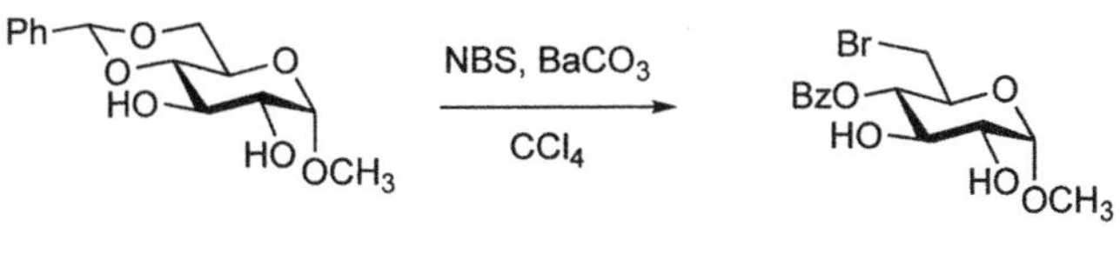

Ph O O HO— HO OCH₃   →(NBS, BaCO₃, CCl₄)   Br— BzO— HO— HO OCH₃

Also, milder conditions could be an advantage for the introduction of halogen at the anomeric carbon; the product, a glycosyl halide, was exceptionally reactive, being an 'α-halo ether'. Many of the reagents for the introduction of halogen now rely on triphenylphosphine. The discussion that follows will, by necessity, separate the introduction of halogen at the anomeric and non-anomeric positions.

Non-anomeric halogenation: Although the traditional nucleophilic displacements, e.g. of a tosylate by iodide, are still used, the methods currently in vogue for the introduction of chlorine, bromine and iodine all utilize triphenylphosphine. Probably, the most popular choices are triphenylphosphine with carbon tetrachloride, carbon tetrabromide or iodine/imidazole. In fact, this last combination has become the reagent of choice for the regioselective iodination (at C6) of many hexopyranosides:[75–79]

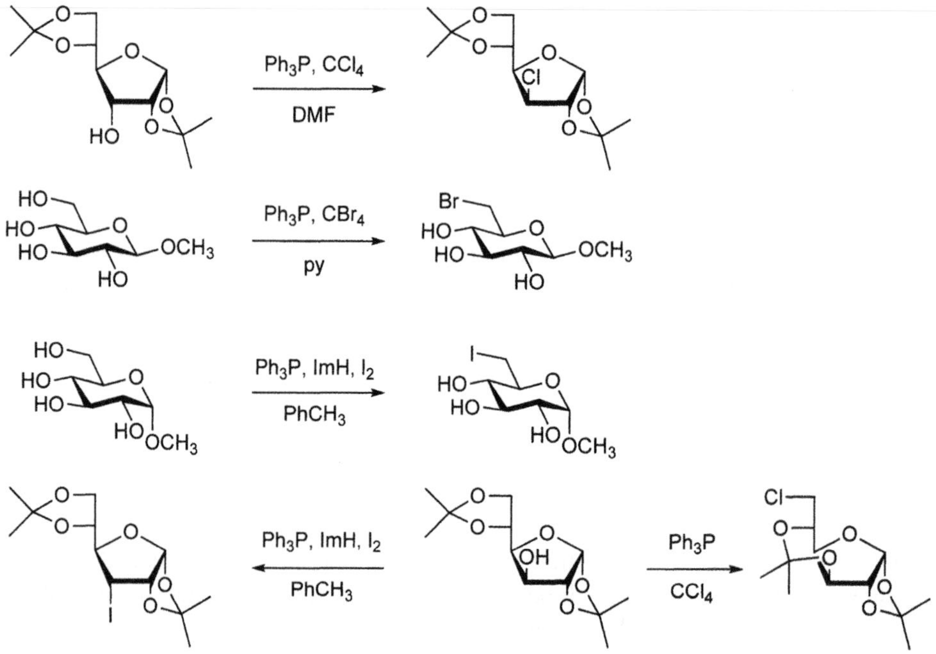

Where applicable, it can be seen that the above halogenations proceed with the inversion of configuration at carbon. This observation is in line with the general principle of nucleophilic displacement reactions in carbohydrates: the preference for S_N2 processes over S_N1. Presumably, carbenium ion intermediates are disfavoured by the presence of so many (electronegative) oxygen atoms. The driving force behind triphenylphosphine-mediated halogenation is the formation of the very stable phosphine oxide:

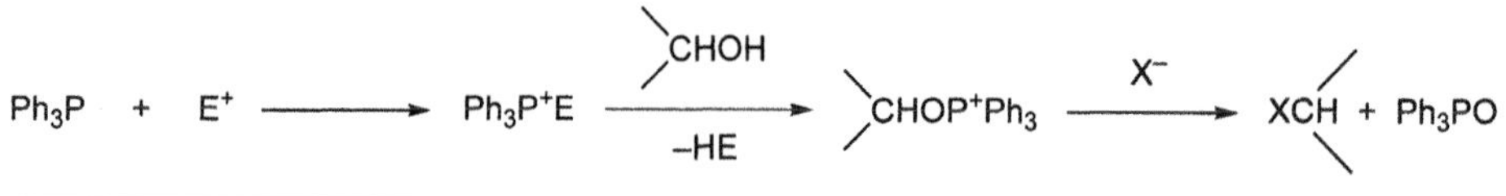

Removal of triphenylphosphine oxide can be a problem, and some solutions have been offered.[80]

Although fluorine can be introduced into a carbohydrate by conventional nucleophilic substitution, often employing a sulfonate and tetrabutylammonium fluoride, the method of choice utilizes (diethylamino)sulfur trifluoride (DAST):[81–83]

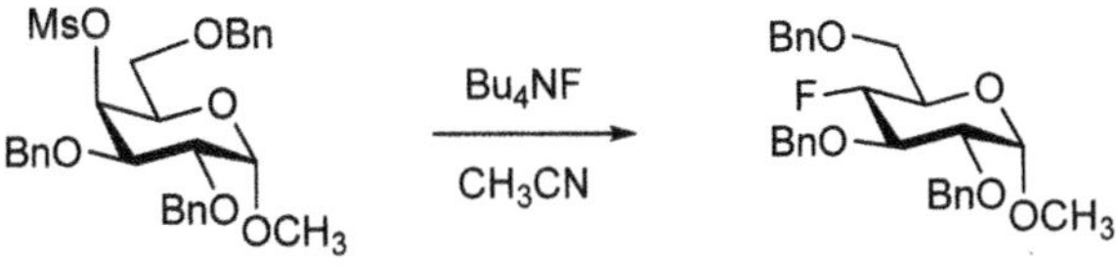

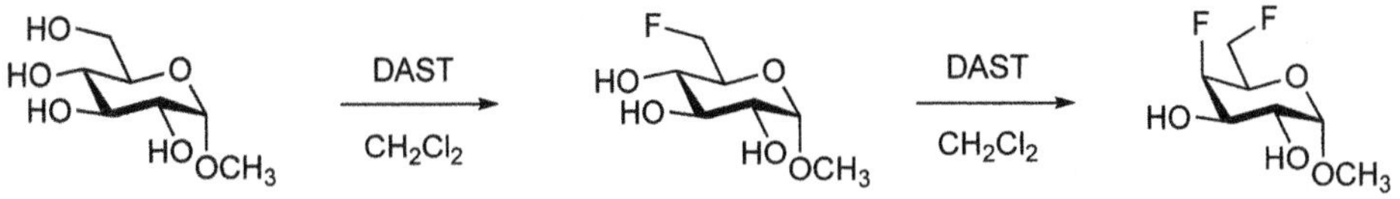

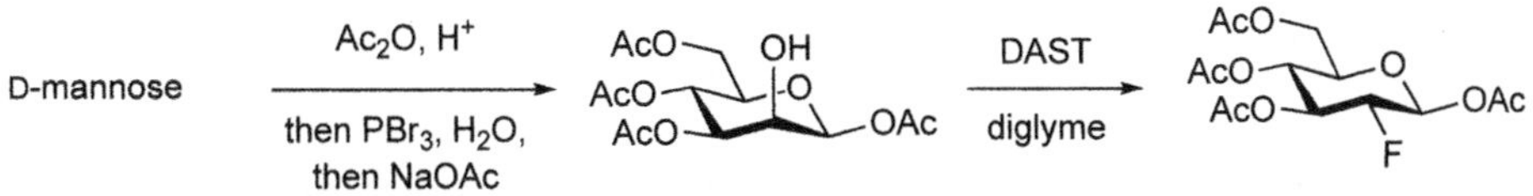

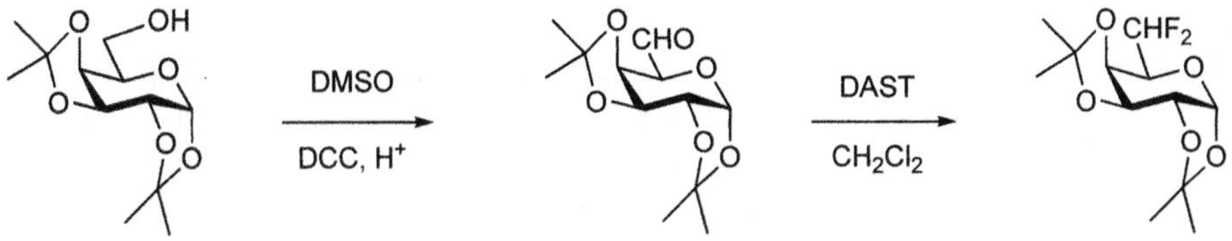

The reagent supplies an active electrophile as well as the fluoride ion but is relatively expensive; the substitution usually proceeds with inversion of configuration:

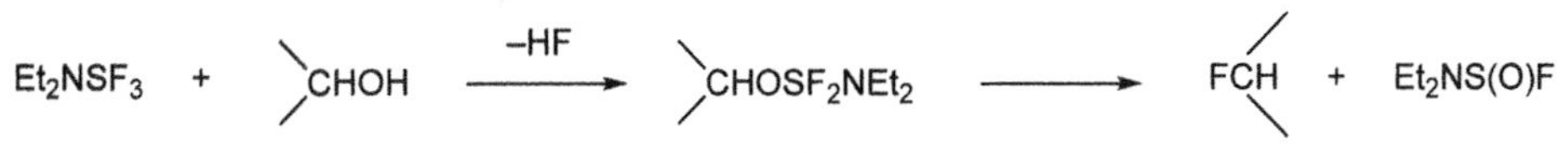

$$Et_2NSF_3 \; + \; {>}CHOH \xrightarrow{-HF} {>}CHOSF_2NEt_2 \longrightarrow FCH{<} \; + \; Et_2NS(O)F$$

Rearrangements are often observed with DAST in susceptible substrates:[84,85]

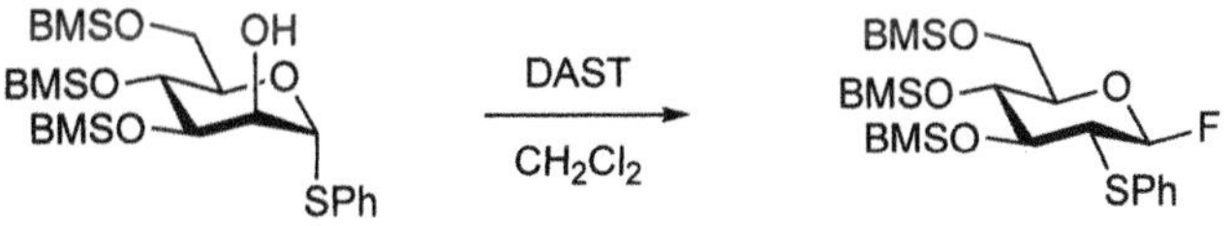

A comprehensive review of methods for the replacement of the hydroxyl group by fluorine has recently appeared.[86]

A novel preparation of 2-deoxy-2-fluoro sugars involves the addition of 'Selectfluor' {1-chloromethyl-4-fluoro-1,4-diazoniabicyclo[2.2.2]octane bis(tetrafluoroborate)}, a rare source of the fluoronium ion, to a 1,2-alkene:[83,87–92]

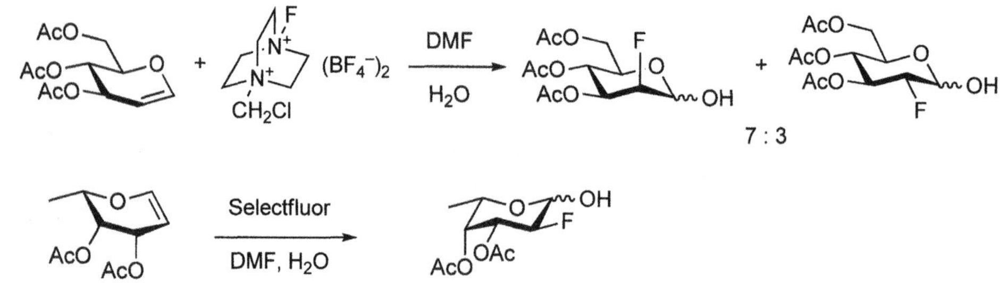

Only with some substrates is any useful diastereoselectivity exhibited in the products. The presence of a fluorine atom in deoxyfluoro sugars adds, of course, another dimension to NMR spectroscopy.[93]

Probably, one of the most useful deoxyfluoro sugars is 2-deoxy-2-[^{18}F]fluoro-D-glucose, [^{18}F]FDG (something of a misnomer), having a half-life of 110 min and therefore ideally suitable for medical imaging using positron emission tomography (PET). The molecule is prepared in 'hot boxes' fed directly by a cyclotron and is sufficiently active for clinical use for eight half-lives (about 15 h).[94–96]

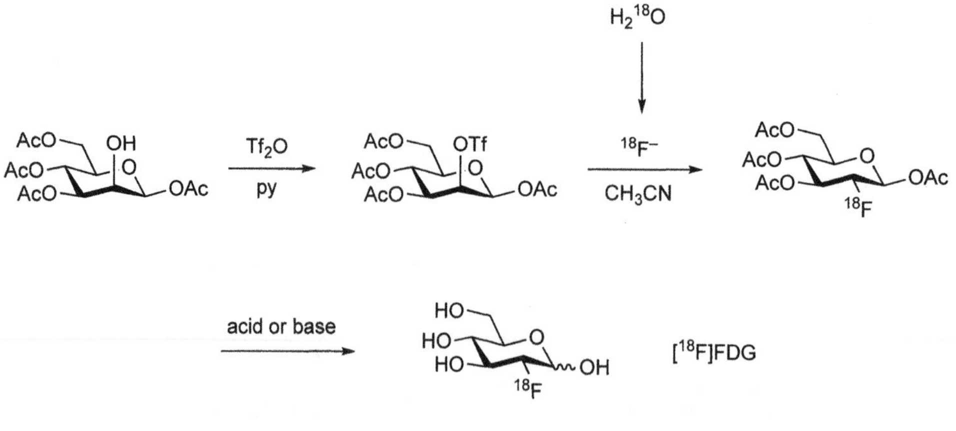

Anomeric halogenation: Glycosyl halides can be viewed as α-halo ethers, and consequently their method of preparation and subsequent chemical reactivity are different from the normal deoxyhalogeno sugars. The first examples of this valuable class were the glycosyl chlorides and bromides:[97–101]

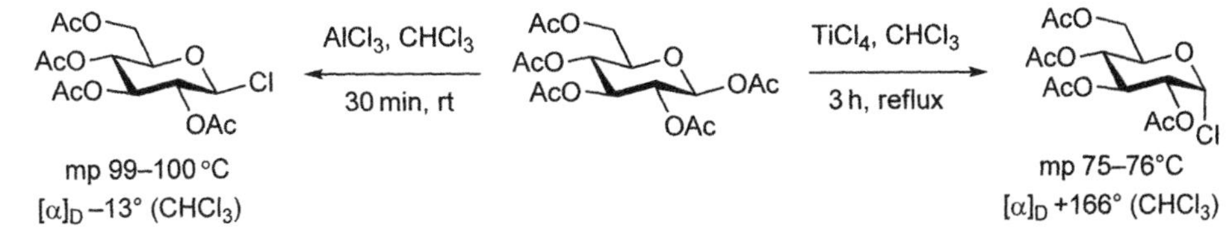

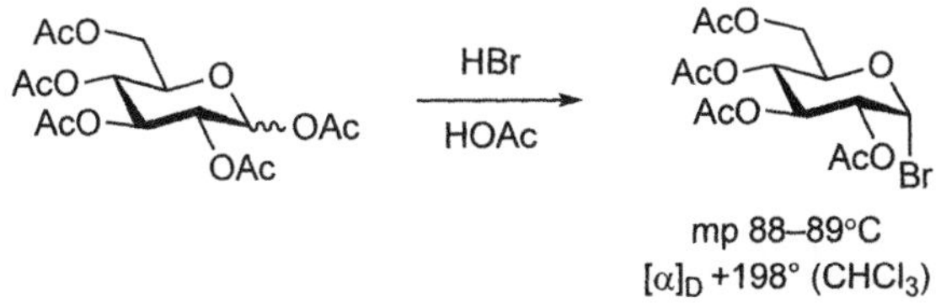

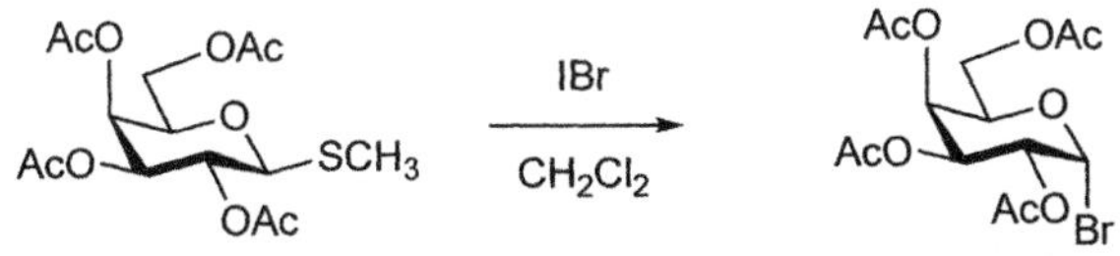

mp 99–100 °C
$[\alpha]_D$ –13° (CHCl₃)

mp 75–76°C
$[\alpha]_D$ +166° (CHCl₃)

mp 88–89°C
$[\alpha]_D$ +198° (CHCl₃)

Generally, Glycosyl chlorides can be obtained in either the α- or the β-pyranose form. The α-anomer is naturally the more stable, but the strength of the carbon–chlorine bond allows the formation of the β-anomer under conditions of kinetic control:[102]

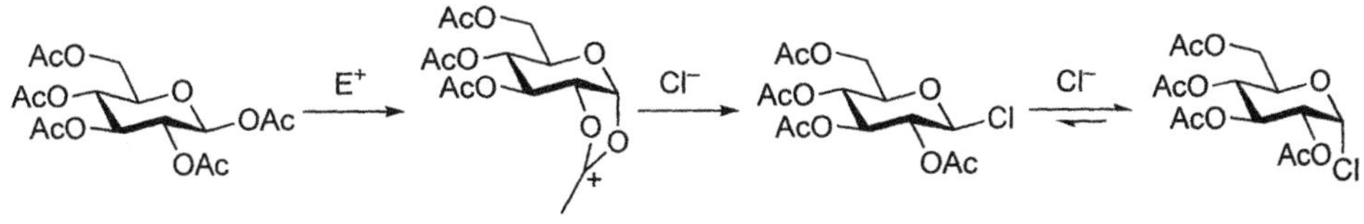

Glycosyl bromides, because of the weaker carbon–bromine bond, are generally isolated only as the α-anomer; this also holds for glycosyl iodides. Although first reported in 1929,[103] glycosyl iodides until recently were deemed too unstable to be of any use in subsequent synthetic transformations:[98,104–111]

References start on page 124

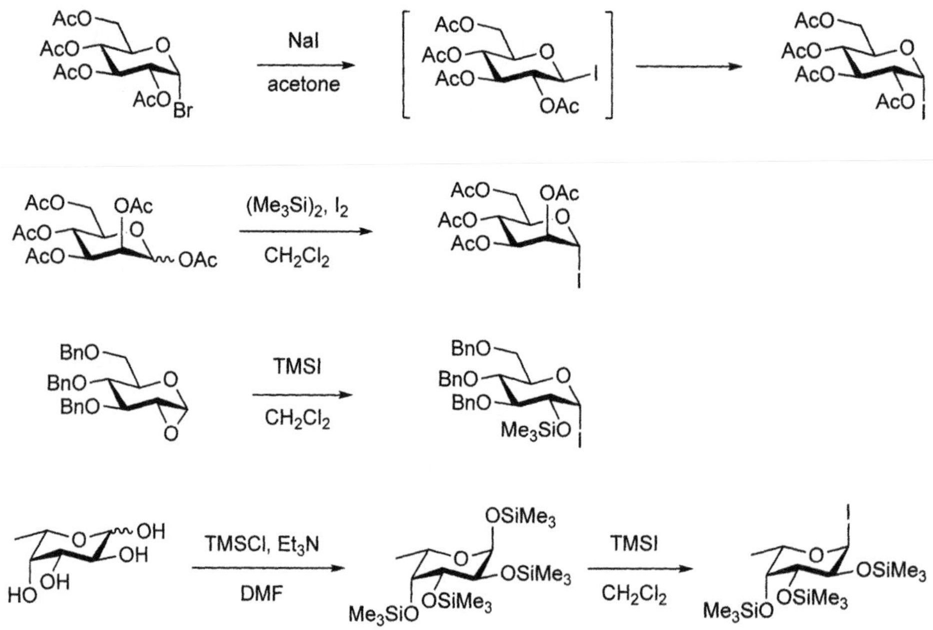

Glycosyl fluorides have been known for many years, but it took the inventiveness of chemists such as Mukaiyama and Nicolaou to realize the synthetic potential of these fluorides and so to stimulate new methods for their preparation:[81,112–117]

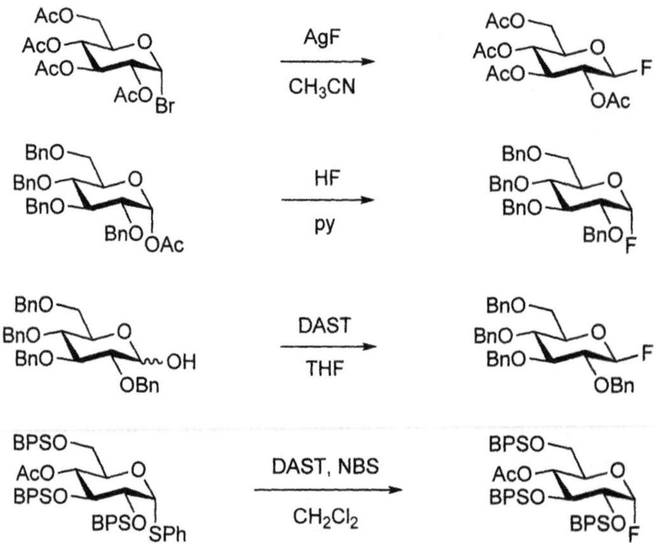

The α-anomer is the more stable, but again owing to the strength of the carbon–fluorine bond, it is possible to prepare the β-anomer in high yield. A unique feature of glycosyl fluorides is that they may be deprotected to give the polyol:[81]

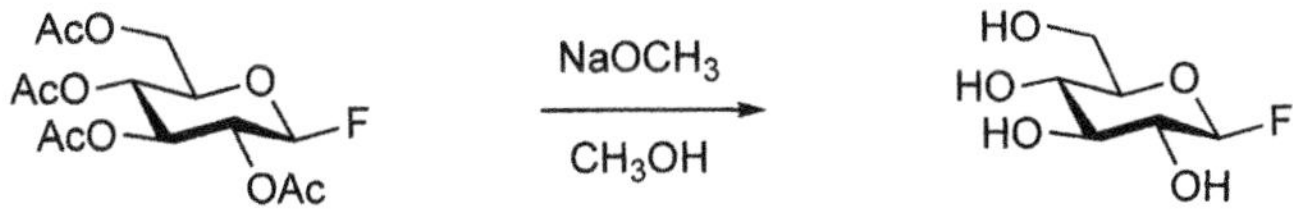

Some interesting and useful molecules result from the combination of various

Anomeric halogenation: Glycosyl halides can be viewed as α-halo ethers, and consequently their method of preparation and subsequent chemical reactivity are different from the normal deoxyhalogeno sugars. The first examples of this valuable class were the glycosyl chlorides and bromides:[97–101]

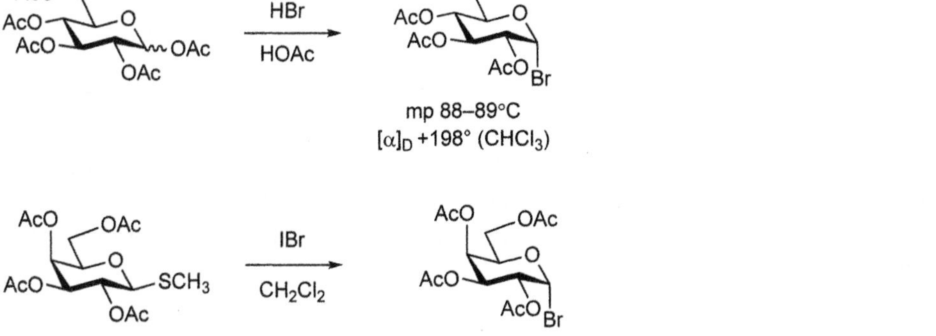

Generally, Glycosyl chlorides can be obtained in either the α- or the β-pyranose form. The α-anomer is naturally the more stable, but the strength of the carbon–chlorine bond allows the formation of the β-anomer under conditions of kinetic control:[102]

Finally, mention is made of a reaction that does not occur at the anomeric centre but generates *another* anomeric centre – Ferrier's 'photobromination':[120]

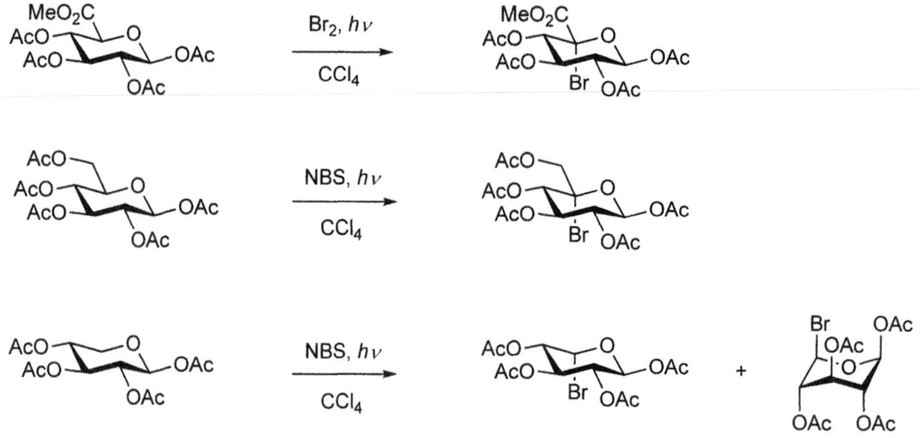

The brominated compounds can be put to good synthetic use:[120,121]

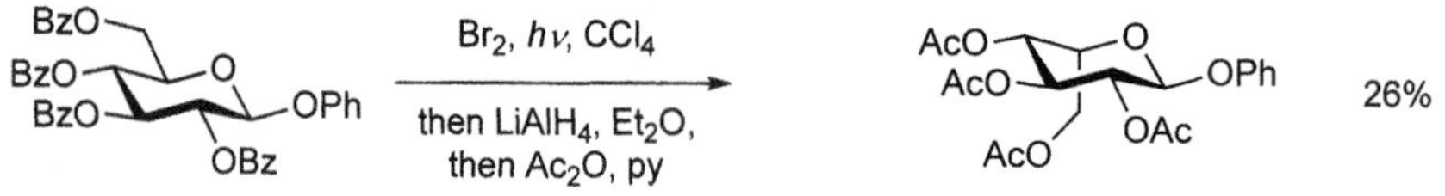

phenyl tetra-O-acetyl-α-L-idopyranoside

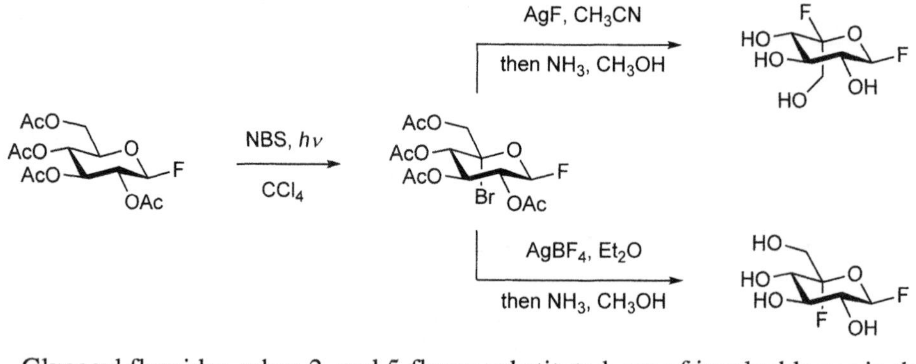

Glycosyl fluorides, when 2- and 5-fluoro substituted, are of invaluable use in the labelling of amino acid residues located in the active site of various glycoside hydrolases (see Chapter 7).[122–124] DAST can also be employed to prepare these interesting difluorides:[125]

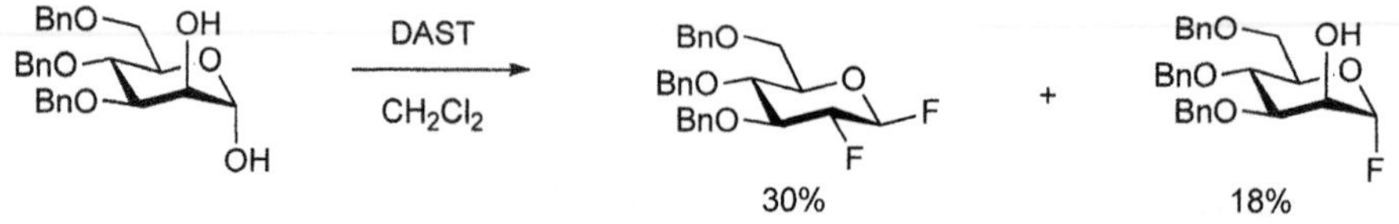

Alkenes and Carbocycles

It is appropriate to discuss alkenes and carbocycles together as the former are sometimes precursors to the latter. In general, alkenes derived from carbohydrates are rarely found naturally and usually result from some chemical transformation in the laboratory, e.g. dehydration, dehydrohalogenation or 'de-dihydroxylation'. Anomeric alkenes, termed 'glycals', are unique and their chemistry will be discussed separately. Carbocycles form the core of several important classes of biologically important molecules, e.g. the cyclitols.[126]

Non-anomeric alkenes:[127,128] In the pyranose series, 2,3-, 3,4-, 4,5- and 5,6-alkenes are all known, with the last type probably being the most common and useful. The generation of a 2,3- or 3,4-alkene can conveniently occur from the appropriate diol, using older (Tipson–Cohen) methodology or an extension of methods that we have already seen as developed by Barton and by Garegg:[129–133]

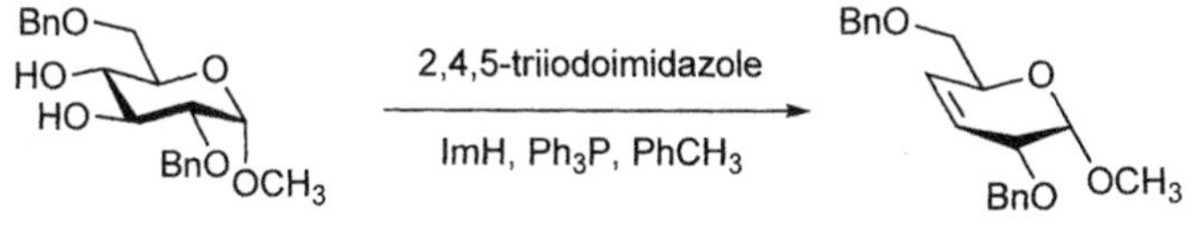

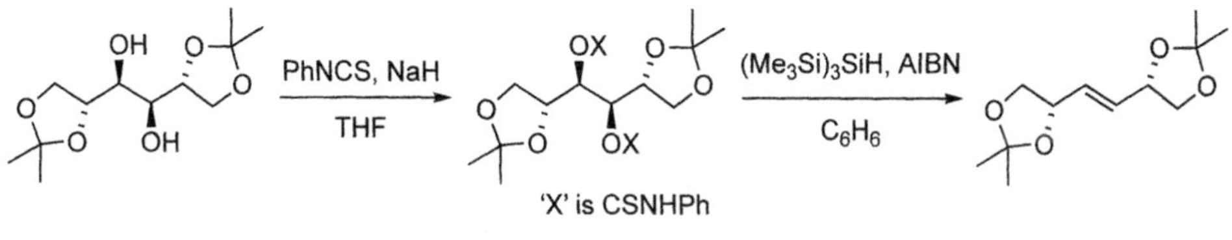

methyl 4,6-O-benzylidene-2,
3-dideoxy-α-D-*erythro*-hex-2-enoside

A rather unusual method for the synthesis of a 2,3-alkene involves the reduction of a cyclic sulfate:[134]

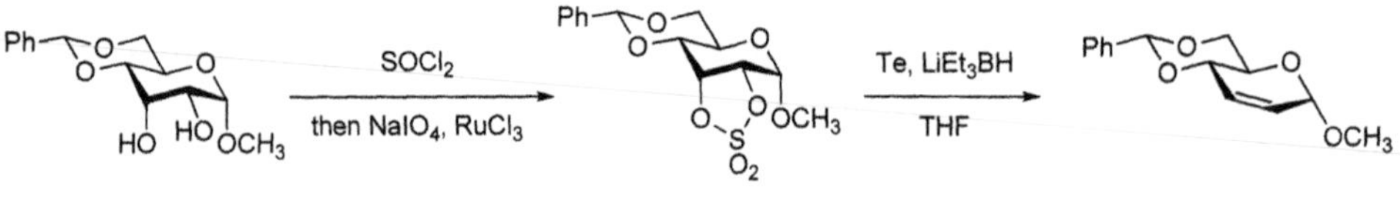

For the synthesis of a 4,5- or 5,6-alkene, a dehydration process may appear obvious, but it is a dehydrohalogenation that is more efficient:[135]

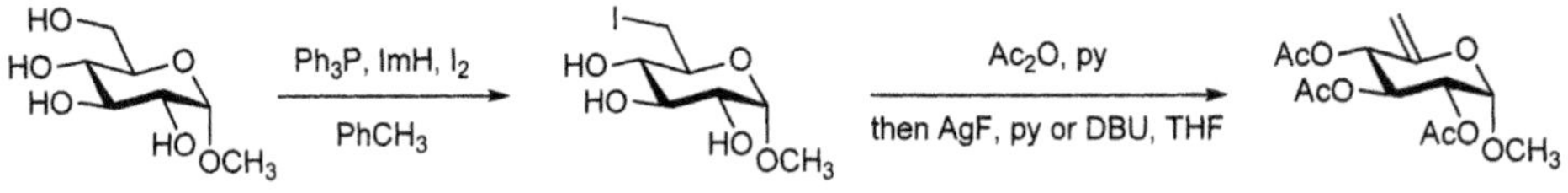

An unusual synthesis, often termed the Bernet–Vasella reaction, of acyclic 5,6-alkenes involves the reduction of a 6-bromo-6-deoxy-D-glucopyranoside:[136–138]

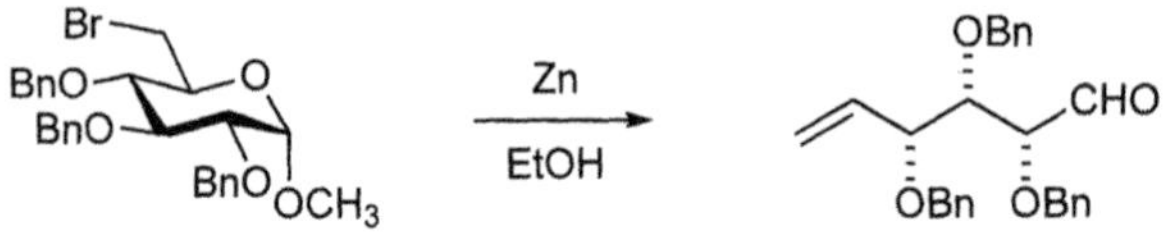

Anomeric alkenes:[127,128] Although derivatives of 1,5-anhydro-hex-1-enitols are known, it is the 2-deoxy derivatives that are much more common and synthetically useful; the name 'glycal' was given to the latter group when the properties of an aldehyde (from hydrolysis of the enol structure) were originally noticed:

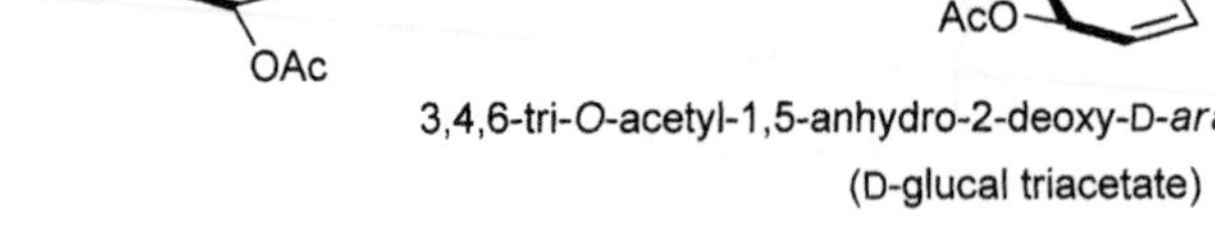

3,4,6-tri-O-acetyl-1,5-anhydro-2-deoxy-D-*arabino*-hex-1-enitol

(D-glucal triacetate)

There are many methods for the formation of glycals, ranging from the original treatment of 'acetobromoglucose' with zinc in acetic acid (by Fischer in 1913) to a comparable reduction with samarium(II) iodide:[48,127,128,139–141]

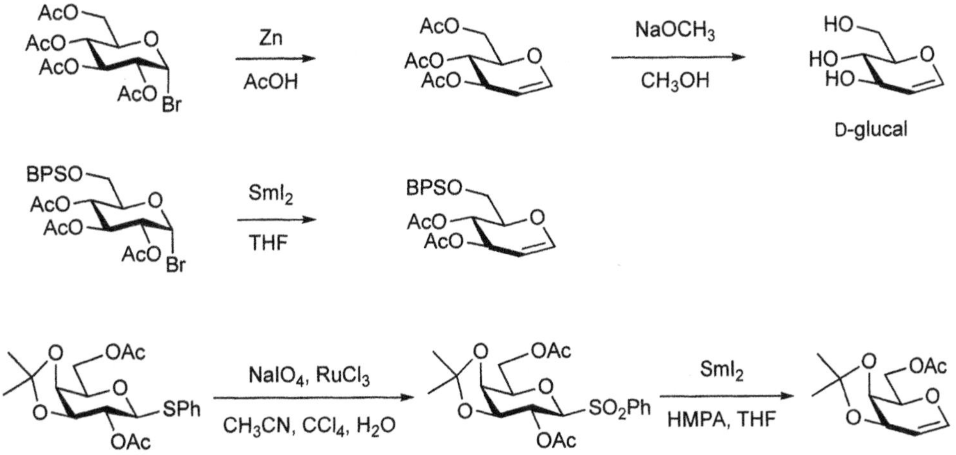

D-glucal

The use of zinc/silver/graphite offers a significant improvement,[142] and a 'one pot' procedure seems to be of advantage when conversion of a free sugar directly into the glycal is required.[143] An aprotic procedure (zinc and a tertiary amine in benzene) has also been developed,[144] and potassium–graphite laminate is a vigorous enough reagent to cause the formation of glycals even from O-alkyl or O,O-alkylidene precursors:[145]

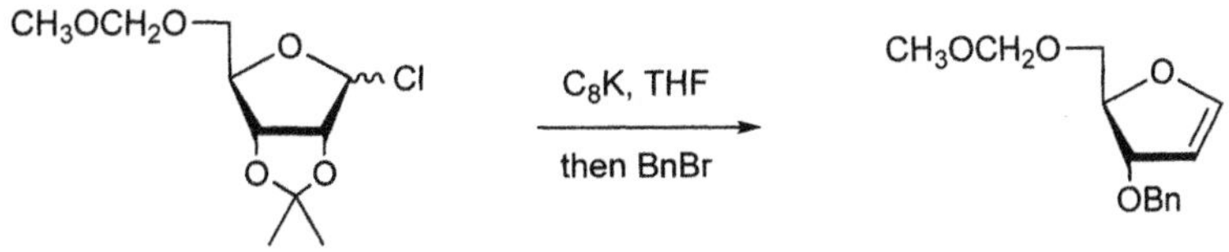

New methods employing other metals, e.g. titanium, vanadium and cobalt (of vitamin B$_{12}$), continue to be announced.[146–149] Furanoid glycals are generally difficult to prepare and are much less stable than their pyranoid counterparts. In general, glycals, being enol ethers, do not particularly enjoy the company of acids; an indirect synthesis of 4,6-O-benzylidene-D-glucal, avoiding such acidic conditions, has recently appeared:[150]

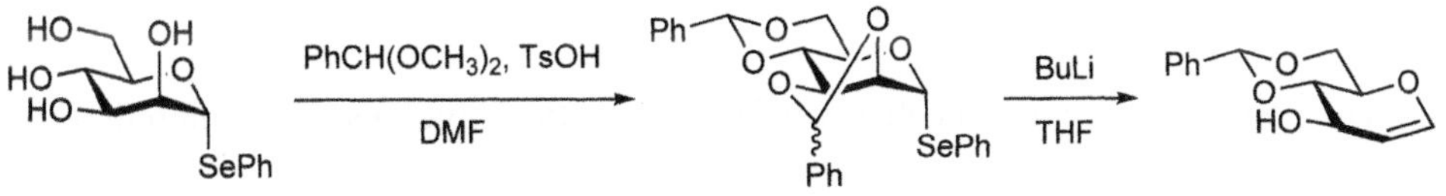

Exo-glycals, which have the double bond at the anomeric carbon external to the ring, are not nearly as common as the parent glycals.[151,152]

Carbocycles:[153–157] For the biosynthesis of the inositols (polyhydroxylated cyclohexanes), nature uses a carbohydrate precursor, D-*xylo*-hexos-5-ulose 6-phosphate, and makes use of the aldol reaction (see Chapter 6):[158]

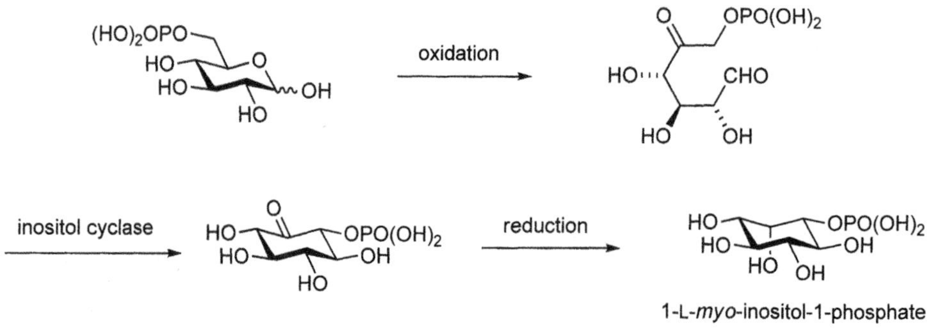

1-L-*myo*-inositol-1-phosphate

This sort of reaction has been mimicked[159] and emulated by chemists for the synthesis of carbocycles from carbohydrates and, without doubt, the procedure developed by Ferrier is one of the most valuable:[135,154,160]

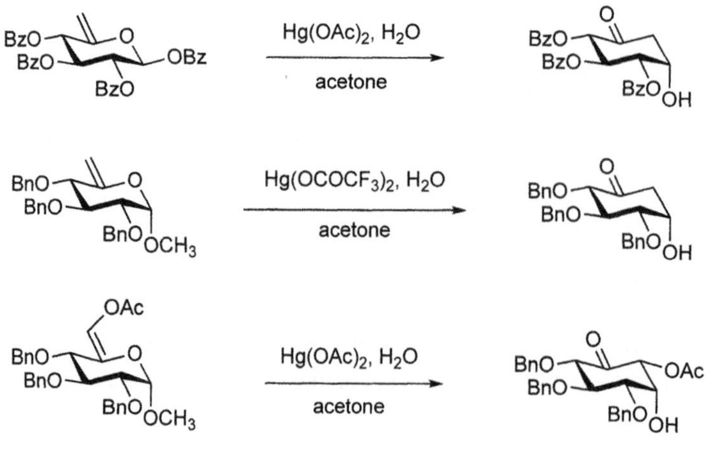

The generally accepted mechanism for this reaction involves an oxymercuration, followed by an intramolecular aldol reaction:[154]

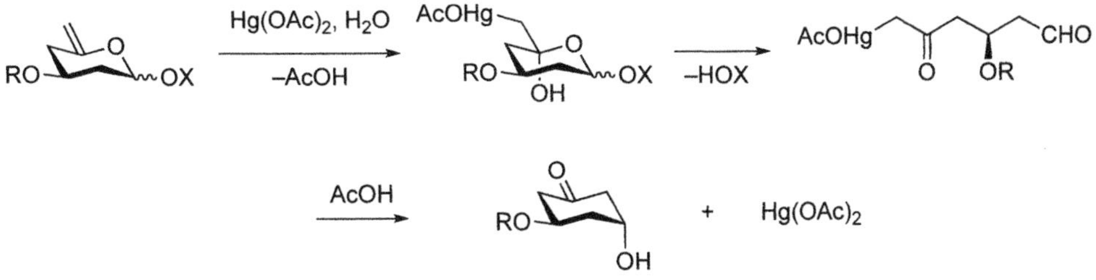

The stereochemistry that is observed in the final product has been commented upon.[154] Certainly, one of the strengths of this transformation is that the initial products can be put to good use:[135,154,160–162]

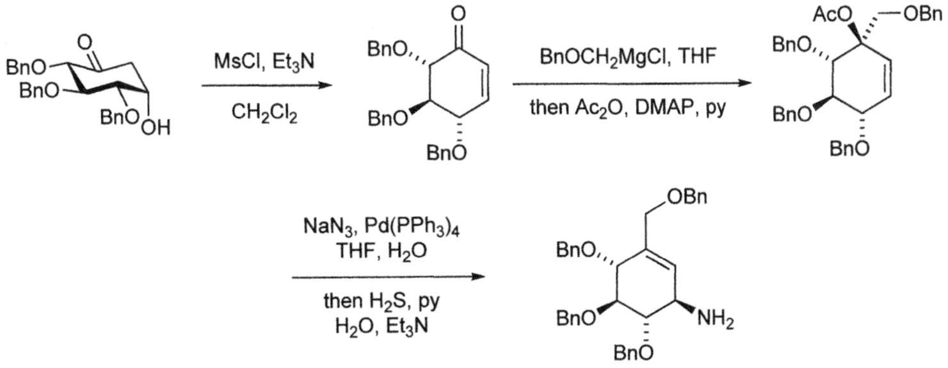

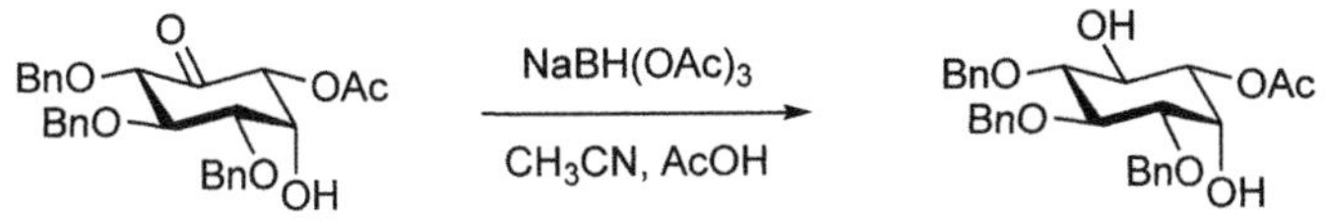

Two variations of this carbocycle synthesis, both of which retain the anomeric substituent, have been reported:[163,164]

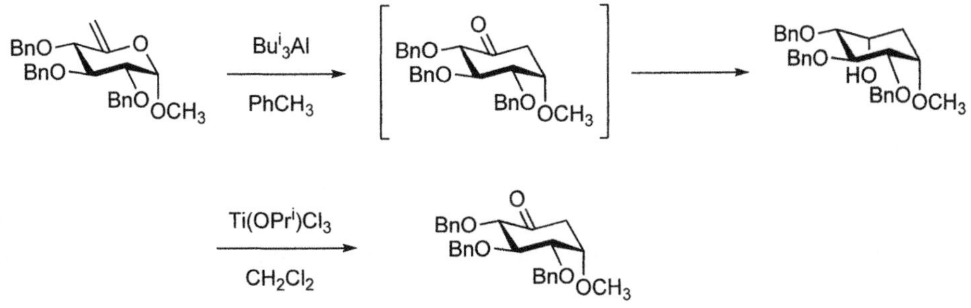

Another approach to the synthesis of carbocycles from carbohydrates involves highly functionalized alkenes:[153,154,165]

These sorts of reactions all involve free radical intermediates, with hex-5-enyl radicals cyclizing more rapidly than their hept-6-enyl counterparts and in accord with the rules as espoused by Baldwin and Beckwith:[153,166–169]

The most recent approach to carbocycles involves olefin metathesis, in particular, ring-closing metathesis employing Grubbs' first-generation catalysts:[170–174]

Anhydro Sugars[175]

As the name implies, anhydro sugars formally result from the loss of water between two hydroxyl groups within a monosaccharide unit, with the concomitant formation of a new carbon–oxygen bond. Because of the number of different types of hydroxyl groups present in a monosaccharide, a whole range of anhydro sugars is possible; a classification based on whether the 'anhydro linkage' involves the anomeric centre or not is convenient.

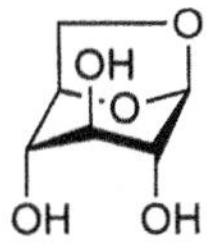

1,6-anhydro-β-D-glucopyranose

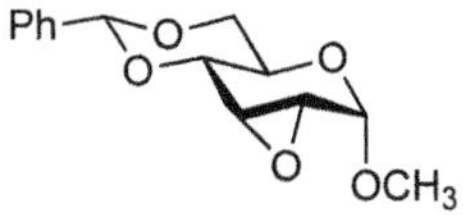

methyl 2,3-anhydro-4,6-O-benzylidene-α-D-alloside

Non-anomeric anhydro sugars:[176] The most important of the non-anomeric anhydro sugars are the epoxides, which can be formed somewhat circuitously by the oxidation of an unsaturated sugar but, more commonly, they result from an internal nucleophilic displacement:[177–180]

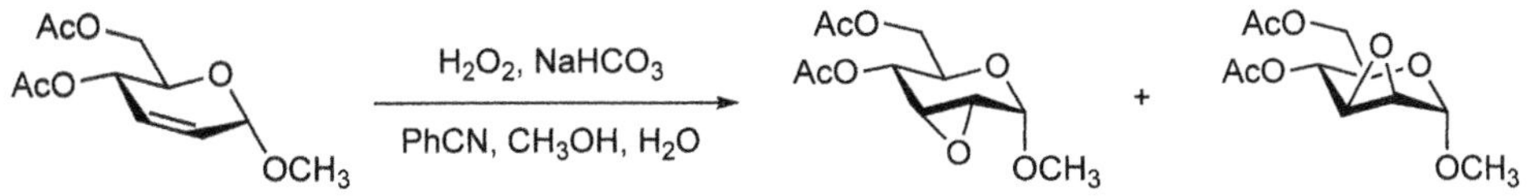

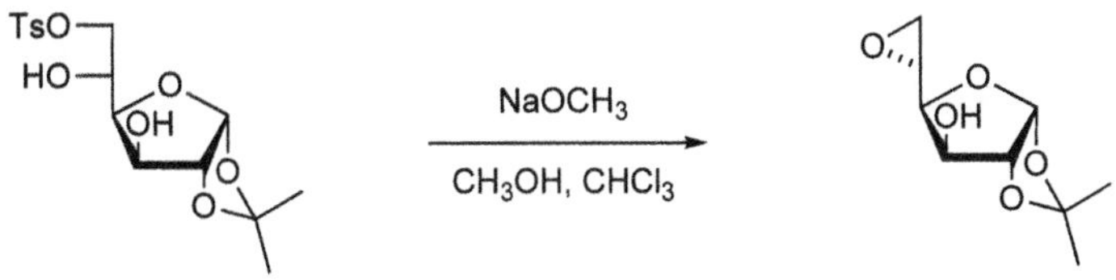

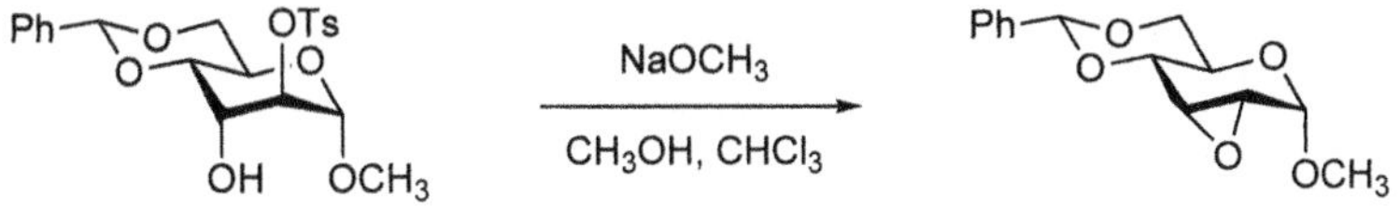

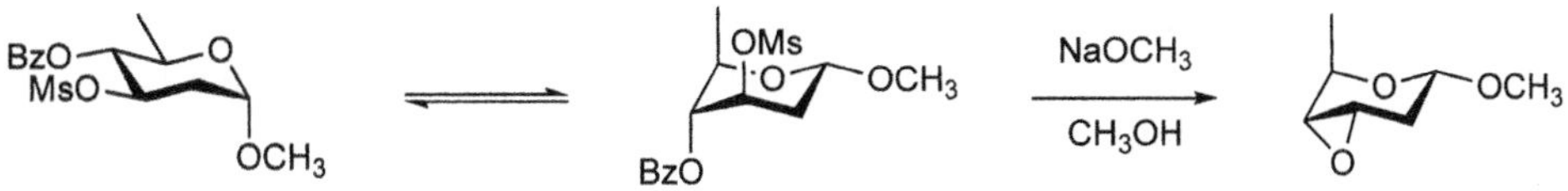

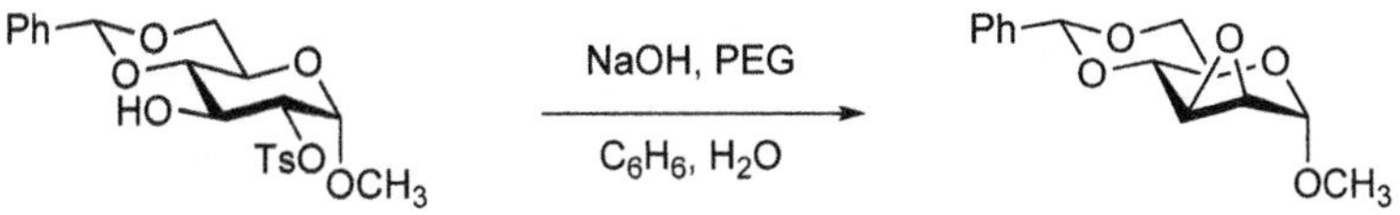

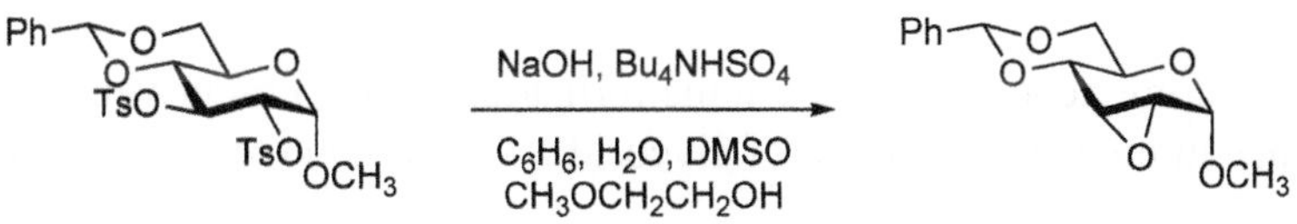

The direct oxidation of an alkene is (generally) not very stereoselective. In addition, the formation of an epoxide by the internal displacement of a sulfonic acid ester is a favoured process when either few conformational constraints exist or an 'anti-periplanar' arrangement of nucleophile (O$^-$) and leaving group (OTs, OMs) can be attained. For substrates where such an arrangement cannot be reached by the normal conformational change ($^4C_1$ to $^1C_4$), it has been suggested that other conformations may be involved.[176,181]

Another approach to sugar epoxides involves an intramolecular Mitsunobu reaction, directly on the diol:[181]

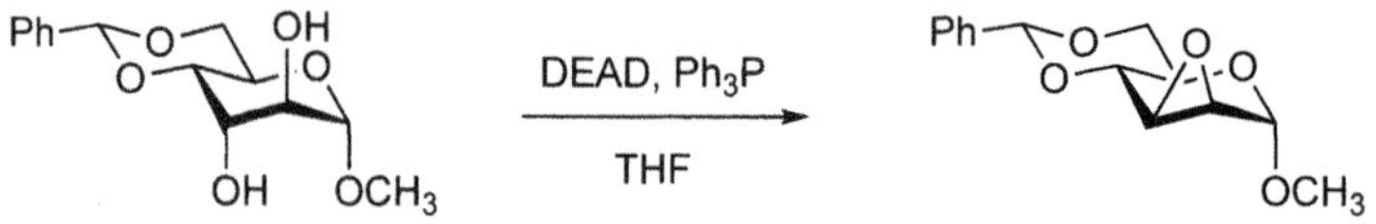

Care must be employed in some cases where an internal 'Payne rearrangement' is possible in an initially formed product:[182,183]

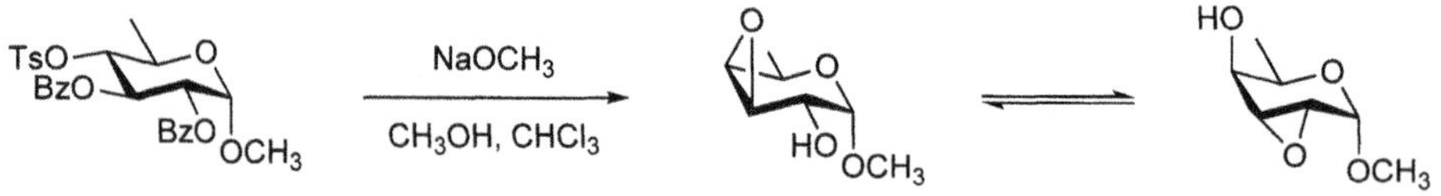

Spiro-epoxides, apart from arising from the epoxidation of an *exo*-cyclic alkene, can be formed from ketones and dimethylsulfoxonium methylide or dimethylsulfonium methylide:[184–186]

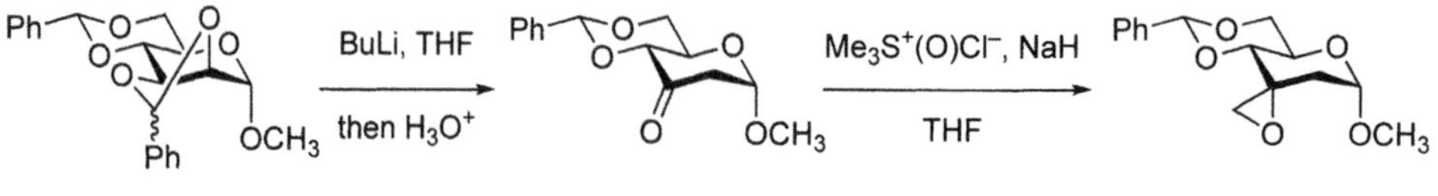

Of the other non-anomeric anhydro sugars, the formation of the 3,6-derivative is a common occurrence in the base treatment of glycosides with a good leaving group at C6:[187]

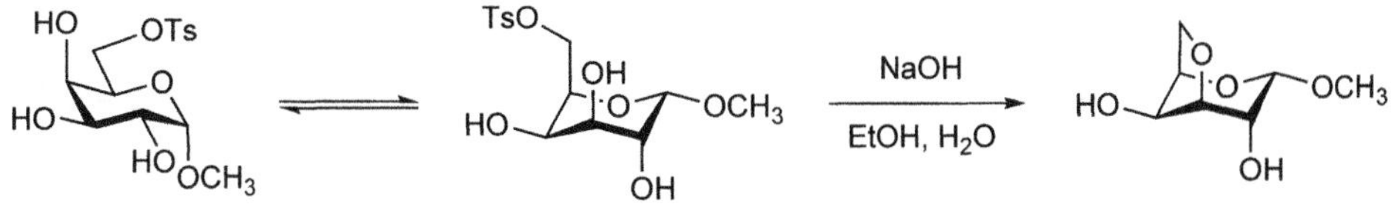

Anomeric anhydro sugars: Anomeric anhydro sugars, in which the new carbon–oxygen bond involves the anomeric centre, may be considered as 'internal' acetals. Again, many possibilities exist, but here only the important 1,6- and 1,2-anhydro hexopyranoses will be discussed:

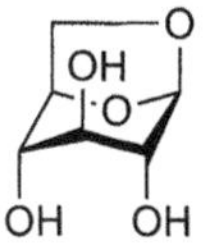

1,6-anhydro-β-D-glucopyranose

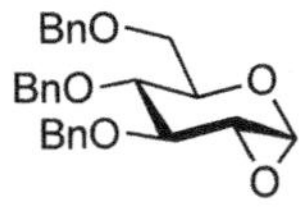

1,2-anhydro-3,4,6-tri-O-benzyl-α-D-glucose

1,6-Anhydro hexopyranoses:[188,189] Some aldohexoses, in aqueous acid at equilibrium, form a large amount of the 1,6-anhydro pyranose derivative:[190]

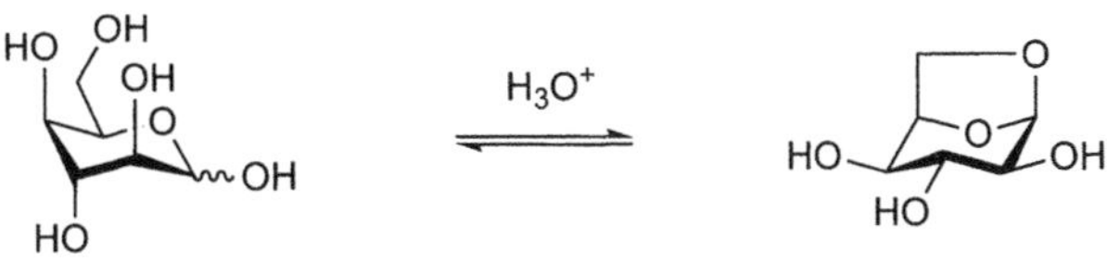

D-idopyranose

1,6-anhydro-β-D-idopyranose

For D-idose, as illustrated, the result is not surprising: the product is stabilized by a favourable anomeric effect and is no longer destabilized by having axial hydroxyl groups on the pyranose ring. In fact, for D-altrose, D-gulose and D-idose, this simple acid-catalysed process represents a viable method for preparing the anhydro sugar. For other aldohexoses, however, alternative methods had to be developed.

One of the oldest ways of producing 1,6-anhydro hexopyranoses is by the pyrolysis of readily available carbohydrates, such as starch, cellulose, mannan and α-lactose. In this way 1,6-anhydro-β-D-glucopyranose, -mannopyranose and -galactopyranose are produced:[191–193]

References start on page 124

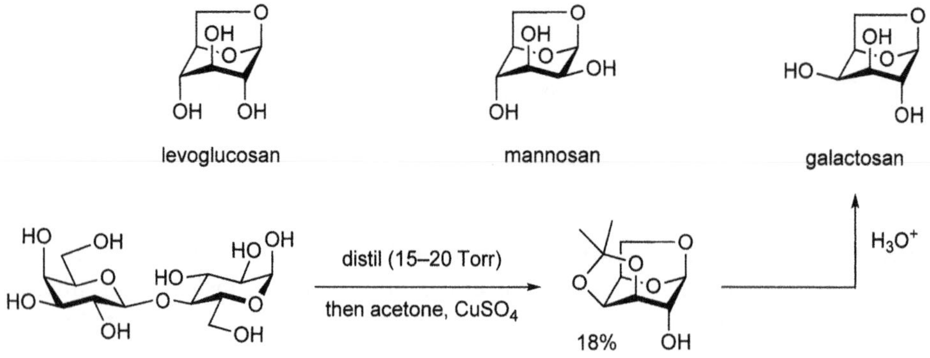

levoglucosan   mannosan   galactosan

The trivial names arise from 'glycosan' (*glycos*e *an*hydride), and 'levoglucosan' from the fact that an early report on glucosan gave a positive value for the optical rotation of what was obviously an impure sample.

Other methods are commonly used for the preparation of glycosans, generally involving some sort of activation of either the C1–O1 or the C6–O6 bond:[194–199]

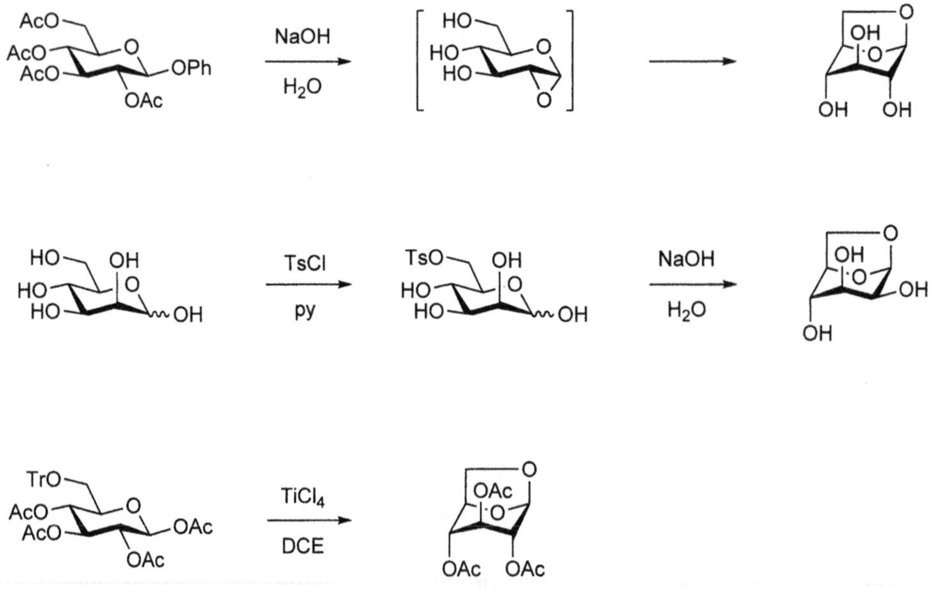

Glycals can be convenient sources of functionalized anhydro sugars:[200–204]

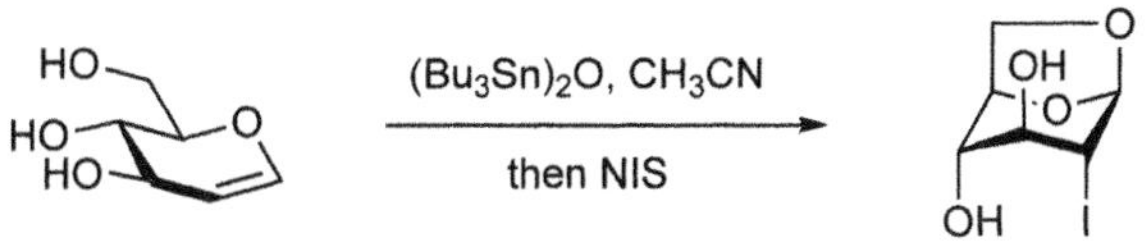

One of the reasons that so much effort has been put on the synthesis of the glycosans is that they offer so much to synthetic chemists in subsequent transformations. Included in the gift are protection of both C1 and C6, an unusual conformation ($^1C_4$), an array of hydroxyl groups of unusual and differing reactivity, and an anhydro bridge that, when finally broken, allows a return to the normal $^4C_1$ conformation of the pyranose ring. Making use of most of these features are the 'Čěrný' 2,3- and 3,4-epoxides:[188,203-208]

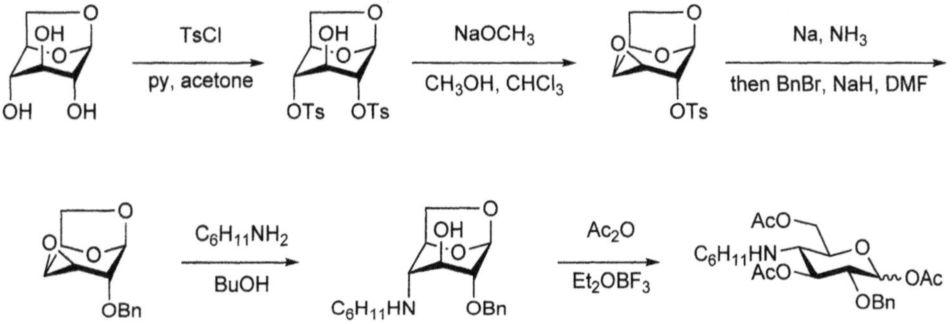

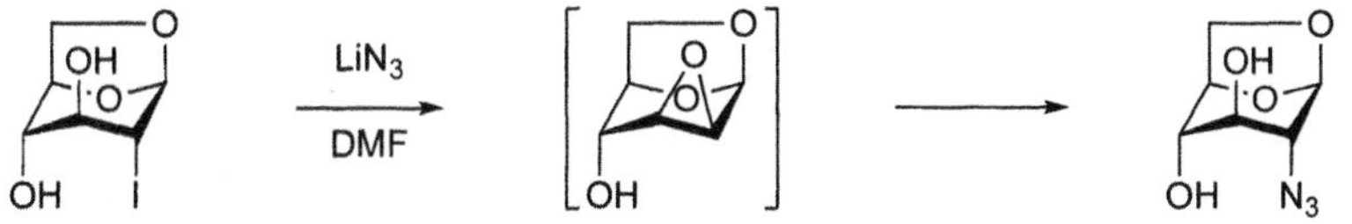

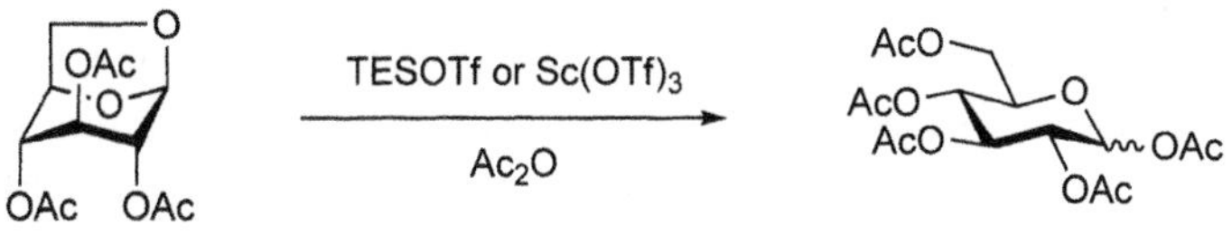

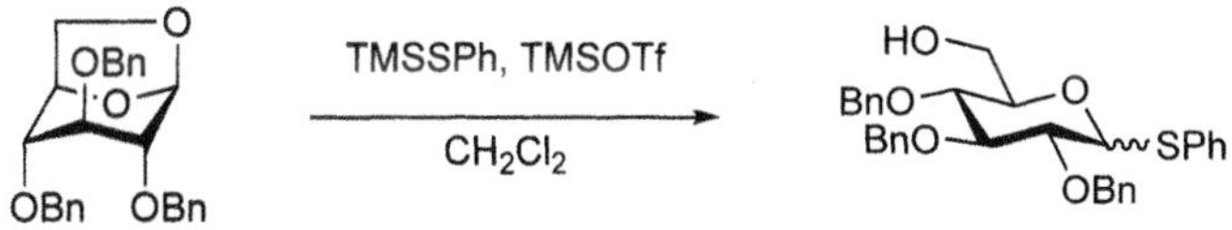

1,2-Anhydro hexopyranoses: The classic example of a 1,2-anhydro hexopyranose was prepared by Brigl, many years ago:[209]

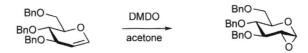

'Brigl's anhydride' took on a special role in the history of carbohydrate chemistry when it was used in the first chemical synthesis of sucrose by Lemieux.[210] Nowadays, this method of preparation of anomeric epoxides (the displacement of some good leaving group at C1 by an oxygen nucleophile at C2) has given way to more general procedures.

The most convenient and operationally simple method for the synthesis of 1,2-anhydro hexopyranoses involves the oxidation of glycals with dimethyldioxirane:[211,212]

As we have seen, all sorts of glycals are readily prepared and the reagent dimethyldioxirane is easily generated in acetone solution, free of moisture.[213,214] The work-up procedure after the oxidation involves nothing more than a simple evaporation of the solvent, and generally, one diastereoisomer is either formed exclusively or predominates. The use of more traditional oxidizing agents, e.g. 3-chloroperbenzoic acid, produces acidic by-products that destroy the desired epoxide. More will be said about these anomeric epoxides in the next chapter, but it will suffice to say that they are generous precursors of glycosides, glycosyl amines, thioacetals and glycosyl fluorides, as one would expect of an epoxide that is also part of an acetal:[215]

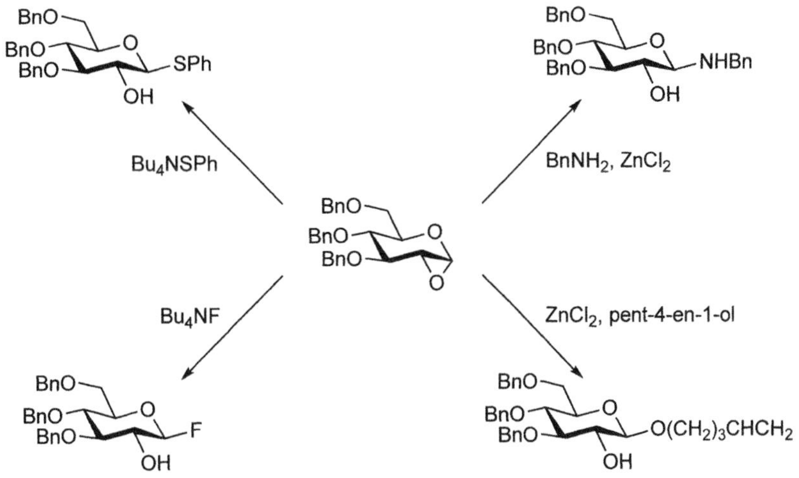

Deoxy, Amino Deoxy and Branched-chain Sugars

Deoxy, amino deoxy and branched-chain sugars are all characterized by some sort of modification of the carbohydrate: loss of a hydroxyl group, replacement of a hydroxyl group by an amino group or addition of an alkyl group. These changes are not without purpose, as the resulting molecules often show unique physical characteristics that are associated with some sort of biological role or response.

Deoxy sugars:[216,217] Deoxy sugars occur commonly in nature, the most abundant being 'deoxy ribose' (2-deoxy-β-D-*erythro*-pentofuranose) as found in deoxyribonucleic acid. Although deoxygenation is possible at any position of a carbohydrate, it is the 6-deoxy sugars that are widespread – D-quinovose (6-deoxy-D-glucopyranose), L-rhamnose (6-deoxy-L-mannopyranose) and L-fucose (6-deoxy-L-galactopyranose) are all constituents of molecules that are found, among other places, in plants and bacteria:

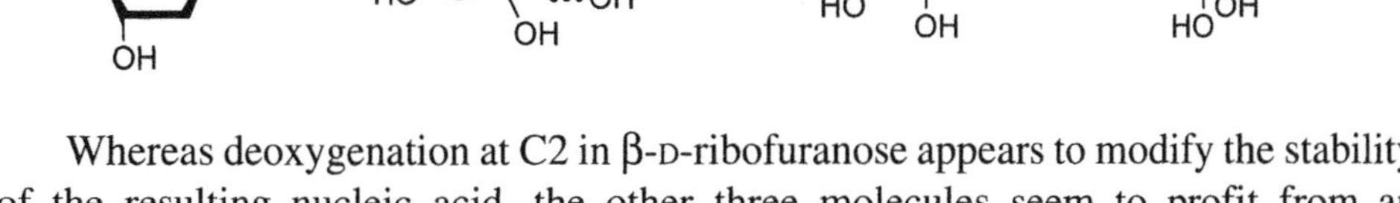

Whereas deoxygenation at C2 in β-D-ribofuranose appears to modify the stability of the resulting nucleic acid, the other three molecules seem to profit from an improved hydrophobic interaction at C6 (CH$_3$) with appropriate receptor sites in biological molecules.

For the synthesis of 3-, 4- and 6-deoxy sugars,[218] it is generally simply a matter of manipulating the appropriate sugar so that only the hydroxyl group of interest is unprotected; a simple reduction or deoxygenation then provides a precursor to the desired deoxy sugar:[219–221]

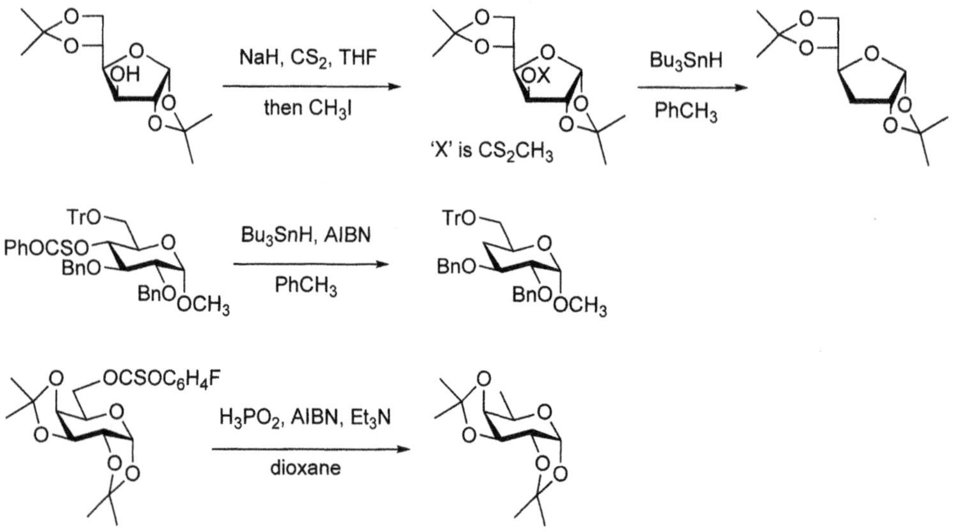

Glycals offer a convenient starting point for the synthesis of 2-deoxy sugars:[222,223]

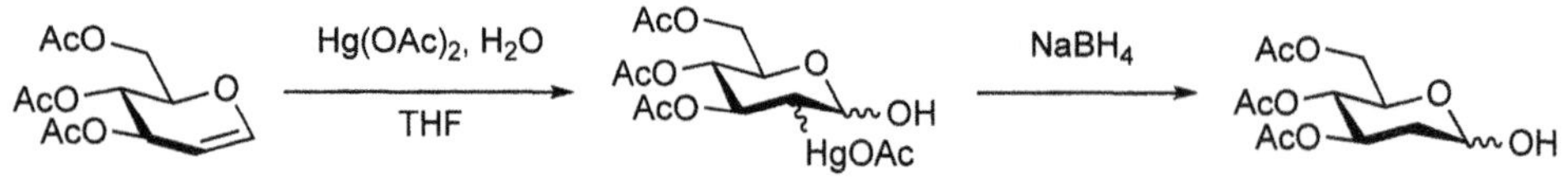

Somewhat rarer but of no less importance are the dideoxy sugars, which are often found at the termini of oligosaccharides attached to the surface protein of various bacteria:

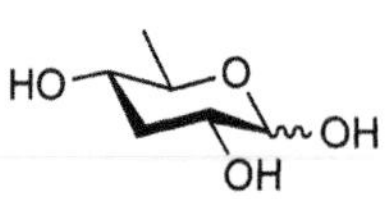

paratose

3,6-dideoxy-D-*ribo*-hexopyranose

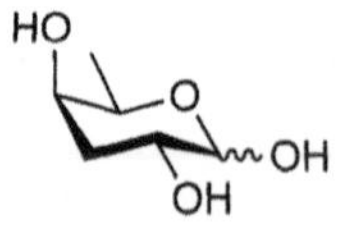

abequose

3,6-dideoxy-D-*xylo*-hexopyranose

A simple synthesis of paratose follows from the deoxygenation of 'diacetone glucose' (at C3) above:[219]

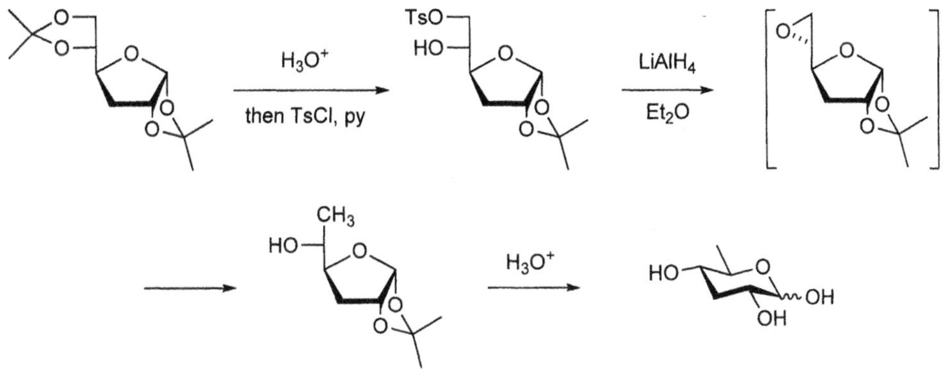

One final point: deoxy sugars are more electron rich than their parent sugars and so are more prone to hydrolysis. For example, the hydrolysis of methyl 2-deoxy-α-D-*arabino*-hexopyranoside is about 2000 times faster than that of methyl α-D-glucopyranoside.[224]

Amino deoxy sugars:[225] Amino deoxy sugars, usually referred to as just the 'amino sugars', are no doubt the most important of the modified sugars under discussion here. The most common members are 2-amino-2-deoxy- and 2-acetamido-2-deoxy-D-glucopyranose and -D-galactopyranose:

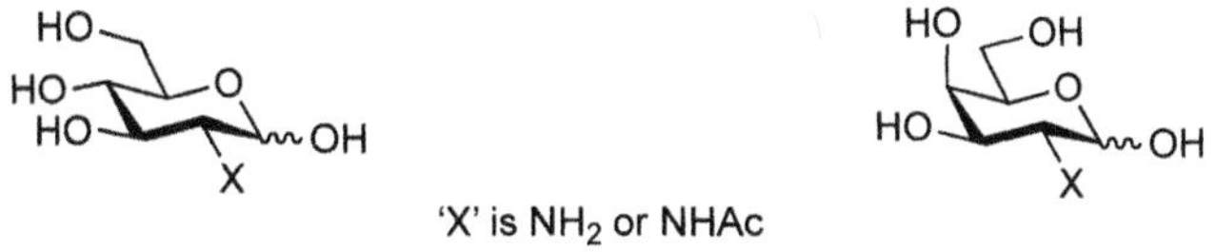

'X' is NH$_2$ or NHAc

These amino sugars are crucial to the well-being of most organisms, including humans, because they play pivotal roles in the structure and function of biologically important oligosaccharides, polysaccharides and glycoproteins. The presence of a free amine in a monosaccharide residue allows, of course, for protonation and the generation of a positively charged ammonium ion; the acetamido group, although far less basic than an amine, imparts a good deal of polarity to the molecule in question.

'*N*-Acetyl-D-glucosamine' is an inexpensive and abundant amino sugar and therefore attracts little synthetic interest as such. However, derivatives of

References start on page 124

N-acetyl-D-glucosamines, much sought after as intermediates in the synthesis of oligosaccharides, are a little more difficult to access:[226]

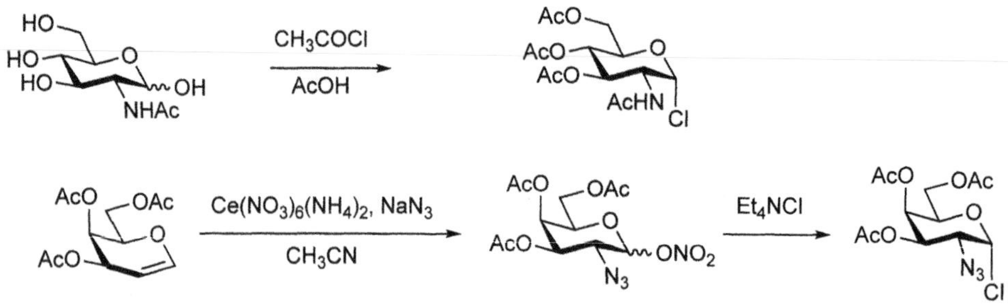

The latter transformation involves an 'azidonitration', invented by Lemieux, which readily provides a masked amino sugar in the form of a glycosyl halide.[227–231] The azido group, of course, can be introduced by diazotransfer onto an already present amine and avoids the expensive cerium(IV) ammonium nitrate (see Chapter 2).

Whereas *N*-acetyl-D-glucosamine is plentiful and cheap, such is not the case with *N*-acetyl-D-galactosamine. Normally obtained with difficulty from natural sources such as mammalian tissue and cartilage, or as a surrogate from the azidonitration of D-galactal derivatives (above), a simple derivative of *N*-acetyl-D-galactosamine can now be prepared conveniently in a direct sequence:[232]

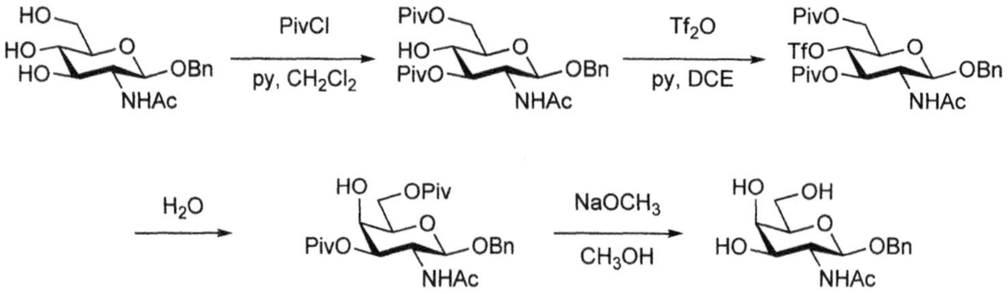

Another common amino sugar found in glycoproteins is *N*-acetylneuraminic acid (5-acetamido-3,5-dideoxy-D-*glycero*-D-*galacto*-non-2-ulopyranosonic acid), a member of the general class of sialic acids:[233]

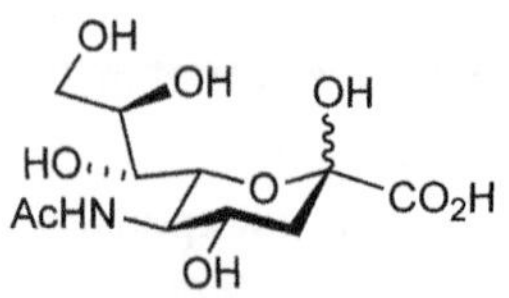

This 'higher homologue' amino sugar is the precursor of zanamavir, a drug used to combat infection by the influenza virus (see Chapter 7).[234] Consequently, efficient methods for the large-scale production of N-acetylneuraminic acid have been devised:[235–239]

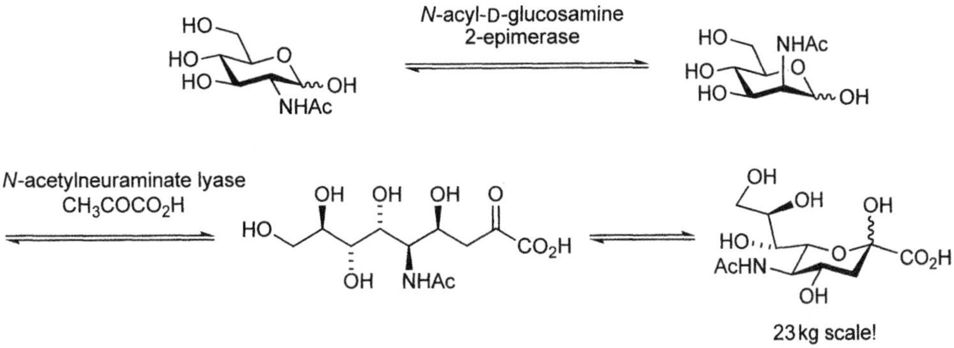

N-acyl-D-glucosamine 2-epimerase

N-acetylneuraminate lyase
CH_3COCO_2H

23 kg scale!

Although certainly not an amino sugar but obviously related to N-acetylneuraminic acid is 'Kdn' (3-deoxy-D-glycero-D-galacto-non-2-ulopyranosonic acid), prepared in a remarkable way that bears some resemblance to the biosynthesis of the molecule (see Chapter 6):[240]

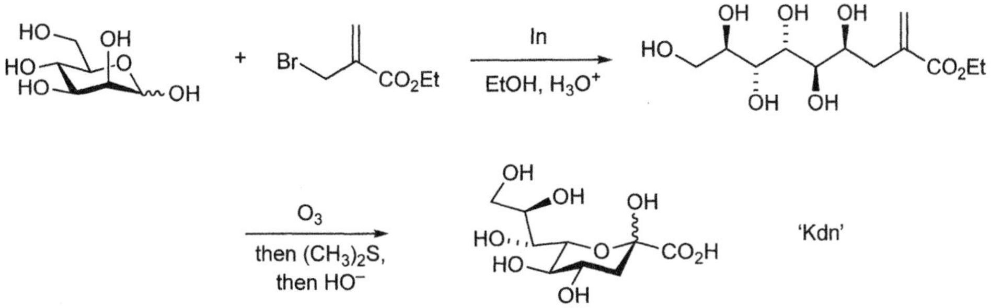

In
EtOH, H_3O^+

O_3
then $(CH_3)_2S$,
then HO^-

'Kdn'

Finally, there are amino sugars that are best described as 'glycosylamines' – the amino group is attached directly to the anomeric carbon atom:

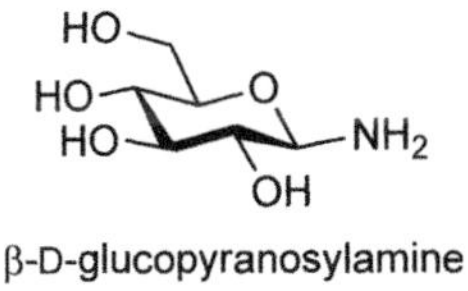

β-D-glucopyranosylamine

Although such a molecule can be prepared from D-glucose, it is not particularly stable under aqueous conditions:[241]

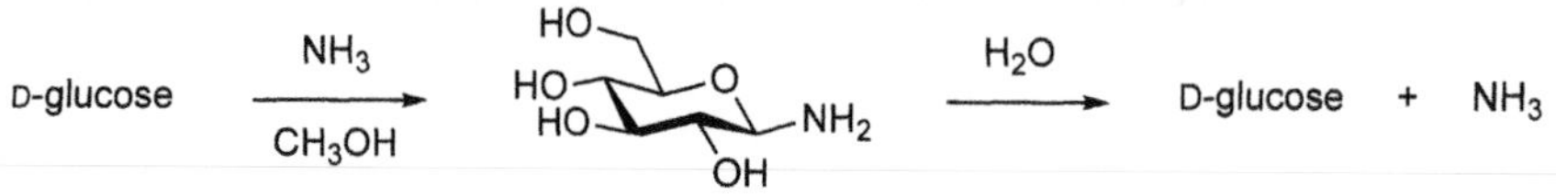

Ammonium carbamate in methanol gives a better result and has been applied to the synthesis of β-chitobiosylamine.[242] However, with aromatic amines and amino acids, the so formed glycosylamines are both more stable and more crystalline. When necessary, a glycosyl azide can be a latent precursor of a glycosylamine:[243,244]

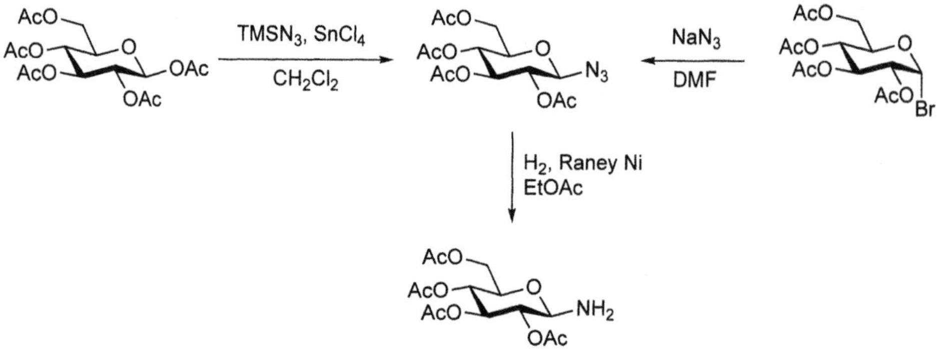

Glycosylamines are implicated as intermediates in the *Amadori rearrangement*, the formation of 1-amino-1-deoxy-2-uloses when aldoses are treated with amines:[241,245]

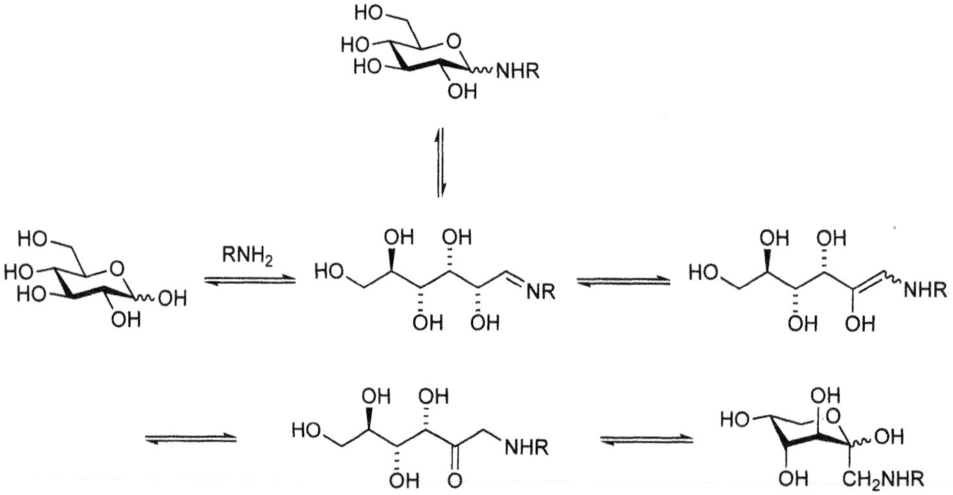

There is a certain similarity here to the Lobry de Bruyn–Alberda van Ekenstein rearrangement discussed in Chapter 1, and some steps of the Amadori rearrangement are no doubt common to the 'Maillard reaction', a complex series of reactions that cause the browning and darkening of sugars when heated in the presence of proteins or their hydrolysis products.[241,246] The 'Heyns rearrangement' is a related and useful reaction that converts ketoses into 2-amino-2-deoxy aldoses:[247]

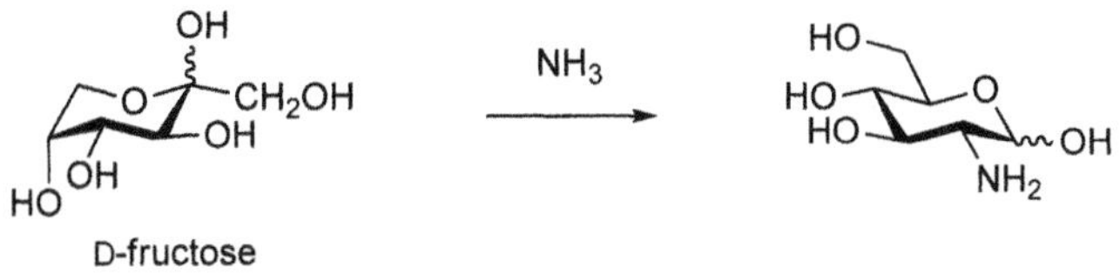

D-fructose

Branched-chain sugars:[248–250] As the name implies, branched-chain sugars have a carbon substituent at one of the non-terminal carbon atoms of the sugar chain; such substituents may replace either a hydrogen atom or a hydroxyl group (deoxy). Examples of both sorts of branched sugars are found in molecules that often exhibit biological activity:

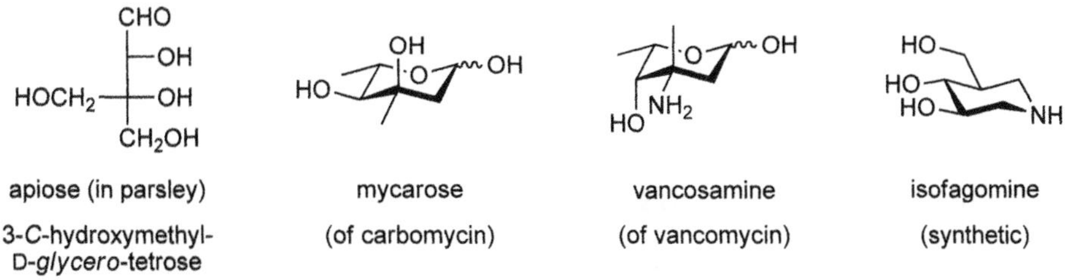

apiose (in parsley)

3-C-hydroxymethyl-
D-*glycero*-tetrose

mycarose

(of carbomycin)

vancosamine

(of vancomycin)

isofagomine

(synthetic)

For the synthesis of both types of branched sugars, uloses can be useful starting materials:

References start on page 124

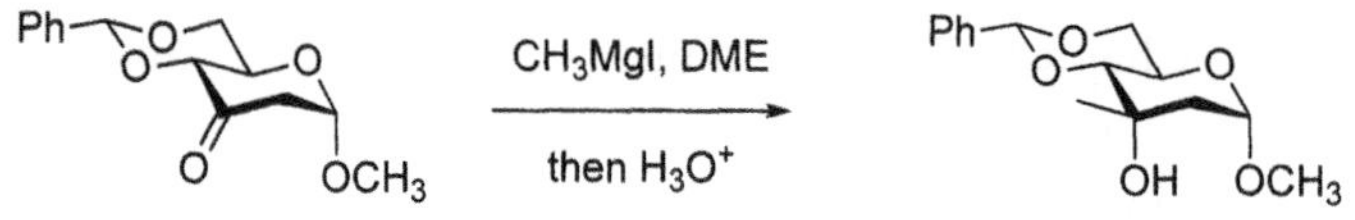

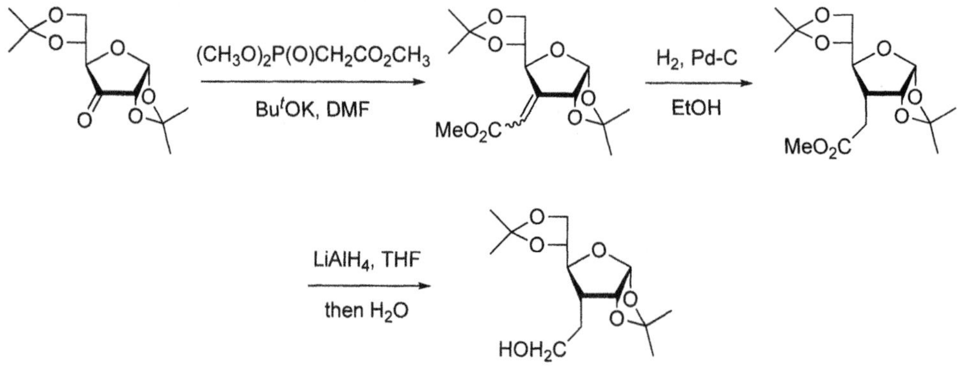

Otherwise, the opening of epoxides and free radical chemistry can be put to good use:

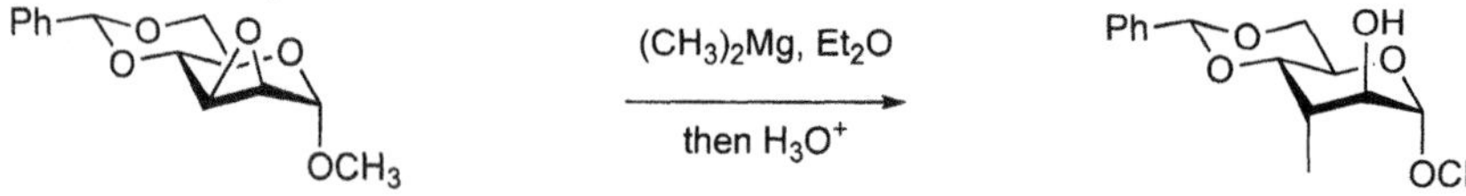

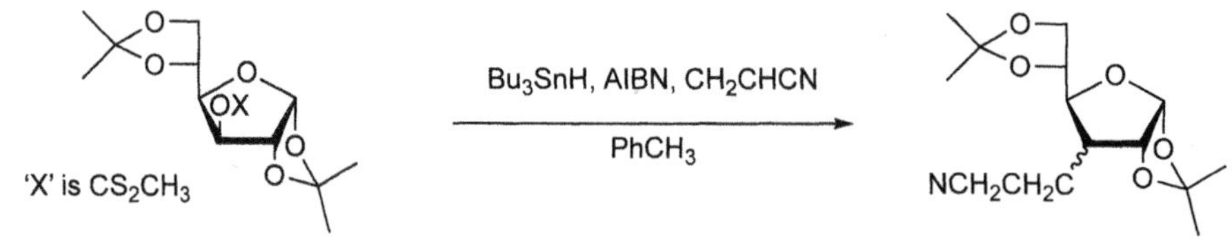

Miscellaneous Reactions

What follows is a selection of chemical reactions that, while important and in common usage in the field of carbohydrates, do not easily fit into any of the sections discussed previously.

Wittig reaction: The Wittig reaction[251–253] and the closely related Horner–Wadsworth–Emmons variants[254,255] have found much use in carbohydrate synthesis. The two reactions have the potential either to extend the carbohydrate chain (for a hexose, at C1 or C6) or to cause a branching of the chain:[256–260]

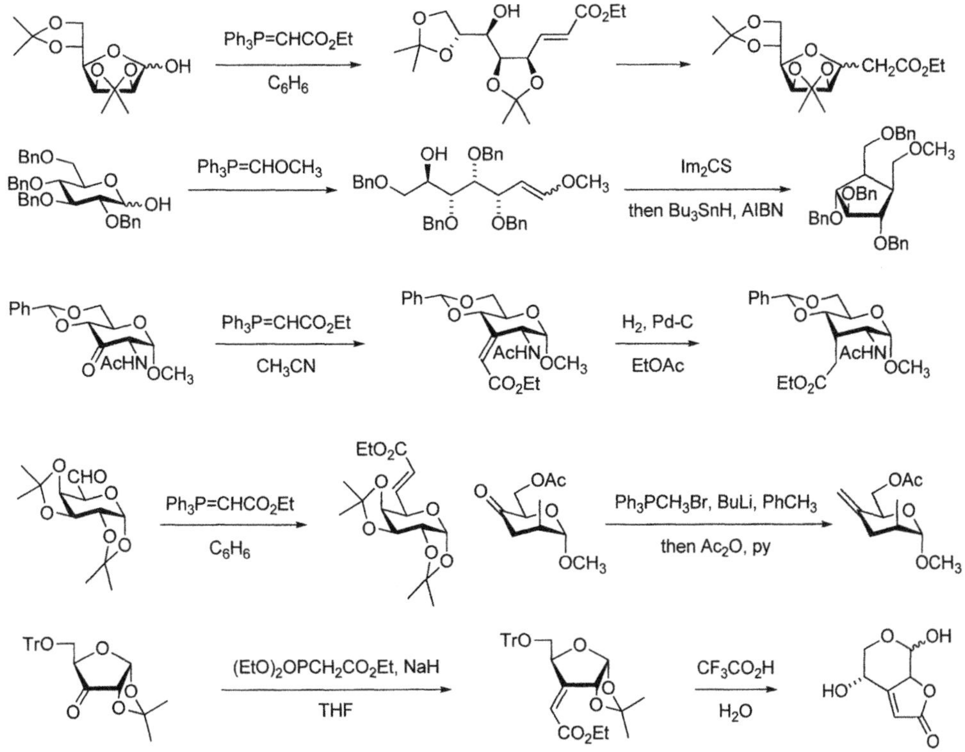

Thiazole-based homologation:[261–263] The Kiliani–Fischer cyanohydrin synthesis has served early and contemporary chemists equally well. However, many improvements have been noted, and one of the best involves the use of 2-trimethylsilylthiazole for the one-carbon extension of, mainly, aldoses:[264]

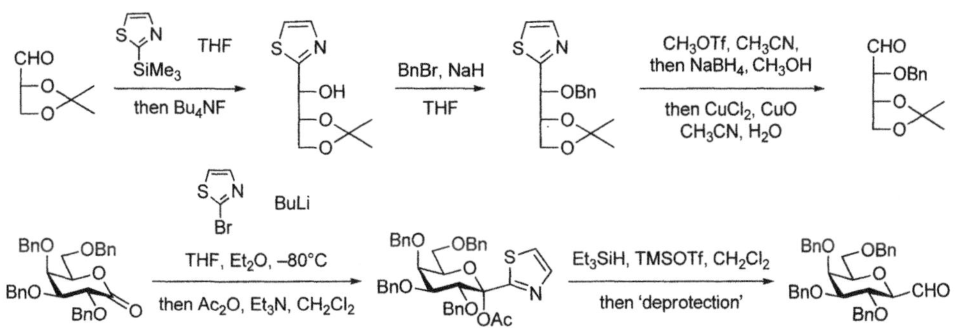

Mitsunobu reaction:[265-267] The Mitsunobu reaction, involving the activation of a hydroxyl group with the combination of triphenylphosphine and a dialkyl azodicarboxylate, and its subsequent replacement with a nucleophile, has been widely used in carbohydrates for the introduction of halogen, oxygen, sulfur and nitrogen:[268-270]

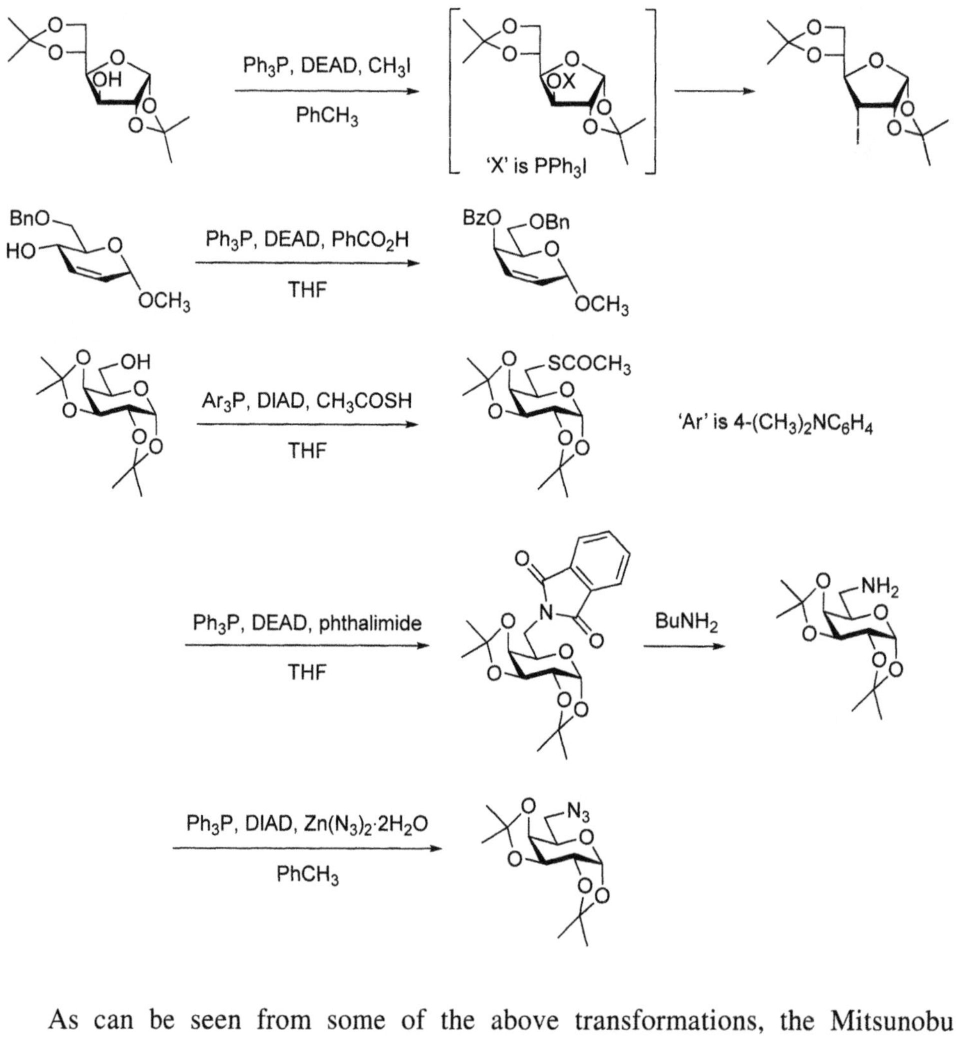

As can be seen from some of the above transformations, the Mitsunobu reaction normally proceeds with inversion of configuration at carbon (an S_{N2} process).

A nice synthesis of a methyl uronate, necessary for the construction of a larger molecule, incorporates many of the reactions discussed here and previously:[271]

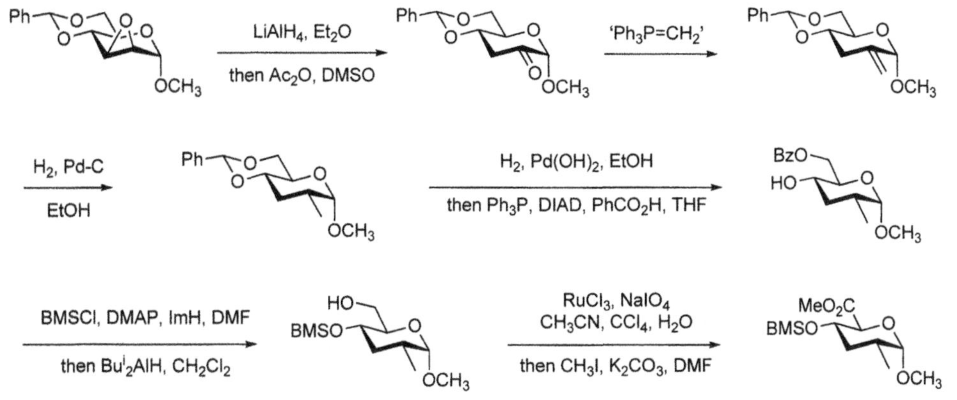

Orthoesters: Orthoesters are encountered not infrequently with carbohydrates and play valuable synthetic roles at both non-anomeric and anomeric positions.

For the synthesis of an orthoester, a diol is generally treated with a simpler orthoester (a transesterification reaction):[272]

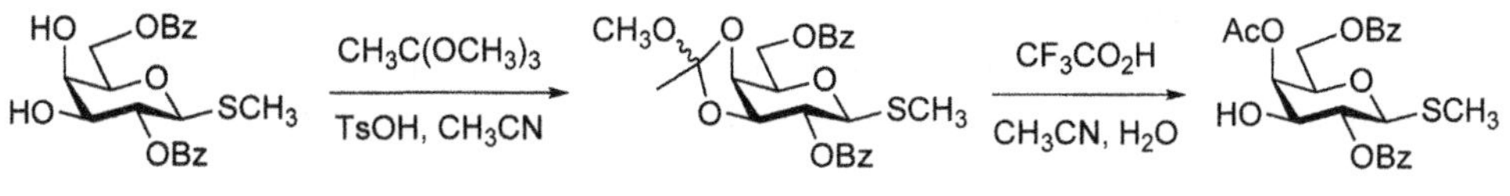

The partial hydrolysis of such an orthoester is a very useful reaction, generally leading to a selective protection of the original diol.[273] Although a rationale for the regioselectivity is not obvious, it is particularly successful with orthoacetates.

For the synthesis of orthoesters at the anomeric position, the conditions devised by Lemieux are in general use:[274–276]

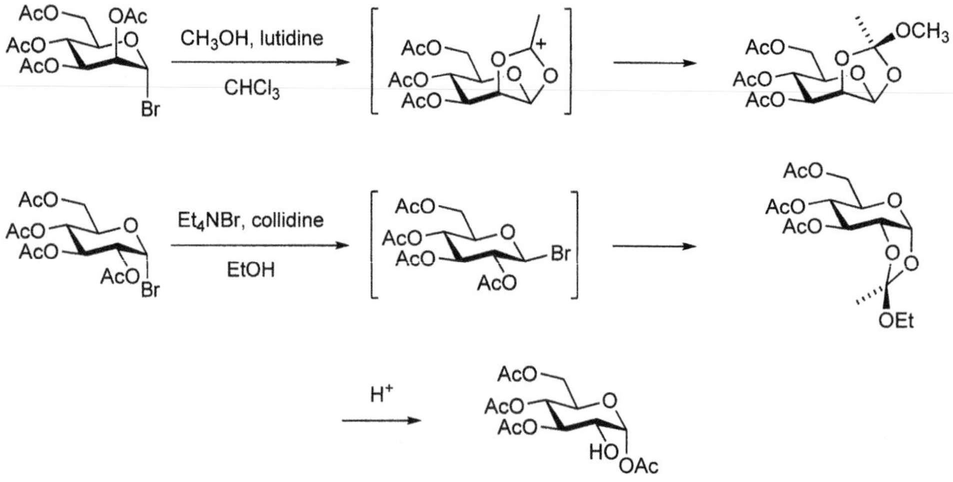

'Amide acetals' can also be used for the synthesis of orthoesters:[277]

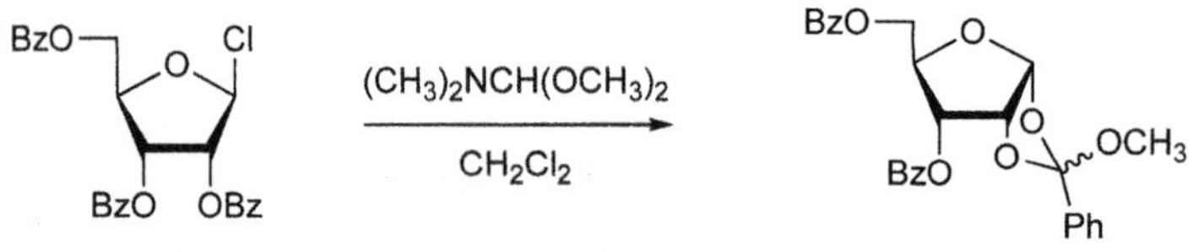

Orthoesters can also be obtained under conditions of kinetic control utilizing 1,1-dimethoxyethene, a reaction somewhat reminiscent of the conversion of diols into acetals utilizing 2-methoxypropene.[278]

A reaction related to orthoester formation, in that it also proceeds through cyclic 1,3-dioxolenium ions, is the transformation of penta-O-acetyl-β-D-glucopyranose into a D-idopyranose salt, a precursor of penta-O-acetyl-α-D-idopyranose:[279]

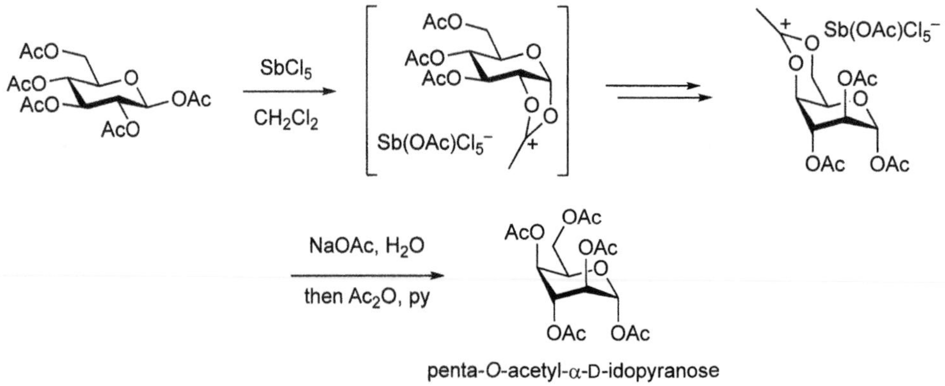

penta-O-acetyl-α-D-idopyranose

In the same vein is the conversion of penta-*O*-acetyl-β-D-galactopyranose into the very useful 1,3,4,6-tetra-*O*-acetyl-α-D-galactopyranose:[280]

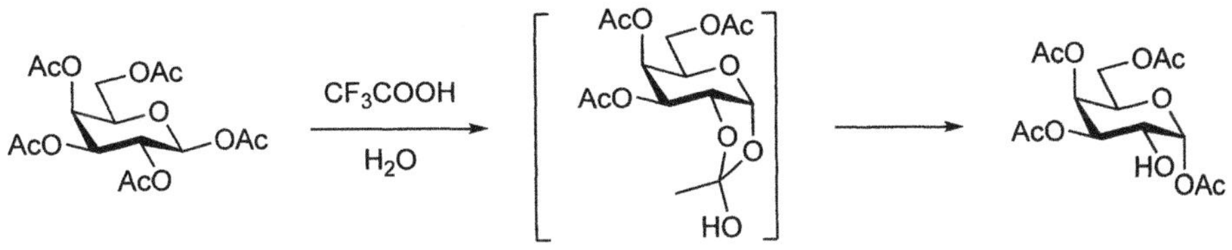

Industrially Important Ketoses[281]

This chapter has so far mainly used aldoses and their derivatives to illustrate the reactions of monosaccharides. It would now seem appropriate to spend a little time on the 'cousins' of the aldoses, the ketoses, and an important aspect of this chemistry is their use in industry. The comments and examples that follow have been drawn mainly from a timely article by Lichtenthaler,[282] a chemist who has been intimately involved with industrial research on carbohydrates in Germany for decades.

There are four ketoses that are each produced annually in excess of 10^7 kg, D-fructose, L-sorbose, isomaltulose and lactulose:

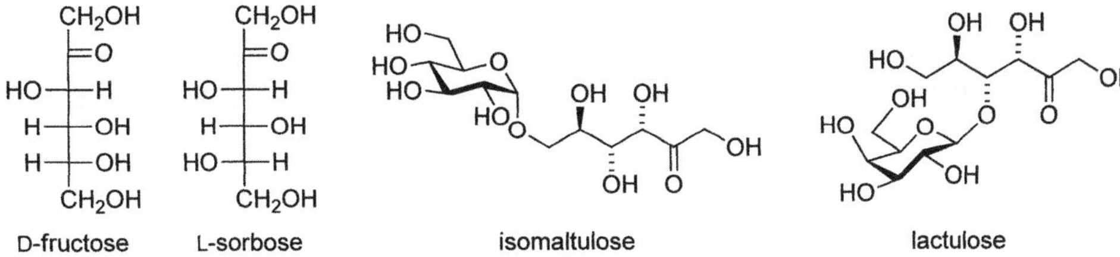

These ketoses are cheap (of the same order as some of the common solvents in use for organic reactions) and as a result are attractive starting materials for industrial processes; however, they are also polyfunctional, polar and water soluble, features not suitable to many chemical transformations. Chemists, therefore, have been forced to design 'entry points'[282] so that these polar molecules can be tamed into becoming suitable substrates.

D-Fructose: This is the most common ketose, available as a natural product or by the isomerization of D-glucose utilizing a D-xylose isomerase ('high fructose corn syrup').[283] In aqueous solution at room temperature, D-fructose rapidly forms anomeric mixtures of both the pyranose and the furanose form (β-*p*:α-*p*:β-*f*: α-*f* = 73:2:20:5):

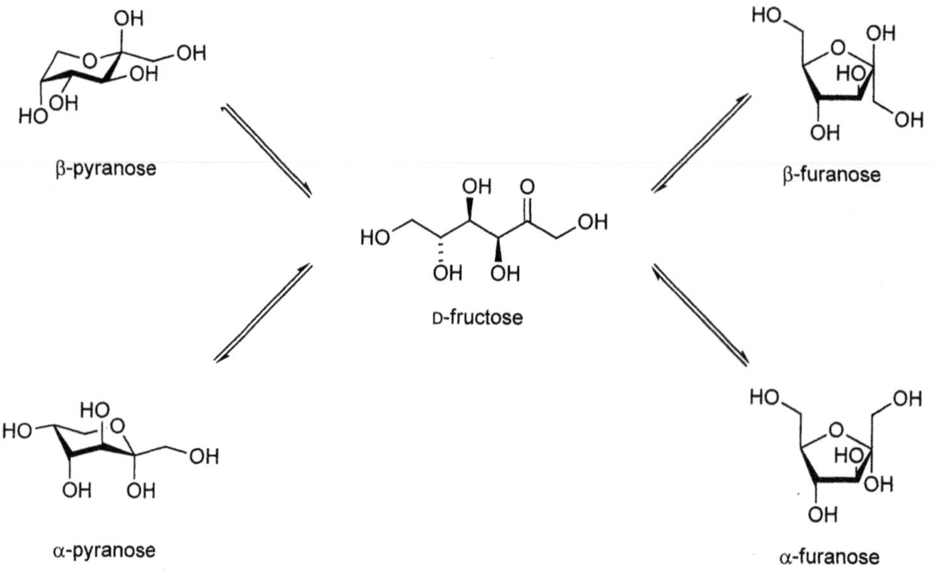

β-pyranose

β-furanose

D-fructose

α-pyranose

α-furanose

The situation changes a little in pyridine (54:5:30:11) and dimethyl sulfoxide (27:4:48:20), but these tautomeric mixtures still pose problems for the high-yielding synthesis of derivatives as entry points for industrial processes:

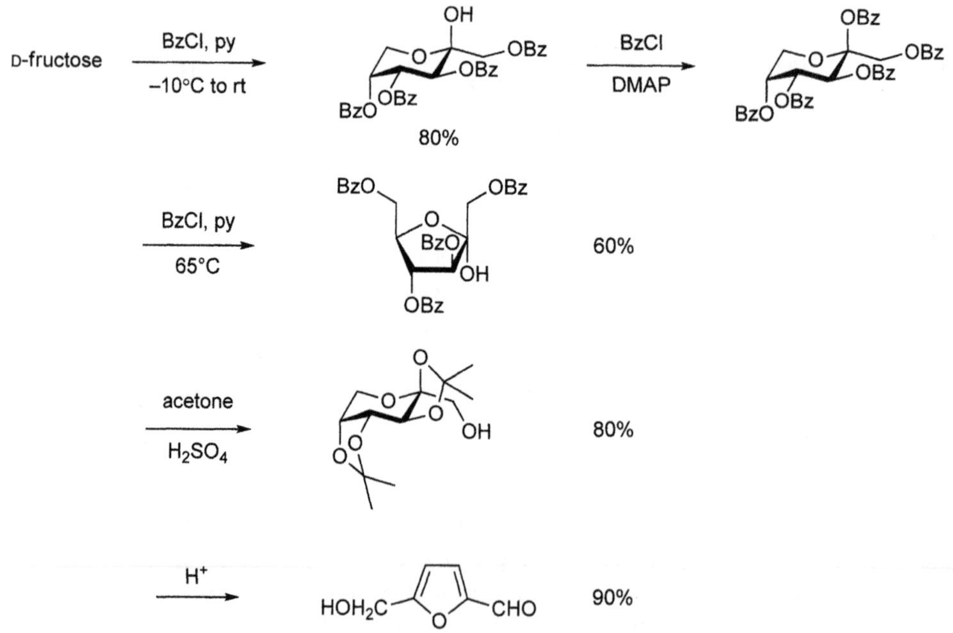

D-fructose $\xrightarrow[\text{−10°C to rt}]{\text{BzCl, py}}$ 80% $\xrightarrow[\text{DMAP}]{\text{BzCl}}$

$\xrightarrow[\text{65°C}]{\text{BzCl, py}}$ 60%

$\xrightarrow[\text{H}_2\text{SO}_4]{\text{acetone}}$ 80%

$\xrightarrow{\text{H}^+}$ HOH$_2$C — CHO   90%

L-Sorbose: This ketohexose is the 5-epimer of D-fructose (i.e. L-sorbose and D-fructose are diastereoisomers at C5) and is readily available from D-sorbitol (D-glucitol), an intermediate in the production of vitamin C (from D-glucose). Fortunately, L-sorbose is present in most solutions as the α-L-pyranose form, and this is generally responsible for the high yields of its derivatives:

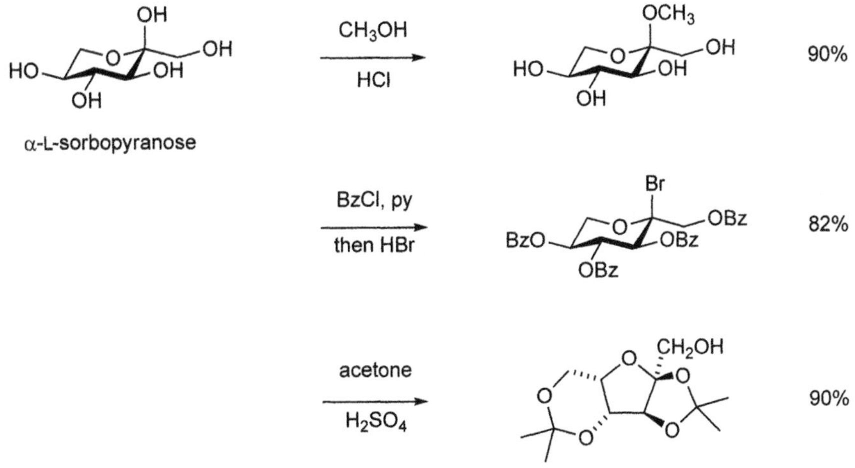

α-L-sorbopyranose

CH_3OH / HCl → 90%

BzCl, py / then HBr → 82%

acetone / H_2SO_4 → 90%

Isomaltulose: One of the degradation products of starch is a disaccharide, isomaltose, the 2-ulose of which takes the name of isomaltulose. However, it is a remarkable bacterium-induced rearrangement of another disaccharide, the ubiquitous sucrose, which is the industrial source of isomaltulose (see Chapter 9):

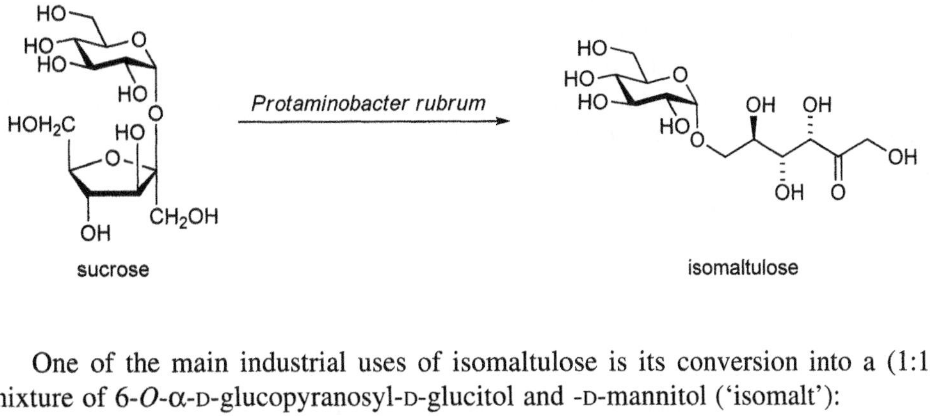

Protaminobacter rubrum

sucrose → isomaltulose

One of the main industrial uses of isomaltulose is its conversion into a (1:1) mixture of 6-*O*-α-D-glucopyranosyl-D-glucitol and -D-mannitol ('isomalt'):

References start on page 124

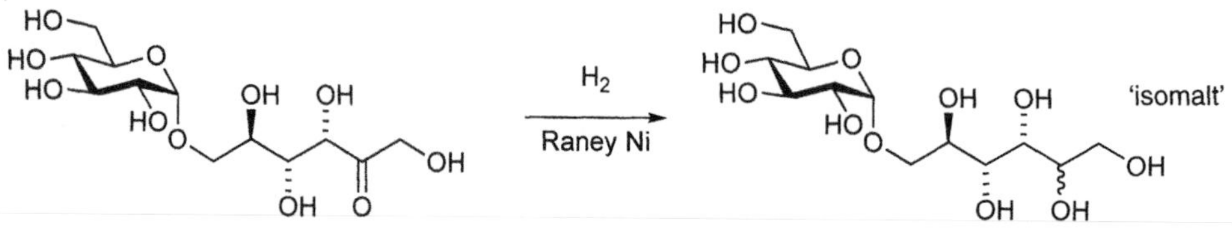

This mixture is used as a non-cariogenic artificial sweetener, e.g. in the hard coating of chewing gum.

Lactulose: This is the last ketose to be discussed here and, perhaps appropriately, it is the one that has received the least attention from industry. Lactose (see Chapter 9) is obtained from whey, a by-product of the cheese industry; the base-catalysed isomerization of lactose then easily produces lactulose:

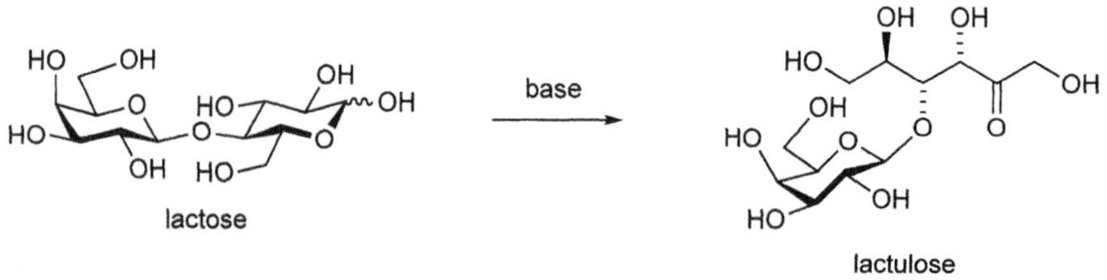

Lactulose,[b] through the Heyns rearrangement, has recently been converted into *N*-acetyllactosamine, a disaccharide that is present in many biologically important oligosaccharides.[284] The key to success was the use of a novel pyrimidine protecting group (see Chapter 2) for nitrogen; in the words of the corresponding author of the paper, 'to see a white, amorphous solid form in the flask upon the addition of the reagent, from what is obviously a complex mixture at equilibrium, makes you believe that your birthday and Christmas have come together'.

---

[b] Lactulose is currently available as 'Laevulac', an osmotic laxative.

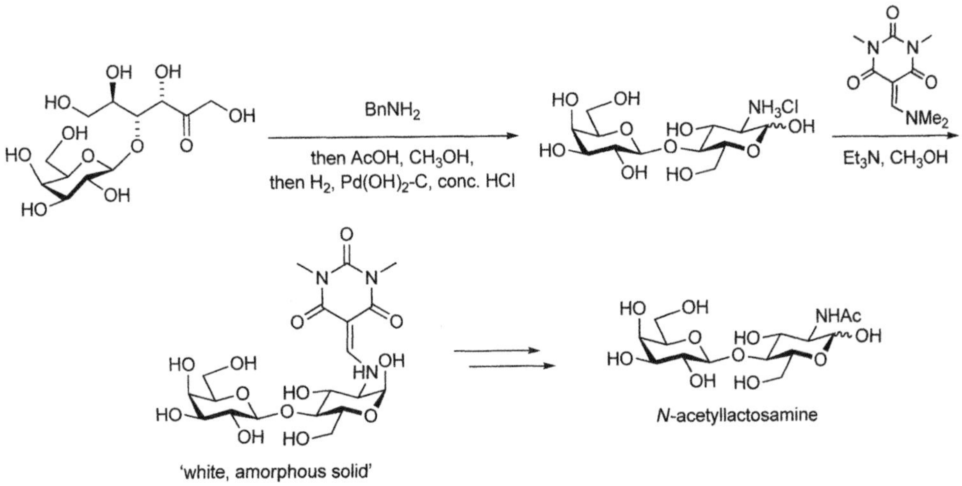

'white, amorphous solid'

N-acetyllactosamine

The Heyns rearrangement can also be used for the efficient conversion of D-tagatose into various *N*-alkyl derivatives of D-galactosamine.[285]

Aza and Imino Sugars[c,286–290]

This chapter will be completed by a discussion about molecules in which the oxygen of the pyranose or furanose ring is absent and a nitrogen atom has been incorporated, replacing either the ring oxygen (an imino sugar) or a carbon atom of the ring (an aza sugar); these molecules are thus polyhydroxylated piperidines or pyrrolidines. However, the ability of these molecules to inhibit the action of many carbohydrate-processing enzymes swiftly attracted the attention of carbohydrate chemists.[291] Some structural and synthetic aspects of these molecules will be discussed here; we will turn to their properties and uses in subsequent chapters.

Nojirimycin was the first example of an imino sugar, isolated at one stage from *Streptomyces nojiriensis*; the molecule is rather unstable, being a hemiaminal. In addition to this example of a polyhydroxylated piperidine, examples have also been found of polyhydroxylated pyrrolidines, indolizidines (swainsonine), pyrrolizidines (alexine) and nortropanes:

---

[c] According to IUPAC-IUBMB nomenclature recommendations for carbohydrates, 'Use of the terms 'aza sugar', 'phospha sugar', etc. should be restricted to structures where carbon, not oxygen, is replaced by a heteroatom. The term 'imino sugar' may be used as a class name for cyclic sugar derivatives in which the ring oxygen atom has been replaced by nitrogen.'

References start on page 124

nojirimycin   2-amino-2,5-anhydro-2-deoxy-D-mannitol   swainsonine

(2,5-dideoxy-2,5-imino-D-mannitol)

alexine

Some years later, the D-*manno* and D-*galacto* isomers of nojirimycin were isolated, also from *Streptomyces* sp.:

mannojirimycin   galactonojirimycin

During the early investigations on nojirimycin, Paulsen had prepared '1-deoxynojirimycin', a molecule that is obviously a lot more stable than nojirimycin. This imino sugar showed excellent activity in the inhibition of glycosidases, and it probably came as no great surprise to Paulsen to see it later appear as a natural product (isolated from the roots of mulberry trees):

1-deoxynojirimycin   1-deoxymannojirimycin

1-Deoxymanno(no)jirimycin is also a natural product but not 1-deoxygalacto-nojirimycin (as yet).

Another natural imino sugar is 1,2-dideoxynojirimycin (fagomine). A synthetic isomer, isofagomine (an aza sugar), is an excellent inhibitor of certain glycosidases:

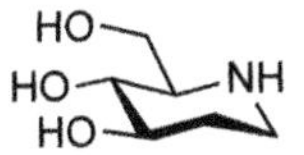

fagomine

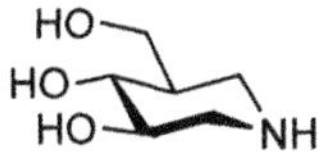

isofagomine

There has been an enormous amount of effort devoted to the synthesis of aza and imino sugars.[287,288] For example, 1-deoxynojirimycin is prepared in four steps from L-sorbose:

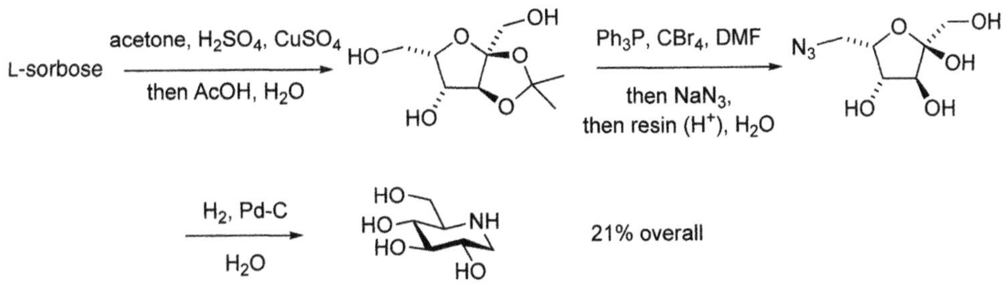

L-sorbose

acetone, H$_2$SO$_4$, CuSO$_4$
then AcOH, H$_2$O

Ph$_3$P, CBr$_4$, DMF
then NaN$_3$,
then resin (H$^+$), H$_2$O

H$_2$, Pd-C
H$_2$O

21% overall

1-Deoxymannojirimycin can be accessed in high yield from either D-fructose[292] or the now readily available 1,5-anhydro-D-fructose:[293]

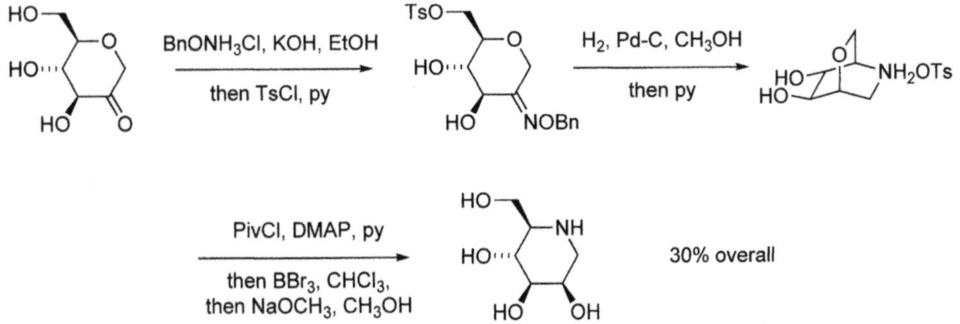

BnONH$_3$Cl, KOH, EtOH
then TsCl, py

H$_2$, Pd-C, CH$_3$OH
then py

PivCl, DMAP, py
then BBr$_3$, CHCl$_3$,
then NaOCH$_3$, CH$_3$OH

30% overall

Isofagomine is available in good amounts from short syntheses starting with either L-xylose[294,295] or D-arabinose:[296]

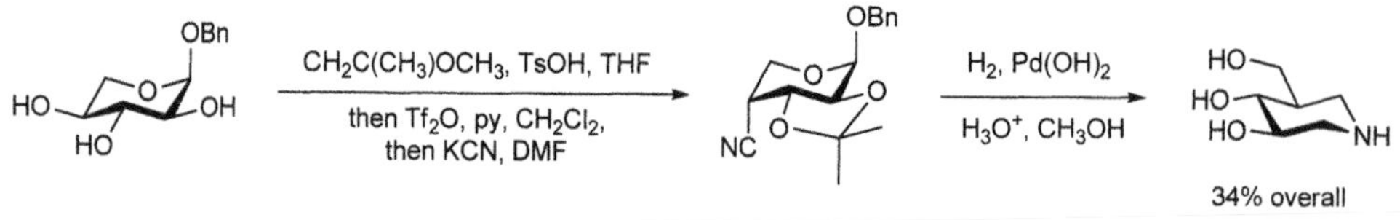

34% overall

Finally, a remarkable transformation converts a methyl α-D-glucopyranoside into a vinyl pyrrolidine, a precursor to any number of different aza sugars:[297,298]

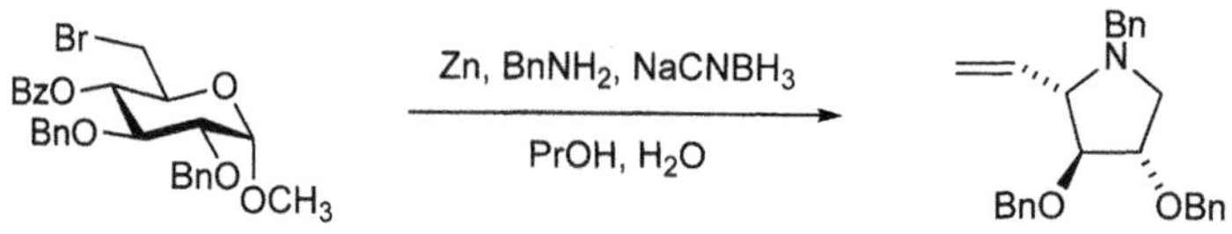

References

1. Levy, D.E. (2006). In *The Organic Chemistry of Sugars* (D.E. Levy and P. Fügedi, eds) p. 225. CRC Press.
2. Muzart, J. (1992). *Chem. Rev.*, **92**, 113.
3. Hudlický, M. (1990). *Oxidations in Organic Chemistry*. ACS Monograph 186. American Chemical Society: Washington DC.
4. Haines, A.H. (1988). *Methods for the Oxidation of Organic Compounds*. Academic Press: London.
5. Arterburn, J.B. (2001). *Tetrahedron*, **57**, 9765.
6. Varela, O. (2003). *Adv. Carbohydr. Chem. Biochem.*, **58**, 308.
7. Barton, D.H.R. and Liu, W. (1997). *Tetrahedron*, **53**, 12067.
8. van Boeckel, C.A.A., Delbressine, L.P.C. and Kaspersen, F.M. (1985). *Recl. Trav. Chim. Pays-Bas*, **104**, 259.
9. Garegg, P.J. and Samuelsson, B. (1978). *Carbohydr. Res.*, **67**, 267.
10. Piancatelli, G., Scettri, A. and D'Auria, M. (1982). *Synthesis*, 245.
11. Herscovici, J., Egron, M.-J. and Antonakis, K. (1982). *J. Chem. Soc., Perkin Trans. 1*, 1967.
12. Andersson, F. and Samuelsson, B. (1984). *Carbohydr. Res.*, **129**, C1.
13. Tidwell, T.T. (1990). *Synthesis*, 857.
14. Pojer, P.M. and Angyal, S.J. (1978). *Aust. J. Chem.*, **31**, 1031.
15. Jones, G.H. and Moffatt, J.G. (1972). *Methods Carbohydr. Chem.*, **6**, 315.
16. Lindberg, B. (1972). *Methods Carbohydr. Chem.*, **6**, 323.
17. Mancuso, A.J. and Swern, D. (1981). *Synthesis*, 165.
18. Harris, J.M., Liu, Y., Chai, S., Andrews, M.D. and Vederas, J.C. (1998). *J. Org. Chem.*, **63**, 2407.
19. Crich, D. and Neelamkavil, S. (2001). *J. Am. Chem. Soc.*, **123**, 7449.
20. Nishide, K., Ohsugi, S.-i., Fudesaka, M., Kodama, S. and Node, M. (2002). *Tetrahedron Lett.*, **43**, 5177.
21. Meyer, S.D. and Schreiber, S.L. (1994). *J. Org. Chem.*, **59**, 7549.

22. Speicher, A., Bomm, V. and Eicher, T. (1996). *J. prakt. Chem.*, **338**, 588.
23. Boeckman, R.K., Jr., Shao, P. and Mullins, J.J. (2000). *Org. Synth.*, **77**, 141.
24. Wirth, T. (2001). *Angew. Chem. Int. Ed.*, **40**, 2812.
25. Haines, A.H. (1988). *Methods for the Oxidation of Organic Compounds*, p. 54. Academic Press: London.
26. Baker, D.C., Horton, D. and Tindall, C.G., Jr. (1972). *Carbohydr. Res.*, **24**, 192.
27. Morris, P.E., Jr., Kiely, D.E. and Vigee, G.S. (1990). *J. Carbohydr. Chem.*, **9**, 661.
28. Ley, S.V., Norman, J., Griffith, W.P. and Marsden, S.P. (1994). *Synthesis*, 639.
29. Hinzen, B., Lenz, R. and Ley, S.V. (1998). *Synthesis*, 977.
30. De Mico, A., Margarita, R., Parlanti, L., Vescovi, A. and Piancatelli, G. (1997). *J. Org. Chem.*, **62**, 6974.
31. Melvin, F., McNeill, A., Henderson, P.J.F. and Herbert, R.B. (1999). *Tetrahedron Lett.*, **40**, 1201.
32. Dijksman, A., Arends, I.W.C.E. and Sheldon, R.A. (2000). *Chem. Commun.*, 271.
33. Huang, L., Teumelsan, N. and Huang, X. (2006). *Chem. Eur. J.*, **12**, 5246.
34. Haines, A.H. (1988). *Methods for the Oxidation of Organic Compounds*, p. 10. Academic Press: London.
35. de Nooy, A.E.J., Besemer, A.C. and van Bekkum, H. (1996). *Synthesis*, 1153.
36. Zou, W. and Szarek, W.A. (1994). *Carbohydr. Res.*, **254**, 25.
37. den Drijver, L., Holzapfel, C.W., Koekemoer, J.M., Kruger, G.J. and van Dyk, M.S. (1986). *Carbohydr. Res.*, **155**, 141.
38. Kong, X. and Grindley, T.B. (1993). *J. Carbohydr. Chem.*, **12**, 557.
39. Hanessian, S. and Roy, R. (1985). *Can. J. Chem.*, **63**, 163.
40. Perlin, A.S. (1980). In *The Carbohydrates* (W. Pigman and D. Horton, eds) Vol. IB, p. 1167. Academic Press.
41. Patroni, J.J. and Stick, R.V. (1985). *Aust. J. Chem.*, **38**, 947.
42. Pigman, W. and Horton, D., eds (1980). *The Carbohydrates*. Vol IB, p. 1299. Academic Press.
43. Gorin, P.A.J. (1981). *Adv. Carbohydr. Chem. Biochem.*, **38**, 13.
44. Tvaroska, I. and Taravel, F.R. (1995). *Adv. Carbohydr. Chem. Biochem.*, **51**, 15.
45. Dais, P. and Perlin, A.S. (1987). *Adv. Carbohydr. Chem. Biochem.*, **45**, 125.
46. Lemieux, R.U. and Stick, R.V. (1975). *Aust. J. Chem.*, **28**, 1799.
47. Monneret, C., Conreur, C. and Khuong-Huu, Q. (1978). *Carbohydr. Res.*, **65**, 35.
48. Roth, W. and Pigman, W. (1963). *Methods Carbohydr. Chem.*, **2**, 405.
49. Johnstone, R.A.W., Wilby, A.H. and Entwistle, I.D. (1985). *Chem. Rev.*, **85**, 129.
50. Paryzek, Z., Koenig, H. and Tabaczka, B. (2003). *Synthesis*, 2023.
51. Thiem, J. and Meyer, B. (1980). *Chem. Ber.*, **113**, 3067.
52. Binkley, R.W. (1985). *J. Org. Chem.*, **50**, 5646.
53. McAuliffe, J.C. and Stick, R.V. (1997). *Aust. J. Chem.*, **50**, 197.
54. Chang, C.-W.T., Hui, Y. and Elchert, B. (2001). *Tetrahedron Lett.*, **42**, 7019.
55. Neumann, W.P. (1987). *Synthesis*, 665.
56. Baguley, P.A. and Walton, J.C. (1998). *Angew. Chem. Int. Ed.*, **37**, 3072.
57. Barton, D.H.R. and McCombie, S.W. (1975). *J. Chem. Soc., Perkin Trans. 1*, 1574.
58. Lee, A.W.M., Chan, W.H., Wong, H.C. and Wong, M.S. (1989). *Synth. Commun.*, **19**, 547.
59. Hartwig, W. (1983). *Tetrahedron*, **39**, 2609.
60. Barton, D.H.R. (1992). *Tetrahedron*, **48**, 2529.
61. Barton, D.H.R., Ferreira, J.A. and Jaszberenyi, J.C. (1997). In *Preparative Carbohydrate Chemistry* (S. Hanessian, ed.) p. 151. Marcel Dekker, Inc.
62. Robins, M.J., Wilson, J.S. and Hansske, F. (1983). *J. Am. Chem. Soc.*, **105**, 4059.
63. Oba, M. and Nishiyama, K. (1994). *Synthesis*, 624.
64. Oba, M. and Nishiyama, K. (1994). *Tetrahedron*, **50**, 10193.

65. Barton, D.H.R. and Jacob, M. (1998). *Tetrahedron Lett.*, **39**, 1331.
66. Barton, D.H.R., Blundell, P., Dorchak, J., Jang, D.O. and Jaszberenyi, J.Cs. (1991). *Tetrahedron*, **47**, 8969.
67. Berge, J.M. and Roberts, S.M. (1979). *Synthesis*, 471.
68. Neumann, W.P. and Peterseim, M. (1992). *Synlett*, 801.
69. Boussaguet, P., Delmond, B., Dumartin, G. and Pereyre, M. (2000). *Tetrahedron Lett.*, **41**, 3377.
70. Lesage, M., Chatgilialoglu, C. and Griller, D. (1989). *Tetrahedron Lett.*, **30**, 2733.
71. Schummer, D. and Höfle, G. (1990). *Synlett*, 705.
72. Jang, D.O., Cho, D.H. and Barton, D.H.R. (1998). *Synlett*, 39.
73. Barton, D.H.R. and Stick, R.V. (1975). *J. Chem. Soc., Perkin Trans. 1*, 1773.
74. Barton, D.H.R. and Subramanian, R. (1977). *J. Chem. Soc., Perkin Trans. 1*, 1718.
75. Szarek, W.A. (1973). *Adv. Carbohydr. Chem. Biochem.*, **28**, 225.
76. Castro, B.R. (1983). *Org. React.*, **29**, 1.
77. Haylock, C.R., Melton, L.D., Slessor, K.N. and Tracey, A.S. (1971). *Carbohydr. Res.*, **16**, 375.
78. Whistler, R.L. and Anisuzzaman, A.K.M. (1980). *Methods Carbohydr. Chem.*, **8**, 227.
79. Garegg, P.J. (1984). *Pure Appl. Chem.*, **56**, 845.
80. Skaanderup, P.R., Poulsen, P.S., Hyldtoft, L., Jørgensen, M.R. and Madsen, R. (2002). *Synthesis*, 1721.
81. Tsuchiya, T. (1990). *Adv. Carbohydr. Chem. Biochem.*, **48**, 91.
82. Taylor, N.F., ed. (1988). *Fluorinated Carbohydrates. Chemical and Biochemical Aspects.* ACS Symposium Series 374. Washington DC: American Chemical Society.
83. Dax, K., Albert, M., Ortner, J. and Paul, B.J. (1999). *Curr. Org. Chem.*, **3**, 287.
84. Nicolaou, K.C., Ladduwahetty, T., Randall, J.L. and Chucholowski, A. (1986). *J. Am. Chem. Soc.*, **108**, 2466.
85. Barrena, M.I., Matheu, M.I. and Castillón, S. (1998). *J. Org. Chem.*, **63**, 2184.
86. Dax, K. (2005). In *Science of Synthesis, Houben-Weyl Methods of Molecular Transformations.* Vol. 34, p. 71. Thieme Verlag.
87. Banks, R.E. (1998). *J. Fluorine Chem.*, **87**, 1.
88. Vincent, S.P., Burkart, M.D., Tsai, C.-Y., Zhang, Z. and Wong, C.-H. (1999). *J. Org. Chem.*, **64**, 5264.
89. Albert, M., Dax, K. and Ortner, J. (1998). *Tetrahedron*, **54**, 4839.
90. Ortner, J., Albert, M., Weber, H. and Dax, K. (1999). *J. Carbohydr. Chem.*, **18**, 297.
91. Taylor, S.D., Kotoris, C.C. and Hum, G. (1999). *Tetrahedron*, **55**, 12431.
92. Lamberth, C. (1998). *Recent Res. Devel. Synth. Org. Chem.*, 1.
93. Csuk, R. and Glänzer, B.I. (1988). *Adv. Carbohydr. Chem. Biochem.*, **46**, 73.
94. Mock, B.H., Vavrek, M.T. and Mulholland, G.K. (1996). *Nucl. Med. Biol.*, **23**, 497.
95. Gillies, J.M., Prenant, C., Chimon, G.N., Smethurst, G.J., Perrie, W., Hamblett, I., Dekker, B. and Zweit, J. (2006). *Appl. Rad. Isotop.*, **64**, 325.
96. Schirrmacher, R., Wängler, C. and Schirrmacher, E. (2007). *Mini-Rev. Org. Chem.*, **4**, 317.
97. Lemieux, R.U. (1963). *Methods Carbohydr. Chem.*, **2**, 223, 224.
98. Haynes, L.J. and Newth, F.H. (1955). *Adv. Carbohydr. Chem.*, **10**, 207.
99. Kartha, K.P.R. and Field, R.A. (1997). *Tetrahedron Lett.*, **38**, 8233.
100. Hunsen, M., Long, D.A., D'Ardenne, C.R. and Smith, A.L. (2005). *Carbohydr. Res.*, **340**, 2670.
101. Presser, A., Kunert, O. and Pötschger, I. (2006). *Monatsh. Chem.*, **137**, 365.
102. Ibatullin, F.M. and Selivanov, S.I. (2002). *Tetrahedron Lett.*, **43**, 9577.
103. Helferich, B. and Gootz, R. (1929). *Ber. Dtsch. Chem. Ges.*, **62**, 2788.
104. Kartha, K.P.R. and Field, R.A. (1998). *Carbohydr. Lett.*, **3**, 179.
105. Gervay, J. (1998). In *Organic Synthesis: Theory and Applications* (T. Hudlicky, ed.) Vol. 4, p. 121. JAI Press, Inc.

106. Hadd, M.J. and Gervay, J. (1999). *Carbohydr. Res.*, **320**, 61.
107. Uchiyama, T. and Hindsgaul, O. (1996). *Synlett*, 499.
108. Uchiyama, T. and Hindsgaul, O. (1998). *J. Carbohydr. Chem.*, **17**, 1181.
109. Caputo, R., Kunz, H., Mastroianni, D., Palumbo, G., Pedatella, S. and Solla, F. (1999). *Eur. J. Org. Chem.*, 3147.
110. Bickley, J., Cottrell, J.A., Ferguson, J.R., Field, R.A., Harding, J.R., Hughes, D.L., Kartha, K.P.R, Law, J.L., Scheinmann, F. and Stachulski, A.V. (2003). *Chem. Commun.*, 1266.
111. Mukhopadhyay, B., Kartha, K.P.R., Russell, D.A. and Field, R.A. (2004). *J. Org. Chem.*, **69**, 7758.
112. Shimizu, M., Togo, H. and Yokoyama, M. (1998). *Synthesis*, 799.
113. Miethchen, R. and Defaye, J., eds (2000). *Carbohydr. Res.*, **327**, 1–218.
114. Shoda, S.-i., Izumi, R. and Fujita, M. (2003). *Bull. Chem. Soc. Jpn.*, **76**, 1.
115. Kobayashi, S., Yoneda, A., Fukuhara, T. and Hara, S. (2004). *Tetrahedron*, **60**, 6923.
116. Huang, K.-T. and Winssinger, N. (2007). *Eur. J. Org. Chem.*, 1887.
117. López, J.C., Bernal-Albert, P., Uriel, C., Valverde, S. and Gómez, A.M. (2007). *J. Org. Chem.*, **72**, 10268.
118. Ernst, B. and Winkler, T. (1989). *Tetrahedron Lett.*, **30**, 3081.
119. Baeschlin, D.K., Chaperon, A.R., Charbonneau, V., Green, L.G., Ley, S.V., Lücking, U. and Walther, E. (1998). *Angew. Chem. Int. Ed.*, **37**, 3423.
120. Somsák, L. and Ferrier, R.J. (1991). *Adv. Carbohydr. Chem. Biochem.*, **49**, 37.
121. Wong, A.W., He, S. and Withers, S.G. (2001). *Can. J. Chem.*, **79**, 510.
122. Davies, G.J., Mackenzie, L., Varrot, A., Dauter, M., Brzozowski, A.M., Schülein, M. and Withers, S.G. (1998). *Biochemistry*, **37**, 11707.
123. Howard, S., He, S. and Withers, S.G. (1998). *J. Biol. Chem.*, **273**, 2067.
124. Vocadlo, D.J. and Withers, S.G. (2005). *Carbohydr. Res.*, **340**, 379.
125. Street, I.P. and Withers, S.G. (1986). *Can. J. Chem.*, **64**, 1400.
126. Sureshan, K.M., Shashidhar, M.S., Praveen, T. and Das, T. (2003). *Chem. Rev.*, **103**, 4477.
127. Ferrier, R.J. (1969). *Adv. Carbohydr. Chem. Biochem.*, **24**, 199.
128. Ferrier, R.J. and Hoberg, J.O. (2003). *Adv. Carbohydr. Chem. Biochem.*, **58**, 55.
129. Baptistella, L.H.B., Neto, A.Z., Onaga, H. and Godoi, E.A.M. (1993). *Tetrahedron Lett.*, **34**, 8407.
130. Barton, D.H.R., Jang, D.O. and Jaszberenyi, J.Cs. (1991). *Tetrahedron Lett.*, **32**, 2569.
131. Garegg, P.J. and Samuelsson, B. (1979). *Synthesis*, 813.
132. Liu, Z., Classon, B. and Samuelsson, B. (1990). *J. Org. Chem.*, **55**, 4273.
133. Oba, M., Suyama, M., Shimamura, A. and Nishiyama, K. (2003). *Tetrahedron Lett.*, **44**, 4027.
134. Chao, B., McNulty, K.C. and Dittmer, D.C. (1995). *Tetrahedron Lett.*, **36**, 7209.
135. McAuliffe, J.C. and Stick, R.V. (1997). *Aust. J. Chem.*, **50**, 193.
136. Bernet, B. and Vasella, A. (1979). *Helv. Chim. Acta*, **62**, 2400.
137. Tanaka, K., Yamano, S. and Mitsunobu, O. (2001). *Synlett*, 1620.
138. Skaanderup, P.R., Hyldtoft, L. and Madsen, R. (2002). *Monatsh. Chem.*, **133**, 117.
139. Helferich, B. (1952). *Adv. Carbohydr. Chem.*, **7**, 209.
140. Ferrier, R.J. (1965). *Adv. Carbohydr. Chem.*, **20**, 67.
141. de Pouilly, P., Chénedé, A., Mallet, J.-M. and Sinaÿ, P. (1992). *Tetrahedron Lett.*, **33**, 8065.
142. Csuk, R., Fürstner, A., Glänzer, B.I. and Weidmann, H. (1986). *J. Chem. Soc., Chem. Commun.*, 1149.
143. Shull, B.K., Wu, Z. and Koreeda, M. (1996). *J. Carbohydr. Chem.*, **15**, 955.
144. Somsák, L. and Németh, I. (1993). *J. Carbohydr. Chem.*, **12**, 679.
145. Fürstner, A. and Weidmann, H. (1988). *J. Carbohydr. Chem.*, **7**, 773.
146. Forbes, C.L. and Franck, R.W. (1999). *J. Org. Chem.*, **64**, 1424.
147. Hansen, T., Krintel, S.L., Daasbjerg, K. and Skrydstrup, T. (1999). *Tetrahedron Lett.*, **40**, 6087.
148. Spencer, R.P., Cavallaro, C.L. and Schwartz, J. (1999). *J. Org. Chem.*, **64**, 3987.

149. Stick, R.V., Stubbs, K.A., Tilbrook, D.M.G. and Watts, A.G. (2002). *Aust. J. Chem.*, **55**, 83; **55** (4), iv.
150. Osborn, H.M.I., Meo, P. and Nijjar, R.K. (2005). *Tetrahedron:* Asymm., **16**, 1935.
151. Taillefumier, C. and Chapleur, Y. (2004). *Chem. Rev.*, **104**, 263.
152. Zhu, X., Jin, Y. and Wickham, J. (2007). *J. Org. Chem.*, **72**, 2670.
153. Hanessian, S., ed. (1977). *Preparative Carbohydrate Chemistry*, pp. 543–591. Marcel Dekker, Inc.
154. Ferrier, R.J. and Middleton, S. (1993). *Chem. Rev.*, **93**, 2779.
155. Ferrier, R.J. (2001). *Top. Curr. Chem.*, **215**, 277.
156. Zhou, J., Wang, G., Zhang, L.-H. and Ye, X.-S. (2006). *Curr. Org. Chem.*, **10**, 625.
157. Arjona, O., Gómez, A.M., López, J.C. and Plumet, J. (2007). *Chem. Rev.*, **107**, 1919.
158. Barnett, J.E.G., Rasheed, A. and Corina, D.L. (1973). *Biochem. J.*, **131**, 21.
159. Kiely, D.E. and Sherman, W.R. (1975). *J. Am. Chem. Soc.*, **97**, 6810.
160. Bender, S.L. and Budhu, R.J. (1991). *J. Am. Chem. Soc.*, **113**, 9883.
161. Sato, K.-i., Sakuma, S., Muramatsu, S. and Bokura, M. (1991). *Chem. Lett.*, 1473.
162. Takahashi, H., Kittaka, H. and Ikegami, S. (1998). *Tetrahedron Lett.*, **39**, 9703.
163. Dalko, P.I. and Sinaÿ, P. (1999). *Angew. Chem. Int. Ed.*, **38**, 773.
164. Dalko, P.I. and Sinaÿ, P. (2003). In *Organic Synthesis Highlights* V (H.-G. Schmalz and T. Wirth, eds) p. 1. Wiley-VCH: Weinheim.
165. Marco-Contelles, J., Pozuelo, C., Jimeno, M.L., Martínez, L. and Martínez-Grau, A. (1992). *J. Org. Chem.*, **57**, 2625.
166. Baldwin, J.E. (1976). *J. Chem. Soc., Chem. Commun.*, 734.
167. Beckwith, A.L.J., Easton, C.J. and Serelis, A.K. (1980). *J. Chem. Soc., Chem. Commun.*, 482.
168. Giese, B. (1986). *Radicals in Organic Synthesis: Formation of Carbon-Carbon Bonds.* Pergamon Press.
169. Curran, D.P., Porter, N.A. and Giese, B. (1996). *Stereochemistry of Radical Reactions.* Wiley-VCH: Weinheim.
170. Trnka, T.M. and Grubbs, R.H. (2001). *Acc. Chem. Res.*, **34**, 18.
171. Roy, R. and Das, S.K. (2000). *Chem. Commun.*, 519.
172. Leeuwenburgh, M.A., Appeldoorn, C.C.M., van Hooft, P.A.V., Overkleeft, H.S., van der Marel, G.A. and van Boom, J.H. (2000). *Eur. J. Org. Chem.*, 873.
173. Madsen, R. (2007). *Eur. J. Org. Chem.*, 399.
174. Plumet, J., Gómez, A.M. and López, J.C. (2007). *Mini-Rev. Org. Chem.*, **4**, 201.
175. Cěrný, M. (2003). *Adv. Carbohydr. Chem. Biochem.*, **58**, 122.
176. Williams, N.R. (1970). *Adv. Carbohydr. Chem. Biochem.*, **25**, 109.
177. Stewart, A.O. and Williams, R.M. (1984). *Carbohydr. Res.*, **135**, 167.
178. Wiggins, L.F. (1963). *Methods Carbohydr. Chem.*, **2**, 188.
179. Szeja, W. (1986). *Carbohydr. Res.*, **158**, 245.
180. Szeja, W. (1988). *Carbohydr. Res.*, **183**, 135.
181. Guthrie, R.D. and Jenkins, I.D. (1981). *Aust. J. Chem.*, **34**, 1997.
182. Behrens, C.H., Ko, S.Y., Sharpless, K.B. and Walker, F.J. (1985). *J. Org. Chem.*, **50**, 5687.
183. Buchanan, J.G., Fletcher, R., Parry, K. and Thomas, W.A. (1969). *J. Chem. Soc.* B, 377.
184. Horton, D. and Weckerle, W. (1975). *Carbohydr. Res.*, **44**, 227.
185. King, R.D., Overend, W.G., Wells, J. and Williams, N.R. (1967). *J. Chem. Soc., Chem. Commun.*, 726.
186. Hagen, S., Anthonsen, T. and Kilaas, L. (1979). *Tetrahedron*, **35**, 2583.
187. Lewis, B.A., Smith, F. and Stephen, A.M. (1963). *Methods Carbohydr. Chem.*, **2**, 172.
188. Cěrný, M. and Stanek, J., Jr. (1977). *Adv. Carbohydr. Chem. Biochem.*, **34**, 23.
189. Bols, M. (1996). In *Carbohydrate Building Blocks*, pp. 43–48. John Wiley and Sons.
190. Angyal, S.J. and Dawes, K. (1968). *Aust. J. Chem.*, **21**, 2747.
191. Ward, R.B. (1963). *Methods Carbohydr. Chem.*, **2**, 394.
192. Shafizadeh, F., Furneaux, R.H., Stevenson, T.T. and Cochran, T.G. (1978). *Carbohydr. Res.*, **67**, 433.
193. Hann, R.M. and Hudson, C.S. (1942). *J. Am. Chem. Soc.*, **64**, 2435.

194. Coleman, G.H. (1963). *Methods Carbohydr. Chem.*, **2**, 397.
195. Boons, G.-J., Isles, S. and Setälä, P. (1995). *Synlett*, 755.
196. Zottola, M.A., Alonso, R., Vite, G.D. and Fraser-Reid, B. (1989). *J. Org. Chem.*, **54**, 6123.
197. Lafont, D., Boullanger, P., Cadas, O. and Descotes, G. (1989). *Synthesis*, 191.
198. Rao, M.V. and Nagarajan, M. (1987). *Carbohydr. Res.*, **162**, 141.
199. Caron, S., McDonald, A.I. and Heathcock, C.H. (1996). *Carbohydr. Res.*, **281**, 179.
200. Czernecki, S., Leteux, C. and Veyrières, A. (1992). *Tetrahedron Lett.*, **33**, 221.
201. Mereyala, H.B. and Venkataramanaiah, K.C. (1991). *J. Chem. Res., Synop.*, 197.
202. Haeckel, R., Lauer, G. and Oberdorfer, F. (1996). *Synlett*, 21.
203. Černý, M. (1994). In *Frontiers in Biomedicine and Biotechnology. Levoglucosenone and Levoglucosans, Chemistry and Applications* (Z.J. Witczak, ed.) Vol. 2, pp. 121–146. ATL Press: Shrewsbury.
204. McAuliffe, J.C., Skelton, B.W., Stick, R.V. and White, A.H. (1997). *Aust. J. Chem.*, **50**, 209.
205. Xue, J. and Guo, Z. (2001). *Tetrahedron Lett.*, **42**, 6487.
206. Burgey, C.S., Vollerthun, R. and Fraser-Reid, B. (1994). *Tetrahedron Lett.*, **35**, 2637.
207. Wang, L.-X., Sakairi, N. and Kuzuhara, H. (1990). *J. Chem. Soc., Perkin Trans. 1*, 1677.
208. Lee, J.-C., Tai, C.-A. and Hung, S.-C. (2002). *Tetrahedron Lett.*, **43**, 851.
209. Lemieux, R.U. and Howard, J. (1963). *Methods Carbohydr. Chem.*, **2**, 400.
210. Lemeiux, R.U. and Huber, G. (1956). *J. Am. Chem. Soc.*, **78**, 4117.
211. Halcomb, R.L. and Danishefsky, S.J. (1989). *J. Am. Chem. Soc.*, **111**, 6661.
212. Seeberger, P.H., Bilodeau, M.T. and Danishefsky, S.J. (1997). *Aldrichimica Acta*, **30**, 75.
213. Adam, W., Chan, Y.-Y., Cremer, D., Gauss, J., Scheutzow, D. and Schindler, M. (1987). *J. Org. Chem.*, **52**, 2800.
214. Cheshev, P., Marra, A. and Dondoni, A. (2006). *Carbohydr. Res.*, **341**, 2714.
215. Danishefsky, S.J. and Bilodeau, M.T. (1996). *Angew. Chem. Int. Ed. Engl.*, **35**, 1380.
216. Hanessian, S. (1966). *Adv. Carbohydr. Chem.*, **21**, 143.
217. Butterworth, R.F. and Hanessian, S. (1971). *Adv. Carbohydr. Chem. Biochem.*, **26**, 279.
218. Kirschning, A., Jesberger, M. and Schöning, K.-U. (2001). *Synthesis*, 507.
219. Patroni, J.J. and Stick, R.V. (1978). *Aust. J. Chem.*, **31**, 445.
220. Czernecki, S., Horns, S. and Valéry, J.-M. (1995). *J. Carbohydr. Chem.*, **14**, 157.
221. Barton, D.H.R., Jang, D.O. and Jaszberenyi, J.Cs. (1992). *Tetrahedron Lett.*, **33**, 5709.
222. Bettelli, E., Cherubini, P., D'Andrea, P., Passacantilli, P. and Piancatelli, G. (1998). *Tetrahedron*, **54**, 6011.
223. Costantino, V., Imperatore, C., Fattorusso, E. and Mangoni, A. (2000). *Tetrahedron Lett.*, **41**, 9177.
224. Capon, B. (1969). *Chem. Rev.*, **69**, 407.
225. Pigman, W. and Horton, D., eds (1980). *The Carbohydrates*, Vol. IB, p. 643. Academic Press.
226. McLaren, J.M., Stick, R.V. and Webb, S. (1977). *Aust. J. Chem.*, **30**, 2689.
227. Lemieux, R.U. and Ratcliffe, R.M. (1979). *Can. J. Chem.*, **57**, 1244.
228. Kinzy, W. and Schmidt, R.R. (1985). *Liebigs Ann. Chem.*, 1537.
229. Bertozzi, C.R. and Bednarski, M.D. (1992). *Tetrahedron Lett.*, **33**, 3109.
230. Gauffeny, F., Marra, A., Shun, L.K.S., Sinaÿ, P. and Tabeur, C. (1991). *Carbohydr. Res.*, **219**, 237.
231. Seeberger, P.H., Roehrig, S., Schell, P., Wang, Y. and Christ, W.J. (2000). *Carbohydr. Res.*, **328**, 61.
232. Rochepeau-Jobron, L. and Jacquinet, J.-C. (1998). *Carbohydr. Res.*, **305**, 181.
233. Banaszek, A. and Mlynarski, J. (2005). In *Studies in Natural Products Chemistry* (Atta-ur-Rahman, ed.) Vol. 30, p. 419. Elsevier.
234. von Itzstein, M., Wu, W.-Y., Kok, G.B., Pegg, M.S., Dyason, J.C., Jin, B., Phan, T.V., Smythe, M.L., White, H.F., Oliver, S.W., Colman, P.M., Varghese, J.N., Ryan, D.M., Woods, J.M., Bethell, R.C., Hotham, V.J., Cameron, J.M. and Penn, C.R. (1993). *Nature*, **363**, 418.
235. Hemeon, I. and Bennet, A.J. (2007). *Synthesis*, 1899.

236. Maru, I., Ohnishi, J., Ohta, Y. and Tsukada, Y. (1998). *Carbohydr. Res.*, **306**, 575.
237. Maru, I., Ohnishi, J., Ohta, Y. and Tsukada, Y. (2002). *J. Biosc. Bioeng.*, **93**, 258.
238. Lee, J.-O., Yi, J.-K., Lee, S.-G., Takahashi, S. and Kim, B.-G. (2004). *Enzyme and Microbial Technology*, **35**, 121.
239. Yu, H., Chokhawala, H.A., Huang, S. and Chen, X. (2006). *Nat. Protoc.*, **1**, 2485.
240. Warwel, M. and Fessner, W.-D. (2000). *Synlett*, 865.
241. Pigman, W. and Horton, D., eds (1980). *The Carbohydrates.* Vol. IB, p. 881. Academic Press.
242. Hackenberger, C.P.R., O'Reilly, M.K. and Imperiali, B. (2005). *J. Org. Chem.*, **70**, 3574.
243. Pfleiderer, W. and Bühler, E. (1966). *Chem. Ber.*, **99**, 3022.
244. Paulsen, H., Györgydeák, Z. and Friedmann, M. (1974). *Chem. Ber.*, **107**, 1568.
245. Wrodnigg, T.M. and Eder, B. (2001). *Top. Curr. Chem.*, **215**, 115.
246. Ledl, F. and Schleicher, E. (1990). *Angew. Chem. Int. Ed. Engl.*, **29**, 565.
247. Wrodnigg, T.M. and Stütz, A.E. (1999). *Angew. Chem. Int. Ed.*, **38**, 827.
248. Pigman, W. and Horton, D., eds (1980). *The Carbohydrates.* Vol. IB, p. 761. Academic Press.
249. Yoshimura, J. (1984). *Adv. Carbohydr. Chem. Biochem.*, **42**, 69.
250. Chapleur, Y. and Chrétien, F. (1997). In *Preparative Carbohydrate Chemistry* (S. Hanessian, ed.) p. 207. Marcel Dekker, Inc.
251. Zhdanov, Yu.A., Alexeev, Yu.E. and Alexeeva, V.G. (1972). *Adv. Carbohydr. Chem. Biochem.*, **27**, 227.
252. Maryanoff, B.E. and Reitz, A.B. (1989). *Chem. Rev.*, **89**, 863.
253. RajanBabu, T.V. (1997). In *Preparative Carbohydrate Chemistry* (S. Hanessian, ed.) p. 545. Marcel Dekker, Inc.
254. Horner, L., Hoffmann, H., Wippel, H.G. and Klahre, G. (1959). *Chem. Ber.*, **92**, 2499.
255. Wadsworth, W.S., Jr. and Emmons, W.D. (1961). *J. Am. Chem. Soc.*, **83**, 1733.
256. Collins, P.M., Overend, W.G. and Shing, T.S. (1982). *J. Chem. Soc., Chem. Commun.*, 297.
257. RajanBabu, T.V. (1991). *Acc. Chem. Res.*, **24**, 139.
258. Brimacombe, J.S., Hanna, R. and Bennett, F. (1985). *Carbohydr. Res.*, **135**, C17.
259. Calvo-Mateo, A., Camarasa, M.J. and de las Heras, F.G. (1984). *J. Carbohydr. Chem.*, **3**, 475.
260. Xavier, N.M. and Rauter, A.P. (2007). *Org. Lett.*, **9**, 3339.
261. Dondoni, A. and Marra, A. (1997). In *Preparative Carbohydrate Chemistry* (S. Hanessian, ed.) p. 173. Marcel Dekker, Inc.
262. Dondoni, A. (1998). *Synthesis*, 1681.
263. Dondoni, A. and Perrone, D. (1997). *Aldrichimica Acta*, **30**, 35.
264. Dondoni, A., Marra, A. and Perrone, D. (1993). *J. Org. Chem.*, **58**, 275.
265. Mitsunobu, O. (1981). *Synthesis*, 1.
266. Hughes, D.L. (1992). *Organic Reactions*, **42**, 335.
267. Herr, R.J. and Albany Molecular Research, Inc. (1999). *Technical Reports*, **3** http://www.albmolecular.com/chemlinks/vol3.html.
268. Kunz, H. and Schmidt, P. (1982). *Liebigs Ann. Chem.*, 1245.
269. von Itzstein, M., Jenkins, M.J. and Mocerino, M. (1990). *Carbohydr. Res.*, **208**, 287.
270. Viaud, M.C. and Rollin, P. (1990). *Synthesis*, 130.
271. Smith, A.B., III and Hale, K.J. (1989). *Tetrahedron Lett.*, **30**, 1037.
272. Garegg, P.J. and Oscarson, S. (1985). *Carbohydr. Res.*, **136**, 207.
273. Mukhopadhyay, B. and Field, R.A. (2003). *Carbohydr. Res.*, **338**, 2149.
274. Mazurek, M. and Perlin, A.S. (1965). *Can. J. Chem.*, **43**, 1918.
275. Lemieux, R.U. and Morgan, A.R. (1965). *Can. J. Chem.*, **43**, 2199.
276. Shoda, S.-i., Moteki, M., Izumi, R. and Noguchi, M. (2004). *Tetrahedron Lett.*, **45**, 8847.
277. Hanessian, S. and Banoub, J. (1975). *Carbohydr. Res.*, **44**, C14.
278. Bouchra, M. and Gelas, J. (1998). *Carbohydr. Res.*, **305**, 17.

279. Paulsen, H. (1971). *Adv. Carbohydr. Chem. Biochem.*, **26**, 127.

280. Chittenden, G.J.F. (1988). *Carbohydr. Res.*, **183**, 140.

281. Lichtenthaler, F.W., ed. (1991). *Carbohydrates as Organic Raw Materials*. Wiley-VCH: Weinheim.

282. Lichtenthaler, F.W. (1998). *Carbohydr. Res.*, **313**, 69.

283. Häusler, H. and Stütz, A.E. (2001). *Top. Curr. Chem.*, **215**, 77.

284. Stütz, A.E., Dekany, G., Eder, B., Illaszewicz, C. and Wrodnigg, T.M. (2003). *J. Carbohydr. Chem.*, **22**, 253.

285. Wrodnigg, T., Lundt, I. and Stütz, A. (2006). *J. Carbohydr. Chem.*, **25**, 33.

286. Stütz, A.E., ed. (1999). *Iminosugars as Glycosidase Inhibitors. Nojirimycin and Beyond.* Wiley-VCH: Weinheim.

287. Afarinkia, K. and Bahar, A. (2005). *Tetrahedron:* Asymm., **16**, 1239.

288. Pearson, M.S.M., Mathé-Allainmat, M., Fargeas, V. and Lebreton, J. (2005). *Eur. J. Org. Chem.*, 2159.

289. Lillelund, V.H., Jensen, H.H., Liang, X. and Bols, M. (2002). *Chem. Rev.*, **102**, 515.

290. Asano, N. (2003). *Curr. Top. Med. Chem.*, **3**, 471.

291. Häusler, H., Kawakami, R.P., Mlaker, E., Severn, W.B., Wrodnigg, T.M. and Stütz, A.E. (2000). *J. Carbohydr. Chem.*, **19**, 435.

292. Spreitz, J., Stütz, A.E. and Wrodnigg, T.M. (2002). *Carbohydr. Res.*, **337**, 183.

293. Maier, P., Andersen, S.M. and Lundt, I. (2006). *Synthesis*, 827.

294. Best, W.M., Macdonald, J.M., Skelton, B.W., Stick, R.V. and Tilbrook, D.M.G. (2002). *Can. J. Chem.*, **80**, 857.

295. Zhu, X., Sheth, K.A., Li, S., Chang, H.-H. and Fan, J.-Q. (2005). *Angew. Chem. Int. Ed.*, **44**, 7450.

296. Goddard-Borger, E.D. and Stick, R.V. (2007). *Aust. J. Chem.*, **60**, 211.

297. Liotta, L.J. and Ganem, B. (1990). *Synlett*, 503.

298. Suryawanshi, S.N., Dhami, T.S. and Bhakuni, D.S. (1994). *Nat. Prod. Lett.*, **4**, 141.

Chapter 4

Formation of the Glycosidic Linkage[1–26]

We have reached a watershed in the monosaccharides, where to discuss any more of their chemistry would be informative but somewhat trivial and certainly repetitive. It is now time to move on and to consider the broader role of *acetals* in carbohydrate chemistry.

Let us again look at the familiar synthesis of an acetal as performed by Fischer in 1893:[27]

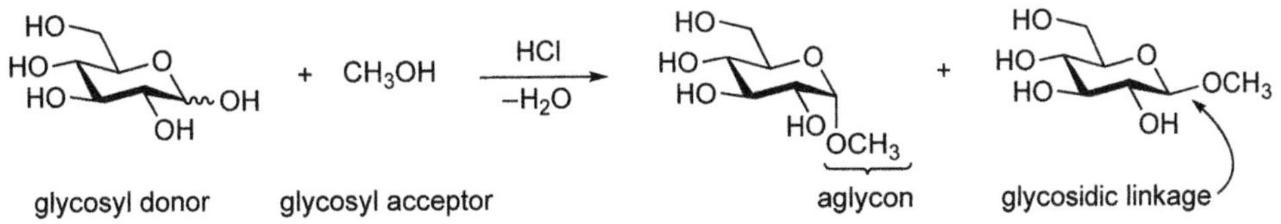

glycosyl donor   glycosyl acceptor   aglycon   glycosidic linkage

The two products, while certainly acetals, are more commonly referred to as *glycosides*: here, more specifically, methyl α- and β-D-glucopyranoside. The carbohydrate (glycon or glycosyl unit) portion of the molecule is distinguished from the non-carbohydrate *aglycon*. Indeed, the *glycosidic linkage* is formed from a *glycosyl donor* and a *glycosyl acceptor*.

Glycosides are commonly found throughout the plant and animal kingdoms and in microorganisms:

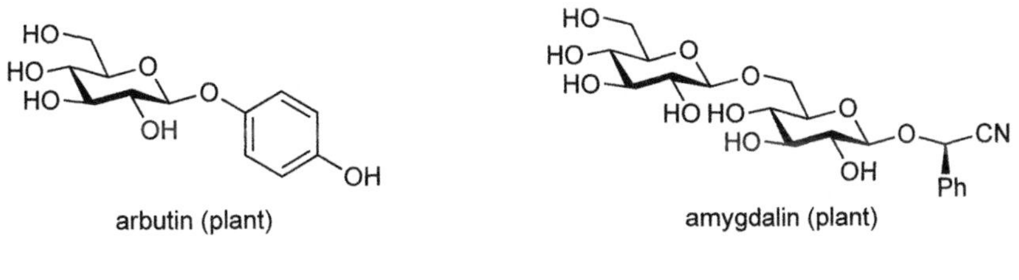

arbutin (plant)   amygdalin (plant)

References start on page 191

The latter example, amygdalin, contains *two* glycosidic linkages, one being involved with the aglycon and the other holding two D-glucopyranose units together; enzymatic hydrolysis removes the aglycon and forms the disaccharide, gentiobiose:

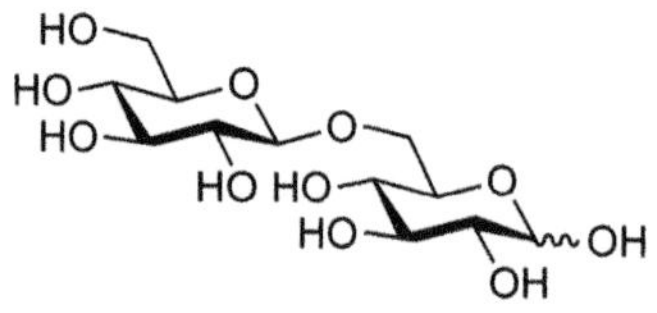

6-O-β-D-glucopyranosyl-D-glucopyranose (gentiobiose)

It is the formation of the glycosidic linkage in molecules such as gentiobiose that is central to the discussion here.

Consider the general formation, now involving a 1,4-linkage, of a disaccharide from two sugars in their pyranose forms:

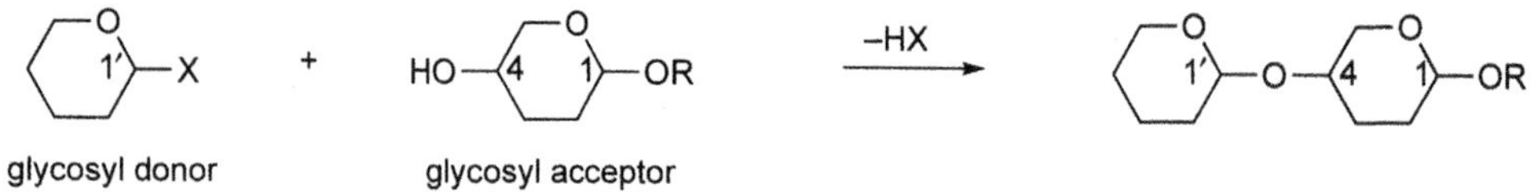

glycosyl donor   glycosyl acceptor

A glycosyl donor, of either the α- or the β-configuration, is treated with a glycosyl acceptor to form, by the elimination of HX, the disaccharide containing the new glycosidic linkage, of either the α- or the β-configuration at C1′. In the process, there is *no change in configuration* at C4 in the glycosyl acceptor.

In less common circumstances, the glycosyl acceptor may react through the hydroxyl group of the anomeric (hemiacetal) centre:

glycosyl donor   glycosyl acceptor

In this case the formation of the glycosidic linkages (there are now two!) results in the α- or β-configuration at both C1 and C1′; the product is a non-reducing disaccharide. An example of such a disaccharide is trehalose:

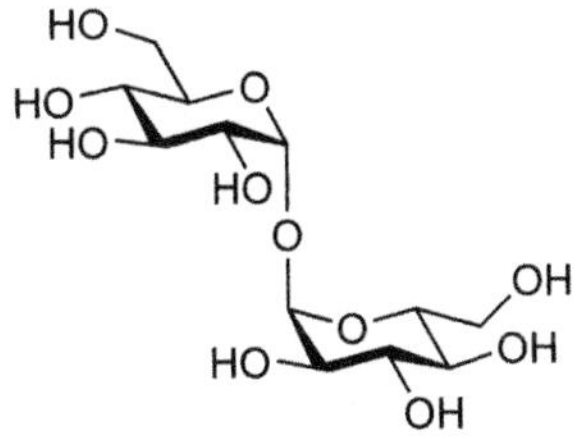

α-D-glucopyranosyl α-D-glucopyranoside (trehalose)

It will be apparent, even at this stage, that the formation of a glycosidic linkage will not be an easy task. Apart from the activation of the glycosyl donor, there are problems of the stereoselectivity (α- or β-) of the process and the access to just the desired hydroxyl group of the glycosyl acceptor (protecting group chemistry). Nature, of course, circumvents all of these problems with the use of enzymes (see Chapter 8), but for the synthetic carbohydrate chemist, much ingenuity, creativity and hard work are necessary to match the rewards of evolution!

It is probably safe to say that there have been no great advances in glycosidation methodology and technology in the past 5 years or so. There is certainly a better understanding of the underlying mechanisms of the various glycosidation methods, thanks in part to Crich and his seminal work on glycosyl sulfoxides and thioglycosides; Fraser-Reid has built on earlier observations by Paulsen and has set us thinking about 'matching' glycosyl donor and acceptor; Boons has presented a new twist to controlling the stereoselectivity of glycosidation; and, finally, the use of (engineered) enzymes has increased in the synthesis of the glycosidic linkage. All of these topics and others will be treated in this and subsequent chapters.

General

The different glycosidic linkages: In forming the glycosidic linkage, close attention must be paid to the orientation of the hydroxyl group at C2 of the glycosyl donor; there are several common scenarios (shown here for a pyranoside):

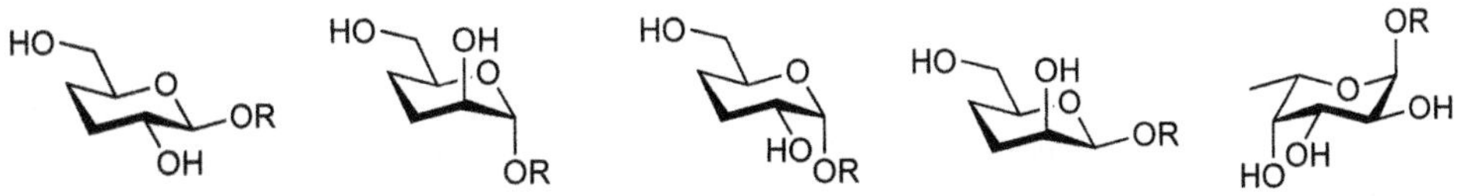

β-D-*gluco*, -*galacto*   α-D-*manno*   α-D-*gluco*, -*galacto*   β-D-*manno*   α-L-*fuco*

EASY (1,2-*trans*)   DIFFICULT (1,2-*cis*)

References start on page 191

As will be seen, it is relatively easy to form 1,2-*trans* linkages but somewhat harder to form 1,2-*cis* linkages. Also, the hydroxyl group at C2 must normally be protected during the actual glycosidation step (otherwise unwanted intra- or inter-molecular reactions could occur), and the nature of this protecting group can have a profound effect on the stereochemistry of the newly introduced aglycon at C1. The whole situation changes when there is no substituent or an amide group at C2 (a 2-deoxy sugar and a 2-acetamido-2-deoxy sugar, respectively); mention will be made of these exceptions later.

The mechanism of glycosidation:[28] The majority of methods available for the formation of the glycosidic linkage utilize a glycosyl donor that is a precursor of either an intermediate oxocarbenium[a] ion (as part of an ion pair) or, at least, a species that has significant positive charge at the anomeric carbon atom:

These oxocarbenium ions are subject to all of the normal factors that stabilize/destabilize such short-lived, high-energy species, and it is appropriate to mention some of them here. Some of these factors also affect the nucleophilicity and, hence, the reactivity of the glycosyl acceptor.

Ion pairs and the solvent: The ionization of a glycosyl donor at the anomeric carbon generates a salt that, depending on the solvent, can have characteristics ranging from a contact (tight) ion pair to a solvent-separated ion pair. The anion of the ion pair may shield one face of the oxocarbenium ion from the approach of the glycosyl acceptor or, if anion exchange can occur, the opposite face may be shielded:

---

[a] The first edition of this book used 'ox*a*carbenium' to describe the intermediate ion; we are now advocating a change to 'ox*o*carbenium'. Both descriptors are used in the carbohydrate literature, seemingly at random and with little justification. Although calculations suggest that most of the positive charge resides on C-1 of the ion, the cyclic oxonium ion possessing a double bond is no doubt a major contributor to the resonance hybrid. Based on the universal acceptance of 'glycosyl', we could see an argument for 'glycosylium' as the descriptor but it is, perhaps unfortunately, not in common use.

The role of the solvent may be passive or active. Solvents of high dielectric constant obviously can stabilize a positive charge very well; those solvents with a basic lone pair of electrons (ethyl ether, tetrahydrofuran) can do so as well, and reversibly.[29,30] One solvent that has dual characteristics of moderate polarity and basicity is acetonitrile; the results of many glycosidation reactions performed in acetonitrile can be explained only by solvent intervention:

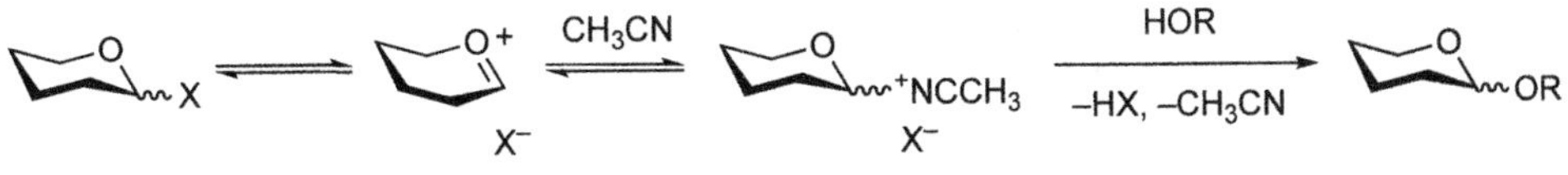

The substituent at C2:[31] The substituent at C2 is usually a functionalized (protected) hydroxyl group, and the choices are generally limited to an ether or an ester, the common groups being a benzyl, 4-methoxybenzyl or silyl ether and an acetic, benzoic or pivalic (2,2-dimethylpropanoic) acid ester. The role of an ether is unique; it is an inert group that provides a degree of steric hindrance to any incoming nucleophile (the glycosyl acceptor). The ester, on the contrary, not only provides a degree of steric encumbrance (especially for the pivalic acid ester) but also can influence the stereochemical outcome of glycoside formation; an orthoester can also be formed:

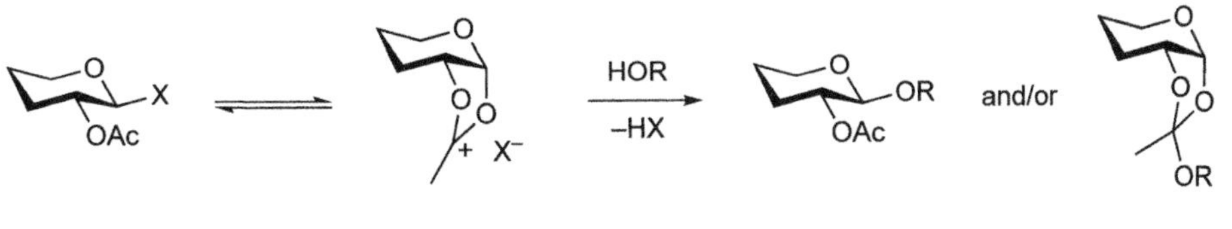

We discussed orthoesters in Chapter 3 and will do so again soon.

The 'armed/disarmed' concept: It is an experimental fact that tetra-O-acetyl-α-D-glucopyranosyl bromide is a colourless, crystalline solid that can be stored in cold, dry conditions for several months. On the contrary, tetra-O-benzyl-α-D-glucopyranosyl bromide is an unstable compound that is normally not isolated or purified but is simply used as generated.

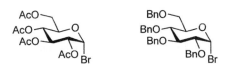

It was Paulsen[b,3] who first noted these differences in stability, but some years passed before Fraser-Reid, in his incisive and wide-ranging studies on 4-pentenyl glycosides, formalized the result with the 'armed/disarmed' concept.[32] Put simply, benzyl ethers are electronically passive groups that do not discourage the development of positive charge in the pyranose ring, most commonly at C1. Conversely, acetate groups (especially when located at C2) are electron-withdrawing, and so disfavour any build-up of positive charge in the ring (at C1). The result is 'arming' of the benzylated bromide and 'disarming' of the ester counterpart; other functional groups show similar effects.[33]

The 'armed/disarmed' concept can also be applied, to a much lesser degree, to the glycosyl acceptor in that electronically passive groups, e.g. ethers, will have little effect on the nucleophilicity of the hydroxyl group in question and so will make for a faster glycosidation; the reverse is true for electron-withdrawing groups, e.g. esters.

less reactive

Bols has recently introduced the concept of 'super armed' glycosyl donors, those that bear bulky silyl protecting groups.[34,35]

The 'torsional control' concept: Again, it was Fraser-Reid who noted that another ring fused to the pyranose ring could have an influence on events occurring at the anomeric centre; for example, the presence of a 4,6-O-benzylidene ring on an otherwise armed donor slows the process of glycosidation.[36] The explanation suggested at the time was that such a ring hinders the necessary flattening of the pyranose ring that accompanies oxocarbenium ion formation:

In a subsequent and elegant investigation, Bols confirmed this explanation but also showed that there is an electronic contribution to the rate reduction in that a 4,6-benzylidene ring locks the substituents on the C5–C6 bond in an unfavourable (less reactive) *tg* conformation.[37,38]

---

[b] Hans Paulsen (1922-), student of Heyns at Hamburg. A contemporary of Lemieux and a powerful force in carbohydrates in the twentieth century.

This 'torsional control' concept was very timely for Ley in his work on the protection of *trans* vicinal diols with diacetal reagents. Such diacetals again impart a degree of rigidity on the pyranose ring and discourage the ready formation of oxocarbenium ions; Ley has introduced the term 'reactivity tuning' to describe the process:[39-41]

less reactive as a glycosyl donor

Kong has recently made related observations with glycosyl donors and acceptors that contain a 4,6-benzylidene protecting group.[42,43]

The 'latent/active' concept: Some 'latent' glycosyl donors are characterized by having a group at the anomeric carbon that is stable under most of the conditions employed in glycosidation reactions, yet can be manipulated later in a synthetic sequence to provide an 'active' donor.[44] For example, but-3-en-2-yl 3,4,6-tri-*O*-benzyl-β-D-glucoside can act only as a glycosyl acceptor, whereas the related but-2-en-2-yl β-D-glucoside (obtained from the former by acetylation and isomerization with Wilkinson's catalyst) is an active glycosyl donor owing to the presence of a reactive enol ether:[45]

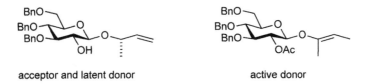

acceptor and latent donor   active donor

Activation of the glycosyl acceptor: There is no point in having a good glycosyl donor if the acceptor is too unreactive for glycosidation to occur. In order to increase the reactivity of some acceptors, and to impart a degree of chemoselectivity to others, various modifications have been made to the hydroxyl group(s) of the acceptor. For example, trityl,[46] silyl[47] and stannyl[48] ethers all seem to polarize the bonding electrons towards the oxygen atom, the first because of the stability of any positive charge that develops on carbon and the other two because of the electropositive nature of silicon and tin. Whatever the reason, the consequence is a more nucleophilic oxygen atom in the acceptor (any steric effect seems to be outweighed by the electronic effect):

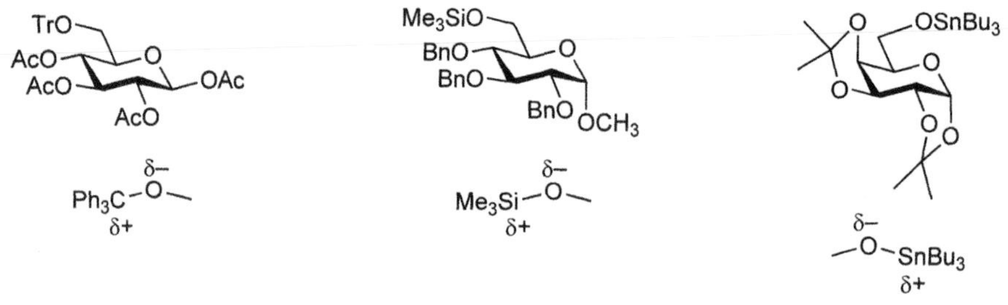

The concept of 'orthogonality': We encountered the concept of orthogonality in Chapter 2, while discussing protecting groups; this was an internal matter for the molecule in question, having two or more hydroxyl groups each protected in a unique manner. The concept of orthogonality as it pertains to glycosyl donors, an external matter, was formalized by Tomoya Ogawa and rapidly embraced by the carbohydrate community.[49] In essence, orthogonality requires two separate donors to have different groups at the anomeric carbon, each capable of activation in a unique and discrete manner. In its simplest form, orthogonality allows for the production of different glycosidic linkages:

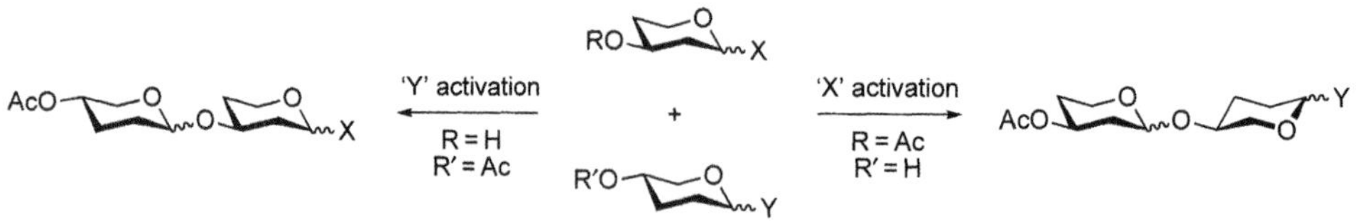

'Reciprocal donor/acceptor selectivity':[50] This is a concept formalized by Fraser-Reid, but one that has its origins in seminal observations (again) made by Paulsen. Essentially, donors and acceptors must ideally be matched: a 2-*O*-benzyl donor matched the C2 hydroxyl group of an inositol acceptor, whereas the 2-*O*-benzoyl counterpart showed complete preference for the C6 hydroxyl group:

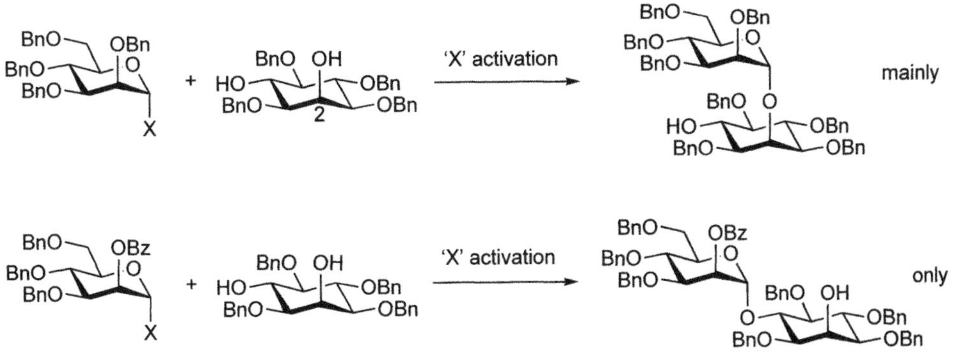

The origins of these not uncommon observations continue to stimulate Fraser-Reid's fertile mind.[51,52] Related effects can also influence the regioselectivity and stereoselectivity of glycosidation.[53–59]

This now brings us to the stage where we can discuss the various methods that are used to form the glycosidic linkage. Much of the discussion will centre around the 'big three' glycosyl donors (halides, trichloroacetimidates and thioglycosides) but will include a handful of less common (but still important) luminaries. Finally, a listing will be given of the less important or 'still rising' donors. Where possible, attention to mechanistic rationalization will be given.[60]

Hemiacetals

A hemiacetal might appear to be the ultimate in terms of convenience when selecting a glycosyl donor, but it can be a 'wolf in sheep's clothing'; considerable synthesis is needed to access even a relatively simple molecule such as 2,3,4,6-tetra-O-benzyl-D-glucose.[61] However, such is not the case with the parent aldose, D-glucose, and a final few words on the Fischer glycosidation are warranted. The strength of this method is its simplicity: a monosaccharide is placed in a large volume of the appropriate alcohol, a small amount of acetyl chloride is added (to generate hydrogen chloride) and the mixture is heated at reflux.[62] Depending on the nature of the monosaccharide and of the desired outcome, reaction times may be short (to generate products of kinetic control, e.g. furanosides) or long (products of thermodynamic control, pyranosides). The main limitation to the process is that, because of the low reactivity of the hemiacetal, only simple alcohols can be employed. Some examples are given below, together with a generally accepted mechanistic rationalization:[63,64]

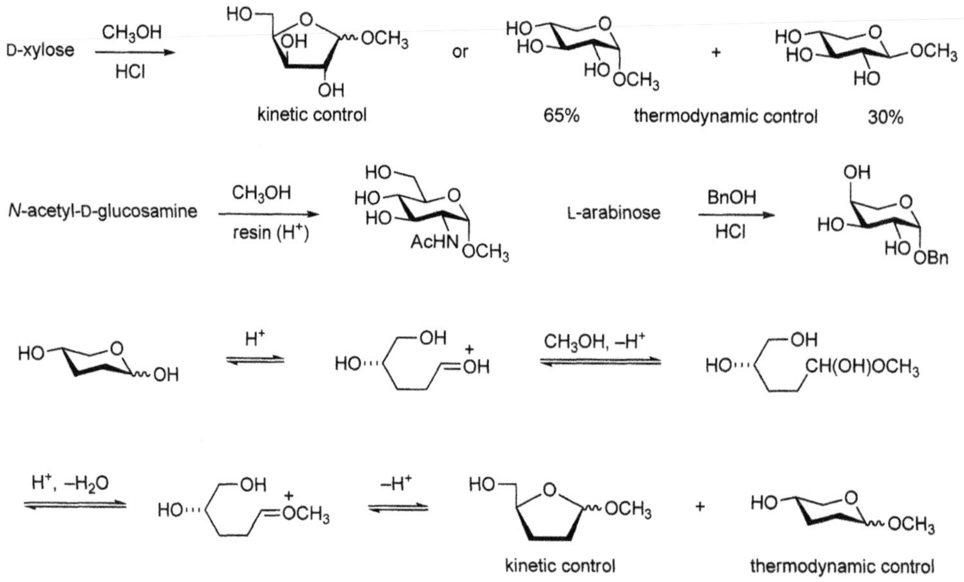

Microwave heating certainly accelerates the glycosidation process, but large-scale reactors are not commonly available.[65]

Basic conditions can also be employed in the synthesis of glycosides by the direct alkylation of the hemiacetal hydroxyl group:

The free sugar forms an equilibrating mixture of anions derived by deprotonation of the anomeric hydroxyl group; preferential alkylation of the oxyanion derived from the β-anomer of the free sugar leads directly to the formation of β-D-glycoside. This method of glycosidation is unique in that the oxygen atom of the glycosyl 'donor' is retained and there is ample opportunity for inversion of configuration at carbon in the alkylating agent:[66–69]

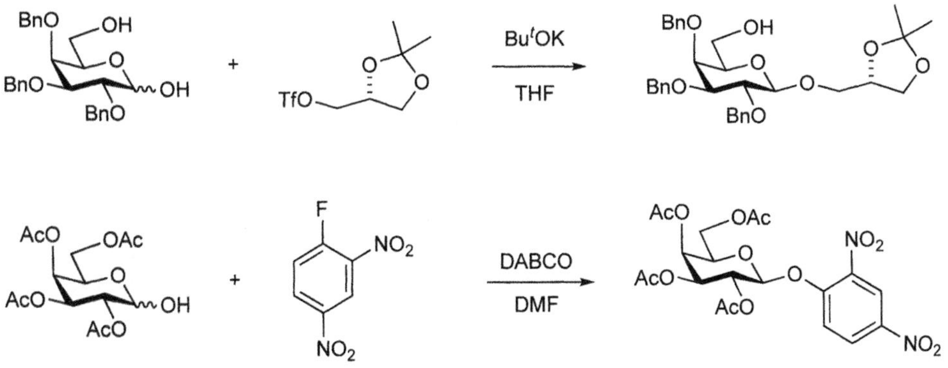

Dibutylstannylene acetals can assist in the process:[70]

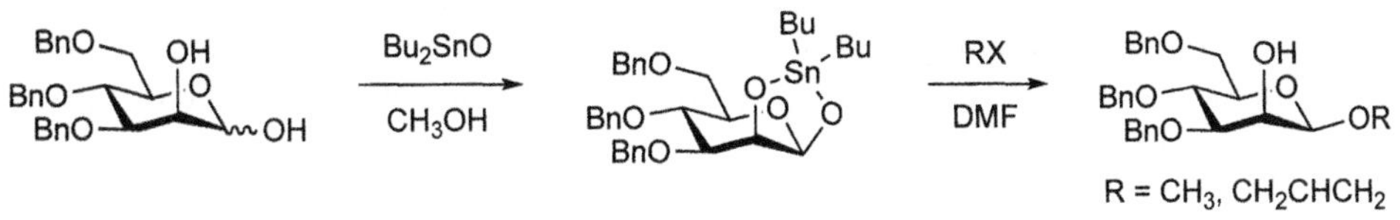

The final method available for the conversion of a hemiacetal into a glycoside, 'dehydrative glycosidation', is a distillation of years of results from the laboratories of Koto and Gin. In the examples below, a hemiacetal is converted in situ into a more reactive donor that then glycosylates the acceptor:[71–74]

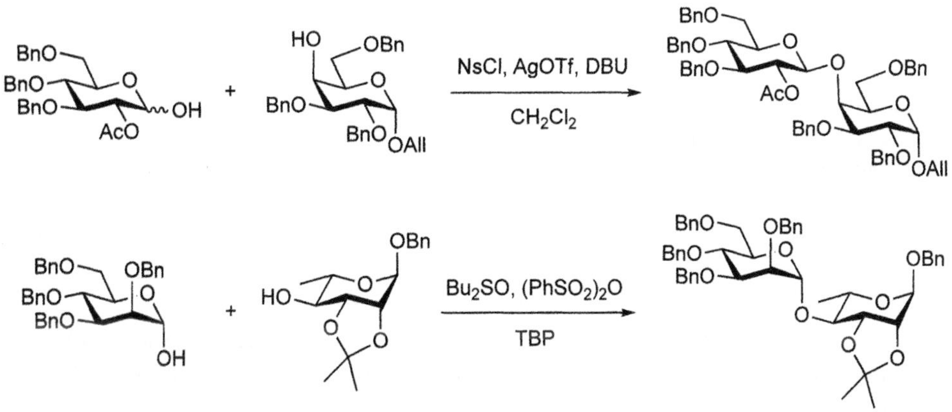

Gin has presented a mechanistic rationale for his procedure:

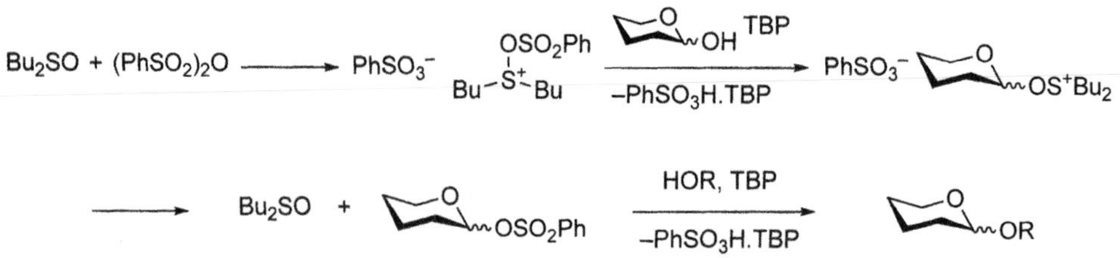

$$Bu_2SO + (PhSO_2)_2O \longrightarrow PhSO_3^-$$

Glycosyl Esters

We have seen that an ester of acetic acid is one of the most common in protecting the hydroxyl groups of a sugar. When this ester is present at the anomeric position, it presents a way of activating the molecule, with the aid of a Lewis acid, into an effective glycosyl donor:[75]

Glycosyl acetates should be considered as cheap and effective glycosyl donors and certainly one of the first 'off-the-shelf' choices (often in combination with boron trifluoride diethyl etherate):[76–78]

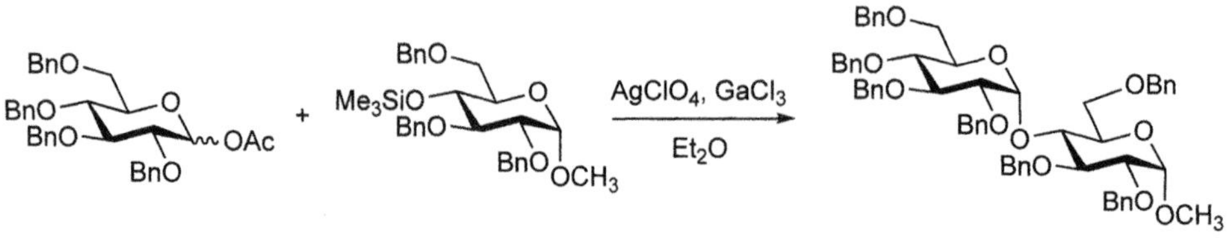

Other sorts of esters have seen increasing use as glycosyl donors, including xanthates (O-alkyl S-glycosyl dithiocarbonates),[7,79–81] phosphites[82–88] and phosphates:[89–91]

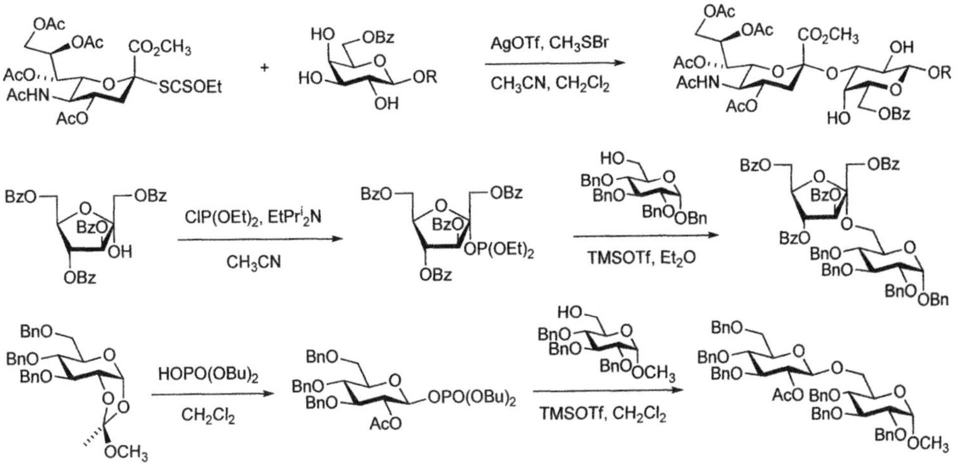

Glycosyl Halides and Orthoesters

Glycosyl halides played a historical role in the development of synthetic approaches to the glycosidic linkage in that it was a glycosyl chloride that was used in the first synthesis (1879) of a glycoside:[75,92]

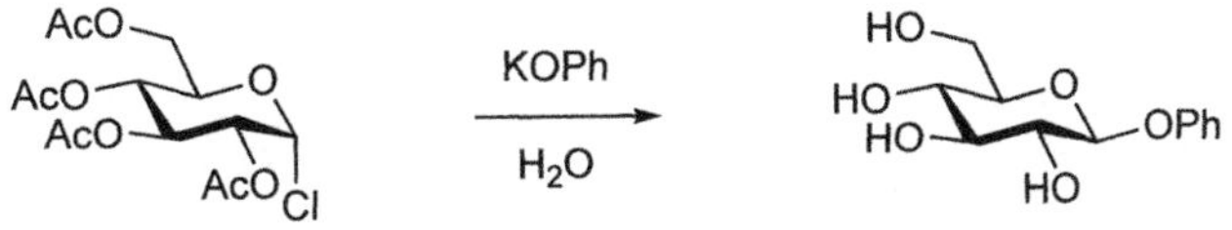

Some years later (1901), Koenigs and Knorr[c,d] extended the approach by treating 'acetobromoglucose' with alcohols in the presence of silver(I) carbonate:[94]

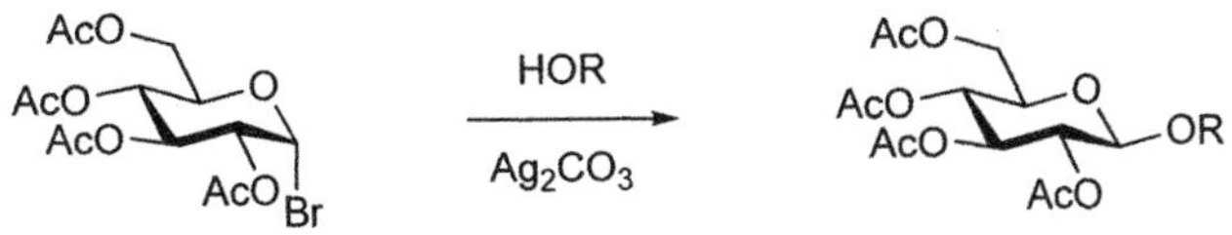

Since these two seminal announcements, glycosyl halides have been at the forefront of new methodology for the synthesis of the glycosidic linkage. Glycosyl chlorides and bromides are used routinely in glycoside synthesis; glycosyl iodides, which were for long considered too unstable to be of any use, are now finding their

---

[c] William Koenigs, or Königs (1851-1906), student of von Baeyer and fellow student with Emil Fischer; Edward Knorr, student of Koenigs.

[d] It is not generally known that Fischer and Armstrong published very similar findings just months after the paper by Koenigs and Knorr.[93]

References start on page 191

rightful place; glycosyl fluorides, the most stable of the glycosyl halides, have broadened the whole approach towards the formation of the glycosidic linkage.

What follows will be a 'cameo' discussion of the various methods that utilize glycosyl halides, with some attention to mechanistic rationale. It is fair to say that, in the interest of the environment, the use of heavy metal salts for the activation of glycosyl halides (chlorides and bromides, in particular) has markedly decreased in the last few decades.

The Koenigs–Knorr reaction (1,2-*trans*):[2,95] Since the inception of the Koenigs–Knorr reaction, there has been an enormous effort to improve the process by utilizing a co-solvent (for acceptors more complex than simple alcohols), adding a desiccant (to absorb any liberated water),[96] adding powdered molecular sieves (to absorb both the liberated hydrogen halide and adventitious water), adding a trace of elemental iodine (to suppress side reactions)[97] and distinguishing the roles of 'promoter' and 'acid acceptor'. In this last regard, heavy metal salts other than those based on silver(I) have been of great use: the combination of mercury(II) bromide/mercury(II) oxide seems to offer the advantages of promotion ($HgBr_2$, a synergistic effect) and acid acceptor (HgO).[98] Soluble promoters such as mercury(II) cyanide, in acetonitrile or nitromethane and introduced by Helferich, and silver(I) triflate are commonly used in the Koenigs–Knorr reaction.[99–101] Some examples of successful glycosidations under Koenigs–Knorr conditions follow:[102–104]

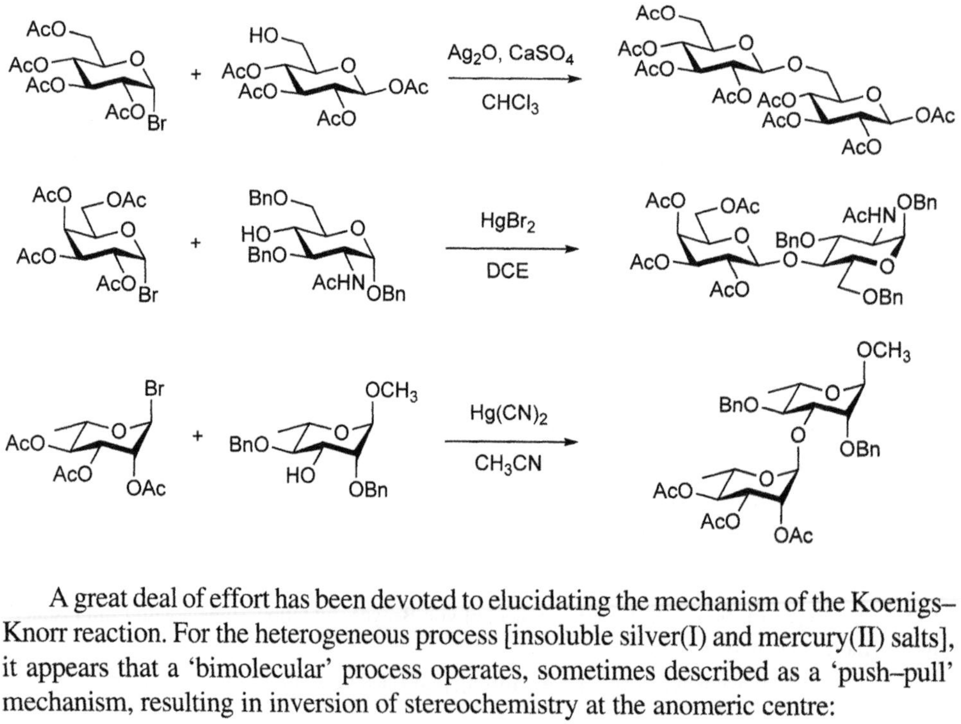

A great deal of effort has been devoted to elucidating the mechanism of the Koenigs–Knorr reaction. For the heterogeneous process [insoluble silver(I) and mercury(II) salts], it appears that a 'bimolecular' process operates, sometimes described as a 'push–pull' mechanism, resulting in inversion of stereochemistry at the anomeric centre:

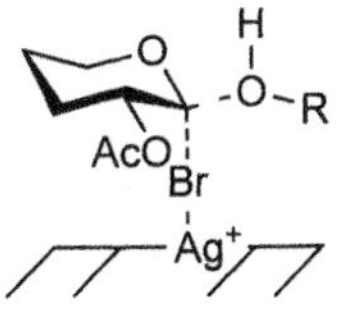

For the homogeneous process [soluble silver(I) and mercury(II) compounds], it has been suggested that a 'unimolecular' mechanism operates and heterolysis of the carbon–halogen bond results in an oxocarbenium ion that undergoes a subsequent reaction with the neighbouring ester group at C2:

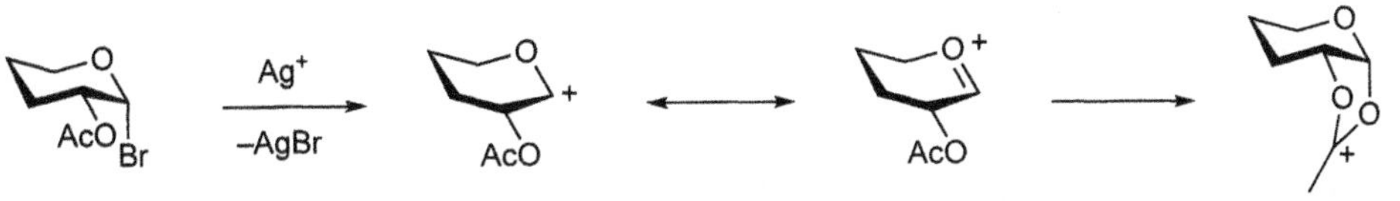

The new cyclic carbenium (1,3-dioxolenium) ion now suffers two fates: approach of the alcohol at the anomeric carbon leads directly to the *trans*-glycoside, whereas attack at the carbon atom of the carbenium ion results in the formation of an orthoester. With the acidic conditions (both protic and Lewis) that prevail in a Koenigs–Knorr reaction, a rearrangement of any intermediate orthoester into the observed glycoside is probable:[105]

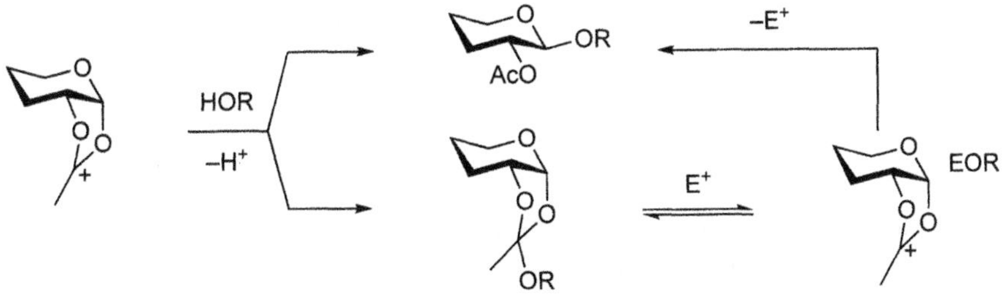

In light of recent studies and the observation that orthoesters are often significant by-products, it is quite possible that the preparation of 1,2-*trans* glycosides using the Koenigs–Knorr reaction proceeds via an orthoester intermediate.[106,107]

Use of the Koenigs–Knorr reaction for the large-scale preparation of glycosides is not recommended, owing to the accumulation of toxic, heavy metal wastes. Although other promoters of the reaction are available,[108] their use is limited; this has resulted in the evolution, and refinement, of other methods for the synthesis of 1,2-*trans* glycosides.

References start on page 191

The orthoester procedure (1,2-*trans*):[2,109,110] We have seen on several occasions that simple orthoesters result most conveniently from the treatment of a glycosyl halide with an alcohol in the presence of a base and, where necessary, a source of halide ions. These simple orthoesters can give rise to more complex orthoesters by a process of acid-catalysed *trans*-esterification; during this process, distillation or the addition of the appropriate molecular sieve usually removes the volatile alcohol. The subsequent rearrangement of the more complex orthoesters into 1,2-*trans* glycosides was initially performed with Lewis acids such as mercury(II) bromide; more recently, trimethylsilyl triflate has appeared as the reagent of choice:[111]

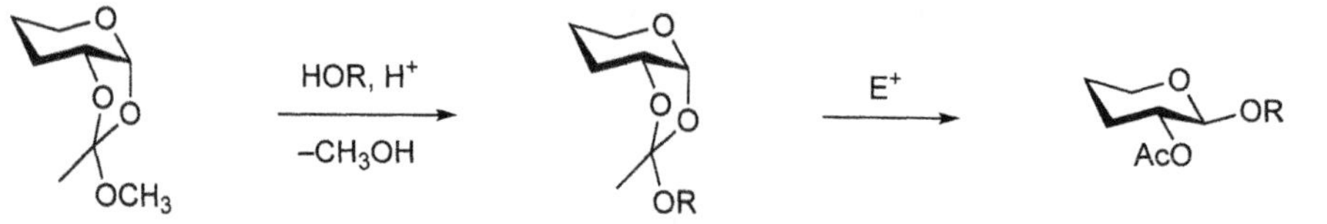

On occasions, especially with orthoacetates, a major and unwanted side reaction is transesterification to give as a by-product the acetic acid ester of the acceptor alcohol; this complication can essentially be eliminated by the use of orthobenzoates:[112,113]

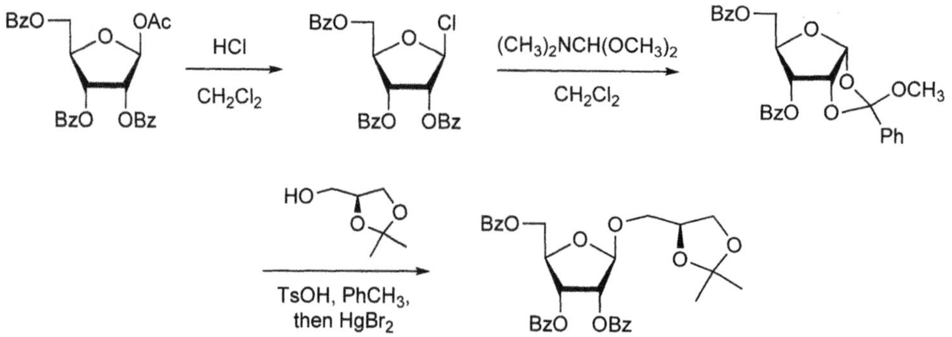

Kochetkov made the orthoester approach virtually his own. In a lifetime of publications, subtle changes such as the use of *tert*-butyl orthoesters,[114] or more major developments involving the introduction of cyano- or thio-ethylidene analogues,[115] have left his name indelibly stamped on the method. For a cyanoethylidene donor, the ideal procedure involves a trityl ether as the acceptor and trityl perchlorate as the catalyst:

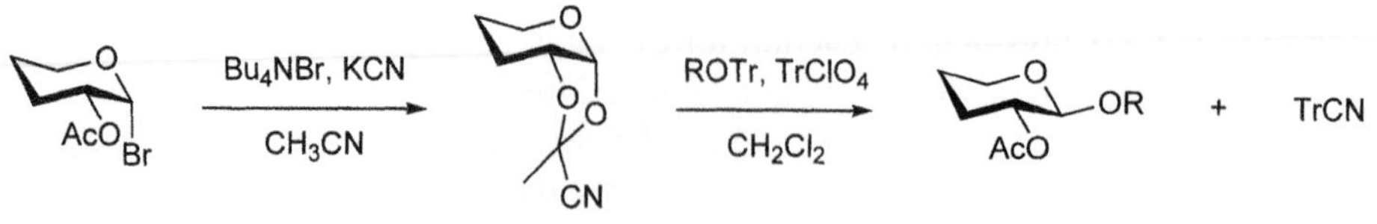

An interesting extension of Kochetkov's work involves the preparation of a tritylated cyanoethylidene derivative of α-D-glucopyranose, capable of polymerization in the presence of trityl perchlorate:

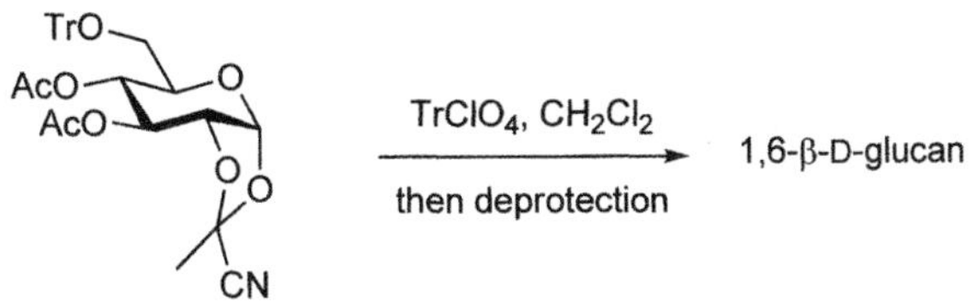

The mechanisms of the orthoester procedures have been thoroughly studied and the various outcomes rationalized:

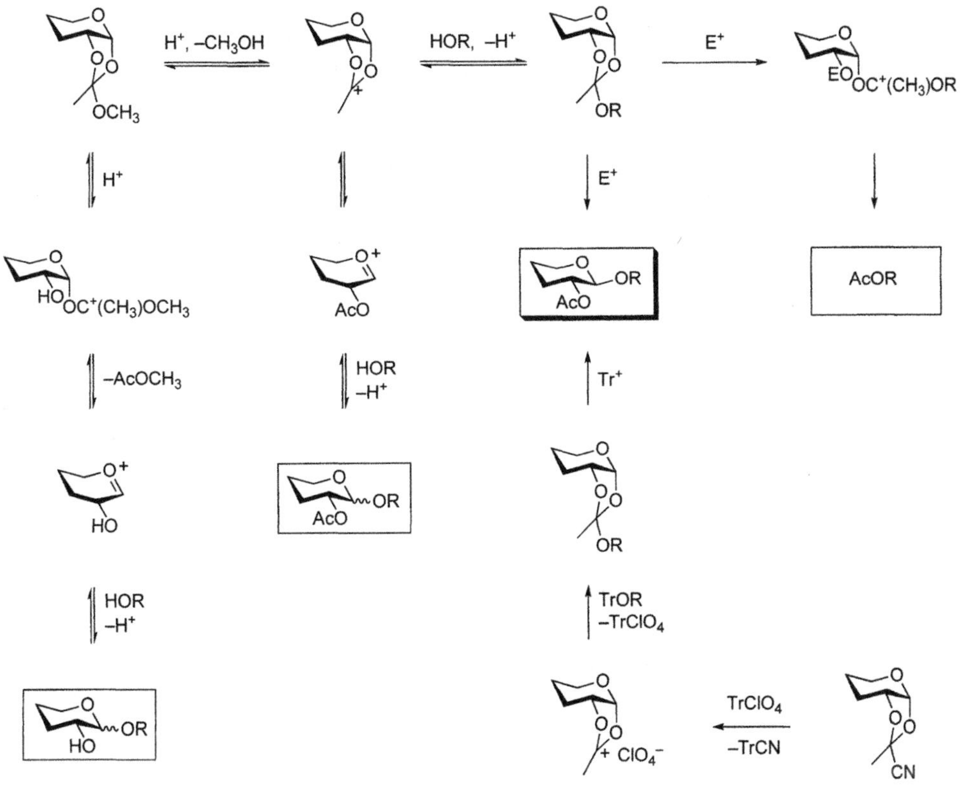

The stereoselective reduction of an anomeric orthoester can give rise to a β-1, 4-linked disaccharide:[116]

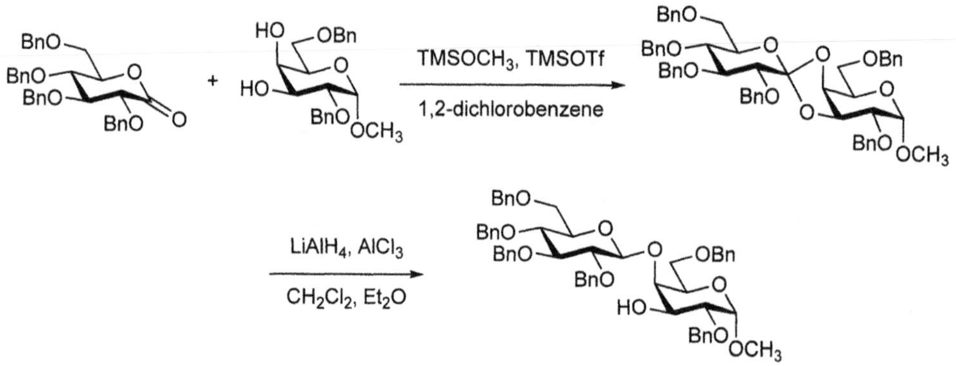

Halide catalysis (1,2-*cis*): It is a credit to the ingenuity and creativity of Lemieux that, some 30 years ago, he was solely responsible for the development of a method for the synthesis of 1,2-*cis* glycosides, termed 'halide catalysis', that is still in common use today.[117] Lemieux, based on an analysis of the anomerization and subsequent reactions of a variety of glycosyl halides, reasoned that a 2-*O*-benzyl protected α-D-glycosyl bromide could be rapidly equilibrated with the β-D-anomer by catalysis involving a tetraalkylammonium bromide. This highly reactive β-D-anomer, or more probably an ion pair derived from it, then proceeds on to the observed product, the α-D-glycoside:

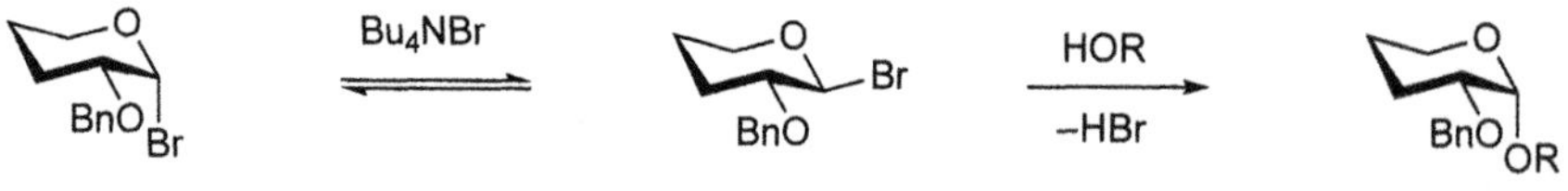

Later workers dubbed the method 'in situ anomerization';[118] Lemieux, in light of the beautiful mechanistic rationalization that he originally proposed,[117] quite rightly objected to this over-simplification of the process. Some recent mechanistic studies have indicated that the 'halide catalysis' procedure is first order in both the donor and the acceptor, and that the tetraalkylammonium ion may accelerate the process by providing electrostatic assistance to the ionization of the carbon–bromine bond.[28]

Many 1,2-*cis* glycosides have been prepared since the announcement of the method in 1975: α-D-glucopyranosides and α-D-galactopyranosides are examples of such linkages, with α-L-fucopyranosides, a common component of many naturally occurring oligosaccharides, being a special bonus:

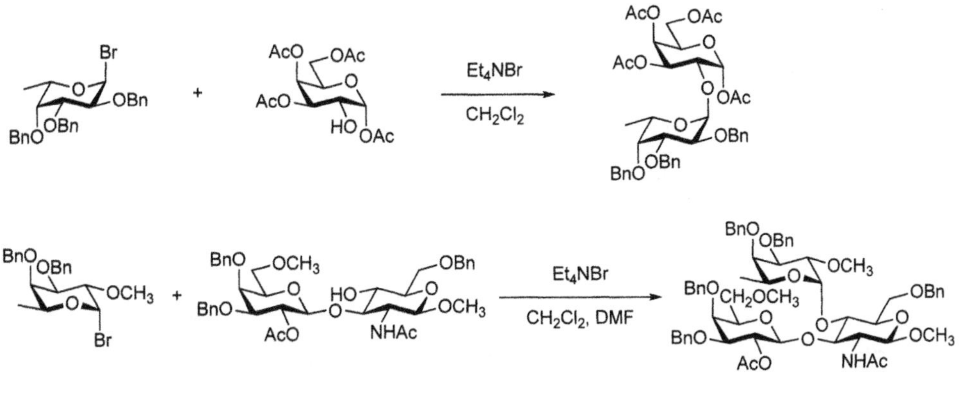

Adding both molecular sieves (to absorb the liberated hydrogen bromide) and dimethylformamide (an apparent catalyst for the reaction) was found to be advantageous.[119–122]

In the early years of the 'halide catalysis' procedure, the glycosyl bromides were generated from the treatment of a 4-nitrobenzoate with hydrogen bromide; nowadays, more convenient procedures are available:[117,123]

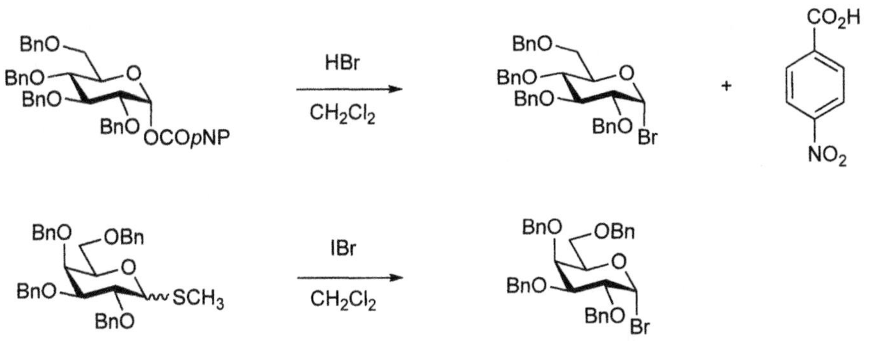

Gervay(-Hague) has developed methods for the synthesis of α-D-glycosyl iodides and found them to be 'superior substrates' for the 'halide catalysis'

protocol;[124,125] other advances followed,[126] none more useful than Mukaiyama's in situ formation of glycosyloxytriphenylphosphonium iodides:[127,128]

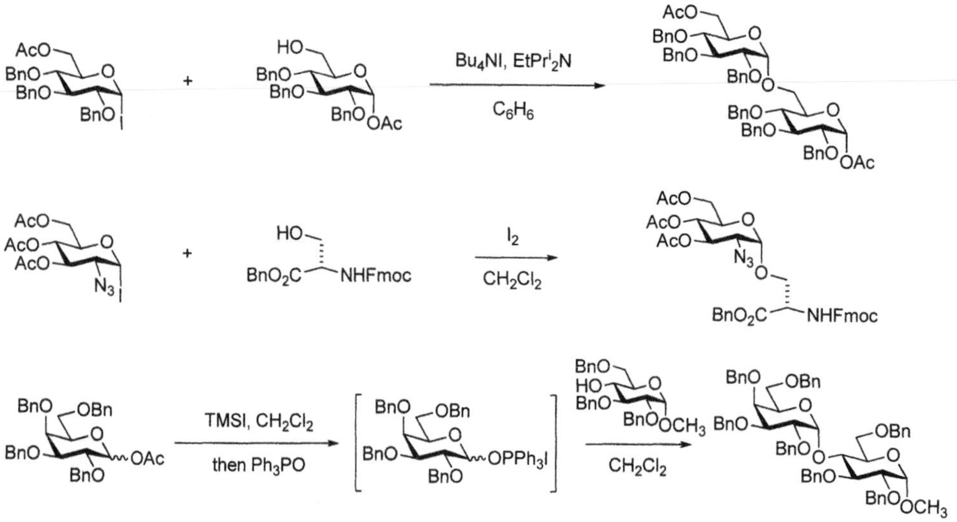

A 3-*O*-acetyl substituent has recently been found to maximize the formation of the α-anomer in glycosidations with various D-glucopyranosyl donors.[129]

Glycosyl fluorides (1,2-*cis* and 1,2-*trans*):[130–133] So far, we have seen that the relatively stable glycosyl chlorides have been of limited use in glycoside synthesis. On the contrary, glycosyl bromides show a versatility that allows their use in the Koenigs–Knorr, orthoester and 'halide catalysis' procedures. Glycosyl iodides, which are of limited stability, show promise for an improvement in the 'halide catalysis' procedure. It seems that the outcast of glycosyl halides is the glycosyl fluoride, and this was to remain so until about 1980. Before then, methods of preparation were limited, β-D-glycosyl fluorides were relatively stable (and the α-D-anomer even more so!) and there existed no 'promoter' to encourage a glycosyl fluoride to act as a donor.

Mukaiyama changed things[134] with the introduction of a range of (exotic) promoters, often as mixtures, to induce glycosyl fluorides to act as respectable glycosyl donors:[135]

SnCl$_2$, AgClO$_4$	Et$_2$OBF$_3$
SnCl$_2$, TrClO$_4$	SiF$_4$
SnCl$_2$, AgB(C$_6$F$_5$)$_4$	SnF$_4$
Cp$_2$MCl$_2$ (M = Ti, Zr, Hf), AgClO$_4$	TiF$_4$
Me$_2$GaCl	TMSOTf
La(ClO$_4$)$_3$.nH$_2$O	Tf$_2$O
TrB(C$_6$F$_5$)$_4$	Yb(OTf)$_3$
HClO$_4$, TfOH or HB(C$_6$F$_5$)$_4$	H$_2$SO$_4$/ZrO$_2$
LiClO$_4$	

Nicolaou became involved in this renaissance of glycosyl fluorides when he showed that a rather benign thioacetal (thioglycoside) could be converted into a fluoride for subsequent use as a glycosyl donor:[136,137]

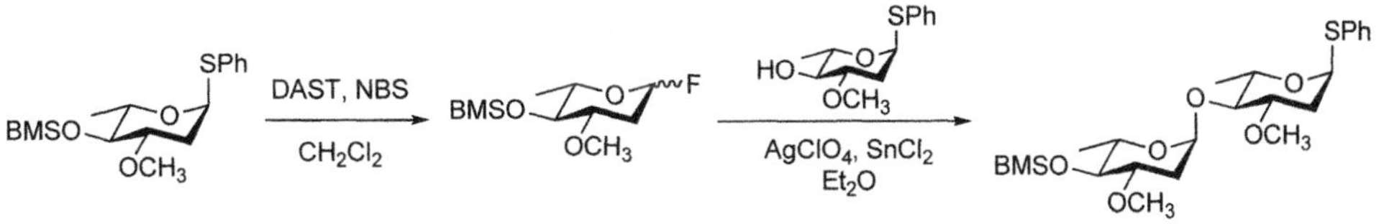

In general, glycosyl fluorides can be used for the synthesis of both 1,2-*cis* and 1,2-*trans* glycosides. For donors having a participating (ester) group at C2, the outcome usually follows the principles of the Koenigs–Knorr reaction, and a 1,2-*trans* glycoside results:

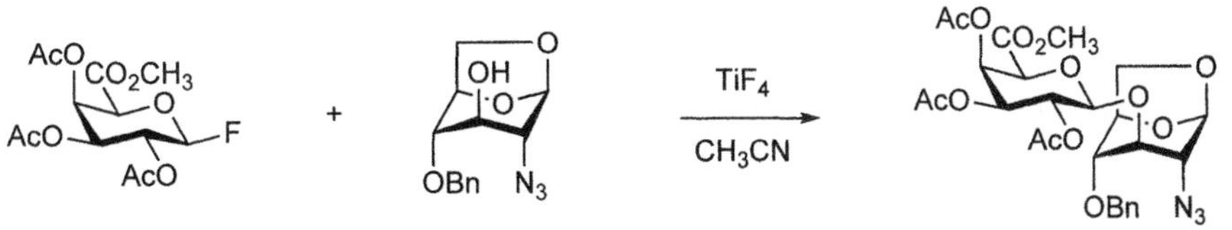

With a non-participating (ether) group at C2, inversion of configuration at the anomeric carbon generally results, which is presumably the consequence of a bimolecular process. Again, the choice of solvent can play a critical role in the outcome of a particular glycosidation.[138,139] Ley has described the application of 1,2-diacetal protecting groups in 'reactivity tuning' of glycosyl fluorides in glycoside synthesis.[140]

Glycosyl Imidates (1,2-*cis* and 1,2-*trans*)[66,67,141]

Sinaÿ was the first to use an imidate for the synthesis of a glycoside,[142] but it was Schmidt who extended the method and made the use of trichloroacetimidates (TCAs) a rival to the well-established Koenigs–Knorr procedure. The TCA method is now, arguably, the preferred method for the synthesis of a 1,2-*trans* glycoside, partly because it does not involve the use of heavy metal reagents in the promotion step but mainly because glycosyl TCAs are often shelf-stable reagents.

Treatment of a hemiacetal with trichloroacetonitrile in dichloromethane in the presence of a suitable base gives rise to anomerically pure, stable TCAs:

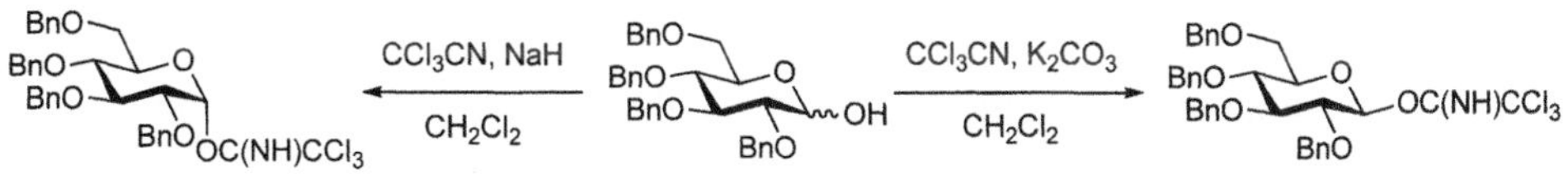

The use of potassium carbonate favours the formation of the β-anomer, whereas sodium hydride or DBU favours the α-anomer; prolonged reaction times also result in the formation of the more stable α-anomer.[143] Although Schmidt invokes kinetic/thermodynamic control to explain these results, it may be that, in the presence of the weaker base (potassium carbonate), a sufficiently high rate of mutarotation of the free sugar and low rate of reversion of the β-TCA will suffice as an alternative explanation:

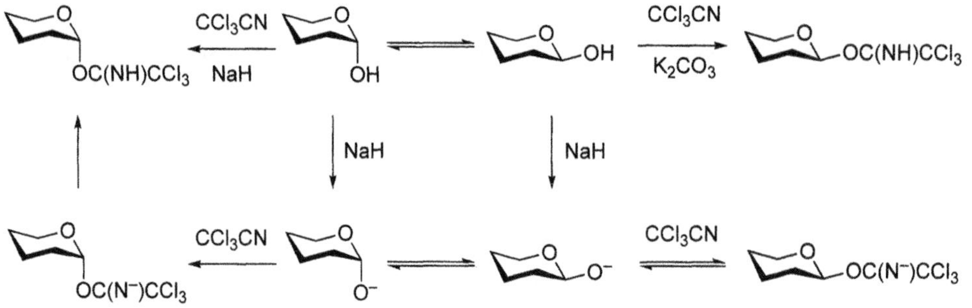

The versatility of the TCA method was obvious from the start:

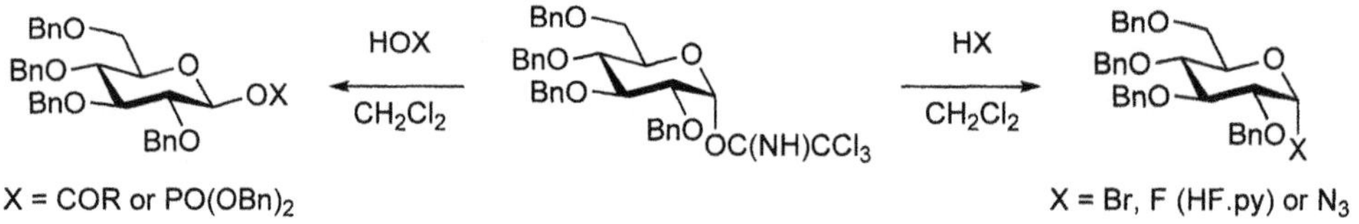

X = COR or PO(OBn)$_2$

X = Br, F (HF.py) or N$_3$

The stronger acids give thermodynamically more stable products, whether through the conventional oxocarbenium ion intermediates or through the isomerization of any initially formed β-D-anomer; the weaker acids give the products of inversion.

The real strength of the TCA method lies in the synthesis of the glycosidic linkage:

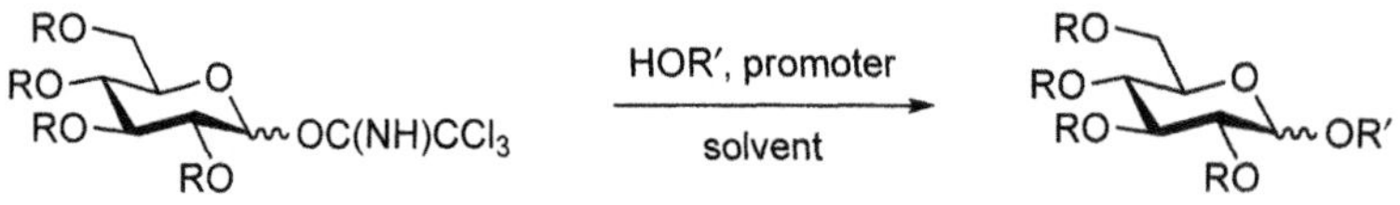

The promoter is generally boron trifluoride diethyl etherate or trimethylsilyl triflate (in catalytic amounts), but zinc bromide,[143] silver(I) triflate[144] and dibutylboron triflate[145] have also been used; new promoters continue to be announced:[146–151]

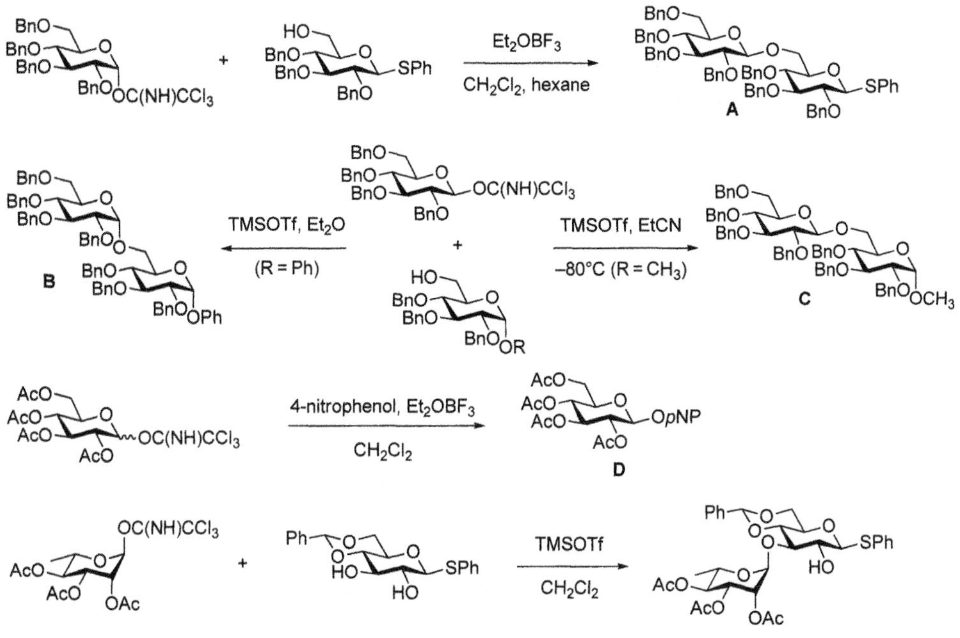

For TCA donors having a non-participating group at C2, treatment with a mild promoter in the presence of an acceptor alcohol generally results in the formation of a glycoside with inversion of configuration at the anomeric carbon (entry **A**), presumably the result of an S_N2-type process. For glycosidations that are promoted by the stronger trimethylsilyl triflate, the fate of the initially formed oxocarbenium ion is regulated by the solvent. In dichloromethane, the stable α-D-glycoside is formed; in ether, the same α-D-glycoside is formed, probably enhanced by the intermediacy of a highly reactive β-D-oxonium ion (entry **B**); in acetonitrile or propionitrile, at low temperature, the rapid formation of an 'α-nitrilium' ion is favoured, leading to the formation of the β-D-glycoside (entry **C**):

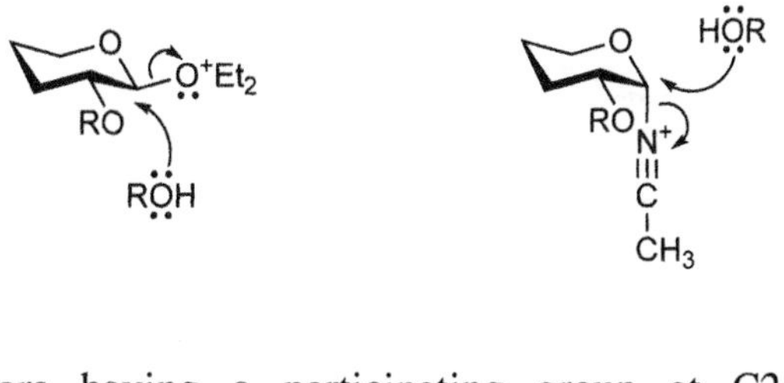

For TCA donors having a participating group at C2, the products of glycosidation are generally those expected (entry **D**); a degree of kinetic control may be exerted in the actual glycosidation step with TCA donors.[152]

Several other aspects of the TCA method warrant comment:

- The incidence of competing orthoester formation with TCA donors having a participating group at C2 is low, presumably owing to the acidic (BF_3, TMSOTf or TfOH) nature of the reaction medium (which encourages the rearrangement of unwanted orthoester into the desired glycoside).
- The incidence of acyl group transfer from a TCA donor having a participating group at C2 is low. Although the reason for this remains unclear, it may be that both boron trifluoride and trimethylsilyl triflate are incapable of activating O2 to initiate the transfer, or that TCA donors are well 'matched' in reactivity to most acceptors.
- For very unreactive acceptor alcohols, for which rearrangement of the TCA donor into an N-glycosyl trichloroacetamide can be a problem, Schmidt has developed an 'inverse procedure': the promoter and alcohol are first mixed and then added to the TCA donor.[153–155]
- The method is well suited to the synthesis of furanosides.[156]

Finally, mention should be made of other glycosyl imidates as potential donors, e.g. N-phenyltrifluoroacetimidates (also useful for highly reactive donors in which rearrangement to the amide can be a problem)[149,157] and thioimidates;[158] the recently introduced glycosyl N-trichloroacetylcarbamate seems somewhat related.[159]

Thioglycosides (1,2-*cis* and 1,2-*trans*)[160–162]

Thioglycosides, in which a sulfur atom replaces the oxygen of the aglycon, are stable derivatives of carbohydrates; we have already addressed the preparation of such molecules in Chapter 2, and other methods exist.[163] We have also had a glimpse of the versatility of thioglycosides in their conversion into glycosyl bromides and glycosyl fluorides. In fact, glycosyl bromides may be generated from thioglycosides in situ and used directly for the conventional synthesis of 1,2-*cis* (halide catalysis) and 1,2-*trans* (Koenigs–Knorr) glycosides:[164,165]

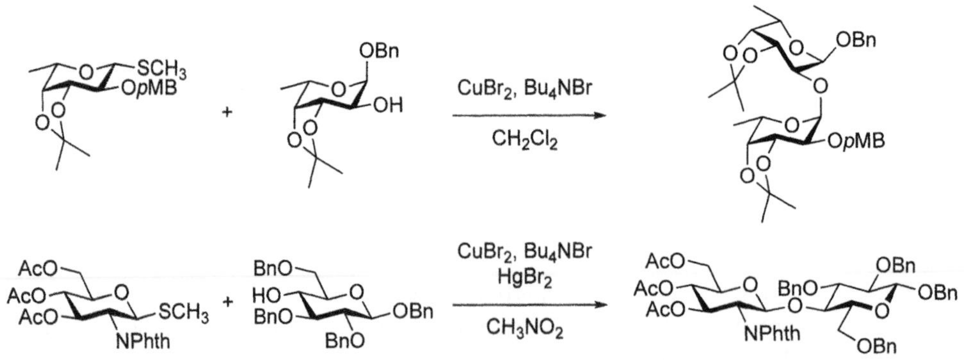

Another reagent, dimethyl(methylthio)sulfonium triflate, can also initiate the 'halide catalysis' cascade on a thioglycoside:[166]

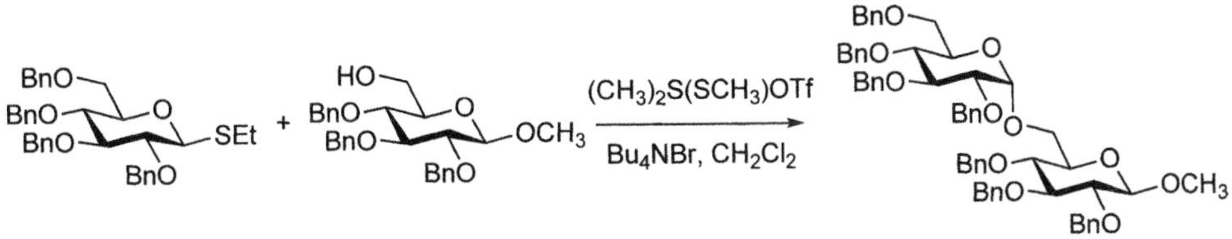

However, the real strength of thioglycosides is that, because of their weak basicity and low reactivity, they are capable of surviving the effects of most of the promoters that are used for other glycosidation protocols and the reagents that are used for protecting group manipulations. Thioglycosides thus offer a measure of 'temporary' (orthogonal) protection to one anomeric centre. Indeed, after the initial observation by Ferrier,[167] a whole range of promoters has been developed specifically for the activation of thioglycosides as glycosyl donors:

CH$_3$OTf	NIS, TfOH [168]
(CH$_3$)$_2$S(SCH$_3$)OTf	IDCClO$_4$ or IDCOTf
CH$_3$SOTf	PhSePhth or PhIO, Mg(ClO$_4$)$_2$ [169]
PhSeOTf	(4-BrC$_6$H$_4$)$_3$N$^{+\bullet}$SbCl$_6^-$ [7,170,171]
4-CH$_3$OC$_6$H$_4$SS(O)Ph	(CH$_3$)$_2$S$_2$, Tf$_2$O [176]
Ph$_2$SO, Tf$_2$O [175]	N-Phenylthio-ε-caprolactam [172]
	1-Benzenesulfinylpiperidine, Tf$_2$O [173,174]

Of these promoters, methyl triflate is a powerful alkylating agent and, therefore, a potential carcinogen; its use should be avoided.

The general process of glycosidation is straightforward and follows the principles established with glycosyl halides and the TCA method:

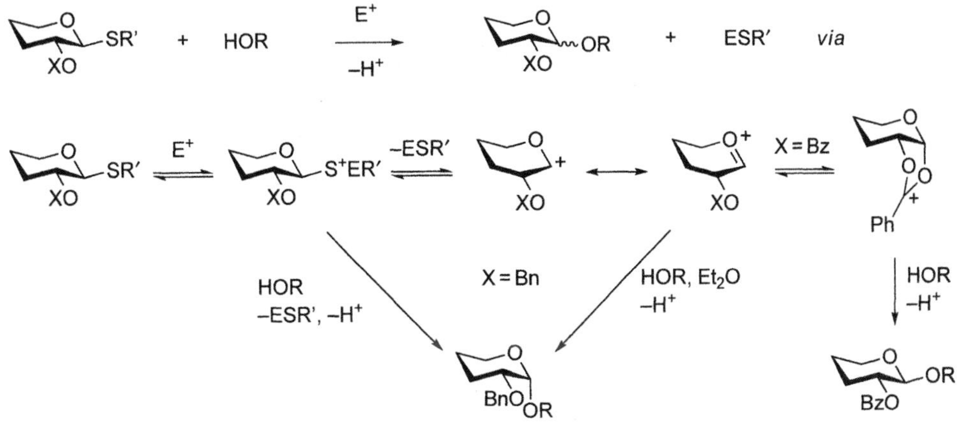

Thioglycosides with a non-participating (ether) group at C2 generally yield the 1,2-*cis* glycoside, whereas a participating (ester) group gives the 1,2-*trans* glycoside; solvents such as ether and acetonitrile can again influence the stereochemical outcome of such events. Acetyl group transfer from the donor to the acceptor can again be a problem, as can aglycon transfer from a thioglycoside acceptor to a donor.[177] For the most recently developed promoters, 1-benzenesulfinylpiperidine[173,174] and diphenyl sulfoxide,[175] both in combination with triflic anhydride, the formation of intermediate glycosyl triflates has been suggested. These glycosyl triflates are stable at low temperature but may anomerize at higher temperatures, leading to a different stereochemical outcome in glycoside formation:

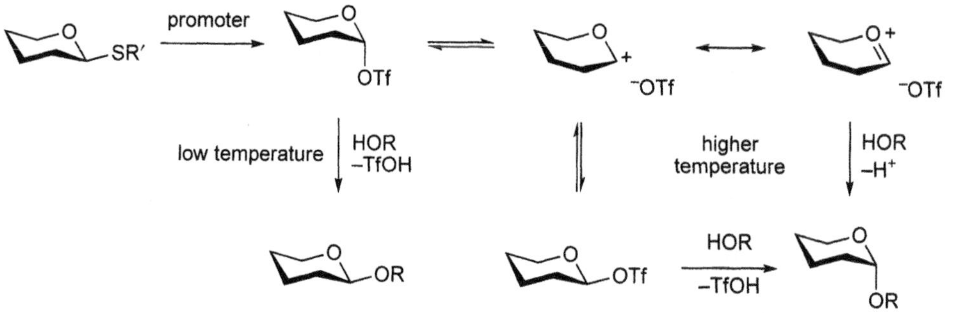

The diphenyl sulfoxide/triflic anhydride combination is a more powerful promoter than the 1-benzenesulfinylpiperidine counterpart;[175] *N*-benzenesulfinylmorpholine/triflic anhydride is the latest variation of the promoter combination.[178] The combination of dimethyl disulfide and triflic anhydride shows promise as a new promoter for thioglycosides.[176]

During van Boom's work on thioglycosides, it became apparent that ether-protected donors could be activated with iodonium dicollidine perchlorate, but ester-protected donors remained inert; for the latter type of donor, a 'stronger' promoter, *N*-iodosuccinimide/triflic acid was necessary:

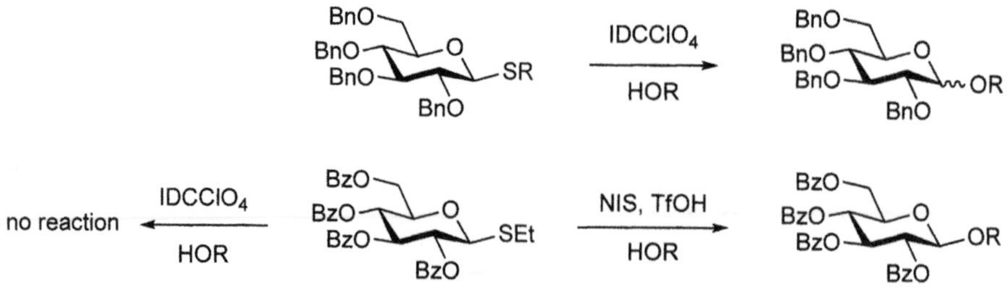

Drawing on Fraser-Reid's earlier concept, van Boom referred to the 'ethers' as 'armed' thioglycosides and to the 'esters' as being correspondingly 'disarmed'. This concept has proven to be of great value to the synthetic chemist:[179,180]

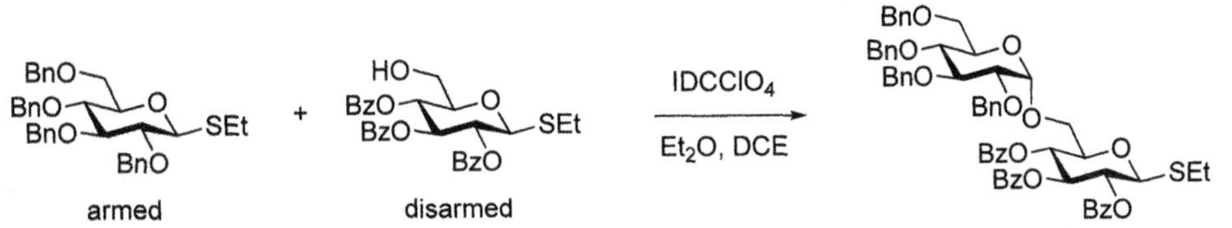

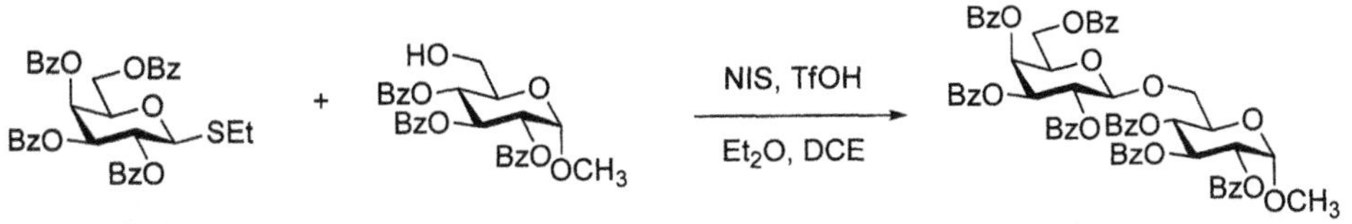

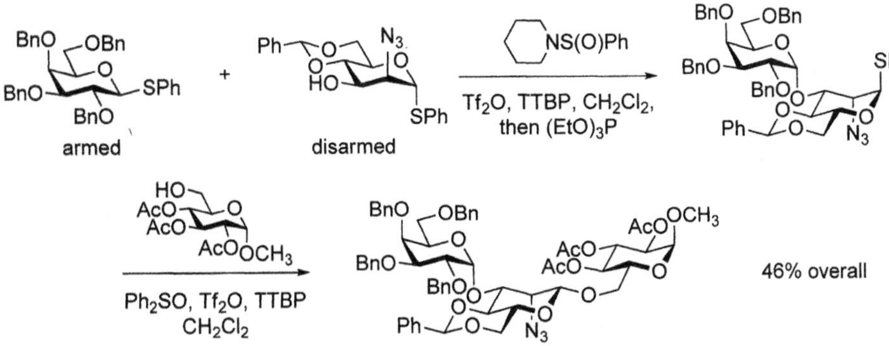

It has been suggested that iodine may be a useful promoter of 'armed' thioglycosides[181] and that 2-*O*-pivalyl thioglycosides seem particularly suited for the preparation of 1,2-*trans* glycosides.[182,183]

Other factors may be brought into play to adjust the reactivity and selectivity of a thioglycoside, namely the choice of solvent,[184] the size of the alkyl groups attached to the sulfur,[185] the anomeric configuration of the thioglycoside,[185] the presence of cyclic protecting groups and the nature of aromatic groups attached to the sulfur:[186–188]

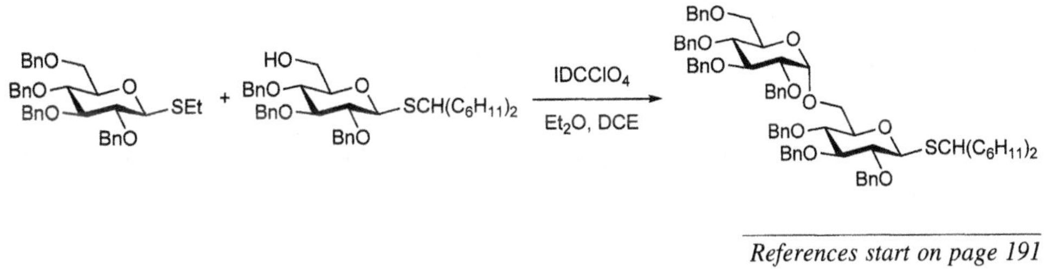

References start on page 191

The 2,6-dimethylphenyl group has been proposed as a convenient and improved aglycon for thioglycosides;[189] 5-*tert*-butyl-2-methylthiophenol avoids the repulsive smell associated with the preparation of many thioglycosides.[190]

In the matter of cyclic protecting groups, Ley has made something of an art form out of the use of various glycosyl donors, including thioglycosides, and the concepts of 'torsional control/reactivity tuning' for glycoside synthesis;[39–41,191] other workers have used cyclic esters and cyclic acetals, across O2/O3 and O4/O6, in a similar manner.[192–194]

This section on thioglycosides ends with a novel 'twist':[195] Boons has recently reported the use of a chiral auxiliary at O2 that, in one sense, forces the formation of a '*trans*-decalin-like' intermediate that leads to the formation of the α-glycoside; alternatively, a '*cis*-decalin-like' intermediate can be arranged that leads to the β-glycoside.[196–198] Only time will tell whether this novel approach will lead to improved stereocontrol in some glycosidations.

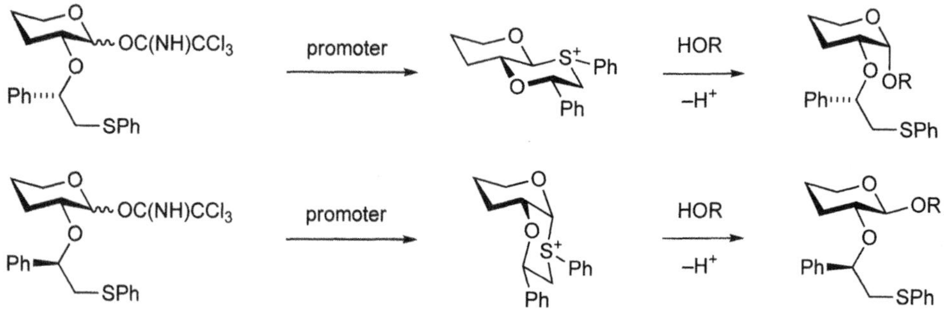

Seleno- and Telluroglycosides[199,200]

Selenoglycosides are prepared from modified carbohydrates in much the same way as are thioglycosides:[201] an anomeric acetate is treated with a Lewis acid and the selenol (sometimes a smelly affair!),[202,203] or a glycosyl halide is treated with a metal selenide:[204,205]

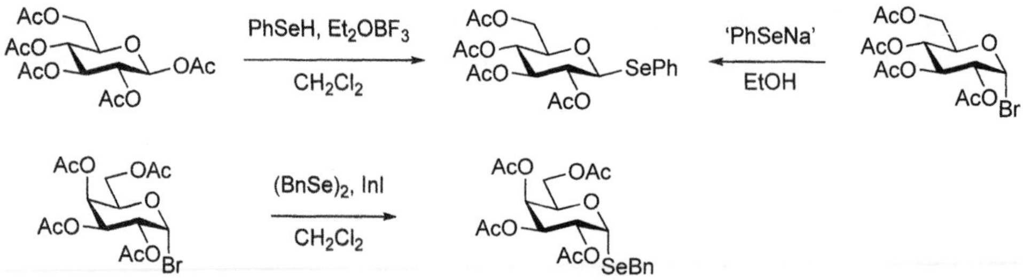

For the corresponding telluroglycosides, it is generally more convenient to proceed via the alkali metal telluride, normally generated from the ditelluride:[204]

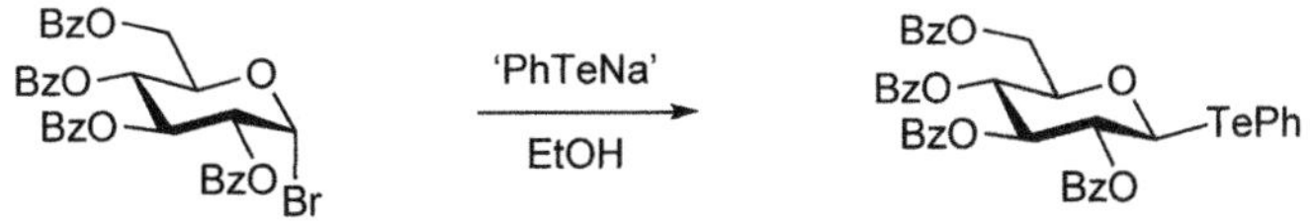

Both seleno- and (aryl)telluroglycosides are generally stable, crystalline solids that can be activated into the role of glycosyl donor by treatment with very mild promoters:

AgOTf, K$_2$CO$_3$ [202]   IDCClO$_4$ [207]

NIS [206]   NIS, TfOH [207]

This activation is chemoselective: telluroglycosides react in preference to selenoglycosides that, again, are more reactive than thioglycosides.[208] Ley, in concert with the concept of 'reactivity tuning', has introduced three different levels of reactivity with the following molecules:[39–41,191]

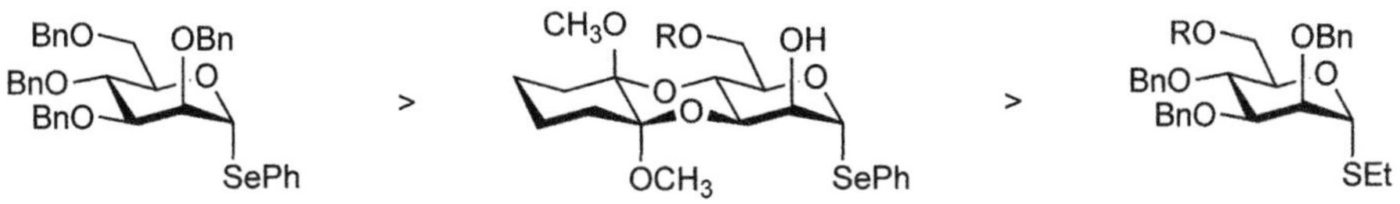

Some examples show the versatility of seleno- and telluroglycosides in glycoside synthesis:[202,208–210]

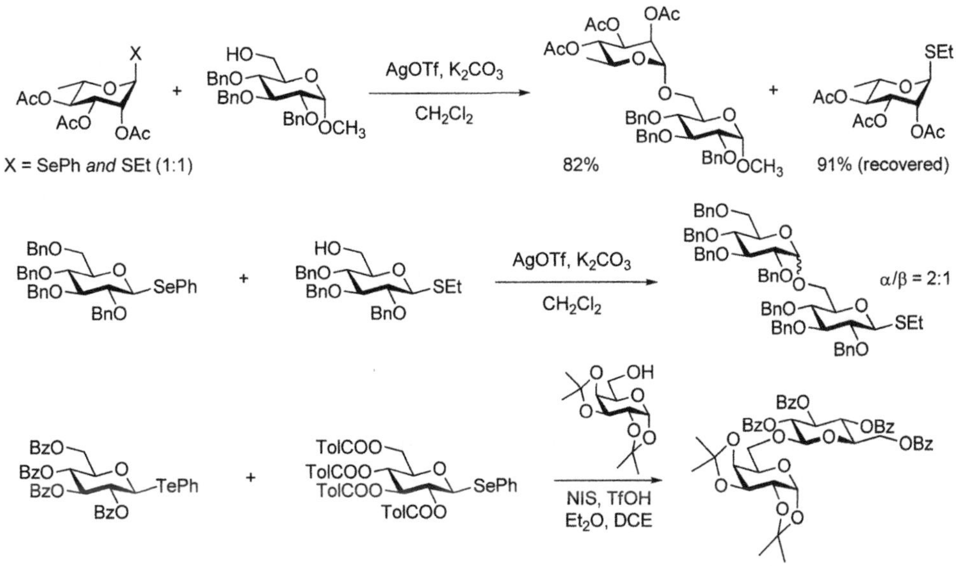

Glycosyl Sulfoxides (sulfinyl glycosides; 1,2-*cis* and 1,2-*trans*)[211]

Thioglycosides, being thioacetals, are naturally amenable to oxidation to produce either the sulfoxide (sulfinyl glycoside) or the sulfone (sulfonyl glycoside):

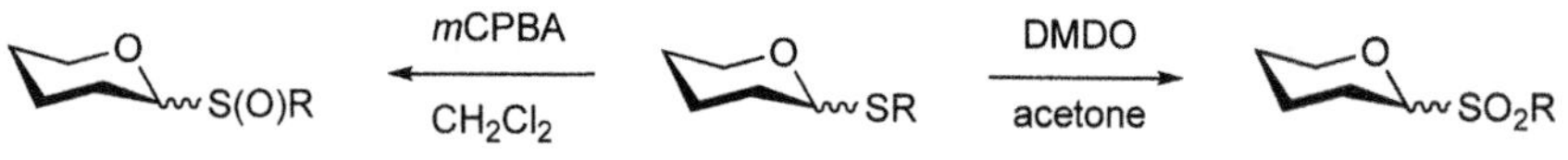

The reagent of choice for preparation of the sulfone is dimethyldioxirane; the less reactive 3-chloroperbenzoic acid is preferred for the preparation of the sulfoxide.[212–214] In fact, the α-D-thioglycoside may be selectively oxidized to form just one stereoisomer of the sulfoxide (ostensibly owing to a combination of steric and anomeric effects); the β-D-anomer does not show this proclivity:[215]

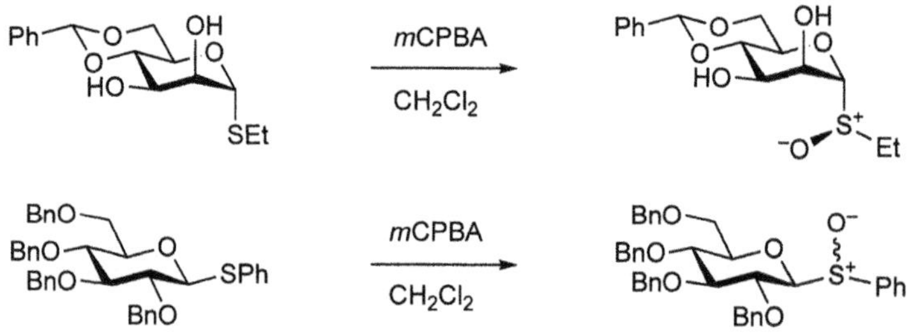

With glycosyl sulfoxides being so available, it was not surprising that they should be investigated as glycosyl donors. However, it took until 1989 for Kahne to realize their potential;[216] the strength of the method lies in the fact that some very unreactive acceptors may be glycosylated under mild conditions:[217–220]

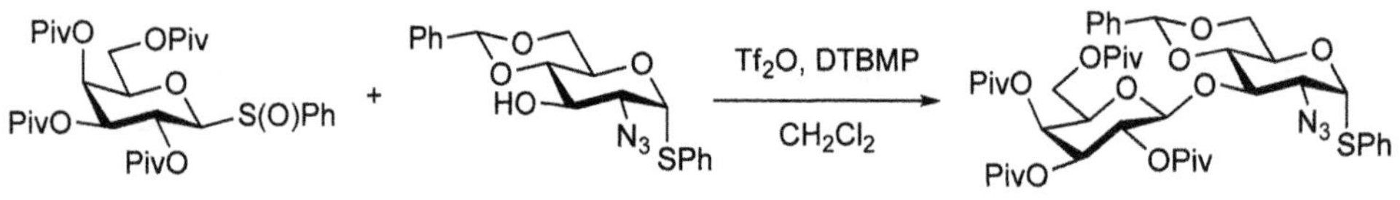

The method seems to rely greatly on the presence of a pivalyl group at O2 of the donor, and this gives rise to the formation of 1,2-*trans* glycosides.[221] However, donors with an ether at O2 react sufficiently well to provide a synthesis of 1,2-*cis* glycosides.[217]

With triflic anhydride as the promoter, the by-products of the 'sulfoxide method' are triflic acid and the very 'thiophilic' phenylsulfenyl triflate; a hindered base is often added to negate the effect of this highly reactive by-product. As an alternative, Kahne introduced the use of a catalytic amount of triflic acid as the promoter; methyl propiolate was then added as a 'scavenger' of the stoichiometric amount of phenylsulfenic acid formed.[219] Other promoter/scavenger combinations have been suggested.[222–224]

All of the general principles of glycoside synthesis can be applied to the sulfoxide method, and Kahne has been able to 'tune' the reactivity of the sulfoxide donor by a careful choice of the substituent on sulfur:[220]

$$Ar = 4\text{-}CH_3OC_6H_4 > C_6H_5 > 4\text{-}O_2NC_6H_4$$

In addition, the mechanism of the method has received much attention:[211,225,226]

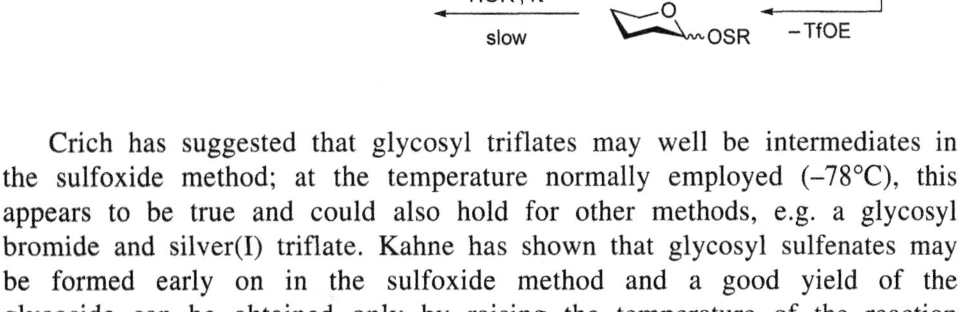

Crich has suggested that glycosyl triflates may well be intermediates in the sulfoxide method; at the temperature normally employed (−78°C), this appears to be true and could also hold for other methods, e.g. a glycosyl bromide and silver(I) triflate. Kahne has shown that glycosyl sulfenates may be formed early on in the sulfoxide method and a good yield of the glycoside can be obtained only by raising the temperature of the reaction mixture later.

It is safe to say that glycosidations with thioglycosides, activated by triflic anhydride and either 1-benzenesulfinylpiperidine or diphenyl sulfoxide, and those with glycosyl sulfoxides, activated by triflic anhydride or triflic acid, while operationally distinct, are mechanistically similar. Anomeric 'sulfimides' have also been suggested as glycosyl donors.[227]

Glycals[228–232]

In 1969, Ferrier and Prasad reported that the treatment of a suitable glycal with a Lewis acid in the presence of a reactive alcohol gave rise to glycosides having unsaturation in the ring:

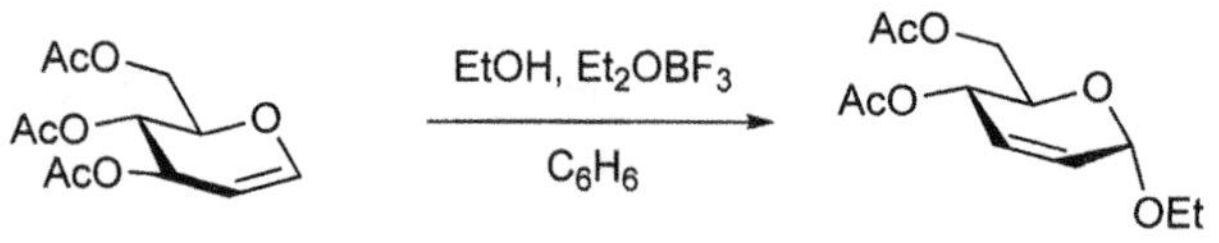

The Lewis acid coordinates with the leaving group at C3, and a resonance-stabilized cation results:

This cation is attacked by the alcohol to yield the 2,3-unsaturated glycoside; the stereoselectivity is normally good as, with no substituents at C2 and C3, the more stable α-glycoside predominates. It is safe to say that more 'improvements' have been made to this method of glycosidation than to any other method; virtually any Lewis[233] or protic[234,235] acid will do the job, and changes can be made even to the leaving group.[236] An interesting variation has recently been reported:[237]

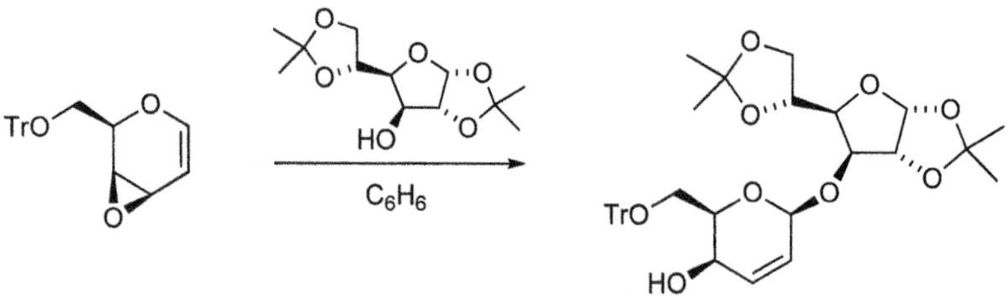

Although the chemistry of glycals was to flourish in the 20 years after Ferrier's discovery, it took two significant events to establish these unsaturated sugars as glycosyl donors in their own right: first, the availability of dimethyldioxirane as a laboratory reagent and, second, Samuel Danishefsky's attraction to the field of carbohydrates.

At first glance, a glycal does not seem to be the ideal glycosyl donor. Apart from the problem of stereoselectivity at the anomeric carbon (α- or β-), an oxygen atom must be reinstated at C2, again stereoselectively; the generation of an intermediate epoxide solved both of these problems:[238,239]

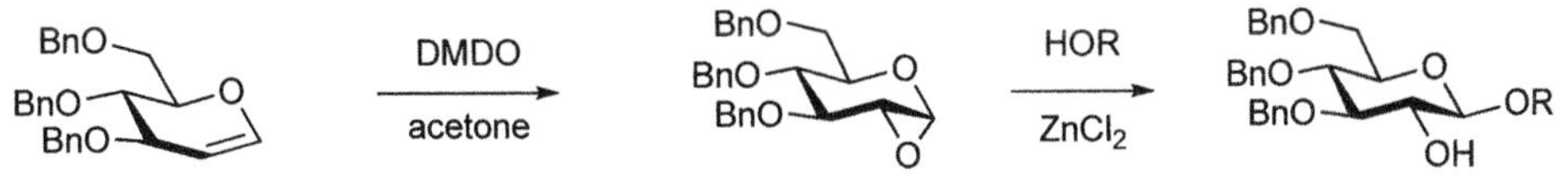

One of the many advantages of this method was that the newly formed glycoside possessed a free hydroxyl group at C2, available for further elaboration (see Chapter 12):

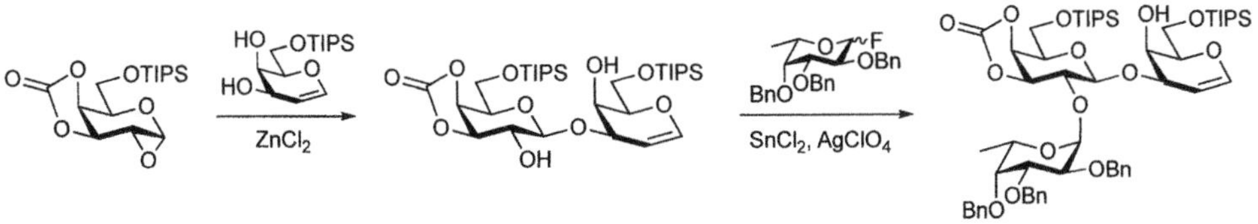

Several points need to be made about this 'glycal epoxide' methodology:

- Only dimethyldioxirane and related dioxiranes can be used to generate the epoxide from the glycal; other reagents, such as 3-chloroperbenzoic acid, cause decomposition of the epoxide.
- The epoxide is usually formed with high stereoselectivity, being installed in a '*trans*' sense to the substituent at C3.
- Normally, an α-D-epoxide upon treatment with an alcohol and a Lewis acid will yield a β-D-glycoside. Sometimes, α-D-glycosides result from the use of less reactive alcohols:

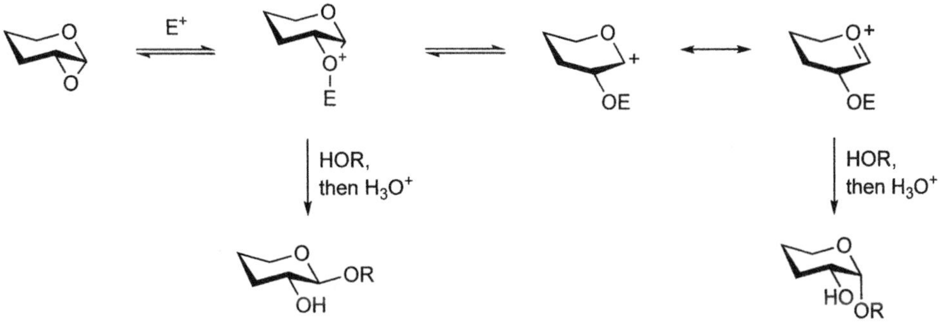

- When an anomeric epoxide is unsatisfactory as a glycosyl donor, conventional transformations may yield a more suitable donor:

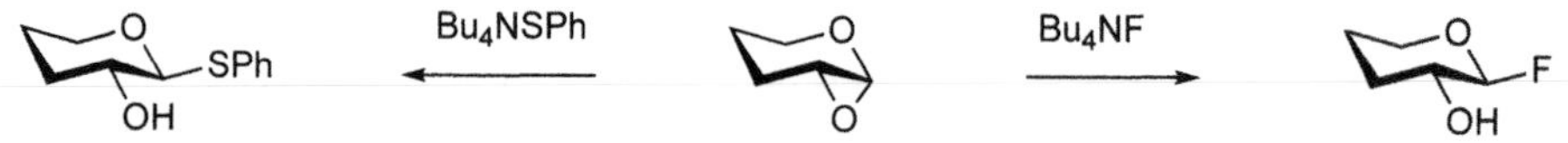

4-Pentenyl Activation (1,2-*cis* and 1,2-*trans*)[240–244]

In the late 1980s, Fraser-Reid noted an interesting transformation of a 'higher' sugar derivative upon treatment with N-bromosuccinimide:

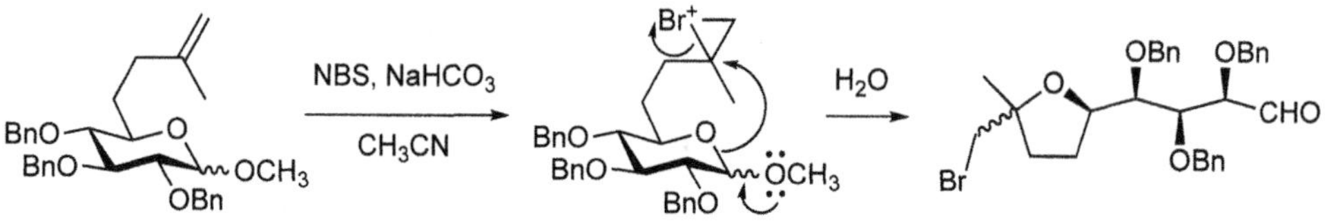

This chance observation led to the idea that 4-pentenyl glycosides would respond similarly to N-bromosuccinimide, thus providing a new type of protecting group for the anomeric centre:

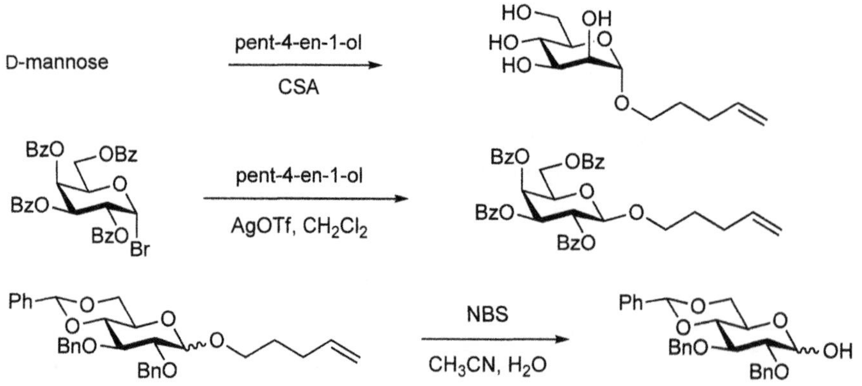

In addition, the 4-pentenyl glycosides were found to be effective glycosyl donors:

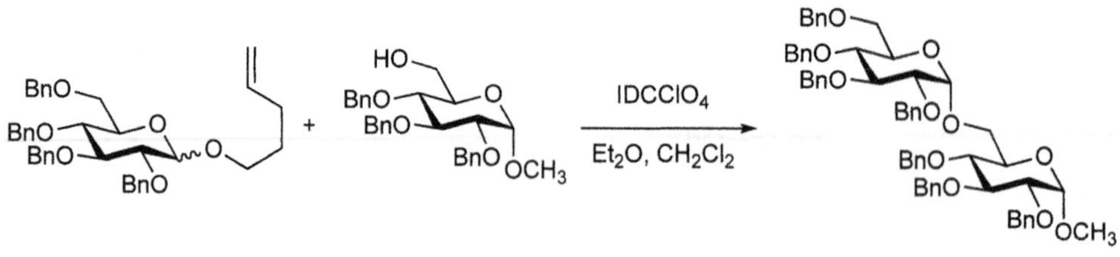

Thus was born the *n*-pentenyl glycoside (NPG) method of glycosidation, one that has gained in popularity because of the ease of installation of the 4-pentenyl glycoside, the stability of the group to most reagents (much akin to a thioglycoside) and the easy promotion by oxidizing agents such as *N*-bromosuccinimide and iodonium dicollidine perchlorate.

The NPG method was to be a fertile area for Fraser-Reid during the next decade; out of it was to come the concept of 'armed/disarmed' glycosyl donors and the necessity for the corresponding promoters:

armed – IDCClO₄ promotion

disarmed – NIS, TfOH or NIS, TESOTf promotion

A whole new facet of glycoside synthesis was exposed:

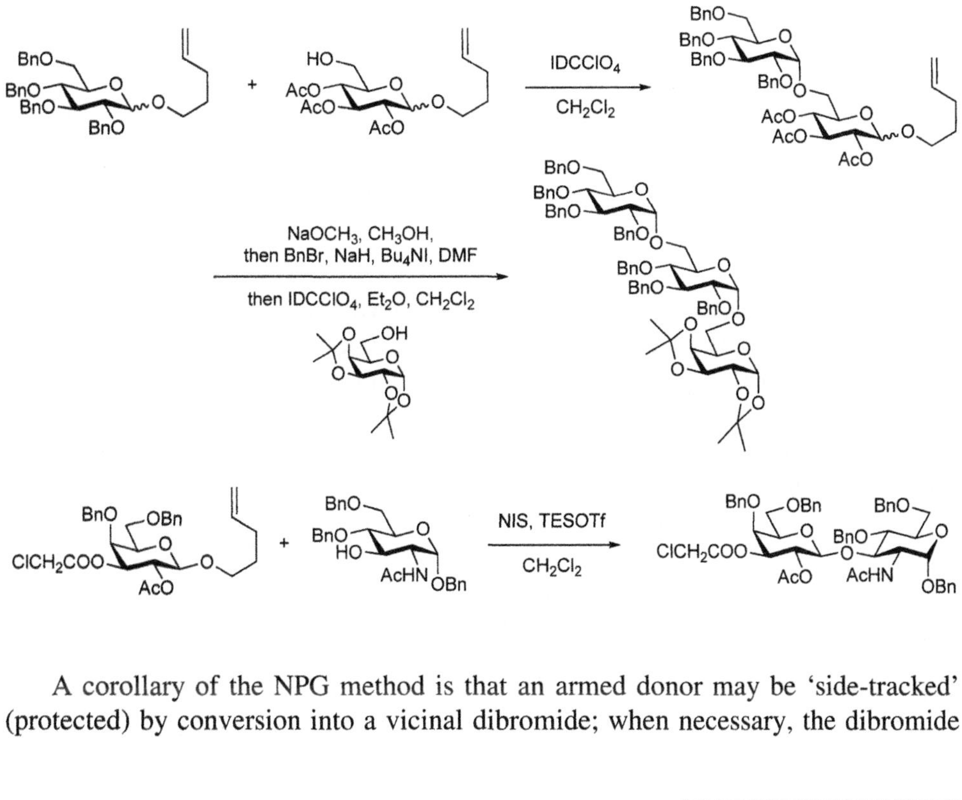

A corollary of the NPG method is that an armed donor may be 'side-tracked' (protected) by conversion into a vicinal dibromide; when necessary, the dibromide (a 'latent' glycosyl donor) may be treated with zinc metal or sodium iodide to regenerate the armed donor:[245,246]

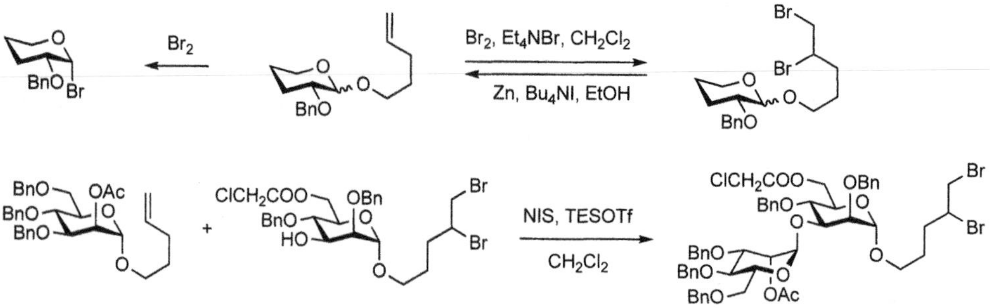

The mechanism of the NPG method has been investigated in some depth and is well understood:

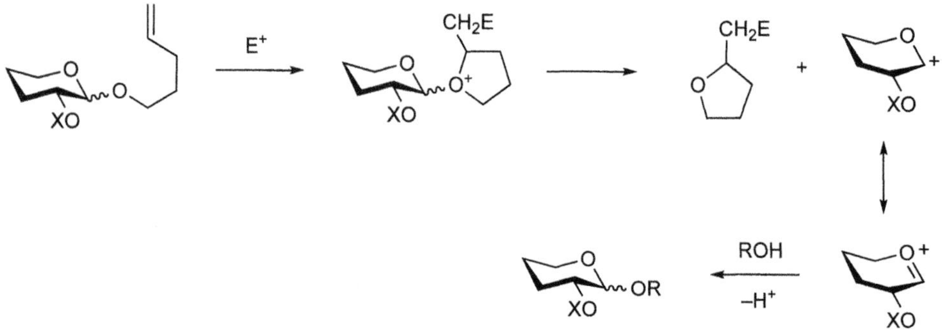

The stereochemical outcome depends, as usual, on the nature of 'X'; ethers generally give 1,2-*cis* glycosides and esters 1,2-*trans* glycosides. Fraser-Reid has continued to probe the underlying mechanism of the NPG method, including a comparison with TCA and thioglycoside donors, linking the results again to 'matching' between donor and acceptor.[247] Also, quite remarkably, it is possible to activate an *n*-pentenyl orthoester in preference to an armed NPG, with ytterbium(III) triflate:[248-250]

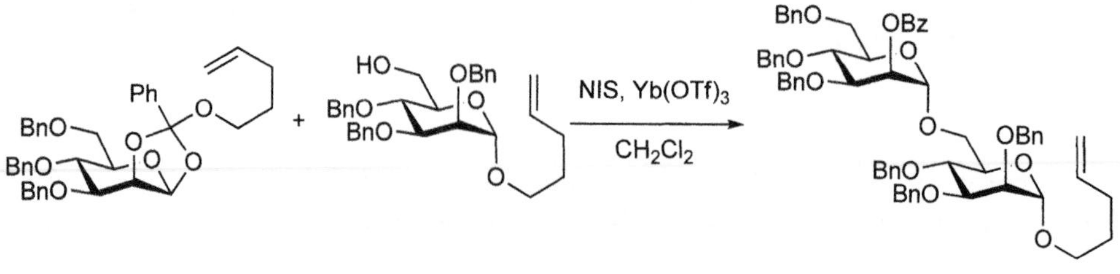

β-D-**Mannopyranosides (1,2-*cis*)**[251–253]

D-Mannose is a common constituent of many naturally occurring oligosaccharides, some of which are attached to proteins and all of which play a biological role in the host organism. Although the common linkage in such oligosaccharides is α-D-*manno*, there is a frequent enough occurrence of the β-D-counterpart. For example, the pentasaccharide core of N-linked glycoproteins is invariant and contains both types of linkages:

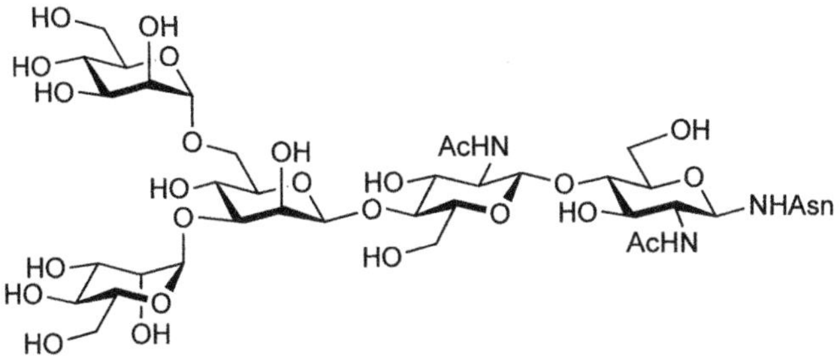

Whilst we have discussed many methods for the synthesis of α-D-mannopyranosides (1,2-*trans* and the thermodynamically favoured arrangement), only very few of these are adaptable to form β-D-mannopyranosides. Here, we shall indicate these versatile methods and also introduce new ones.

The origin of the problem in any approach to β-D-mannopyranosides is the axial orientation of the group at C2. If this group is an ester, the α-D-mannopyranoside generally results from participation of the ester at the anomeric carbon; if the group is an ether, a favourable anomeric effect again results in the preferred formation of the α-D-anomer:

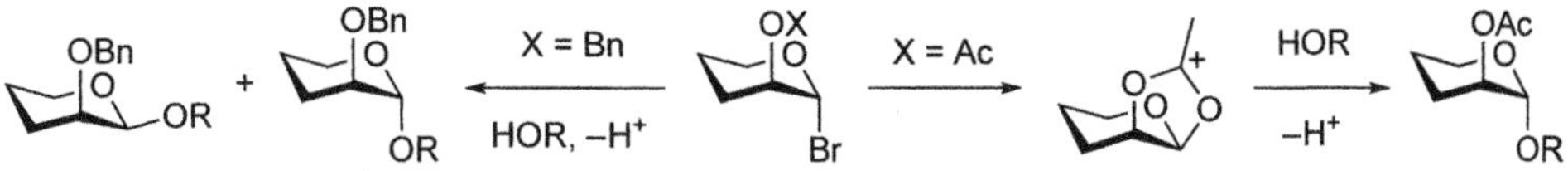

Obviously, to circumvent these problems, some creative thinking was required!

Glycosyl halides: One of the most common methods for the synthesis of β-D-mannopyranosides involves the treatment of a 2-*O*-benzyl-α-D-mannopyranosyl halide with an insoluble promoter such as silver(I) silicate or silver(I) zeolite:

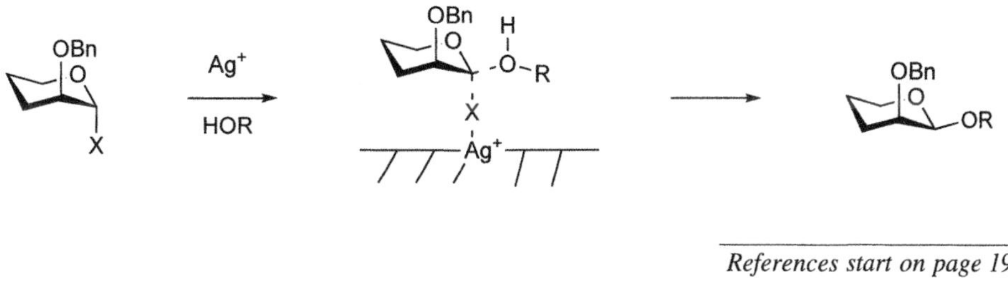

The strength of this Koenigs–Knorr approach lies in its simplicity; it is successful because the 'push–pull' type of mechanism invoked on the surface of the promoter effectively shields the α-face of the donor, forcing the acceptor alcohol to approach in the required manner:[254–256]

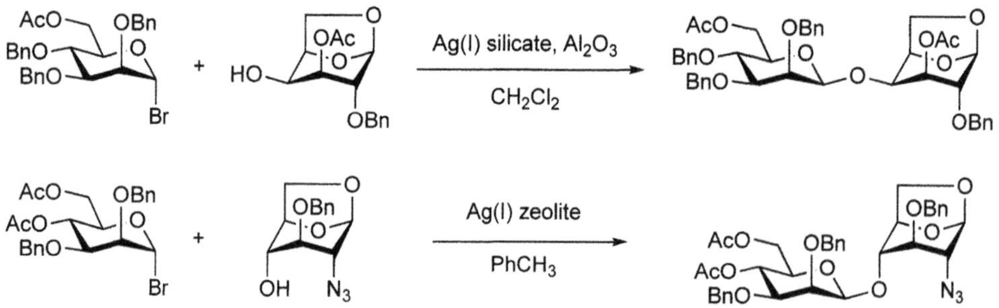

Glycosyl sulfoxides (and thioglycosides): A seminal paper by Crich suggested that glycosyl triflates were intermediates in the sulfoxide method of glycosidation.[225] Indeed, as applied to α-D-mannopyranosyl sulfoxides having non-participating groups at C2 and C3 and a 4,6-O-benzylidene protecting group (presumably to discourage glycosyl cation formation),[253] there now exists a powerful method for the synthesis of β-D-mannopyranosides from virtually any sort of acceptor alcohol:[257–260]

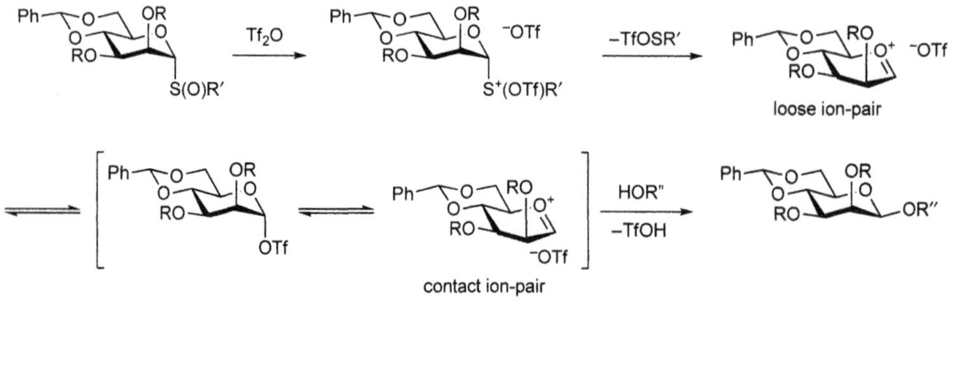

The stereoselectivity of the method is excellent, with any 'leakage' to the α-D-anomer presumably arising via the 'loose' ion pair. However, there were various reasons to try to access these glycosyl triflates from substrates other than sulfoxides and, in an early approach, 1-thio-α-D-mannopyranosides were activated by phenylsulfenyl triflate to provide β-D-mannopyranosides.[261] Even this successful method had limitations and has now been superseded by the aforementioned combinations of 1-benzenesulfinylpiperidine or diphenyl sulfoxide and triflic anhydride:[262–265]

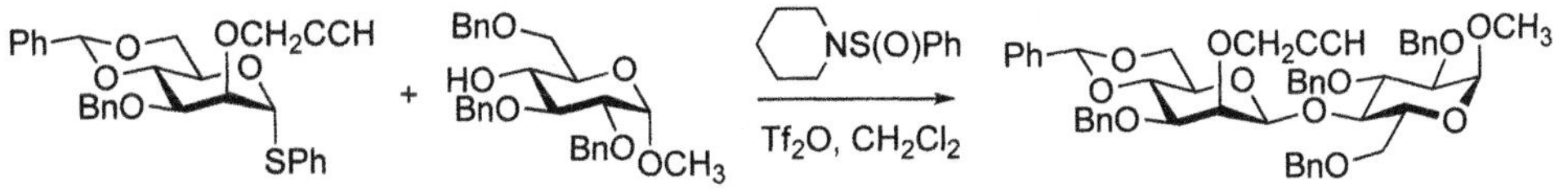

A recent report approaches the synthesis of β-D-mannopyranosides from the hemiacetal, using Gin's dehydrative glycosidation methodology;[266] another uses the ubiquitous TCA approach.[267]

β-D-Glucopyranoside to β-D-mannopyranoside:

In principle, it should be possible to generate a β-D-glucopyranoside by conventional means and, if O2 is differentially protected, invert the stereochemistry by an oxidation–reduction sequence:

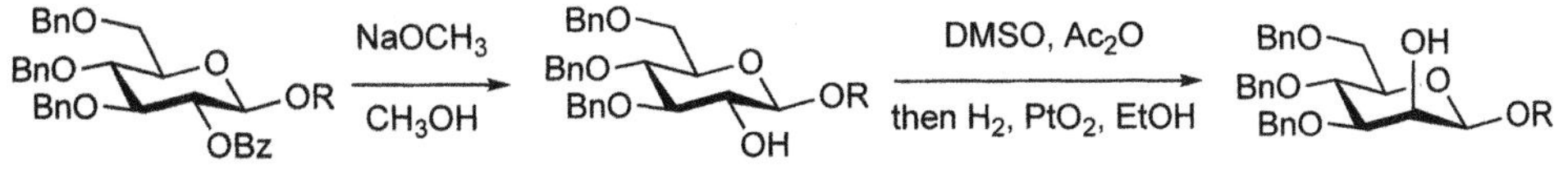

This synthesis of β-D-mannopyranosides was first announced in 1972 and has been used on many occasions, but it obviously suffers from a drawback: the preparation of the 2-*O*-acyl glycosyl donor.[268] A related approach, which inverts the configuration at C2 of the β-D-glucopyranoside by nucleophilic displacement, again suffers a little from the early manipulations in the synthesis of the glycosyl donor:[269–271]

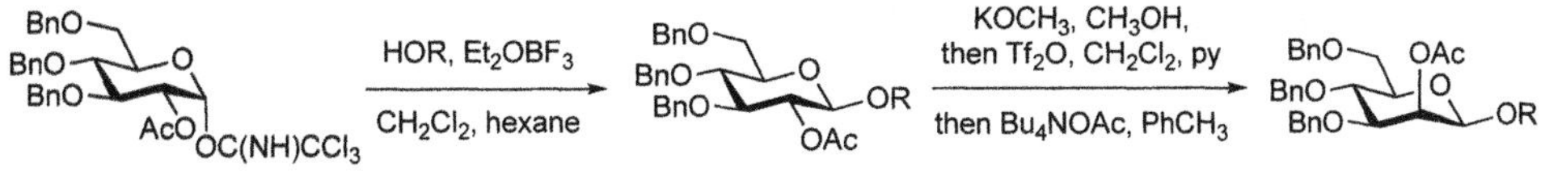

An intramolecular version of this process, which somewhat curiously requires the presence of a 4,6-*O*-benzylidene group, has been developed by Kunz:[272]

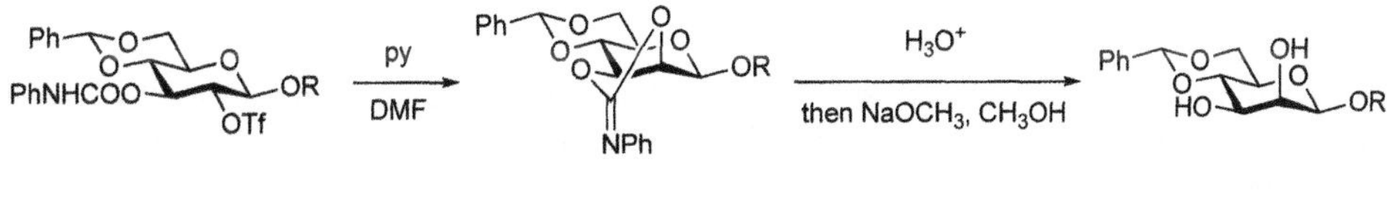

Lichtenthaler has developed a procedure that utilizes glycosyl halides in tandem with a stereoselective reduction for the synthesis of β-D-mannopyranosides:[273]

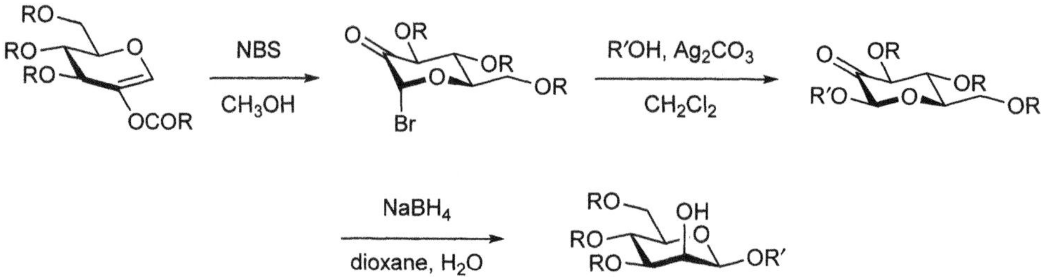

Central to this procedure are the easy preparation of the 2-ulosyl α-D-glycosyl bromide, the β-specific glycosidation (presumably successful owing to the suppression of glycosyl cation formation by the adjacent carbonyl group) and the stereoselective reduction.

Intramolecular aglycon delivery:[e,274] This last method is certainly the most elegant; whether the extra early steps are compensated for by the final, completely stereoselective glycosidation is a moot point. The method has been extended to the synthesis of other 1,2-*cis* linkages and is gradually finding acceptance in the carbohydrate community.[275–278]

First announced by Hindsgaul[279] and later modified by Stork[280] and Ogawa and Ito,[281] the method requires the generation of a D-mannopyranosyl donor with the acceptor *attached* as some sort of acetal at O2; the addition of the appropriate promoter then gives β-D-mannopyranoside:[282]

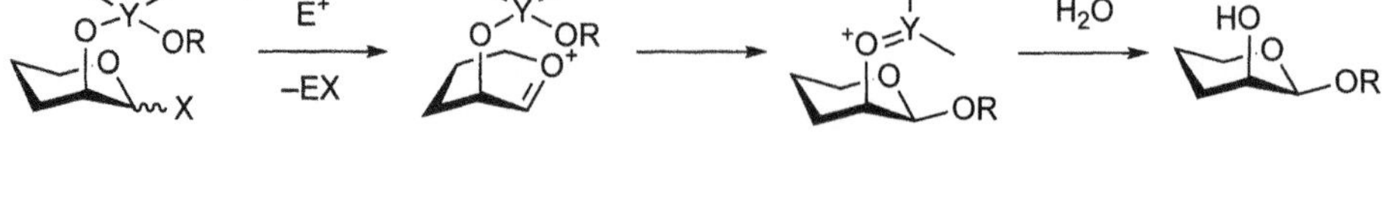

---

[e] To be fair, the orthoester method must be considered the earliest example of intramolecular aglycon delivery; unfortunately, this method generally results in a 1,2-*trans* glycoside, here an α-D-mannopyranoside.

Some details of each modification are presented below:

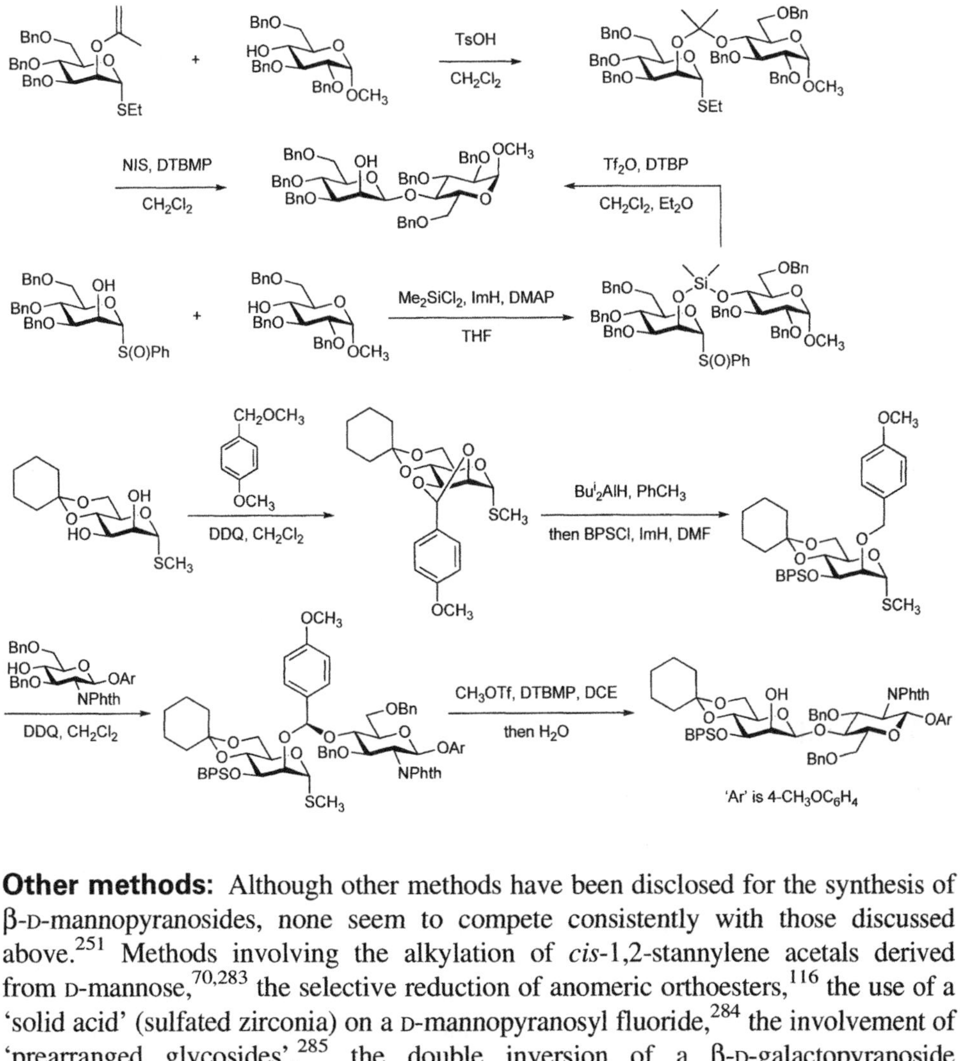

Other methods: Although other methods have been disclosed for the synthesis of β-D-mannopyranosides, none seem to compete consistently with those discussed above.[251] Methods involving the alkylation of *cis*-1,2-stannylene acetals derived from D-mannose,[70,283] the selective reduction of anomeric orthoesters,[116] the use of a 'solid acid' (sulfated zirconia) on a D-mannopyranosyl fluoride,[284] the involvement of 'prearranged glycosides',[285] the double inversion of a β-D-galactopyranoside ditriflate,[286] D-mannopyranosyl phosphites,[287,288] a 6-nitro-2-benzothiazolyl α-D-mannopyranoside[289] and an α-D-mannopyranosyl 4-pentenoate[290] show promise for the synthesis of β-D-mannopyranosides.

One of the rare instances of the Mitsunobu reaction for the synthesis of glycosides was announced by Garegg. The method is best for the preparation of aryl β-D-mannopyranosides, mainly owing to the anomeric purity of the starting hemiacetal, the general 'inversion of configuration' associated with the process and the appropriate acidity of the glycosyl acceptor:[291,292]

The method is general for a whole range of substituents on the aromatic ring.[293]

β-Rhamnopyranosides (1,2-*cis*)

β-L-Rhamnopyranosides (6-deoxy-β-L-mannopyranosides) are common components of bacterial capsular polysaccharides, and their synthesis is relatively straightforward, starting with readily available L-rhamnose and employing one of the methods[283,286,294,295] discussed in this section for the synthesis of a β-mannopyranoside. β-D-Rhamnopyranosides have been encountered more frequently of late, but because of the general unavailability of D-rhamnose, the efficient synthesis of such molecules has generally proceeded via the corresponding β-D-mannopyranoside, with eventual deoxygenation at C6; some recent improvements have been noted.[264,296–298]

2-Acetamido-2-deoxy Glycosides[299–302]

2-Acetamido-2-deoxy glycosides, especially those of the D-*gluco* and D-*galacto* configuration, are common components of many biopolymers (chitin, chondroitin sulfate) and glycoconjugates (glycoproteins, proteoglycans, glycolipids). In fact, over the past two decades, the role of 2-acetamido-2-deoxy-D-glucopyranose (GlcNAc) in many important biological events has become increasingly apparent (see Chapter 11).[303–305]

The presence of an acetamido group changes the polarity and hydrogen-bonding potential of a molecule; also, a 2-acetamido group is capable of participating in events at the anomeric carbon (even to the extent of forming a somewhat unreactive oxazoline). Together, these two features make the synthesis of 2-acetamido-2-deoxy glycosides unique and challenging.

The most obvious approach to the synthesis of 2-acetamido-2-deoxy glycosides utilizes a glycosyl donor with the acetamido group already in place. Indeed, this procedure works well for the synthesis of 1,2-*trans* glycosides derived from reasonably reactive alcohols; the intermediate oxazoline is rather stable and, naturally, controls the stereoselectivity of the event:[306]

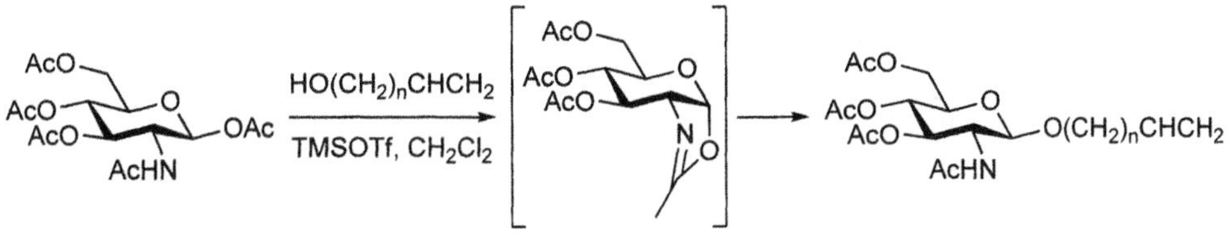

Glycosyl chlorides and glycosyl TCAs may also be used as the donor.[67] On some occasions, the oxazoline may first be prepared and then used in the subsequent glycosidation;[307] particularly striking results have been obtained in enzyme-mediated processes.[270,308] An interesting and easily accessible furanosyl oxazoline provides direct access to simple alkyl 2-acetamido-2-deoxy glycopyranosides:[309]

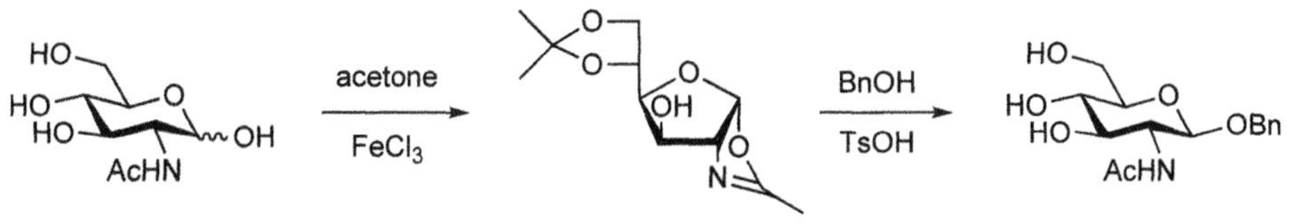

The trichloroethoxycarbonyl and DTPM groups have proven useful for the protection of the primary amine during the glycosidation step.[310,311]

Much better results are obtained when *N,N*-diacyl glycosyl donors are used;[312] the only drawback is that a deprotection (and sometimes subsequent acetylation) step is now necessary. One of the most common diacyl donors is based on phthalimide:[67,313–315]

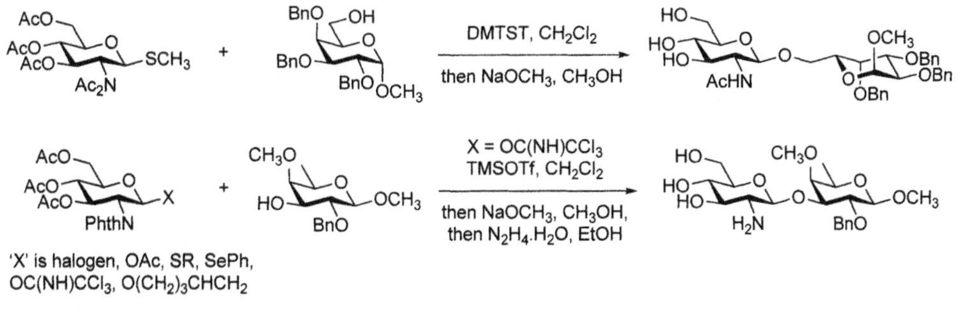

'X' is halogen, OAc, SR, SePh,
OC(NH)CCl₃, O(CH₂)₃CHCH₂

Again, the glycosidations generally produce only the 1,2-*trans* products. Although participation by the diacetylamino group at the anomeric centre is perhaps to be expected, it is a moot point whether the same occurs with the

References start on page 191

phthalimido group (the planar aromatic system is generally orthogonal to the D-glycopyranose ring and thus blocks the α-face).[316]

2-Azido-2-deoxy glycosyl donors are easily accessible from the appropriate glycal by 'azido nitration', 'azido selenation' or direct 'diazo transfer' onto the natural 2-amino-2-deoxy sugar itself[317] and offer convenient methods for the stereoselective synthesis of 2-acetamido-2-deoxy-D-glycopyranosides:[67,300,318–320]

Other 2-azido-2-deoxy glycosyl donors have been used for related syntheses.[321,322]

Glycals have also been used for the synthesis of 2-acetamido-2-deoxy-β-D-glycopyranosides, or their direct precursors;[238,323] 2-nitroglycals add another dimension to the synthesis of such molecules:[324]

Needless to say, both phosphites and phosphates have been suggested as useful donors in the synthesis of 2-acetamido-2-deoxy glycosides.[325,326]

An interesting development of late has been the introduction of oxazolidinones (*trans*-fused 2,3-carbamates derived from 2-amino-2-deoxy sugars) as glycosyl donors. These molecules are easily prepared and, with a judicious choice of donor and reagents, can give rise to both the normal β- and less common α-glycosides.[327–330]

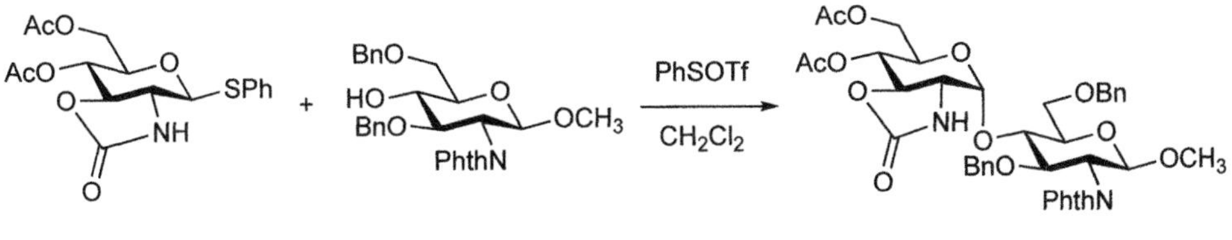

A corollary of this novel cyclic carbamate is that, when located in a glycosyl acceptor, it imparts an increased reactivity to O4;[331] others have commented on this phenomenon.[332]

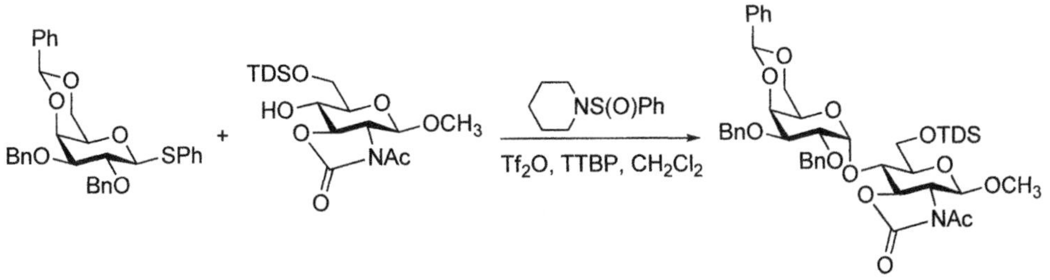

Finally, as with the synthesis of β-D-mannopyranosides, special methods had to be developed for their 2-acetamido-2-deoxy counterparts:[333–335]

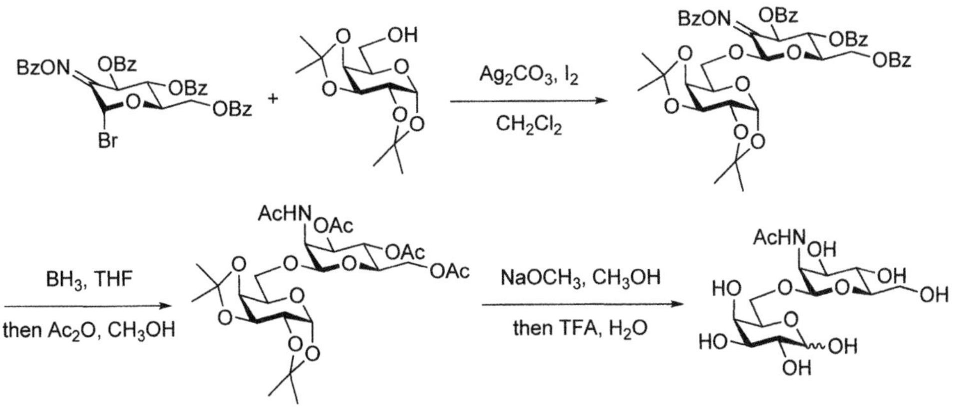

A philosophical comment: everyone is willing to work with D-glucose; some gentle encouragement will see good progress with D-glucosamine; a brave (and wealthy!) person ventures into the 'delights' of synthesis with D-galactosamine.[336]

2-Deoxy Glycosides[337–340]

2-Deoxy glycosides are often found in the oligosaccharide portion of natural products derived from plant and bacterial sources. The lack of a hydroxyl group at C2 makes the associated glycoside more amenable to hydrolysis; also, this absence gives rise to problems in the stereoselectivity of formation of the glycosidic linkage. It is fairly obvious that any synthesis of 2-deoxy glycosides can be approached in just two ways – the donor can be either a 2-deoxy sugar in its own right or can contain a functionality at C2 that, once used to control the stereoselectivity at the anomeric carbon, is easily removed.

Any method of synthesis of a 2-deoxy glycoside that simply delivers a hydrogen atom to C2 of a glycal has the potential for poor stereoselectivity;[341] also, an unwanted by-product can be the 2,3-ene:[342]

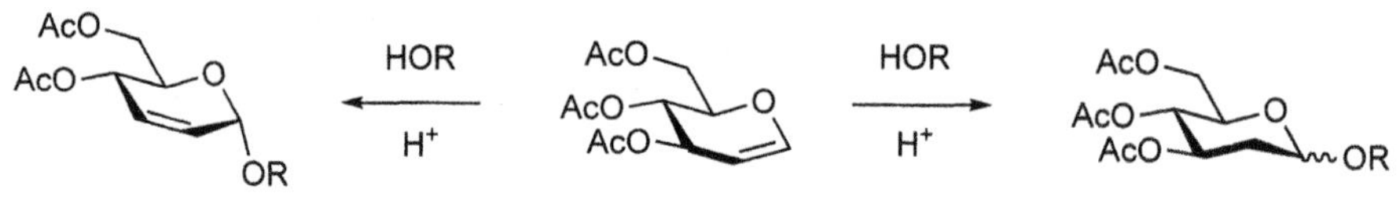

However, there have been some notable successes, with the anomeric effect usually being responsible for the production of the more stable product:[343]

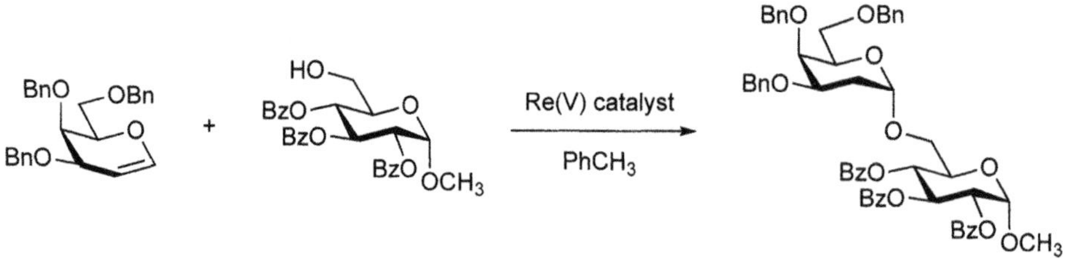

Other 2-deoxy glycosyl donors such as thioglycosides,[344] phosphonodithioates,[345,346] TCAs,[347,348] fluorides,[30] tetrazoles and phosphites[349] have been used successfully for the synthesis of a variety of glycosides. The main product is generally the more stable α-D-glycoside, but a careful choice of the donor or the solvent can change the outcome to produce just the β-anomer:[139,349–351]

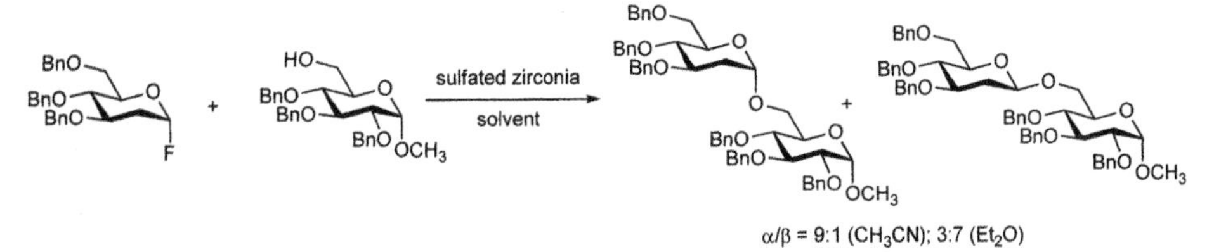

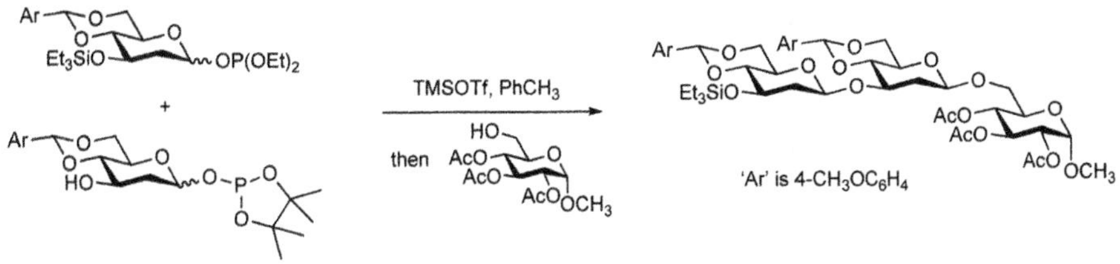

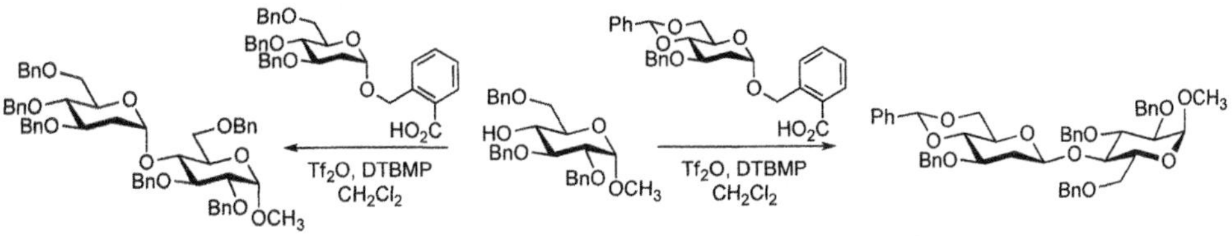

$\alpha/\beta = 9:1$ (CH$_3$CN); 3:7 (Et$_2$O)

'Ar' is 4-CH$_3$OC$_6$H$_4$

For the alternative approach, where the glycosyl donor harbours a temporary function at C2, the anomeric stereoselectivity is generally improved; 2-deoxy-2-iodo glycosyl donors are particularly attractive:[292,352-361]

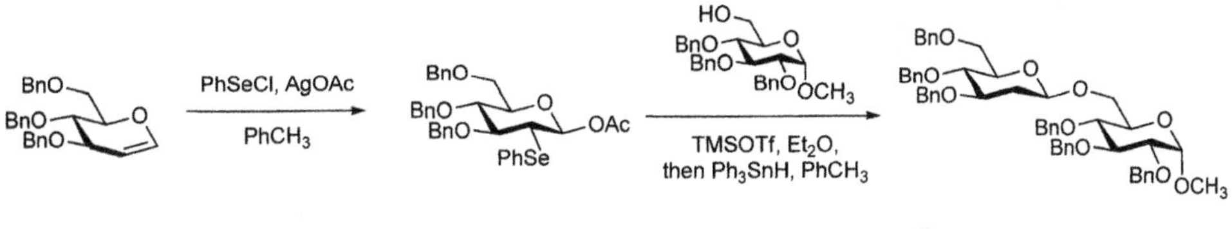

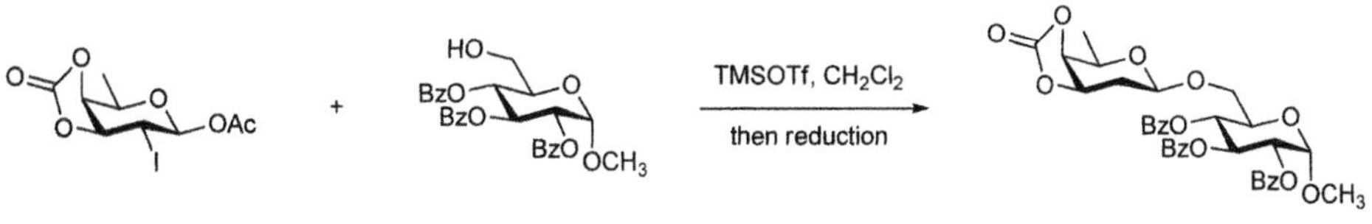

Nicolaou has reported a very neat reaction of a thioglycoside, powered by diethylaminosulfur trifluoride and leading to a glycosyl fluoride, suitable for the synthesis of both 2-deoxy α- and β-glycosides:[362]

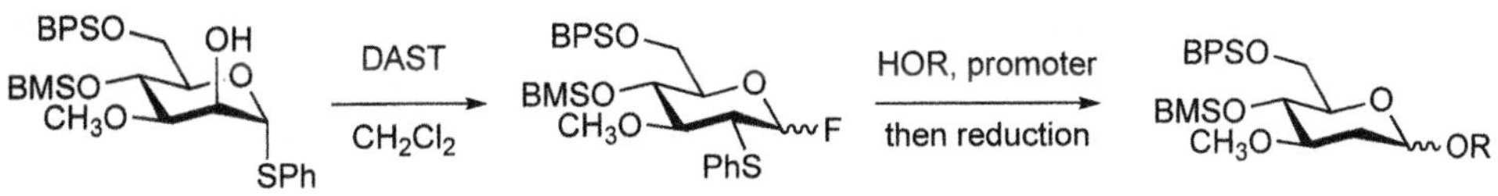

van Boom, in a process that again involves an interesting rearrangement, devised a direct synthesis of the same types of glycosides:[363]

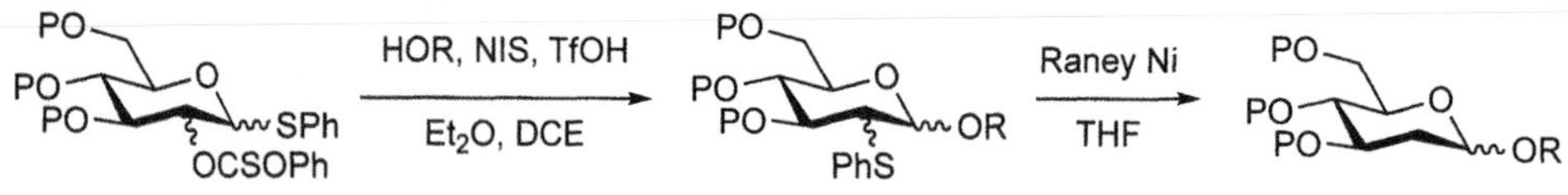

'P' is a protecting group

Related chemistry can lead to an efficient synthesis of 2-deoxy furanosides.[364] Roush has also made many contributions to the stereoselective synthesis of 2-deoxy glycosides.[365–367]

Sialosides[368–370]

The past two decades have highlighted the importance of sialic acids ('2-keto-3-deoxy-nononic acids') in many biological processes, e.g. the sialylation of glycoconjugates, the role of sialyl Lewis^x in inflammation and the involvement of N-acetylneuraminic acid (5-acetamido-3,5-dideoxy-D-*glycero*-D-*galacto*-non-2-ulopyranosonic acid) in infection by influenza virus.[371]

3-deoxy-D-*glycero*-β-D-*galacto*-non-2-ulopyranosonic acid (β-Kdn)   β-Neu5Ac

Most of the linkages to N-acetylneuraminic acid are found with the α-configuration, and so it comes as no surprise to see the majority of the synthetic effort in this direction. We have already discussed the synthesis of a sialoside (utilizing a glycosyl ester as a donor), and some of the common donors discussed earlier in this chapter are also employed, e.g. 2-thio glycosides:[372–374]

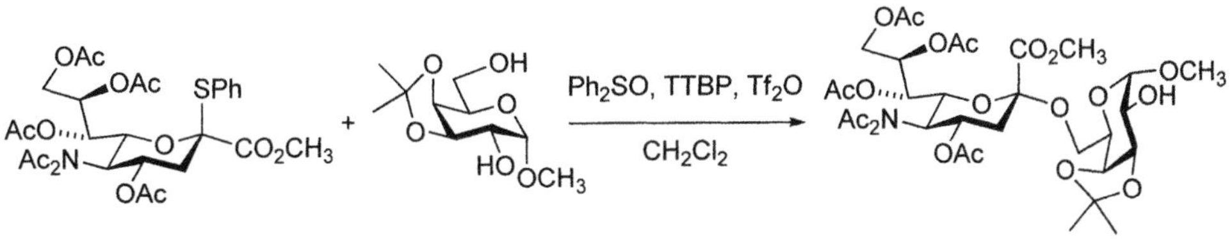

Somewhat surprisingly, a TCA is not a useful donor for sialoside synthesis.[67]

The challenges in any synthesis of an α-sialoside are fairly obvious: the presence of an electron-withdrawing group at C1 slows the whole process, there is no substituent at C3 to control the stereoselectivity of the glycosidation (reminiscent of the discussion above on 2-deoxy sugars), the hydrogen atoms on C3 are always available to participate in unwanted 2,3-ene formation and the gain from any anomeric effect has to be shared between the two substituents (at C2). A judicious choice of solvent (acetonitrile) maximizes the amount of α-anomer,[375] but a concerted effort is still needed to develop better methods for the synthesis of α-sialosides:[376,377]

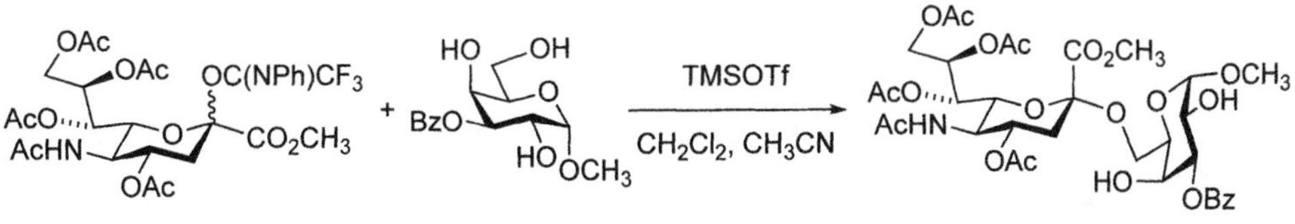

Furanosides[378–380]

The discussion in this chapter has been restricted mainly to the synthesis of pyranosides. However, furanosides (as distinct from nucleosides that are generally considered to be *N*-glycosides) are commonly found in nature with many biologically significant roles, e.g. in the polysaccharides of bacteria, parasites, plants and fungi. An arabinogalactan (a polymer of arabinofuranose and galactofuranose) and a lipoarabinomannan (a polymer of arabinofuranose and mannopyranose), major components of the mycobacterial cell wall, are targets for the treatment of diseases such as tuberculosis. It has been necessary to develop methods for the synthesis of furanosides, both to provide fragment oligosaccharides (of the polysaccharide) and to develop inhibitors that interfere with the assembly of the oligosaccharide.

Hemiacetals, esters, halides, TCAs and 1-thio sugars[381,382] have all been used for the synthesis of furanosides. A recent and effective method uses a 2,3-anhydro-furanose,[383–385] another conformationally locked donor:[386]

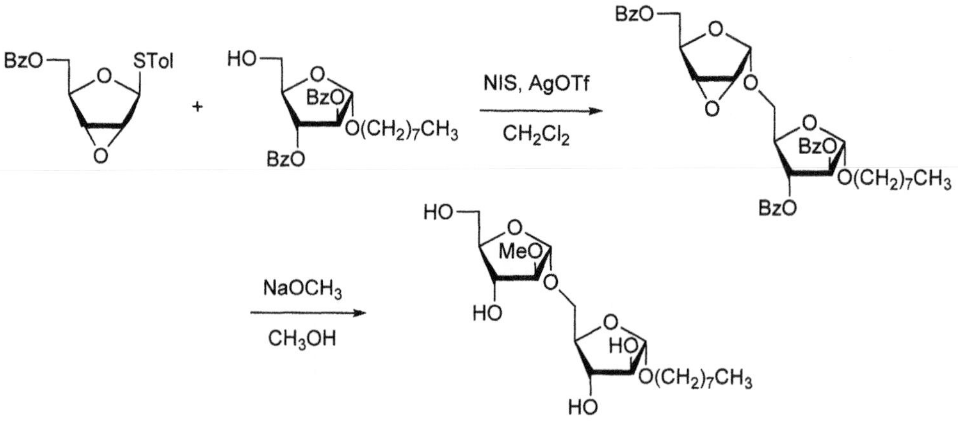

Miscellaneous Methods[387]

The glycosidic linkage has exerted a certain power over chemists for centuries; it has never yielded to just one method of synthesis and has offered a sort of 'fatal attraction' for improvements in its construction. So, in addition to the tried-and-tested procedures discussed so far, there exists an array of other methods that are either in their infancy or, as yet, have not gained general acceptance.

Alkenyl glycosides: We have encountered alkenyl glycosides in our discussion of the 'latent/active' concept as applied to glycosyl donors. In fact, there is a range of such donors that has been used successfully in the synthesis of glycosides:[7,45,81,388–390]

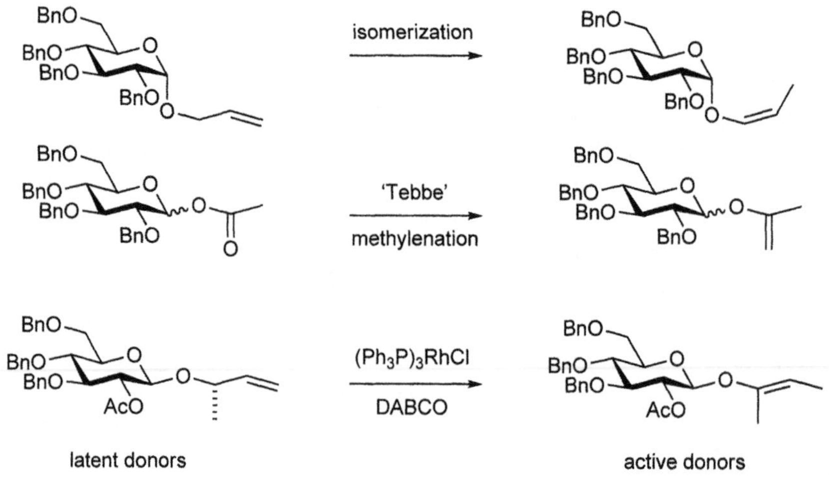

latent donors   active donors

Two interesting variations on the same theme are a glycosyl isopropenyl carbonate, and an isopropenyl ether on the acceptor:[7,81,388]

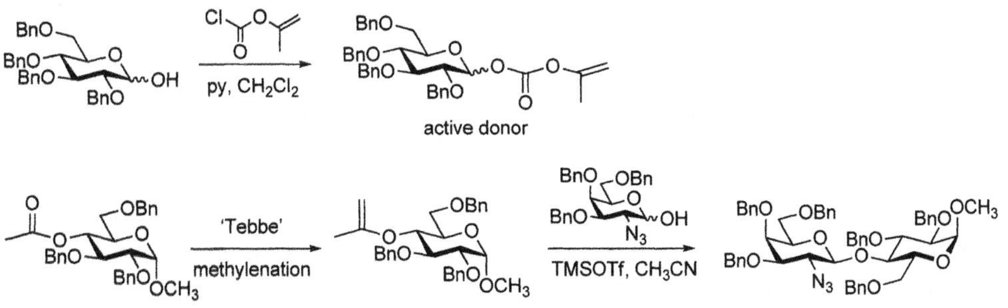

active donor

'Tebbe'
methylenation

TMSOTf, CH$_3$CN

All of the above processes operate on much the same principles:

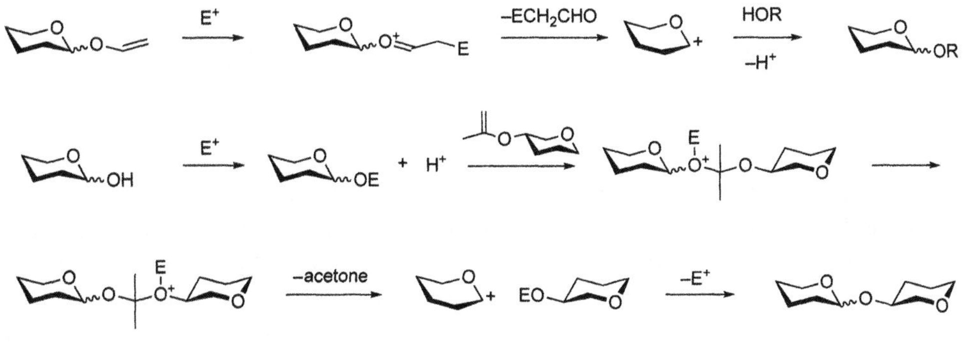

Remote activation: 'Remote activation', as applied to carbohydrates, is essentially the brainchild of Hanessian.[391,392] In essence, many methods of forming the glycosidic bond involve direct activation of the atom attached to the anomeric carbon:

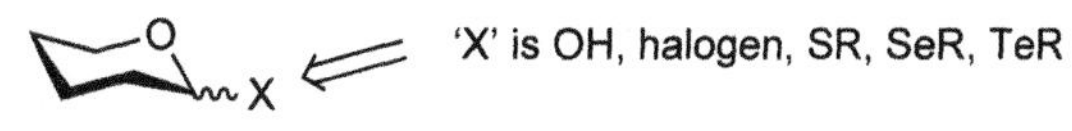

'X' is OH, halogen, SR, SeR, TeR

Other methods, however, rely on the activation of an atom or a group that is remote from the anomeric carbon but is attached to it via another atom or atoms:

References start on page 191

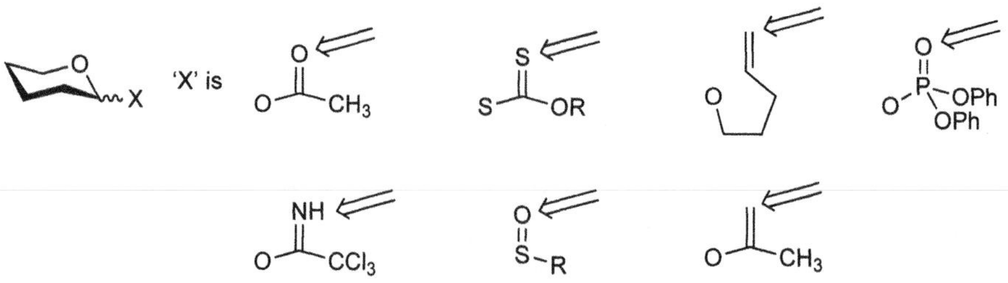

Although this distinction with these two groups of examples may seem somewhat forced and flimsy, the concept of 'remote activation' does take on some significance when one attempts to develop a glycosyl donor *that bears no protecting groups* (as exists in nature).

The first generation of such molecules requiring 'remote activation' were 2-pyridylthio glycosides:[393]

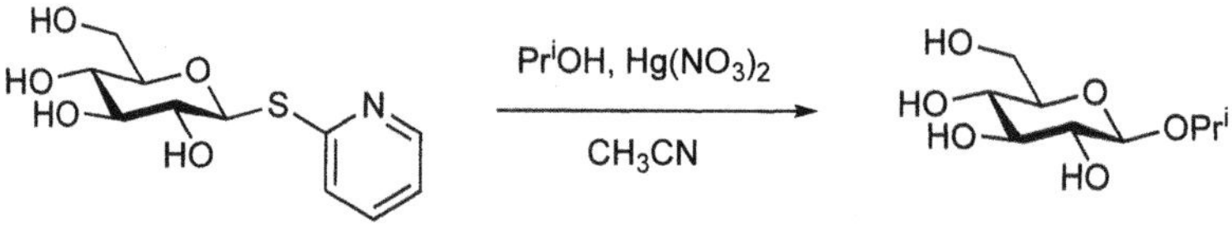

Although mixtures of anomers were generally obtained, the reaction was rapid, presumably initiated by complexation between the metal ion and the bidentate aglycon. Later, the method was extended to include such donors as 3-methoxy-2-pyridyl (MOP) glycosides that, although activated by just a 'catalytic' amount of methyl triflate (and thereafter the liberated triflic acid), still required the use of a large excess of the glycosyl acceptor:

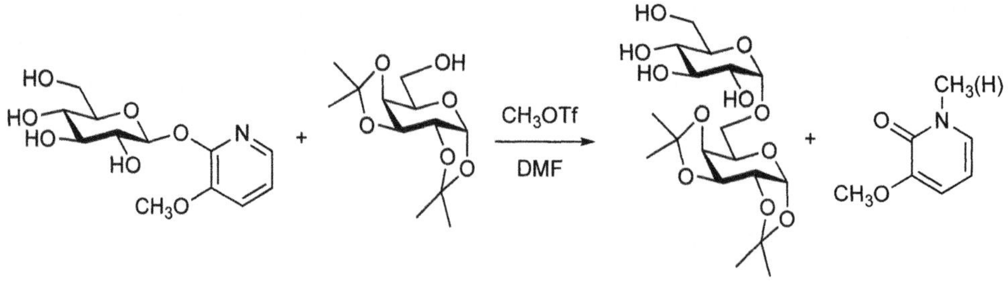

Generally, the S_N2 nature of the process guaranteed the preponderance of 1,2-*cis* glycoside. Further studies by Hanessian showed that *protected* MOP donors, as with most other sorts of glycosyl donors, could be used effectively for the glycosidation of *equimolar* amounts of acceptor alcohols – the ideal promoter was copper(II) triflate. The stereochemical outcome of the various glycosidations depended, predictably, on the nature of the protecting group at O2 and on the solvent.

A further development in the 'remote activation' concept was the use of *O*-glycosyl *S*-(2-pyridyl) thiocarbonates (TOPCAT) as glycosyl donors:[391,392]

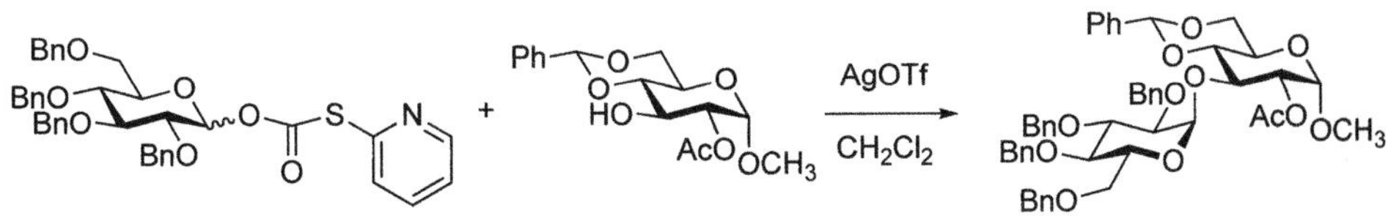

The unique promoter, silver(I) triflate, again probably operating through the formation of some sort of coordinated species, allows for the chemoselective activation of TOPCAT over MOP [copper(II) triflate] groups.

Hanessian has applied the 'MOP glycoside' approach to the synthesis of glycosyl phosphates and their nucleoside derivatives,[394] and Kobayashi has expanded the use of 'glycosyl 2-pyridinecarboxylates' for the preparation of various disaccharides:[395]

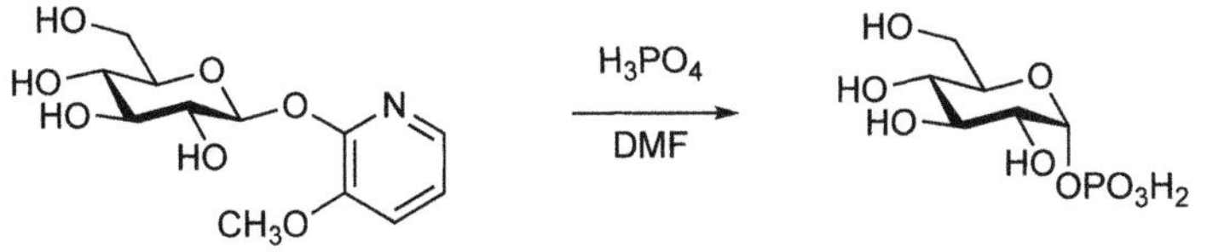

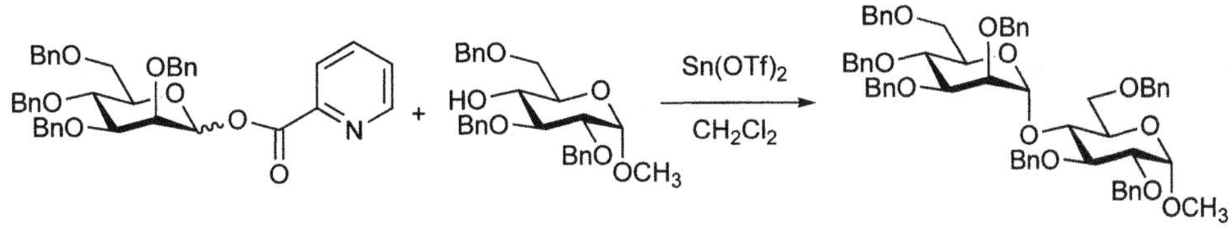

Finally, mention should be made of 2-haloethyl 1-thioglycosides (Redlich),[396] glycosyl disulfides (Davis),[397] glycosyl sulfimides (Rollin),[398] SBox (*S*-benzoxazolyl) and STaz (*S*-thiazolinyl) glycosides (Demchenko),[158,399–402] (2-carboxyphenyl)methyl glycosides (Kim),[403,404] DISAL (methyl 3,5-dinitrosalicylate) glycosides (Jensen)[405,406] and 6-nitro-2-benzothiazolyl glycosides (Mukaiyama)[289] as useful donors for the future. There has been a suggestion that sonication of reaction mixtures vastly improves the outcome of many glycosidation reactions, and even some of the simpler transformations discussed in previous chapters.[407]

References start on page 191

C-Glycosides[408–419]

We have spent a great deal of time discussing glycosides and, *inter alia*, *N*- and *S*-glycosides (thioglycosides):

These three classes of compounds are actually acetals, hemiaminals and thioacetals, all relatively stable to the action of bases but generally unstable to acid, acid and thiophilic reagents, respectively. Another class of compound that we have neglected until now arises from the replacement of the exocyclic oxygen of the acetal by carbon:

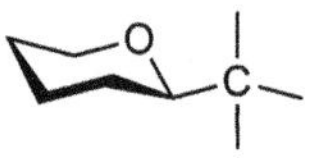

These compounds are known as *C*-glycosides (if you are a carbohydrate chemist); others would view them as tetrahydropyrans, simple cyclic ethers. Consequently, they are inert compounds, essentially stable to the action of both acids and bases.

Why, then, are *C*-glycosides of any interest? First, many natural products contain a *C*-glycosidic linkage: aquayamycin is a *C*-aryl glycoside produced by *Strepto-myces misawanensis* and possesses antibiotic activity,[420] showdomycin is a *C*-nucleoside that possesses antibacterial and antitumour properties,[408] and palytoxin and maitotoxin[421] are marine metabolites that are two of the most toxic chemicals known:

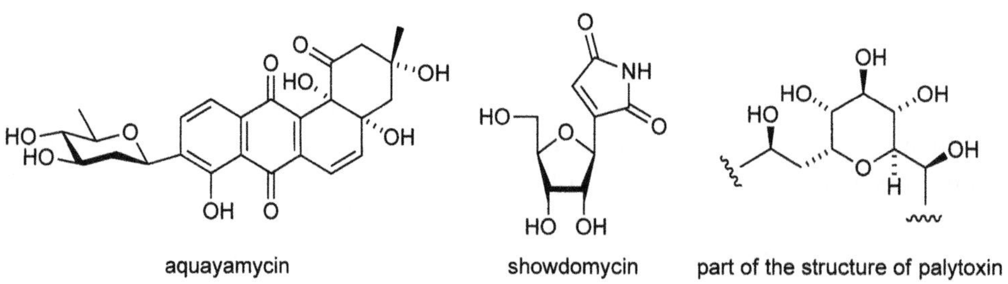

aquayamycin   showdomycin   part of the structure of palytoxin

Second, chemists have long held the notion that *C*-glycosides would be stable 'mimics' of glycosides and so would be potential candidates for the design of molecules with expected biological activity but possessing none of the lability of a glycoside.[413,422] Whatever the reason, there has been an enormous drive over the past two decades for the synthesis of *C*-glycosides.

Whereas the synthesis of glycosides (C–O bond construction), by and large, is restricted to methods that develop a partial positive charge at the anomeric carbon atom, such is not the case with the synthesis of *C*-glycosides; the construction of a C–C bond allows more flexibility to the method. The three main routes followed involve both ionic and free radical processes:

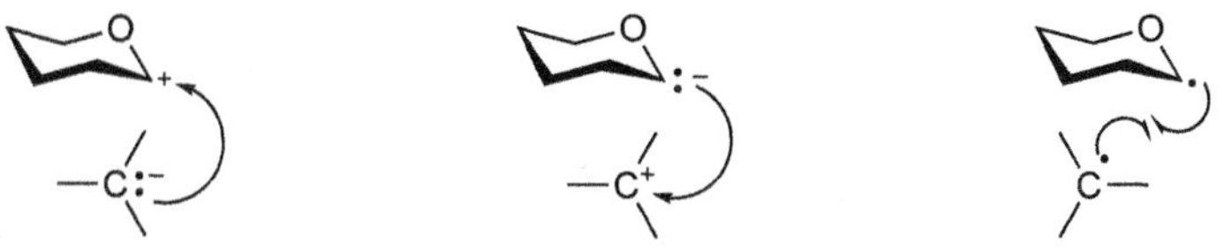

Again, stereoselectivity of bond formation at the anomeric carbon (α- or β-) is of prime importance.

The addition of carbanions to anomeric electrophiles:[423] There is a plethora of methods for the addition of a carbanion to an anomeric electrophile, and only a few will be mentioned here. The substitution by a nucleophilic reagent of a glycosyl halide is one of the older methods, but yields are often low, especially with ester protecting groups in the sugar:

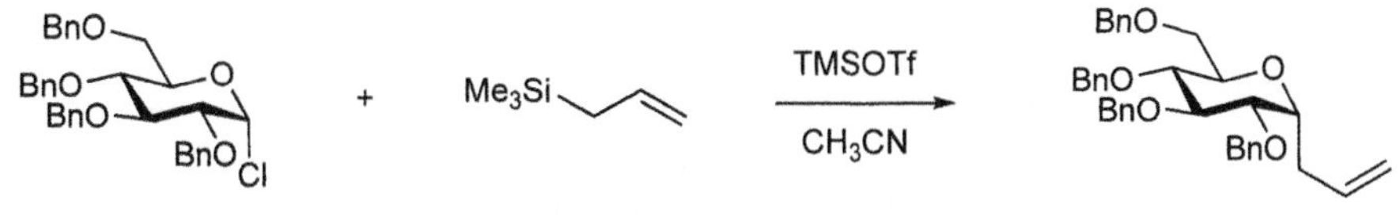

Of general usage is the method developed by Kishi, the addition of an organometallic reagent to an aldonolactone:[424,425]

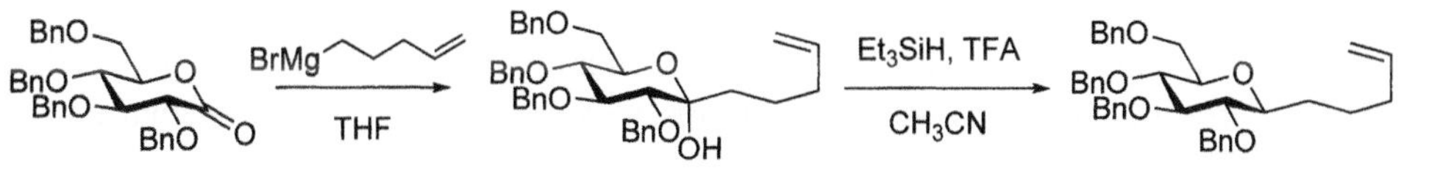

Epoxides and glycals are useful alkylating agents of nucleophilic species:[424,426–428]

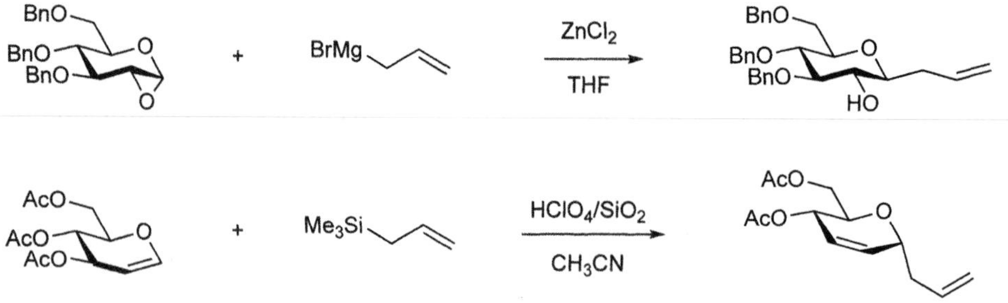

Hemiacetals, upon chain extension with phosphoranes, phosphonates or organometallic species, yield alkenes that cyclize either spontaneously or by treatment with bases or electrophiles:

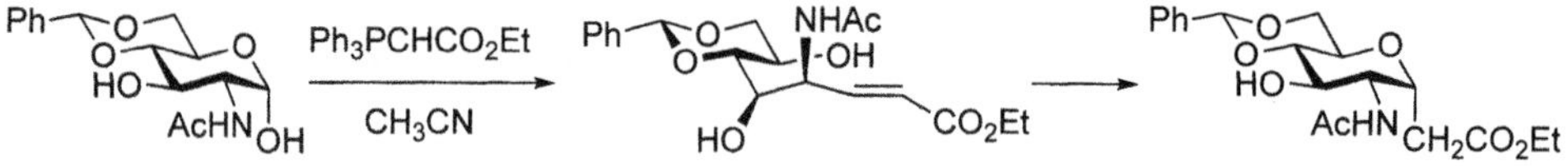

Formyl tetra-O-benzyl-β-D-C-glucopyranoside is a synthetically useful intermediate that is available in large amounts:[429]

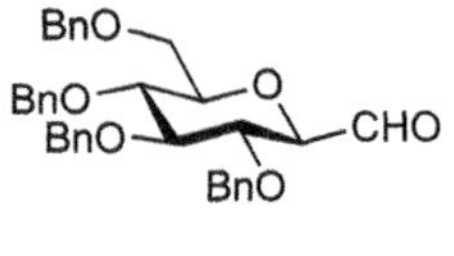

The addition of electrophiles to anomeric carbanions: Again, many methods exist for the addition of an electrophile to an anomeric carbanion, but most typically utilize the presence of an electron-withdrawing group at the anomeric carbon to aid in the generation and stabilization of the negative charge:[430]

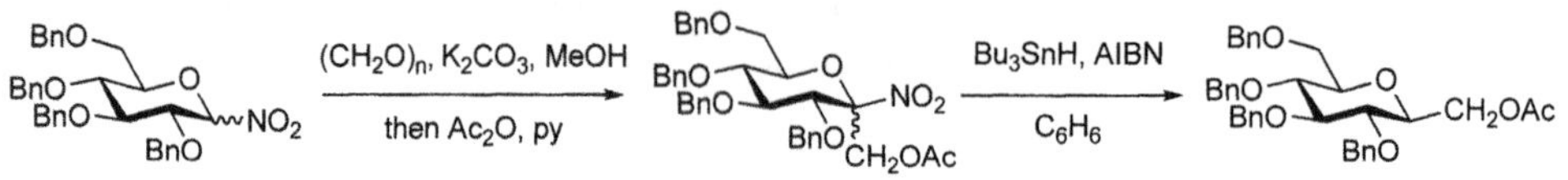

Glycosyl sulfones are particularly useful donors for the synthesis of C-glycosides, offering versatility not found with other groups; they are often activated by samarium(II) iodide:[431–433]

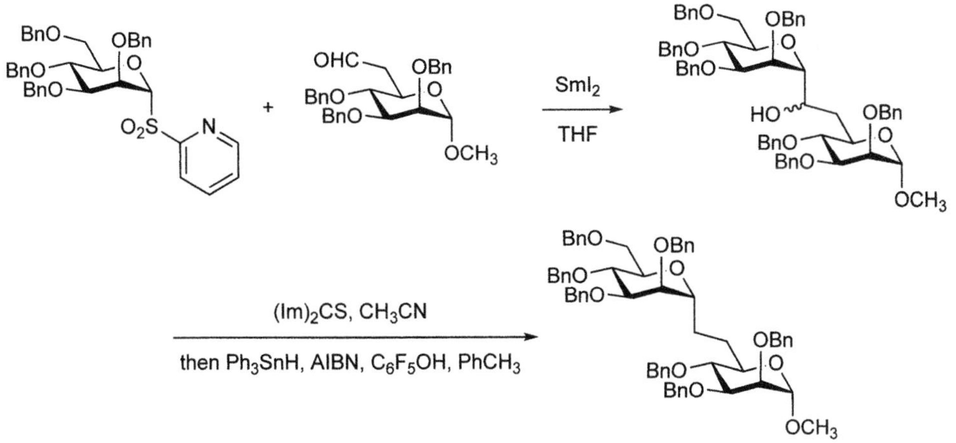

The Ramberg–Bäcklund reaction also makes use of glycosyl sulfones:[434]

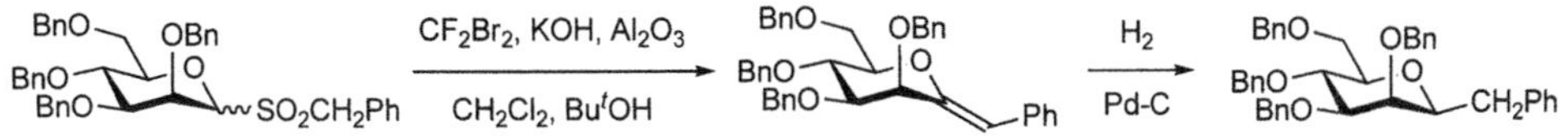

Very often, when the anomeric carbanion is not stabilized by an electron-withdrawing group, the presence of an oxygen atom at C2 will cause a β-elimination; this can either be avoided or put to good use:[435–437]

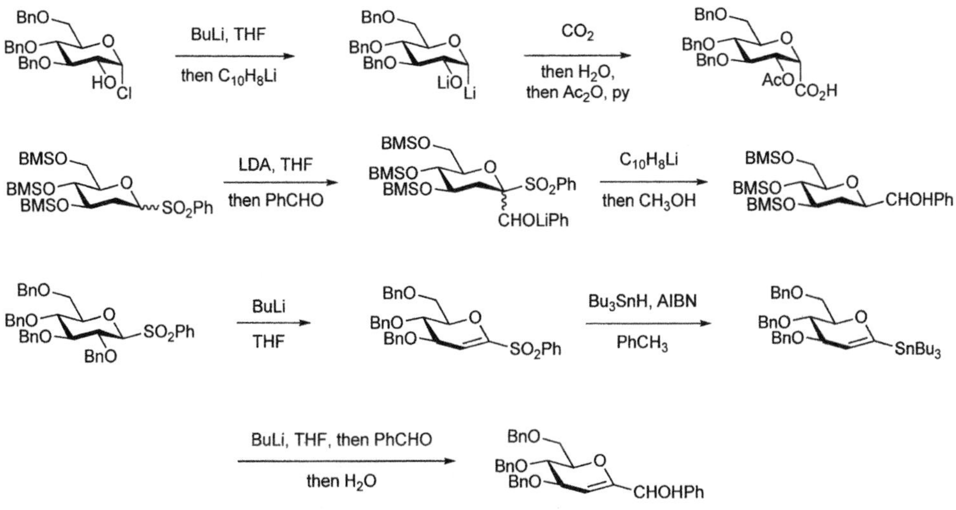

Glycosyl radicals:[438–442] In the early 1980s it was first noticed that anomeric radicals add to electron-deficient alkenes to give axial C-glycosides:

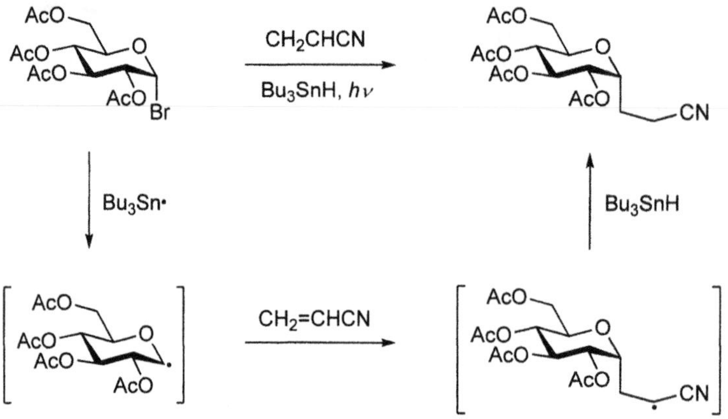

Although the process can sometimes be plagued by side reactions of the highly reactive, intermediate glycosyl radical (reduction, hydrogen atom abstraction), it is this very intermediate that holds the key to the high stereoselectivity usually observed:

favoured ($B_{2,5}$)   disfavoured ($^4C_1$)

Many modifications have been made to the general process in the ensuing 20 years:[443–446]

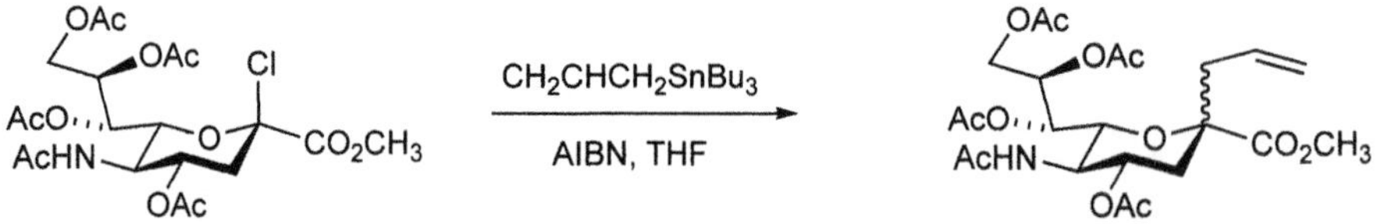

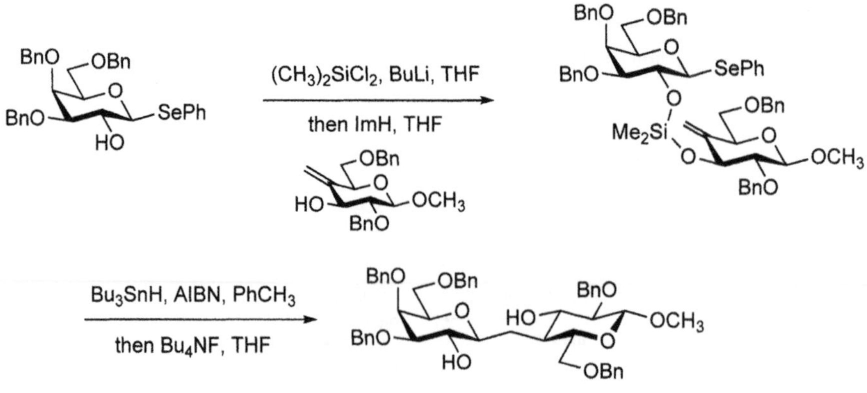

Miscellaneous: Various other methods exist for the synthesis of *C*-glycosides and include substrates such as glycosyl diazirines,[447–448] glycosenes[449] and telluroglycosides;[450] some of the methods employing cross-[451] or ring-closing-[452] metathesis are truly impressive.

References

1. Overend, W.G. (1972). In *The Carbohydrates* (W. Pigman and D. Horton, eds) Vol. IA, p. 279. Academic Press.
2. Bochkov, A.F. and Zaikov, G.E. (1979). *Chemistry of the O-Glycosidic Bond: Formation and Cleavage.* Pergamon Press.
3. Paulsen, H. (1982). *Angew. Chem. Int. Ed. Engl.*, **21**, 155.
4. Paulsen, H. (1984). *Chem. Soc. Rev.*, **13**, 15.
5. Schmidt, R.R. (1986). *Angew. Chem. Int. Ed. Engl.*, **25**, 212.
6. Schmidt, R.R. (1989). *Pure Appl. Chem.*, **61**, 1257.
7. Sinaÿ, P. (1991). *Pure Appl. Chem.*, **63**, 519.
8. Schmidt, R.R. (1991). In *Comprehensive Organic Synthesis* (B.M. Trost and I. Fleming, eds) Vol. 6, p. 33. Pergamon.
9. Lockhoff, O. (1992). In *Methoden der Organischen Chemie* (Houben-Weyl) E14a/3, p. 621. Thieme.
10. Darcy, R. and McCarthy, K. (1993). In *Rodd's Chemistry of Carbon Compounds* (M. Sainsbury, ed.) 2nd edn, 2nd supplements, Vol IE/F/G, p. 437. Elsevier.
11. Toshima, K. and Tatsuta, K. (1993). *Chem. Rev.*, **93**, 1503.
12. Barresi, F. and Hindsgaul, O. (1995). In *Modern Synthetic Methods* (B. Ernst and C. Leumann, eds) p. 281. Verlag Helvetica Chimica Acta.
13. Boons, G.-J. (1996). *Contemp. Org. Synth.*, **3**, 173.
14. Boons, G.-J. (1996). *Tetrahedron*, **52**, 1095.
15. Boons, G.-J. (1996). *Drug Discovery Today*, **1**, 331.
16. Paulsen, H. (1996). In *Modern Methods in Carbohydrate Synthesis* (S.H. Khan and R.A. O'Neill, eds) p. 1. Harwood Academic.
17. Whitfield, D.M. and Douglas, S.P. (1996). *Glycoconjugate J.*, **13**, 5.
18. von Rybinski, W. and Hill, K. (1998). *Angew. Chem. Int. Ed.*, **37**, 1328.
19. Schmidt, R.R., Castro-Palomino, J.-C. and Retz, O. (1999). *Pure Appl. Chem.*, **71**, 729.
20. Davis, B.G., (2000). *J. Chem. Soc., Perkin Trans. 1*, 2137.
21. Nicolaou, K.C. and Mitchell, H.J. (2001). *Angew. Chem. Int. Ed.*, **40**, 1576.
22. Jensen, K.J. (2002). *J. Chem. Soc., Perkin Trans. 1*, 2219.
23. Demchenko, A.V. (2003). *Synlett*, 1225.
24. Demchenko, A.V. (2003). *Curr. Org. Chem.*, **7**, 35.
25. Garegg, P.J. (2004). *Adv. Carbohydr. Chem. Biochem.*, **59**, 70.
26. Fügedi, P. (2006). In *The Organic Chemistry of Sugars* (D.E. Levy and P. Fügedi, eds) p. 89. CRC Press.
27. Fischer, E. (1893). *Ber. Dtsch. Chem. Ges.*, **26**, 2400.
28. Chandrasekhar, S. (2005). *ARKIVOC*, 37.
29. Wulff, G., Schröder, U. and Wichelhaus, J. (1979). *Carbohydr. Res.*, **72**, 280.
30. Jünnemann, J., Lundt, I. and Thiem, J. (1991). *Liebigs Ann. Chem.*, 759.
31. Ionescu, A.R., Whitfield, D.M. and Zgierski, M.Z. (2007). *Carbohydr. Res.*, **342**, 2793.

32. Fraser-Reid, B., Wu, Z., Udodong, U.E. and Ottosson, H. (1990). *J. Org. Chem.*, **55**, 6068.
33. Zhang, Z., Ollmann, I.R., Ye, X.-S., Wischnat, R., Baasov, T. and Wong, C.-H. (1999). *J. Am. Chem. Soc.*, **121**, 734.
34. Pedersen, C.M., Nordstrøm, L.U. and Bols, M. (2007). *J. Am. Chem. Soc.*, **129**, 9222.
35. Jensen, H.H., Pedersen, C.M., and Bols, M. (2007). *Chem. Eur. J.*, **13**, 7576.
36. Fraser-Reid, B., Wu, Z., Andrews, C.W., Skowronski, E. and Bowen, J.P. (1991). *J. Am. Chem. Soc.*, **113**, 1434.
37. Jensen, H.H., Nordstrøm, L.U. and Bols, M. (2004). *J. Am. Chem. Soc.*, **126**, 9205.
38. Crich, D. and Vinogradova, O. (2007). *J. Am. Chem. Soc.*, **129**, 11756.
39. Douglas, N.L., Ley, S.V., Lücking, U. and Warriner, S.L. (1998). *J. Chem. Soc., Perkin Trans. 1*, 51.
40. Green, L., Hinzen, B., Ince, S.J., Langer, P., Ley, S.V. and Warriner, S.L. (1998). *Synlett*, 440.
41. Grice, P., Ley, S.V., Pietruszka, J., Priepke, H.W.M. and Walther, E.P.E. (1995). *Synlett*, 781.
42. Chen, L. and Kong, F. (2003). *Tetrahedron Lett.*, **44**, 3691.
43. Zeng, Y. and Kong, F. (2003). *Carbohydr. Res.*, **338**, 843.
44. Cao, S., Hernández-Matéo, F. and Roy, R. (1998). *J. Carbohydr. Chem.*, **17**, 609.
45. Boons, G.-J. and Isles, S. (1994). *Tetrahedron Lett.*, **35**, 3593.
46. Tsvetkov, Y.E., Kitov, P.I., Backinowsky, L.V. and Kochetkov, N.K. (1996). *J. Carbohydr. Chem.*, **15**, 1027.
47. Ziegler, T. (1998). *J. prakt. Chem.*, **340**, 204.
48. Ogawa, T. and Matsui, M. (1976). *Carbohydr. Res.*, **51**, C13.
49. Kanie, O., Ito, Y. and Ogawa, T. (1994). *J. Am. Chem. Soc.*, **116**, 12073.
50. Fraser-Reid, B., López, J.C., Gómez, A.M. and Uriel, C. (2004). *Eur. J. Org. Chem.*, 1387.
51. Jayaprakash, K.N. and Fraser-Reid, B. (2004). *Org. Lett.*, **6**, 4211.
52. Uriel, C., Gómez, A.M., López, J.C. and Fraser-Reid, B. (2005). *J. Carbohydr. Chem.*, **24**, 665.
53. Masamune, S., Choy, W., Petersen, J.S. and Sita, L.R. (1985). *Angew. Chem. Int. Ed. Engl.*, **24**, 1.
54. Spijker, N.M. and van Boeckel, C.A.A. (1991). *Angew. Chem. Int. Ed. Engl.*, **30**, 180.
55. Ziegler, T. and Lemanski, G. (1998). *Eur. J. Org. Chem.*, 163.
56. Cid, M.B., Alfonso, F. and Martín-Lomas, M. (2005). *Chem. Eur. J.*, **11**, 928.
57. Rising, T.W.D.F., Heidecke, C.D. and Fairbanks, A.J. (2007). *Synlett*, 1421.
58. Litjens, R.E.J.N., van den Bos, L.J., Codée, J.D.C., Overkleeft, H.S. and van der Marel, G.A. (2007). *Carbohydr. Res.*, **342**, 419.
59. Fraser-Reid, B., Jayaprakash, K.N., López, J.C., Gómez, A.M. and Uriel, C. (2007). In *Frontiers in Modern Carbohydrate Chemistry* (A.V. Demchenko, ed.) ACS Symposium Series 960, p. 91. Washington, DC: American Chemical Society.
60. Bülow, A., Meyer, T., Olszewski, T.K. and Bols, M. (2004). *Eur. J. Org. Chem.*, 323.
61. Dinkelaar, J., Witte, M.D., van den Bos, L.J., Overkleeft, H.S. and van der Marel, G.A. (2006). *Carbohydr. Res.*, **341**, 1723.
62. Mowery, D.F., Jr. (1963). *Methods Carbohydr. Chem.*, **2**, 328.
63. Capon, B. (1969). *Chem. Rev.*, **69**, 407.
64. Garegg, P.J., Johansson, K.-J., Konradsson, P. and Lindberg, B. (1999). *J. Carbohydr. Chem.*, **18**, 31.
65. Bornaghi, L.F. and Poulsen, S.-A. (2005). *Tetrahedron Lett.*, **46**, 3485.
66. Schmidt, R.R. (1996). In *Modern Methods in Carbohydrate Synthesis* (S.H. Khan and R.A. O'Neill, eds) p. 20. Harwood Academic.
67. Schmidt, R.R. and Kinzy, W. (1994). *Adv. Carbohydr. Chem. Biochem.*, **50**, 21.
68. Koeners, H.J., de Kok, A.J., Romers, C. and van Boom, J.H. (1980). *Recl. Trav. Chim. Pays-Bas*, **99**, 355.
69. Sharma, S.K., Corrales, G. and Penadés, S. (1995). *Tetrahedron Lett.*, **36**, 5627.
70. Srivastava, V.K. and Schuerch, C. (1979). *Tetrahedron Lett.*, 3269.

71. Gin, D. (2002). *J. Carbohydr. Chem.*, **21**, 645.
72. Boebel, T.A. and Gin, D.Y. (2005). *J. Org. Chem.*, **70**, 5818.
73. Koto, S., Shinoda, Y., Hirooka, M., Sekino, A., Ishizumi, S., Koma, M., Matuura, C. and Sakata, N. (2003). *Bull. Chem. Soc. Jpn.*, **76**, 1603.
74. Nishida, Y., Shingu, Y., Dohi, H. and Kobayashi, K. (2003). *Org. Lett.*, **5**, 2377.
75. Jacobsson, M., Malmberg, J. and Ellervik, U. (2006). *Carbohydr. Res.*, **341**, 1266.
76. Conchie, J., Levvy, G.A. and Marsh, C.A. (1957). *Adv. Carbohydr. Chem.*, **12**, 157.
77. Mukaiyama, T., Katsurada, M. and Takashima, T. (1991). *Chem. Lett.*, 985.
78. Chatterjee, S.K. and Nuhn, P. (1998). *Chem. Commun.*, 1729.
79. Lönn, H. and Stenvall, K. (1992). *Tetrahedron Lett.*, **33**, 115.
80. Marra, A. and Sinaÿ, P. (1990). *Carbohydr. Res.*, **195**, 303.
81. Sinaÿ, P. and Mallet, J.-M. (1996). In *Modern Methods in Carbohydrate Synthesis* (S.H. Khan and R.A. O'Neill, eds) p. 130. Harwood Academic.
82. Kondo, H., Aoki, S., Ichikawa, Y., Halcomb, R.L., Ritzen, H. and Wong, C.-H. (1994). *J. Org. Chem.*, **59**, 864.
83. Müller, T., Schneider, R. and Schmidt, R.R. (1994). *Tetrahedron Lett.*, **35**, 4763.
84. Hashimoto, S.-i., Sano, A., Umeo, K., Nakajima, M. and Ikegami, S. (1995). *Chem. Pharm. Bull.*, **43**, 2267.
85. Tanaka, H., Sakamoto, H., Sano, A., Nakamura, S., Nakajima, M. and Hashimoto, S. (1999). *Chem. Commun.*, 1259.
86. Zhang, Z. and Wong, C.-H. (2000). In *Carbohydrates in Chemistry and Biology* (B. Ernst, G.W. Hart, and P. Sinaÿ, eds) Vol 1, p. 117. Wiley-VCH.
87. Cheng, Y.-P., Chen, H.-T. and Lin, C.-C. (2002). *Tetrahedron Lett.*, **43**, 7721.
88. Nagai, H., Matsumura, S. and Toshima, K. (2002). *Tetrahedron Lett.*, **43**, 847.
89. Hashimoto, S.-i., Yanagiya, Y., Honda, T., Harada, H. and Ikegami, S. (1992). *Tetrahedron Lett.*, **33**, 3523.
90. Hariprasad, V., Singh, G. and Tranoy, I. (1998). *Chem. Commun.*, 2129.
91. Ravidà, A., Liu, X., Kovacs, L. and Seeberger, P.H. (2006). *Org. Lett.*, **8**, 1815.
92. Michael, A. (1879). *American Chemical J.*, **1**, 305; 1885, **6**, 336.
93. Fischer, E. and Armstrong, E.F. (1901). *Ber. Dtsch. Chem. Ges.*, **34**, 2885.
94. Koenigs, W. and Knorr, E. (1901). *Ber. Dtsch. Chem. Ges.*, **34**, 957.
95. Igarashi, K. (1977). *Adv. Carbohydr. Chem. Biochem.*, **34**, 243.
96. Helferich, B. and Gootz, R. (1931). *Ber. Dtsch. Chem. Ges.*, **64**, 109.
97. Helferich, B., Bohn, E. and Winkler, S. (1930). *Ber. Dtsch. Chem. Ges.*, **63**, 989.
98. Flowers, H.M. (1972). *Methods Carbohydr. Chem.*, **6**, 474.
99. Helferich, B. and Zirner, J. (1962). *Chem. Ber.*, **95**, 2604.
100. Hanessian, S. and Banoub, J. (1980). *Methods Carbohydr. Chem.*, **8**, 247.
101. Presser, A., Kunert, O. and Pötschger, I. (2006). *Monatsh. Chem.*, **137**, 365.
102. Reynolds, D.D. and Evans, W.L. (1938). *J. Am. Chem. Soc.*, **60**, 2559.
103. Jacquinet, J.-C., Duchet, D., Milat, M.-L. and Sinaÿ, P. (1981). *J. Chem. Soc., Perkin Trans. 1*, 326.
104. Lipták, A., Nánási, P., Neszmélyi, A. and Wagner, H. (1980). *Tetrahedron*, **36**, 1261.
105. Crich, D., Dai, Z. and Gastaldi, S. (1999). *J. Org. Chem.*, **64**, 5224.
106. Garegg, P.J., Konradsson, P., Kvarnström, I., Norberg, T., Svennson, S.C.T. and Wigilius, B. (1985). *Acta Chem. Scand.*, **B39**, 569.
107. Nukada, T., Berces, A., Zgierski, M.Z. and Whitfield, D.M. (1998). *J. Am. Chem. Soc.*, **120**, 13291.
108. Kartha, K.P.R., Aloui, M. and Field, R.A. (1996). *Tetrahedron Lett.*, **37**, 8807.
109. Pacsu, E. (1945). *Adv. Carbohydr. Chem.*, **1**, 77.
110. Kong, F. (2007). *Carbohydr. Res.*, **342**, 345.

111. Wang, W. and Kong, F. (1998). *J. Org. Chem.*, **63**, 5744.
112. McAdam, D.P., Perera, A.M.A. and Stick, R.V. (1987). *Aust. J. Chem.*, **40**, 1901.
113. Bérces, A., Whitfield, D.M., Nukada, T., do Santos Z., I., Obuchowska, A. and Krepinsky, J.J. (2004). *Can. J. Chem.*, **82**, 1157.
114. Kochetkov, N.K., Bochkov, A.F., Sokolovskaya, T.A. and Snyatkova, V.J. (1971). *Carbohydr. Res.*, **16**, 17.
115. Kochetkov, N.K. (1987). *Tetrahedron*, **43**, 2389.
116. Ohtake, H., Iimori, T. and Ikegami, S. (1998). *Synlett*, 1420.
117. Lemieux, R.U., Hendriks, K.B., Stick, R.V. and James, K. (1975). *J. Am. Chem. Soc.*, **97**, 4056.
118. Khan, S.H. and O'Neill, R.A., eds (1996). *Modern Methods in Carbohydrate Synthesis*, p. xi. Harwood Academic.
119. Lemieux, R.U. and Driguez, H. (1975). *J. Am. Chem. Soc.*, **97**, 4069.
120. Nikrad, P.V., Beierbeck, H. and Lemieux, R.U. (1992). *Can. J. Chem.*, **70**, 241.
121. Lemieux, R.U. and Driguez, H. (1975). *J. Am. Chem. Soc.*, **97**, 4063.
122. Hindsgaul, O., Norberg, T., Le Pendu, J. and Lemieux, R.U. (1982). *Carbohydr. Res.*, **109**, 109.
123. Kartha, K.P.R. and Field, R.A. (1997). *Tetrahedron Lett.*, **38**, 8233.
124. Gervay, J. (1998). In *Organic Synthesis: Theory and Applications* (T. Hudlicky, ed.) Vol. 4, p. 121. JAI Press, Inc.
125. Lam, S.N. and Gervay-Hague, J. (2002). *Org. Lett.*, **4**, 2039.
126. van Well, R.M., Kartha, K.P.R. and Field, R.A. (2005). *J. Carbohydr. Chem.*, **24**, 463.
127. Shingu, Y., Nishida, Y., Dohi, H. and Kobayashi, K. (2003). *Org. Biomol. Chem.*, **1**, 2518.
128. Kobashi, Y. and Mukaiyama, T. (2004). *Chem. Lett.*, **33**, 874.
129. Ustyuzhanina, N., Komarova, B., Zlotina, N., Krylov, V., Gerbst, A., Tsvetkov, Y. and Nifantiev, N. (2006). *Synlett*, 921.
130. Dax, K., Albert, M., Ortner, J. and Paul, B.J. (1999). *Curr. Org. Chem.*, **3**, 287.
131. Tsuchiya, T. (1990). *Adv. Carbohydr. Chem. Biochem.*, **48**, 91.
132. Thiem, J. and Wiesner, M. (1993). *Carbohydr. Res.*, **249**, 197.
133. Mukaiyama, T. and Jona, H. (2002). *Proc. Japan Acad., Ser. B.*, **78**, 73.
134. Mukaiyama, T., Murai, Y. and Shoda, S.-i. (1981). *Chem. Lett.*, 431.
135. Jona, H., Mandai, H., Chavasiri, W., Takeuchi, K. and Mukaiyama, T. (2002). *Bull. Chem. Soc. Jpn.*, **75**, 291.
136. Nicolaou, K.C., Dolle, R.E., Papahatjis, D.P. and Randall, J.L. (1984). *J. Am. Chem. Soc.*, **106**, 4189.
137. Ohtsuka, I., Ako, T., Kato, R., Daikoku, S., Koroghi, S., Kanemitsu, T. and Kanie, O. (2006). *Carbohydr. Res.*, **341**, 1476.
138. Kreuzer, M. and Thiem, J. (1986). *Carbohydr. Res.*, **149**, 347.
139. Toshima, K., Nagai, H. and Kasumi, K.-i. (2004). *Tetrahedron*, **60**, 5331.
140. Baeschlin, D.K., Green, L.G., Hahn, M.G., Hinzen, B., Ince, S.J. and Ley, S.V. (2000). *Tetrahedron: Asymm.*, **11**, 173.
141. Schmidt, R.R. and Jung, K.-H. (2000). In *Carbohydrates in Chemistry and Biology* (B. Ernst, G.W. Hart, and P. Sinaÿ, eds) Vol. 1, p. 5. Wiley-VCH.
142. Pougny, J.-R., Jacquinet, J.-C., Nassr, M., Duchet, D., Milat, M.-L. and Sinaÿ, P. (1977). *J. Am. Chem. Soc.*, **99**, 6762.
143. Urban, F.J., Moore, B.S. and Breitenbach, R. (1990). *Tetrahedron Lett.*, **31**, 4421.
144. Wei, G., Gu, G. and Du, Y. (2003). *J. Carbohydr. Chem.*, **22**, 385.
145. Wang, Z.-G., Douglas, S.P. and Krepinsky, J.J. (1996). *Tetrahedron Lett.*, **37**, 6985.
146. Geiger, J., Barroca, N. and Schmidt, R.R. (2004). *Synlett*, 836.
147. Du, Y., Wei, G., Cheng, S., Hua, Y. and Linhardt, R.J. (2006). *Tetrahedron Lett.*, **47**, 307.
148. Adinolfi, M., Iadonisi, A. and Ravidà, A. (2006). *Synlett*, 583.

149. Adinolfi, M., Iadonisi, A., Ravidà, A. and Valerio, S. (2006). *Tetrahedron Lett.*, **47**, 2595.

150. Tian, Q., Zhang, S., Yu, Q., He, M.-B. and Yang, J.-S. (2007). *Tetrahedron*, **63**, 2142.

151. Yang, J., Cooper-Vanosdell, C., Mensah, E.A. and Nguyen, H.M. (2008). *J. Org. Chem.*, **73**, 794.

152. Kasuya, M.C. and Hatanaka, K. (1998). *Tetrahedron Lett.*, **39**, 9719.

153. Schmidt, R.R. and Toepfer, A. (1991). *Tetrahedron Lett.*, **32**, 3353.

154. Liu, M., Yu, B. and Hui, Y. (1998). *Tetrahedron Lett.*, **39**, 415.

155. Larsen, K., Olsen, C.E. and Motawia, M.S. (2008). *Carbohydr. Res.*, **343**, 383.

156. Gelin, M., Ferrieres, V. and Plusquellec, D. (1997). *Carbohydr. Lett.*, **2**, 381.

157. Yu, B. and Tao, H. (2001). *Tetrahedron Lett.*, **42**, 2405.

158. Ramakrishnan, A., Pornsuriyasak, P. and Demchenko, A.V. (2005). *J. Carbohydr. Chem.*, **24**, 649.

159. Matsuo, J.-i., Shirahata, T. and Omura, S. (2006). *Tetrahedron Lett.*, **47**, 267.

160. Garegg, P.J. (1997). *Adv. Carbohydr. Chem. Biochem.*, **52**, 179.

161. Oscarson, S. (2000). In *Carbohydrates in Chemistry and Biology* (B. Ernst, G.W. Hart, and P. Sinaÿ, eds) Vol. 1, p. 93. Wiley-VCH.

162. Codée, J.D.C., Litjens, R.E.J.N., van den Bos, L.J., Overkleeft, H.S. and van der Marel, G.A. (2005). *Chem. Soc. Rev.*, **34**, 769.

163. Norberg, T. (1996). In *Modern Methods in Carbohydrate Synthesis* (S.H. Khan and R.A. O'Neill, eds) p. 82. Harwood Academic.

164. Ludewig, M. and Thiem, J. (1998). *Synthesis*, 56.

165. Sato, S., Mori, M., Ito, Y. and Ogawa, T. (1986). *Carbohydr. Res.*, **155**, C6.

166. Andersson, F., Fügedi, P., Garegg, P.J. and Nashed, M. (1986). *Tetrahedron Lett.*, **27**, 3919.

167. Ferrier, R.J., Hay, R.W. and Vethaviyasar, N. (1973). *Carbohydr. Res.*, **27**, 55.

168. Mukhopadhyay, B., Collet, B. and Field, R.A. (2005). *Tetrahedron Lett.*, **46**, 5923.

169. Fukase, K., Nakai, Y., Kanoh, T. and Kusumoto, S. (1998). *Synlett*, 84.

170. Zhang, Y.-M., Mallet, J.-M. and Sinaÿ, P. (1992). *Carbohydr. Res.*, **236**, 73.

171. Mehta, S. and Pinto, B.M. (1998). *Carbohydr. Res.*, **310**, 43.

172. Durón, S.G., Polat, T. and Wong, C.-H. (2004). *Org. Lett.*, **6**, 839.

173. Crich, D. and Smith, M. (2001). *J. Am. Chem. Soc.*, **123**, 9015.

174. Crich, D., Banerjee, A., Li, W. and Yao, Q. (2005). *J. Carbohydr. Chem.*, **24**, 415.

175. Codée, J.D.C., van den Bos, L.J., Litjens, R.E.J.N., Overkleeft, H.S., van Boeckel, C.A.A., van Boom, J.H. and van der Marel, G.A. (2004). *Tetrahedron*, **60**, 1057.

176. Tatai, J. and Fügedi, P. (2007). *Org. Lett.*, **9**, 4647.

177. Li, Z. and Gildersleeve, J.C. (2007). *Tetrahedron Lett.*, **48**, 559.

178. Wang, C., Wang, H., Huang, X., Zhang, L.-H. and Ye, X.-S. (2006). *Synlett*, 2846.

179. Veeneman, G.H. and van Boom, J.H. (1990). *Tetrahedron Lett.*, **31**, 275.

180. Veeneman, G.H., van Leeuwen, S.H. and van Boom, J.H. (1990). *Tetrahedron Lett.*, **31**, 1331.

181. Cura, P., Aloui, M., Kartha, K.P.R. and Field, R.A. (2000). *Synlett*, 1279.

182. Sato, S., Nunomura, S., Nakano, T., Ito, Y. and Ogawa, T. (1988). *Tetrahedron Lett.*, **29**, 4097.

183. Knapp, S. and Nandan, S.R. (1994). *J. Org. Chem.*, **59**, 281.

184. Demchenko, A., Stauch, T. and Boons, G.-J. (1997). *Synlett*, 818.

185. Geurtsen, R., Holmes, D.S. and Boons, G.-J. (1997). *J. Org. Chem.*, **62**, 8145.

186. Roy, R., Andersson, F.O. and Letellier, M. (1992). *Tetrahedron Lett.*, **33**, 6053.

187. Sliedregt, L.A.J.M., Zegelaar-Jaarsveld, K., van der Marel, G.A. and van Boom, J.H. (1993). *Synlett*, 335.

188. Lahmann, M. and Oscarson, S. (2002). *Can. J. Chem.*, **80**, 889.

189. Li, Z. and Gildersleeve, J.C. (2006). *J. Am. Chem. Soc.*, **128**, 11612.

190. Collot, M., Savreux, J. and Mallet, J.-M. (2008). *Tetrahedron*, **64**, 1523.

191. Grice, P., Ley, S.V., Pietruszka, J., Osborn, H.M.I., Priepke, H.W.M. and Warriner, S.L. (1997). *Chem. Eur. J.*, **3**, 431.

192. Zhu, T. and Boons, G.-J. (2001). *Org. Lett.*, **3**, 4201.
193. Crich, D. and Jayalath, P. (2005). *J. Org. Chem.*, **70**, 7252.
194. Crich, D., de la Mora, M. and Vinod, A.U. (2003). *J. Org. Chem.*, **68**, 8142.
195. Flitsch, S.L. (2005). *Nature*, **437**, 201.
196. Kim, J.-H., Yang, H., Park, J. and Boons, G.-J. (2005). *J. Am. Chem. Soc.*, **127**, 12090.
197. Kim, J.-H., Yang, H., Khot, V., Whitfield, D. and Boons, G.-J. (2006). *Eur. J. Org. Chem.*, 5007.
198. Kim, J.-H., Yang, H. and Boons, G.-J. (2007). In *Frontiers in Modern Carbohydrate Chemistry* (A.V. Demchenko, ed.) ACS Symposium Series 960, p. 73. Washington, DC: American Chemical Society.
199. Mehta, S. and Pinto, B.M. (1996). In *Modern Methods in Carbohydrate Synthesis* (S.H. Khan and R.A. O'Neill, eds) p. 107. Harwood Academic.
200. Witczak, Z.J. and Czernecki, S. (1998). *Adv. Carbohydr. Chem. Biochem.*, **53**, 143.
201. Ferrier, R.J. and Furneaux, R.H. (1980). *Methods Carbohydr. Chem.*, **8**, 251.
202. Mehta, S. and Pinto, B.M. (1993). *J. Org. Chem.*, **58**, 3269.
203. Tiwari, P. and Misra, A.K. (2006). *Tetrahedron Lett.*, **47**, 2345.
204. Stick, R.V., Tilbrook, D.M.G. and Williams, S.J. (1997). *Aust. J. Chem.*, **50**, 233.
205. Valerio, S., Iadonisi, A., Adinolfi, M. and Ravidà, A. (2007). *J. Org. Chem.*, **72**, 6097.
206. Heskamp, B.M., Veeneman, G.H., van der Marel, G.A., van Boeckel, C.A.A. and van Boom, J.H. (1995). *Tetrahedron*, **51**, 8397.
207. Zuurmond, H.M., van der Meer, P.H., van der Klein, P.A.M., van der Marel, G.A. and van Boom, J.H. (1993). *J. Carbohydr. Chem.*, **12**, 1091.
208. Stick, R.V., Tilbrook, D.M.G. and Williams, S.J. (1997). *Aust. J. Chem.*, **50**, 237.
209. Johnston, B.D. and Pinto, B.M. (1998). *Carbohydr. Res.*, **305**, 289.
210. Yamago, S., Kokubo, K., Murakami, H., Mino, Y., Hara, O. and Yoshida, J.-i. (1998). *Tetrahedron Lett.*, **39**, 7905.
211. Crich, D. and Lim, L.B.L. (2004). *Org. React.*, **64**, 115.
212. Skelton, B.W., Stick, R.V., Tilbrook, D.M.G., White, A.H. and Williams, S.J. (2000). *Aust. J. Chem.*, **53**, 389.
213. Kakarla, R., Dulina, R.G., Hatzenbuhler, N.T., Hui, Y.W. and Sofia, M.J. (1996). *J. Org. Chem.*, **61**, 8347.
214. Agnihotri, G. and Misra, A.K. (2005). *Tetrahedron Lett.*, **46**, 8113.
215. Crich, D., Mataka, J., Zakharov, L.N., Rheingold, A.L. and Wink, D.J. (2002). *J. Am. Chem. Soc.*, **124**, 6028.
216. Kahne, D., Walker, S., Cheng, Y. and Van Engen, D. (1989). *J. Am. Chem. Soc.*, **111**, 6881.
217. Yan, L. and Kahne, D. (1996). *J. Am. Chem. Soc.*, **118**, 9239.
218. Ikemoto, N. and Schreiber, S.L. (1990). *J. Am. Chem. Soc.*, **112**, 9657.
219. Yang, D., Kim, S.-H. and Kahne, D. (1991). *J. Am. Chem. Soc.*, **113**, 4715.
220. Raghavan, S. and Kahne, D. (1993). *J. Am. Chem. Soc.*, **115**, 1580.
221. Thompson, C., Ge, M. and Kahne, D. (1999). *J. Am. Chem. Soc.*, **121**, 1237.
222. Sliedregt, L.A.J.M., van der Marel, G.A. and van Boom, J.H. (1994). *Tetrahedron Lett.*, **35**, 4015.
223. Alonso, I., Khiar, N. and Martín-Lomas, M. (1996). *Tetrahedron Lett.*, **37**, 1477.
224. Gildersleeve, J., Smith, A., Sakurai, K., Raghaven, S. and Kahne, D. (1999). *J. Am. Chem. Soc.*, **121**, 6176.
225. Crich, D. and Sun, S. (1997). *J. Am. Chem. Soc.*, **119**, 11217.
226. Gildersleeve, J., Pascal, R.A., Jr. and Kahne, D. (1998). *J. Am. Chem. Soc.*, **120**, 5961.
227. Cassel, S., Plessis, I., Wessel, H.P. and Rollin, P. (1998). *Tetrahedron Lett.*, **39**, 8097.
228. Ferrier, R.J. and Prasad, N. (1969). *J. Chem. Soc. C*, 570.
229. Ferrier, R.J. (1969). *Adv. Carbohydr. Chem. Biochem.*, **24**, 199.

230. Ferrier, R.J. (2001). *Top. Curr. Chem.*, **215**, 153.
231. Ferrier, R.J. and Hoberg, J.O. (2003). *Adv. Carbohydr. Chem. Biochem.*, **58**, 55.
232. Ferrier, R.J. and Zubkov, O.A. (2003). *Org. React.*, **62**, 569.
233. Takhi, M., Abdel-Rahman, A.A.-H. and Schmidt, R.R. (2001). *Synlett*, 427.
234. Misra, A.K., Tiwari, P. and Agnihotri, G. (2005). *Synthesis*, 260.
235. Yadav, J.S., Satyanarayana, M., Balanarsaiah, E. and Raghavendra, S. (2006). *Tetrahedron Lett.*, **47**, 6095.
236. Abdel-Rahman, A.A.-H., Winterfeld, G.A., Takhi, M. and Schmidt, R.R. (2002). *Eur. J. Org. Chem.*, 713.
237. Di Bussolo, V., Caselli, M., Romano, M.R., Pineschi, M. and Crotti, P. (2004). *J. Org. Chem.*, **69**, 7383.
238. Danishefsky, S.J. and Bilodeau, M.T. (1996). *Angew. Chem. Int. Ed. Engl.*, **35**, 1380.
239. Seeberger, P.H., Bilodeau, M.T. and Danishefsky, S.J. (1997). *Aldrichimica Acta*, **30**, 75.
240. Fraser-Reid, B., Udodong, U.E, Wu, Z., Ottosson, H., Merritt, J.R., Rao, C.S., Roberts, C. and Madsen, R. (1992). *Synlett*, 927.
241. Fraser-Reid, B., Merritt, J.R., Handlon, A.L. and Andrews, C.W. (1993). *Pure Appl. Chem.*, **65**, 779.
242. Fraser-Reid, B. and Madsen, R. (1997). In *Preparative Carbohydrate Chemistry* (S. Hanessian, ed.) p. 339. Marcel Dekker, Inc.
243. Madsen, R. and Fraser-Reid, B. (1996). In *Modern Methods in Carbohydrate Synthesis* (S.H. Kahn and R.A. O'Neill, eds) p. 155. Harwood Academic.
244. Fraser-Reid, B., Anilkumar, G., Gilbert, M.R., Joshi, S. and Kraehmer, R. (2000). In *Carbohydrates in Chemistry and Biology* (B. Ernst, G.W. Hart and P. Sinaÿ, eds) Vol. 1, p. 135. Wiley-VCH.
245. Rodebaugh, R., Debenham, J.S., Fraser-Reid, B. and Snyder, J.P. (1999). *J. Org. Chem.*, **64**, 1758.
246. Merritt, J.R., Debenham, J.S. and Fraser-Reid, B. (1996). *J. Carbohydr. Chem.*, **15**, 65.
247. López, J.C., Gómez, A.M., Uriel, C. and Fraser-Reid, B. (2003). *Tetrahedron Lett.*, **44**, 1417.
248. Mach, M., Schlueter, U., Mathew, F., Fraser-Reid, B. and Hazen, K.C. (2002). *Tetrahedron*, **58**, 7345.
249. Jayaprakash, K.N., Radhakrishnan, K.V. and Fraser-Reid, B. (2002). *Tetrahedron Lett.*, **43**, 6953.
250. Jayaprakash, K.N. and Fraser-Reid, B. (2004). *Synlett*, 301.
251. Barresi, F. and Hindsgaul, O. (1996). In *Modern Methods in Carbohydrate Synthesis* (S.H. Khan and R.A. O'Neill, eds) p. 251. Harwood Academic.
252. Gridley, J.J. and Osborn, H.M.I. (2000). *J. Chem. Soc., Perkin Trans. 1*, 1471.
253. Crich, D. (2002). *J. Carbohydr. Chem.*, **21**, 667.
254. Paulsen, H. and Lockhoff, O. (1981). *Chem. Ber.*, **114**, 3102.
255. van Boeckel, C.A.A., Beetz, T. and van Aelst, S.F. (1984). *Tetrahedron*, **40**, 4097.
256. Garegg, P.J. and Ossowski, P. (1983). *Acta Chem. Scand.*, **B37**, 249.
257. Crich, D. and Sun, S. (1998). *Tetrahedron*, **54**, 8321.
258. Crich, D. and Barba, G.R. (1998). *Tetrahedron Lett.*, **39**, 9339.
259. Crich, D. and Dai, Z. (1999). *Tetrahedron*, **55**, 1569.
260. Crich, D. and Wu, B. (2006). *Org. Lett.*, **8**, 4879.
261. Crich, D., Cai, W. and Dai, Z. (2000). *J. Org. Chem.*, **65**, 1291.
262. Crich, D., Jayalath, P. and Hutton, T.K. (2006). *J. Org. Chem.*, **71**, 3064.
263. Crich, D. and Vinogradova, O. (2006). *J. Org. Chem.*, **71**, 8473.
264. Crich, D. (2007). In *Frontiers in Modern Carbohydrate Chemistry* (A.V. Demchenko, ed.) ACS Symposium Series 960 p. 60. Washington, DC: American Chemical Society.
265. Crich, D. and Li, L. (2007). *J. Org. Chem.*, **72**, 1681.
266. Codée, J.D.C., Hossain, L.H. and Seeberger, P.H. (2005). *Org. Lett.*, **7**, 3251.
267. Abdel-Rahman, A.A.-H., Jonke, S., El Ashry, E.S.H. and Schmidt, R.R. (2002). *Angew. Chem. Int. Ed.*, **41**, 2972.

268. Ekborg, G., Lindberg, B. and Lönngren, J. (1972). *Acta Chem. Scand.*, **26**, 3287.
269. Fürstner, A. and Konetzki, I. (1998). *Tetrahedron Lett.*, **39**, 5721.
270. Rising, T.W.D.F., Claridge, T.D.W., Davies, N., Gamblin, D.P., Moir, J.W.B. and Fairbanks, A.J. (2006). *Carbohydr. Res.*, **341**, 1574.
271. Dong, H., Pei, Z., Angelin, M., Byström, S. and Ramström, O. (2007). *J. Org. Chem.*, **72**, 3694.
272. Günther, W. and Kunz, H. (1992). *Carbohydr. Res.*, **228**, 217.
273. Lichtenthaler, F.W., Kläres, U., Szurmai, Z. and Werner, B. (1998). *Carbohydr. Res.*, **305**, 293.
274. Jung, K.-H., Müller, M. and Schmidt, R.R. (2000). *Chem. Rev.*, **100**, 4423.
275. Fairbanks, A.J. (2003). *Synlett*, 1945.
276. Cumpstey, I., Chayajarus, K., Fairbanks, A.J., Redgrave, A.J. and Seward, C.M.P. (2004). *Tetrahedron*: Asymm., **15**, 3207.
277. Attolino, E., Rising, T.W.D.F., Heidecke, C.D. and Fairbanks, A.J. (2007). *Tetrahedron*: Asymm., **18**, 1721.
278. Pratt, M.R., Leigh, C.D. and Bertozzi, C.R. (2003). *Org. Lett.*, **5**, 3185.
279. Barresi, F. and Hindsgaul, O. (1994). *Can. J. Chem.*, **72**, 1447.
280. Stork, G. and La Clair, J.L. (1996). *J. Am. Chem. Soc.*, **118**, 247.
281. Ito, Y., Ando, H., Wada, M., Kawai, T., Ohnish, Y. and Nakahara, Y. (2001). *Tetrahedron*, **57**, 4123.
282. Attolino, E. and Fairbanks, A.J. (2007). *Tetrahedron Lett.*, **48**, 3061.
283. Hodosi, G. and Kováč, P. (1998). *Carbohydr. Res.*, **308**, 63.
284. Toshima, K., Kasumi, K.-i. and Matsumura, S. (1998). *Synlett*, 643.
285. Ziegler, T. and Lemanski, G. (1998). *Angew. Chem. Int. Ed.*, **37**, 3129.
286. Sato, K.-i. and Yoshitomo, A. (1995). *Chem. Lett.*, 39.
287. Tsuda, T., Sato, S., Nakamura, S. and Hashimoto, S. (2003). *Heterocycles*, **59**, 509.
288. Nagai, H., Matsumura, S. and Toshima, K. (2003). *Carbohydr. Res.*, **338**, 1531.
289. Mandai, H. and Mukaiyama, T. (2006). *Bull. Chem. Soc. Jpn.*, **79**, 479.
290. Baek, J.Y., Choi, T.J., Jeon, H.B. and Kim, K.S. (2006). *Angew. Chem. Int. Ed.*, **45**, 7436.
291. Åkerfeldt, K., Garegg, P.J. and Iversen, T. (1979). *Acta Chem. Scand.*, **B33**, 467.
292. Roush, W.R. and Lin, X.-F. (1991). *J. Org. Chem.*, **56**, 5740.
293. Zechel, D. and Withers, S.G., unpublished results.
294. Lichtenthaler, F.W. and Metz, T. (2003). *Eur. J. Org. Chem.*, 3081.
295. Lee, Y.J., Ishiwata, A. and Ito, Y. (2008). *J. Am. Chem. Soc.*, **130**, 6330.
296. Crich, D. and Bowers, A.A. (2006). *Org. Lett.*, **8**, 4327.
297. Crich, D. and Patel, M. (2006). *Carbohydr. Res.*, **341**, 1467.
298. Fauré, R., Shiao, T.C., Damerval, S. and Roy, R. (2007). *Tetrahedron Lett.*, **48**, 2385.
299. Banoub, J., Boullanger, P. and Lafont, D. (1992). *Chem. Rev.*, **92**, 1167.
300. Veeneman, G.H. (1998). In *Carbohydrate Chemistry* (G.-J. Boons, ed.), p. 98. Blackie.
301. Bongat, A.F.G., Kamat, M.N. and Demchenko, A.V. (2007). *J. Org. Chem.*, **72**, 1480.
302. Bongat, A.F.G. and Demchenko, A.V. (2007). *Carbohydr. Res.*, **342**, 374.
303. Zachara, N.E. and Hart, G.W. (2004). *Trends Cell. Biol.*, **14**, 218.
304. Akimoto, Y., Hart, G.W., Hirano, H. and Kawakami, H. (2005). *Med. Mol. Morphol.*, **38**, 84.
305. Slawson, C., Housley, M.P. and Hart, G.W. (2006). *J. Cell. Biochem.*, **97**, 71.
306. Fairweather, J.K., Stick, R.V. and Tilbrook, D.M.G. (1998). *Aust. J. Chem.*, **51**, 471.
307. Crasto, C.F. and Jones, G.B. (2004). *Tetrahedron Lett.*, **45**, 4891.
308. Zeng, Y., Wang, J., Li, B., Hauser, S., Li, H. and Wang, L.-X. (2006). *Chem. Eur. J.*, **12**, 3355.
309. Cai, Y., Ling, C.-C. and Bundle, D.R. (2005). *Org. Lett.*, **7**, 4021.
310. Meinjohanns, E., Meldal, M. and Bock, K. (1995). *Tetrahedron Lett.*, **36**, 9205.
311. Singh, L. and Seifert, J. (2001). *Tetrahedron Lett.*, **42**, 3133.
312. Suihko, M., Ahlgrén, M., Aulaskari, P. and Rouvinen, J. (2001). *Carbohydr. Res.*, **334**, 337.

313. Castro-Palomino, J.C. and Schmidt, R.R. (1995). *Tetrahedron Lett.*, **36**, 5343, 6871.

314. Silwanis, B.A., El-Sokkary, R.I., Nashed, M.A. and Paulsen, H. (1991). *J. Carbohydr. Chem.*, **10**, 1067.

315. Leung, O.-T., Douglas, S.P., Whitfield, D.M., Pang, H.Y.S. and Krepinsky, J.J. (1994). *New J. Chem.*, **18**, 349.

316. Prahl, I. and Unverzagt, C. (2000). *Tetrahedron Lett.*, **41**, 10189.

317. GoddarD-Borger, E.D. and Stick, R.V. (2007). *Org. Lett.*, **9**, 3797.

318. Grundler, G. and Schmidt, R.R. (1984). *Liebigs Ann. Chem.*, 1826.

319. Mironov, Y.V., Sherman, A.A. and Nifantiev, N.E. (2004). *Tetrahedron Lett.*, **45**, 9107.

320. Park, J., Kawatkar, S., Kim, J.-H. and Boons, G.-J. (2007). *Org. Lett.*, **9**, 1959.

321. Marra, A., Gauffeny, F. and Sinaÿ, P. (1991). *Tetrahedron*, **47**, 5149.

322. Ludewig, M. and Thiem, J. (1998). *Eur. J. Org. Chem.*, 1189.

323. Di Bussolo, V., Liu, J., Huffman, L.G., Jr. and Gin, D.Y. (2000). *Angew. Chem. Int. Ed.*, **39**, 204.

324. Reddy, B.G. and Schmidt, R.R. (2008). *Nat. Protoc.*, **3**, 114.

325. Arihara, R., Nakamura, S. and Hashimoto, S. (2005). *Angew. Chem. Int. Ed.*, **44**, 2245.

326. Tsuda, T., Nakamura, S. and Hashimoto, S. (2004). *Tetrahedron*, **60**, 10711.

327. Boysen, M., Gemma, E., Lahmann, M. and Oscarson, S. (2005). *Chem. Commun.*, 3044.

328. Benakli, K., Zha, C. and Kerns, R.J. (2001). *J. Am. Chem. Soc.*, **123**, 9461.

329. Manabe, S., Ishii, K. and Ito, Y. (2006). *J. Am. Chem. Soc.*, **128**, 10666.

330. Geng, Y., Zhang, L.-H. and Ye, X.-S. (2008). *Chem. Commun.*, 597.

331. Crich, D. and Vinod, A.U. (2005). *J. Org. Chem.*, **70**, 1291.

332. Liao, L. and Auzanneau, F.-I. (2005). *J. Org. Chem.*, **70**, 6265.

333. Kaji, E., Lichtenthaler, F.W., Nishino, T., Yamane, A. and Zen, S. (1988). *Bull. Chem. Soc. Jpn.*, **61**, 1291.

334. Drew, M.G.B., Ennis, S.C., Gridley, J.J., Osborn, H.M.I. and Spackman, D.G. (2001). *Tetrahedron*, **57**, 7919.

335. Litjens, R.E.J.N., van den Bos, L.J., Codée, J.D.C., van den Berg, R.J.B.H.N., Overkleeft, H.S. and van der Marel, G.A. (2005). *Eur. J. Org. Chem.*, 918.

336. Hederos, M. and Konradsson, P. (2005). *J. Carbohydr. Chem.*, **24**, 297.

337. Thiem, J. and Klaffke, W. (1990). *Top. Curr. Chem.*, **154**, 285.

338. Kirschning, A., Bechthold, A.F.-W. and Rohr, J. (1997). *Top. Curr. Chem.*, **188**, 1.

339. Marzabadi, C.H. and Franck, R.W. (2000). *Tetrahedron*, **56**, 8385.

340. Veyrières, A. (2000). In *Carbohydrates in Chemistry and Biology* (B. Ernst, G.W. Hart, and P. Sinaÿ, eds) Vol. 1, p. 367. Wiley-VCH.

341. Toshima, K., Nagai, H., Ushiki, Y. and Matsumura, S. (1998). *Synlett*, 1007.

342. Yadav, J.S., Reddy, B.V.S., Reddy, K.B. and Satyanarayana, M. (2002). *Tetrahedron Lett.*, **43**, 7009.

343. Sherry, B.D., Loy, R.N. and Toste, F.D. (2004). *J. Am. Chem. Soc.*, **126**, 4510.

344. Nicolaou, K.C., Seitz, S.P. and Papahatjis, D.P. (1983). *J. Am. Chem. Soc.*, **105**, 2430.

345. Laupichler, L., Sajus, H. and Thiem, J. (1992). *Synthesis*, 1133.

346. Bielawska, H. and Michalska, M. (1998). *Tetrahedron Lett.*, **39**, 9761.

347. Schene, H. and Waldmann, H. (1998). *Chem. Commun.*, 2759.

348. Tanaka, H., Yoshizawa, A. and Takahashi, T. (2007). *Angew. Chem. Int. Ed.*, **46**, 2505.

349. Pongdee, R., Wu, B. and Sulikowski, G.A. (2001). *Org. Lett.*, **3**, 3523.

350. Kim, K.S., Park, J., Lee, Y.J. and Seo, Y.S. (2003). *Angew. Chem. Int. Ed.*, **42**, 459.

351. Lam, S.N. and Gervay-Hague, J. (2003). *Org. Lett.*, **5**, 4219.

352. Gammon, D.W., Kinfe, H.H., De Vos, D.E., Jacobs, P.A. and Sels, B.F. (2004). *Tetrahedron Lett.*, **45**, 9533.

353. Durham, T.B. and Roush, W.R. (2003). *Org. Lett.*, **5**, 1871.

354. Costantino, V., Fattorusso, E., Imperatore, C. and Mangoni, A. (2002). *Tetrahedron Lett.*, **43**, 9047.

355. Suzuki, K., Sulikowski, G.A., Friesen, R.W. and Danishefsky, S.J. (1990). *J. Am. Chem. Soc.*, **112**, 8895.

356. Thiem, J. and Schöttmer, B. (1987). *Angew. Chem. Int. Ed. Engl.*, **26**, 555.

357. Franck, R.W. and Kaila, N. (1993). *Carbohydr. Res.*, **239**, 71.

358. Perez, M. and Beau, J.-M. (1989). *Tetrahedron Lett.*, **30**, 75.

359. Gervay, J. and Danishefsky, S. (1991). *J. Org. Chem.*, **56**, 5448.

360. Trumtel, M., Veyrières, A. and Sinaÿ, P. (1989). *Tetrahedron Lett.*, **30**, 2529.

361. Tavecchia, P., Trumtel, M., Veyrières, A. and Sinaÿ, P. (1989). *Tetrahedron Lett.*, **30**, 2533.

362. Nicolaou, K.C., Ladduwahetty, T., Randall, J.L. and Chucholowski, A. (1986). *J. Am. Chem. Soc.*, **108**, 2466.

363. Zuurmond, H.M., van der Klein, P.A.M., van der Marel, G.A. and van Boom, J.H. (1993). *Tetrahedron*, **49**, 6501.

364. Hou, D. and Lowary, T.L. (2007). *Org. Lett.*, **9**, 4487.

365. Roush, W.R., Gung, B.W. and Bennett, C.E. (1999). *Org. Lett.*, **1**, 891.

366. Roush, W.R., Narayan, S., Bennett, C.E. and Briner, K. (1999). *Org. Lett.*, **1**, 895.

367. Roush, W.R. and Narayan, S. (1999). *Org. Lett.*, **1**, 899.

368. Boons, G.-J. and Demchenko, A.V. (2000). *Chem. Rev.*, **100**, 4539.

369. Kiso, M., Ishida, H. and Ito, H. (2000). In *Carbohydrates in Chemistry and Biology* (B. Ernst, G.W. Hart, and P. Sinaÿ, eds) Vol. 1, p. 345. Wiley-VCH.

370. Halcomb, R.L. and Chappell, M.D. (2002). *J. Carbohydr. Chem.*, **21**, 723.

371. Dyason, J.C. and von Itzstein, M. (2001). *Aust. J. Chem.*, **54**, 663.

372. Zhao, C., Zhen, X., Zhang, H. and Ding, Y. (2005). *Lett. Org. Chem.*, **2**, 521.

373. Tsvetkov, Y.E. and Nifantiev, N.E. (2005). *Synlett*, 1375.

374. Crich, D. and Li, W. (2006). *Org. Lett.*, **8**, 959.

375. Hasegawa, A., Nagahama, T., Ohki, H., Hotta, K., Ishida, H. and Kiso, M. (1991). *J. Carbohydr. Chem.*, **10**, 493.

376. Cai, S. and Yu, B. (2003). *Org. Lett.*, **5**, 3827.

377. Crich, D. and Li, W. (2007). *J. Org. Chem.*, **72**, 7794.

378. Bogusiak, J. (2002). *Polish J. Chem.*, **76**, 1.

379. Lowary, T.L. (2003). *Curr. Opin. Chem. Biol.*, **7**, 749.

380. Lowary, T.L. (2002). *J. Carbohydr. Chem.*, **21**, 691.

381. Ishiwata, A., Akao, H. and Ito, Y. (2006). *Org. Lett.*, **8**, 5525.

382. Crich, D., Pedersen, C.M., Bowers, A.A. and Wink, D.J. (2007). *J. Org. Chem.*, **72**, 1553.

383. Cociorva, O.M. and Lowary, T.L. (2004). *Tetrahedron*, **60**, 1481.

384. Bai, Y. and Lowary, T.L. (2006). *J. Org. Chem.*, **71**, 9658.

385. Bai, Y. and Lowary, T.L. (2006). *J. Org. Chem.*, **71**, 9672.

386. Zhu, X., Kawatkar, S., Rao, Y. and Boons, G.-J. (2006). *J. Am. Chem. Soc.*, **128**, 11948.

387. Panza, L. and Lay, L. (2000). In *Carbohydrates in Chemistry and Biology* (B. Ernst, G.W. Hart and P. Sinaÿ, eds) Vol. 1, p. 195. Wiley-VCH.

388. Marra, A., Esnault, J., Veyrières, A. and Sinaÿ, P. (1992). *J. Am. Chem. Soc.*, **114**, 6354.

389. Barrett, A.G.M., Bezuidenhoudt, B.C.B., Gasiecki, A.F., Howell, A.R. and Russell, M.A. (1989). *J. Am. Chem. Soc.*, **111**, 1392.

390. Chenault, H.K. and Castro, A. (1994). *Tetrahedron Lett.*, **35**, 9145.

391. Hanessian, S. ed. (1997). *Preparative Carbohydrate Chemistry*, pp. 381–466. Marcel Dekker, Inc.

392. Hanessian, S. and Lou, B. (2000). *Chem. Rev.*, **100**, 4443.

393. Hanessian, S., Bacquet, C. and Lehong, N. (1980). *Carbohydr. Res.*, **80**, C17.

394. Hanessian, S., Lu, P.-P. and Ishida, H. (1998). *J. Am. Chem. Soc.*, **120**, 13296.

395. Furukawa, H., Koide, K., Takao, K.-i. and Kobayashi, S. (1998). *Chem. Pharm. Bull.*, **46**, 1244.

396. Krüger, A., Pyplo-Schnieders, J., Redlich, H. and Winkelmann, P. (2004). *Collect. Czech. Chem. Commun.*, **69**, 1843.

397. Grayson, E.J., Ward, S.J., Hall, A.L., Rendle, P.M., Gamblin, D.P., Batsanov, A.S. and Davis, B.G. (2005). *J. Org. Chem.*, **70**, 9740.

398. Chéry, F., Cassel, S., Wessel, H.P. and Rollin, P. (2002). *Eur. J. Org. Chem.*, 171.

399. Pornsuriyasak, P. and Demchenko, A.V. (2006). *Chem. Eur. J.*, **12**, 6630.

400. Kamat, M.N., De Meo, C. and Demchenko, A.V. (2007). *J. Org. Chem.*, **72**, 6947.

401. Smoot, J.T., Pornsuriyasak, P. and Demchenko, A.V. (2005). *Angew. Chem. Int. Ed.*, **44**, 7123.

402. Pornsuriyasak, P., Kamat, M.N. and Demchenko, A.V. (2007). In *Frontiers in Modern Carbohydrate Chemistry* (A.V. Demchenko, ed.) ACS Symposium Series 960, p. 165. Washington, DC: American Chemical Society.

403. Lee, Y.J., Baek, J.Y., Lee, B.-Y., Kang, S.S., Park, H.-S., Jeon, H.B. and Kim, K.S. (2006). *Carbohydr. Res.*, **341**, 1708.

404. Kim, K.S. and Jeon, H.B. (2007). In *Frontiers in Modern Carbohydrate Chemistry* (A.V. Demchenko, ed.) ACS Symposium Series 960 p. 134. Washington, DC: American Chemical Society.

405. Petersen, L., Laursen, J.B., Larsen, K., Motawia, M.S. and Jensen, K.J. (2003). *Org. Lett.*, **5**, 1309.

406. Worm-Leonhard, K., Larsen, K. and Jensen, K.J. (2007). *J. Carbohydr. Chem.*, **26**, 349.

407. Deng, S., Gangadharmath, U. and Chang, C.-W.T. (2006). *J. Org. Chem.*, **71**, 5179.

408. Hanessian, S. and Pernet, A.G. (1976). *Adv. Carbohydr. Chem. Biochem.*, **33**, 111.

409. Postema, M.H.D. (1992). *Tetrahedron*, **48**, 8545.

410. Jaramillo, C. and Knapp, S. (1994). *Synthesis*, 1.

411. Postema, M.H.D. (1995). *C-Glycoside Synthesis*. CRC Press: Boca Raton.

412. Levy, D.E. and Tang, C. (1995). *The Chemistry of C-Glycosides*. Pergamon: Oxford.

413. Bertozzi, C. and Bednarski, M. (1996). In *Modern Methods in Carbohydrate Synthesis* (S.H. Khan and R.A. O'Neill, eds) p. 316. Harwood Academic.

414. Beau, J.-M. and Gallagher, T. (1997). *Top. Curr. Chem.*, **187**, 1.

415. Nicotra, F. (1997). *Top. Curr. Chem.*, **187**, 55.

416. Du, Y., Linhardt, R.J. and Vlahov, I.R. (1998). *Tetrahedron*, **54**, 9913.

417. Skrydstrup, T., Vauzeilles, B. and Beau, J.-M. (2000). In *Carbohydrates in Chemistry and Biology* (B. Ernst, G.W. Hart, and P. Sinaÿ, eds) Vol. 1, p. 495. Wiley-VCH.

418. Rassu, G., Auzzas, L., Pinna, L., Battistini, L. and Curti, C. (2003). *Studies Nat. Prod. Chem.*, **29**, 449.

419. Levy, D.E. (2006). In *The Organic Chemistry of Sugars* (D.E. Levy and P. Fügedi, eds) p. 269. CRC Press.

420. Sezaki, M., Kondo, S., Maeda, K., Umezawa, H. and Ohno, M. (1970). *Tetrahedron*, **26**, 5171.

421. Kishi, Y. (1998). *Pure Appl. Chem.*, **70**, 339.

422. Espinosa, J.-F., Bruix, M., Jarreton, O., Skrydstrup, T., Beau, J.-M. and Jiménez-Barbero, J. (1999). *Chem. Eur. J.*, **5**, 442.

423. Somsák, L. (2001). *Chem. Rev.*, **101**, 81.

424. Best, W.M., Ferro, V., Harle, J., Stick, R.V. and Tilbrook, D.M.G. (1997). *Aust. J. Chem.*, **50**, 463.

425. Xin, Y.-C., Zhang, Y.-M., Mallet, J.-M., Glaudemans, C.P.J. and Sinaÿ, P. (1999). *Eur. J. Org. Chem.*, 471.

426. Evans, D.A., Trotter, B.W. and Côté, B. (1998). *Tetrahedron Lett.*, **39**, 1709.

427. Takhi, M., Abdel Rahman, A.A.-H. and Schmidt, R.R. (2001). *Tetrahedron Lett.*, **42**, 4053.

428. Tiwari, P., Agnihotri, G. and Misra, A.K. (2005). *Carbohydr. Res.*, **340**, 749.

429. Dondoni, A. and Marra, A. (2003). *Tetrahedron Lett.*, **44**, 13.

430. Baumberger, F. and Vasella, A. (1983). *Helv. Chim. Acta*, **66**, 2210.

431. Palmier, S., Vauzeilles, B. and Beau, J.-M. (2003). *Org. Biomol. Chem.*, **1**, 1097.

432. Mikkelsen, L.M., Krintel, S.L., Jiménez-Barbero, J. and Skrydstrup, T. (2002). *J. Org. Chem.*, **67**, 6297.
433. Belica, P.S. and Franck, R.W. (1998). *Tetrahedron Lett.*, **39**, 8225.
434. Taylor, R.J.K., McAllister, G.D. and Franck, R.W. (2006). *Carbohydr. Res.*, **341**, 1298.
435. Hoffmann, M., Burkhart, F., Hessler, G. and Kessler, H. (1996). *Helv. Chim. Acta*, **79**, 1519.
436. Beau, J.-M. and Sinaÿ, P. (1985). *Tetrahedron Lett.*, **26**, 6185, 6189, 6193.
437. Lesimple, P., Beau, J.-M., Jaurand, G. and Sinaÿ, P. (1986). *Tetrahedron Lett.*, **27**, 6201.
438. Togo, H., He, W., Waki, Y. and Yokoyama, M. (1998). *Synlett*, 700.
439. Giese, B. (1986). *Radicals in Organic Synthesis: Formation of Carbon-Carbon Bonds.* Pergamon Press.
440. Curran, D.P., Porter, N.A. and Giese, B. (1996). *Stereochemistry of Radical Reactions.* Weinheim VCH.
441. Giese, B. and Zeitz, H.-G. (1997). In *Preparative Carbohydrate Chemistry* (S. Hanessian, ed.) p. 507. Marcel Dekker Inc.
442. Praly, J.-P. (2000). *Adv. Carbohydr. Chem. Biochem.*, **56**, 65.
443. Paulsen, H. and Matschulat, P. (1991). *Liebigs Ann. Chem.*, 487.
444. Nagy, J.O. and Bednarski, M.D. (1991). *Tetrahedron Lett.*, **32**, 3953.
445. Sinaÿ, P. (1998). *Pure Appl. Chem.*, **70**, 407, 1495; 1997, **69**, 459.
446. Rubinstenn, G., Mallet, J.-M. and Sinaÿ, P. (1998). *Tetrahedron Lett.*, **39**, 3697.
447. Vasella, A. (1993). *Pure Appl. Chem.*, **65**, 731; 1991, **63**, 507.
448. Vasella, A., Bernet, B., Weber, M. and Wenger, W. (2000). In *Carbohydrates in Chemistry and Biology* (B. Ernst, G.W. Hart, and P. Sinaÿ, eds) Vol. 1, p. 155. Wiley-VCH.
449. Brakta, M., Lhoste, P. and Sinou, D. (1989). *J. Org. Chem.*, **54**, 1890.
450. He, W., Togo, H., Waki, Y. and Yokoyama, M. (1998). *J. Chem. Soc., Perkin Trans. 1*, 2425.
451. Chen, G., Schmieg, J., Tsuji, M. and Franck, R.W. (2004). *Org. Lett.*, **6**, 4077.
452. Piper, J.L. and Postema, M.H.D. (2004). *J. Org. Chem.*, **69**, 7395.

Chapter 5

Oligosaccharide Synthesis[1–4]

Strategies in Oligosaccharide Synthesis

I did it, my way

Frank Sinatra

If you were to enquire about the best way of constructing a particular glycosidic linkage in an oligosaccharide of interest, the answer would depend on the person being asked:

David Gin	'a hemiacetal'
Hans Paulsen	'use a glycosyl halide'
Ray Lemieux	'halide-catalysis, of course'[a]
K. C. Nicolaou and Teruaki Mukaiyama	'a glycosyl fluoride'
Richard Schmidt	'trichloroacetimidates!'
Per Garegg or David Crich	'a thioglycoside'
Dan Kahne	'a glycosyl sulfoxide'
Sam Danishefsky	'glycal assembly'
Bert Fraser-Reid	'NPG activation'

All offer good advice; in fact, there is no one method of glycosidation that is general and reliable enough to guarantee success. The situation is epitomized in the oft-quoted lines of Hans Paulsen:[5]

> *Each oligosaccharide synthesis remains an independent problem, whose resolution requires considerable systematic research and a good deal of know-how. There are no universal reaction conditions for oligosaccharide syntheses.*

---

[a] Raymond U. Lemieux (1920–2000).

References start on page 219

In fact, this statement seems even more profound in light of two recent publications: one advocating precise microwave heating to promote glycosylation[6] and the other accelerated glycosylation under frozen conditions![7]

From the previous chapters we have an appreciation of the various factors that influence the glycosidic bond, as well as a summary of the established and the newer methods for its construction. What is lacking in our knowledge is the *philosophy* behind oligosaccharide synthesis.

Linear syntheses: A *linear synthesis* is one in which a monosaccharide, by the successive addition of single monosaccharides, is transformed into the desired oligosaccharide:[b]

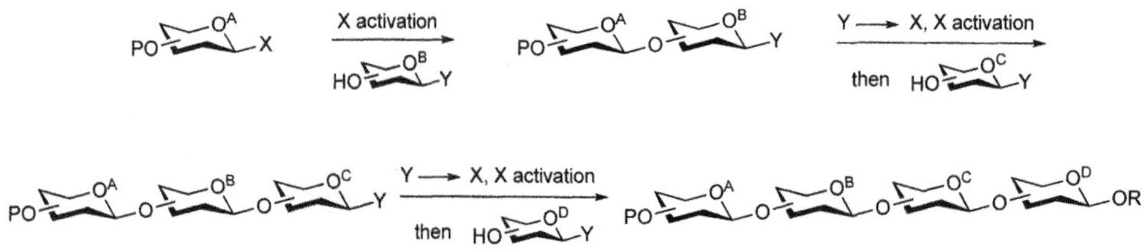

A protected glycosyl donor (A, X activated) is treated with an acceptor alcohol (B, Y inactive) to generate a disaccharide (AB, Y inactive). After the conversion of Y into X, the process is repeated until the final target oligosaccharide is reached. A reverse approach, which benefits from the fact that the glycosyl donor is more readily acquired and can thus be used in excess, is in common use:

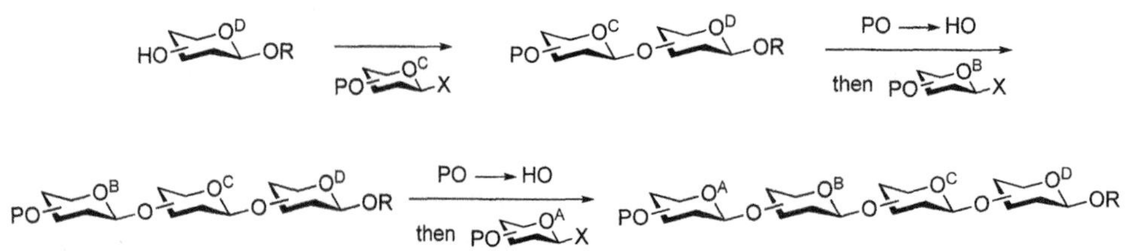

---

[b] In these diagrammatic presentations, no stereochemistry is implied at the anomeric carbon of any generalized structure.

Several modern versions of the linear approach exist, some of which involve *'two-stage activation'* methods:[8–12]

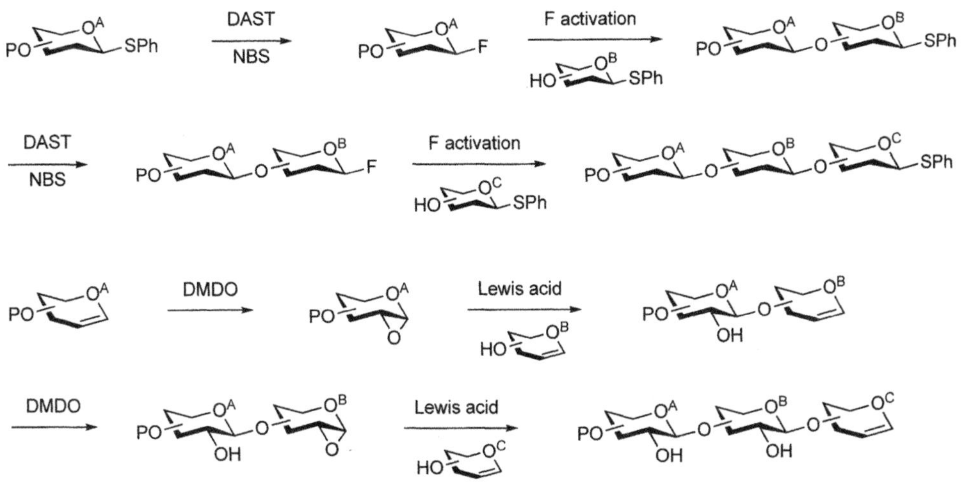

The success of many of these methods obviously relies on the *selective activation* of one glycosyl donor over another (in the above sequences, fluoride over thioglycoside and epoxide over glycal), and there are countless other examples, including bromide over thioglycoside[13] and selenoglycoside over thioglycoside.[14] The 'latent/active' concept discussed earlier is another variant, as is the concept of 'orthogonality'.[1,15,16]

Another useful approach involves the *chemoselective activation* of one glycosyl donor over another and is very much connected with the 'armed/disarmed' concept discussed earlier:

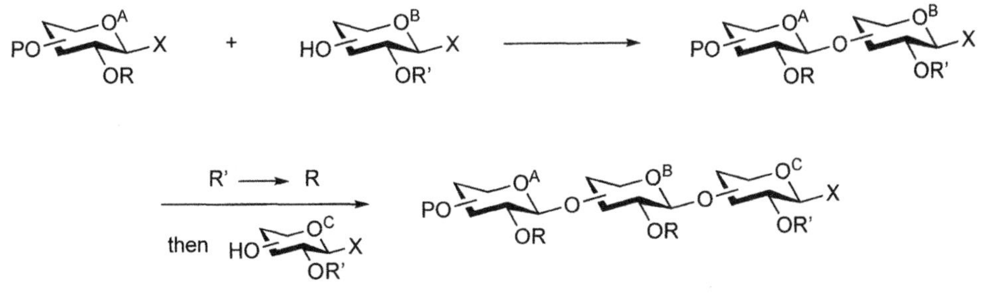

The glycosides A and B possess the same aglycon (X) but act as donor and acceptor, respectively, because of the activating nature of R (A) and the deactivating nature of R′ (B).[1,17,18] In the second step of the above sequence, the protecting group interchange (R′ to R) can sometimes be avoided; for example, in the case of thioglycoside donors, the first step can be achieved with a weak promoter (IDCClO$_4$) and the second with the stronger NIS/TfOH combination.

Convergent syntheses: A *convergent synthesis* assembles the oligosaccharide from smaller, pre-formed components:

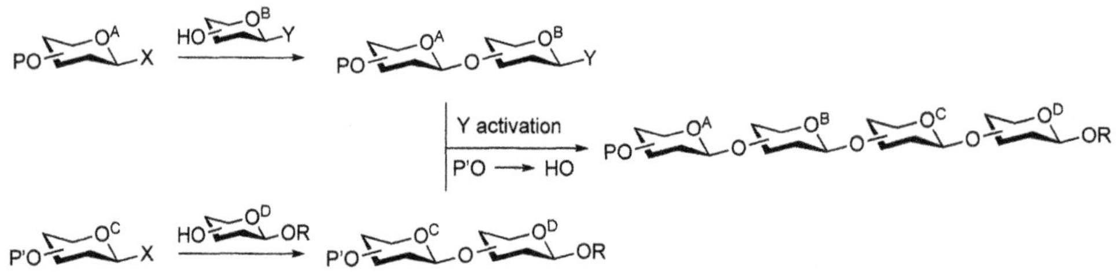

The convergent approach, often dubbed a 'block synthesis' (for obvious reasons), has several advantages over its linear counterpart. The smaller components can often be obtained from readily available disaccharides (lactose, cellobiose), an expensive monosaccharide can be introduced late in the synthetic sequence, and generally, there are fewer overall linear steps. In general, the concepts of two-stage and selective/chemoselective activation, 'latent/active' and orthogonality lend themselves very nicely to convergent syntheses.[1,19]

Two-directional syntheses: Boons has reported a highly convergent approach to oligosaccharide synthesis, whereby a monosaccharide derivative is constructed such that it is capable of acting first as a glycosyl donor and then as a glycosyl acceptor:[20]

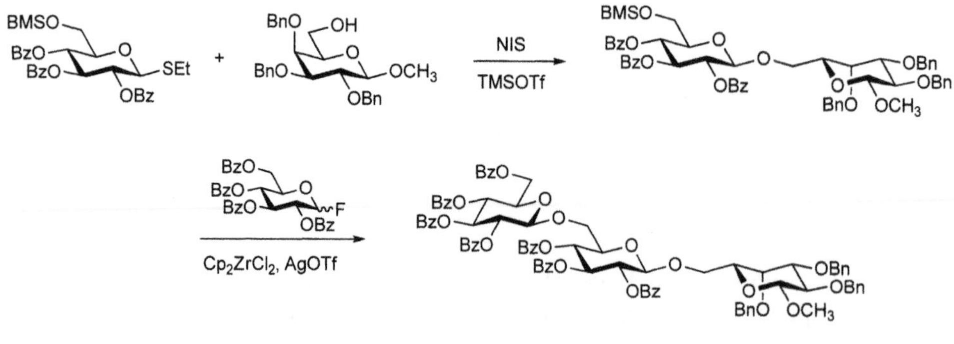

This innovative method, which builds on many of the principles established for glycoside synthesis, has been applied to the synthesis of a pentasaccharide that is associated with the hyperacute rejection response in xenotransplantation from pig to humans.[21]

'One-pot' syntheses: A tantalizing goal in oligosaccharide synthesis is the assembly of the molecule in question in 'one-pot', viz. by the step-wise addition of building blocks to the growing chain, with no need for manipulation of protecting groups and anomeric activating groups and with complete stereochemical control. Such an achievement would rival the successes of polymer-based syntheses (to be discussed next), and notable strides have been made after the original suggestion of the concept by Takahashi:[19,22–31]

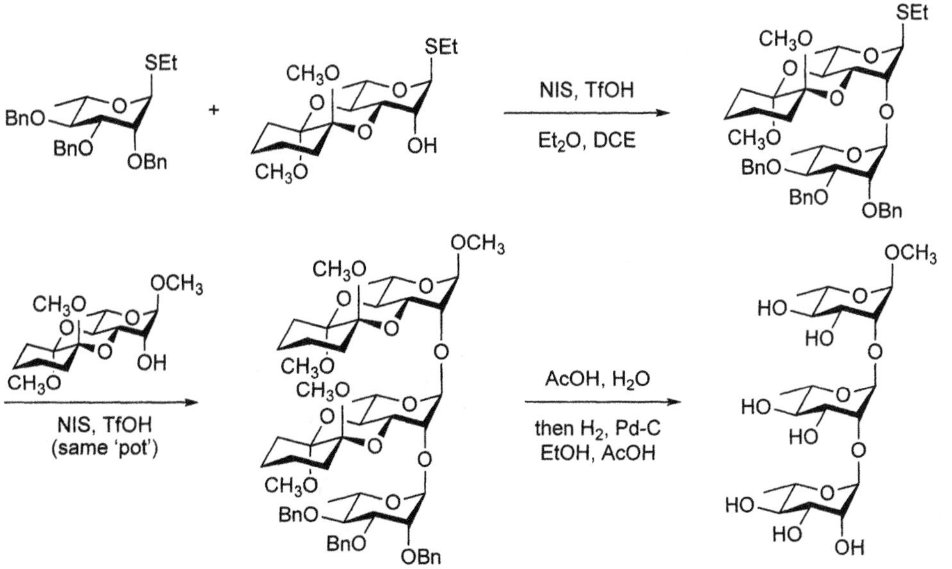

Wong has described 'programmable, one-pot oligosaccharide synthesis' based on a set of thioglycosides arranged in increasing order of reactivity in a database. This valuable database aims to allow a non-specialist to prepare a given oligosaccharide in a controlled manner.[32–37] Another recent approach is to use the pre-activation of tolyl thioglycosides in order to circumvent the varying reactivity of the glycosyl donors.[38]

There have been many interesting syntheses of oligosaccharides in the past decade,[39–66] none of them more so than the syntheses of a complex penta-antennary N-glycan,[67] and a docosanasaccharide arabinan from *Mycobacterium tuberculosis*, the causative agent of tuberculosis:[68]

References start on page 219

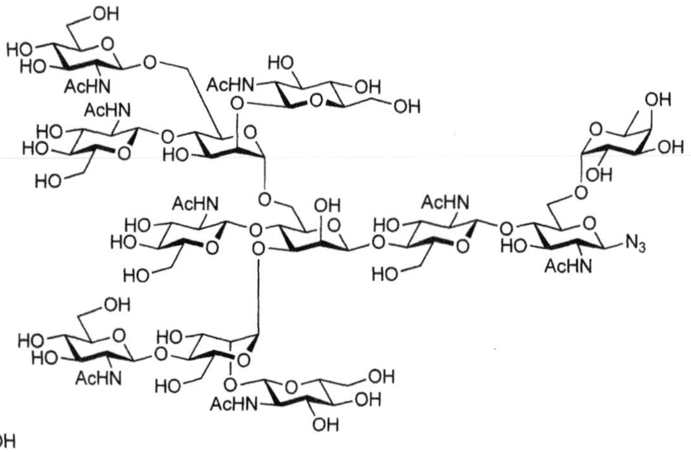

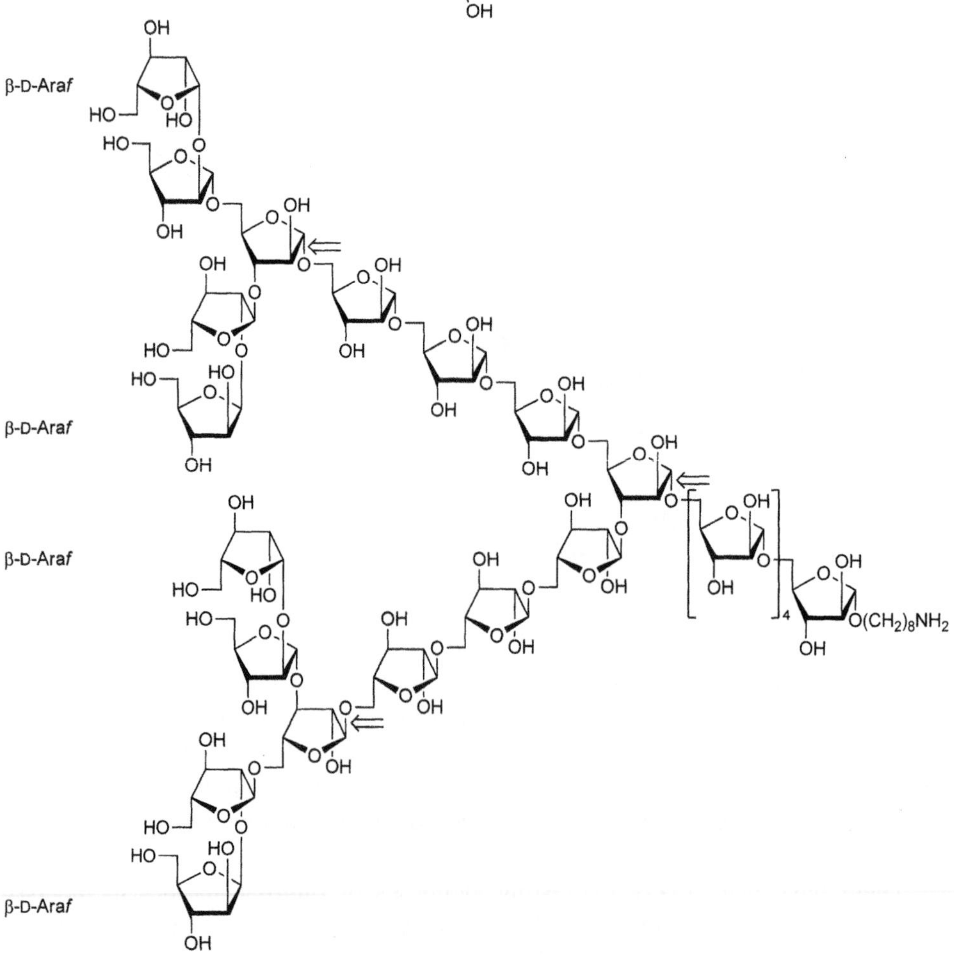

β-D-Ara*f*

β-D-Ara*f*

β-D-Ara*f*

β-D-Ara*f*

The first synthesis is notable in that it uses various protecting groups for hydroxyl, phthalimido and trifluoroacetyl for primary amine, and the ever-reliable thioglycoside and trichloroacetimidate for the construction of the four glycosidic linkages:

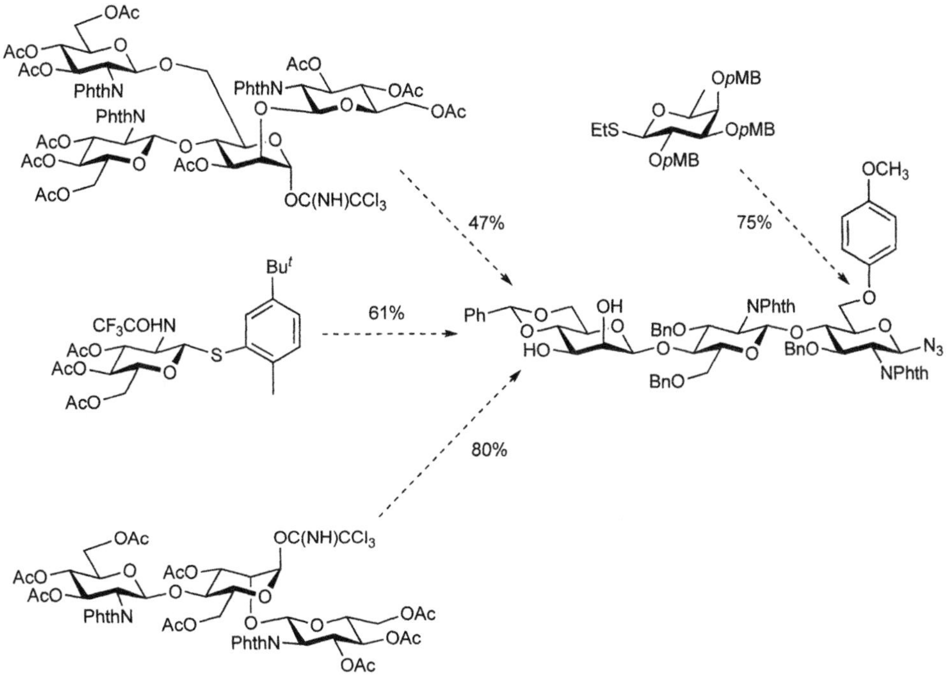

The synthesis of the arabinan is complicated by the presence of four β-D-arabinofuranoside moieties, the construction of which offers problems similar to those encountered in the preparation of β-D-mannopyranosides. However, judicious disconnections (⇐) within the target molecule allow the preparation of three building blocks, assembled mainly with thioglycoside donors and ably protected by resilient benzoyl groups. These building blocks are then joined using the trichloroacetimidate donor derived from **B** (with **A**), followed by desilylation, and then the trichloroacetimidate derived from **C**:

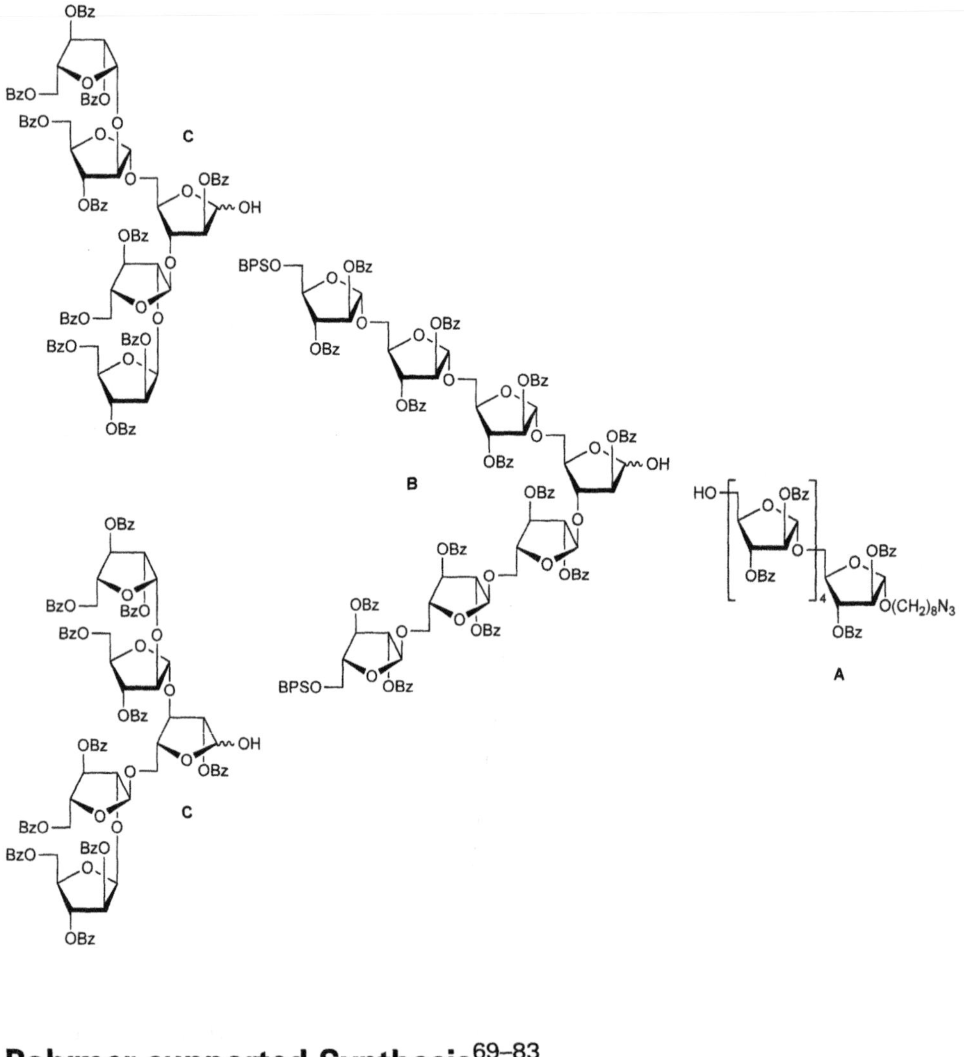

Polymer-supported Synthesis[69–83]

Ever since the invention of the 'solid phase' method for the synthesis of oligopeptides by Merrifield (which ultimately won the Nobel Prize in Chemistry in 1984[84]), and the similar success later enjoyed with oligonucleotides, carbohydrate chemists have dreamt of utilizing the method for the synthesis of oligosaccharides. However, whereas a growing peptide chain on a resin bead presents, after minimal protecting group manoeuvres, a single primary amine (secondary for proline) of predictable reactivity for coupling to a likewise predictable carboxylic acid, such is not the case with carbohydrates. Whether a monosaccharide is attached to a resin bead via an anomeric or a non-anomeric hydroxyl group, the subsequent construction of a glycosidic linkage is plagued by all of the common problems – the influence and removal of protecting groups, the regioselectivity and stereoselectivity of the glycosidation process, the role of the solvent, and the nature of the donor and the associated promoter. In addition, whereas the 20 or so common amino acids are all available with the amino and side-chain functional groups protected in any number of reliable ways, such is not the case with monosaccharides; there is, as yet, not even an orthogonally protected form of D-glucose that could be viewed as a general glycosyl donor.[85,86] However, enormous advances in the use of polymers for the synthesis of oligosaccharides have been made over the last decade, and a broad overview of the area will be given here.

All new methods must stand the test of time. At present, the vast majority of synthetic chemists would turn to conventional methods to synthesize an oligosaccharide. In fact, a rare study compared the solution and the solid-phase synthesis of a hexasaccharide, concluding that the former was generally more efficient;[87] another study, this time concerned with the synthesis of various β-D-mannopyranosides, noticed little difference.[88]

Types of polymers:[89,90] Of the *insoluble polymers*, which are available as discrete beads, the original (Merrifield) cross-linked polystyrene is still used; the chloromethyl group enables the attachment of a 'linker', and then the sugar. 'Wang resin'[91] offers a primary alcohol for attachment, and this can be modified to form an aldehyde;[92] 'TentaGel' has a polystyrene core that is functionalized with poly(ethylene glycol).[93,94] Most of the resins are commercially available with a variety of different functional groups, e.g. Merrifield resin with Cl, OH or NH_2 and TentaGel with Br, OH, NH_2, SH or COOH. Whereas all of the above resins need to 'swell' to achieve good 'loadings' (some macroporous resins, e.g. ArgoPore®, can be used directly), such is not the case with 'controlled-pore glass' – the highly hydroxylated surface is ideal for functionalization by a linker, and then the carbohydrate.[95,96]

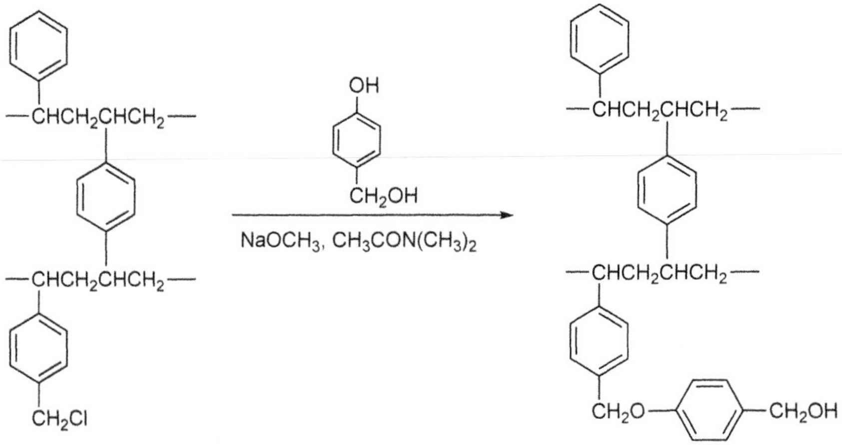

Merrifield resin

Wang resin

cross-linked polystyrene
|
$CH_2O(CH_2CH_2O)_nH$

TentaGel

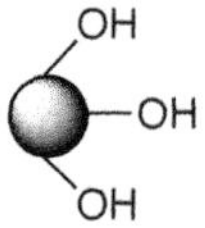

controlled-pore glass

Soluble polymers offer the advantage of being able to conduct the desired chemical reactions in solution, yet display their versatility as insoluble solids during the purification step. The most common soluble polymer that is used in the synthesis of oligosaccharides is *O*-methyl poly(ethylene glycol):[97,98]

$$CH_3O(CH_2CH_2O)_nH$$

Linkers: One sugar of the desired oligosaccharide must, naturally, be first attached to the polymer. This is usually done by employing a *linker* that seems to confer better reactivity on the sugar than if it were attached directly to the polymer itself. The linker, of course, must be stable to all of the synthetic manipulations that will be carried out but must be easily broken to liberate the final product. Because of the lability of the glycosidic linkage towards acid, the useful linkers employed are esters,[94,96] vinylogous amides,[99] benzylic ethers,[100,101] photolabile groups,[69,102,103] silyl ethers,[104] silanes[105] and alkenes,[106,107] all removed by mild and selective reagents.

Attachment of the sugar to the linker/polymer: There are obviously just two discrete ways to attach the first monosaccharide to the linker/polymer, either through the anomeric centre (a glycosidic linkage) or by utilizing one of the other hydroxyl groups on the (pyranose) ring:

For the attachment through a glycosidic linkage, a subsequent deprotection step liberates the hydroxyl group of interest, and another glycosidation can be performed to give a disaccharide now attached to the linked polymer. For the non-anomeric linkage, an acceptor and a promoter are added to the polymer-bound glycosyl donor, and an attached disaccharide again results. There is generally a preference for attachment through a glycosidic linkage, as this allows the use of an excess of the soluble glycosyl donor, ensuring better yields in elongation steps.

The glycosyl donors used: In line with the success of oligosaccharide synthesis in solution, the common glycosyl donors in polymer-supported syntheses are trichloroacetimidates, pentenyl glycosides, glycosyl sulfoxides, thioglycosides and glycals.[108]

Insoluble *versus* soluble polymers: Although much of the early work on the polymer-supported synthesis of oligosaccharides was carried out on insoluble resins, there is something of a modern trend towards the use of soluble polymers. One disadvantage of soluble polymers such as poly(ethyleneglycol) is their low loading capacity; dendritic polymers go some way to avoid this problem.[109,110] An advantage of soluble polymers is that real-time monitoring of the reactions occurring in solution is possible.

In syntheses using insoluble polymers, a commercial resin (purchased as beads) is swollen in a solvent and the chosen linker attached. Next, a monosaccharide is joined to the linker (as either a glycoside or some sort of ether or ester), with complete reaction ensured by the use of an excess of the reagent(s). Following a wash step, any necessary deprotection and glycosyl donor activation, the next monosaccharide is added; the cycle is repeated until the desired oligosaccharide has been assembled. The final deprotection of the many

References start on page 219

hydroxyl groups can precede the liberation of the oligosaccharide from the resin, but this is not always so – it is virtually impossible to remove benzyl ethers 'on resin' by hydrogenolysis.

For the soluble polymer approach, various commercial products are available, which are also attached to a suitable linker, utilizing any solvent apart from ethers. The first monosaccharide is then attached to the linker, via the anomeric or the non-anomeric hydroxyl group (this choice depends on the nature of the linker).[73,98] In the former case, the deprotection of the desired hydroxyl group, followed by treatment with an excess of the next glycosyl donor and promoter, gives the disaccharide attached to the polymer, still in solution. Addition of an ether (usually *tert*-butyl methyl ether) precipitates this product, which may be collected and washed. Repetition of the cycle yields, after deprotection and cleavage from the resin, the desired oligosaccharide.

For such polymer-supported syntheses to be successful, the choice of polymer is important,[96] more active promoters (e.g., dibutylboron triflate with trichloroacetimidates[111]) may be needed to counter the often sluggish glycosylation steps, multiple exposure to reagents is often required,[94] and 'capping' of unreacted residues at each stage of the synthesis helps to improve the purity of the final product.[94,112] Somewhat surprisingly, there is often a marked improvement in the anomeric stereoselectivity for glycosidations performed on a polymer support over those conducted in normal solution.[113]

Trichloroacetimidates[95,114]

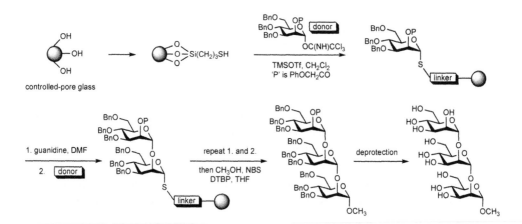

Pentenyl glycosides:[103]

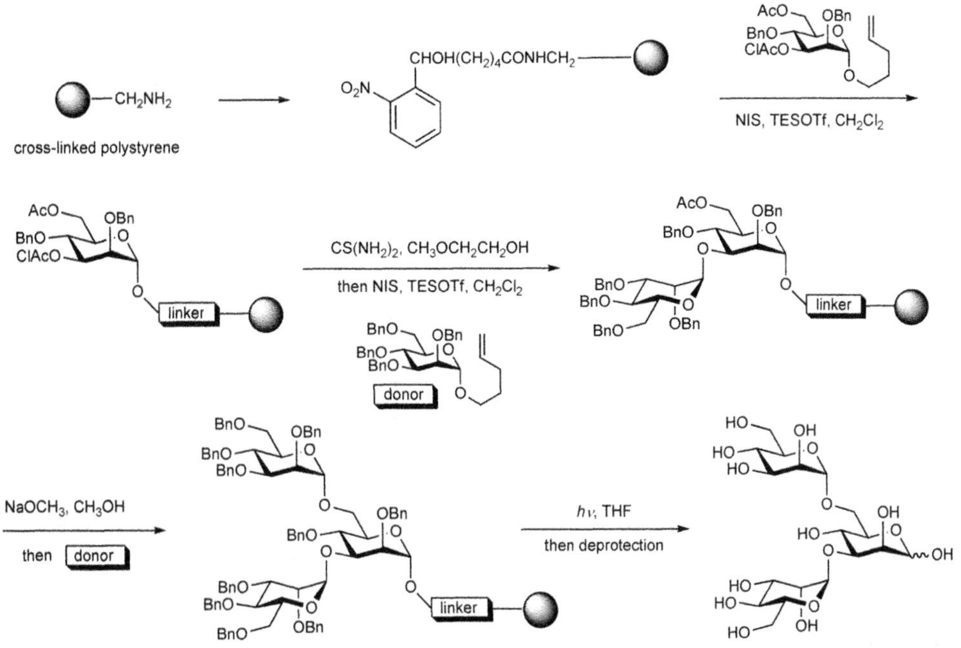

Glycosyl sulfoxides:[115]

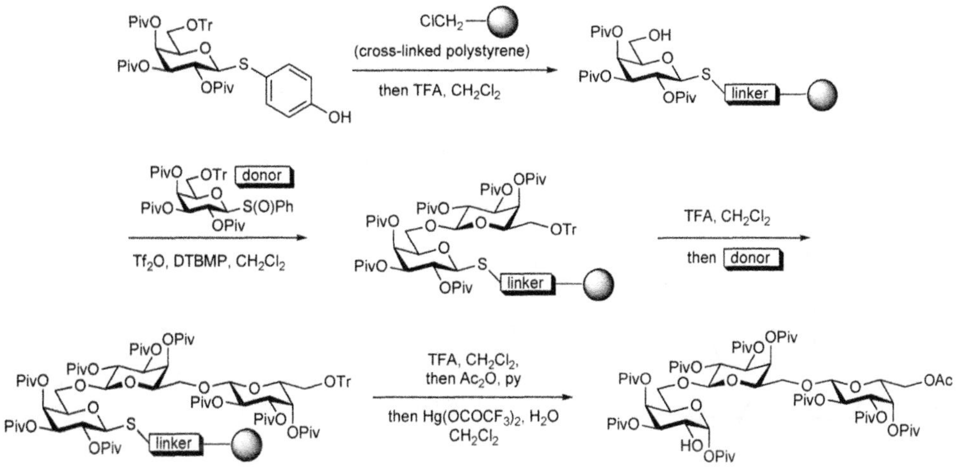

Thioglycosides: Several polymer-supported syntheses utilizing thioglycosides as the donor have been reported; two syntheses of interest utilize O-methyl poly(ethylene glycol) with a succinyl linker,[112] and Merrifield resin, again with a photolabile linker:[116]

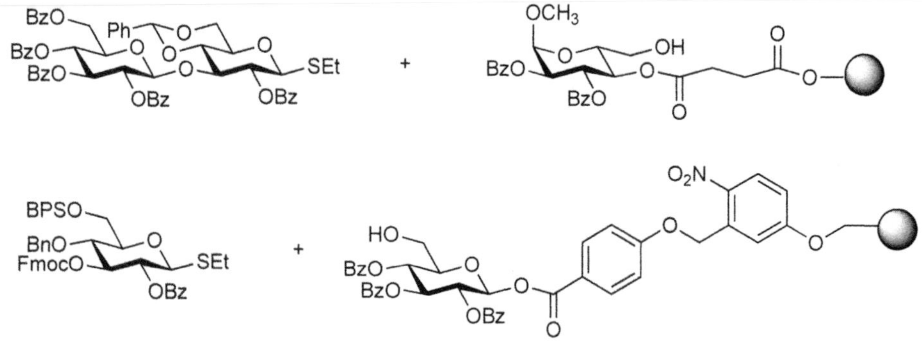

Glycals:[117] Danishefsky, in line with the success of the 'glycal assembly' approach for the synthesis of oligosaccharides in solution, has applied the method to a solid support, generally attaching the first monosaccharide as a glycosyl *donor*. One of the most impressive outcomes is the synthesis of a Lewis[b] blood group determinant glycal:[117,118]

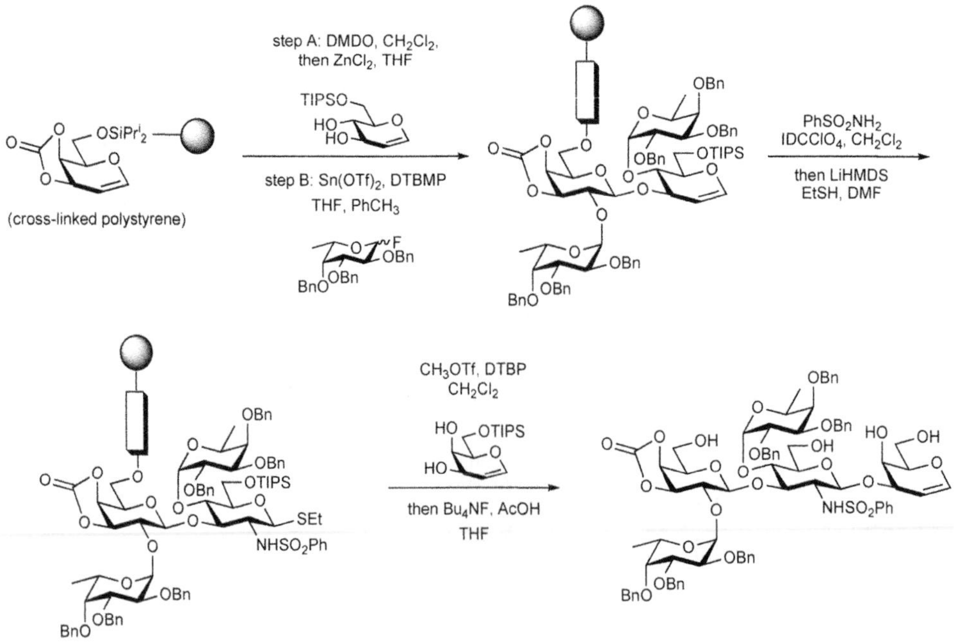

In his work, Danishefsky has never shied away from converting his beloved glycal, when necessary, into a more appropriate donor, e.g. a thioglycoside.[117,119]

Finally, Ito and Ogawa have made significant contributions to polymer-supported oligosaccharide synthesis. Apart from application of the orthogonal glycosylation strategy[120] and intramolecular aglycon delivery,[121] these two collaborators have introduced the concept of a 'hydrophobic tag' that allows for the convenient purification (reversed-phase silica gel) of the oligosaccharide after detachment from the polymer:[122]

Ito has also been responsible for much of the development of 'capture-release' strategies that offer an alternative method of obtaining the desired oligosaccharide from what is usually a complex mixture.[123,124]

Automated oligosaccharide synthesis: The ultimate aim in oligosaccharide synthesis would be to program a 'machine' to do the job, in much the same way as one does with an oligopeptide or oligonucleotide. Perhaps, the previously mentioned work of Wong has gone some way to achieving this aim;[32–36] however, it would be safe to say that Seeberger has been the main driving force behind the automation of oligosaccharide synthesis.[125–133] Much of Seeberger's work uses an octenediol linker that, by olefin cross-metathesis, conveniently liberates the final oligosaccharide from the resin as a pentenyl glycoside; also, glycosyl phosphates tend to be the donors of choice (see Chapter 12).

Combinatorial synthesis and the generation of 'libraries': The past two decades witnessed a change in the philosophy for the discovery of compounds that, many hoped, would eventually lead to novel pharmaceuticals. Traditionally, chemists have worked with botanists, zoologists and even indigenous people to select plants and organisms that, upon chemical extraction and analysis, may yield a chemical with interesting pharmacological properties. Tinkering with the chemical structure of this natural product has often given a better pharmaceutical, one with a broader and/or improved spectrum of activity; a classical example would be that of penicillin.

In these traditional approaches, extensive synthetic effort was invariably required; many have experienced the heartbreak of a long synthesis, only to find that the target molecule exhibited none of the desired properties. The computer-assisted design of target molecules has improved things somewhat but is generally of use only when a deal is known about the mode of action of a disease-causing organism.[134]

A novel approach to 'drug discovery' was suggested, and so was born *combinatorial synthesis*,[69,135–141] the generation of a 'library' of closely related structures by (solid) polymer-supported synthesis. In the 'split and mix' (split and pool) approach, a functionalized bead sample is treated separately with, say, three different but related reagents (A1–A3) to generate three newly functionalized bead samples. These three bead samples are then mixed and subsequently split into three before being separately treated again with three new reagents (B1–B3), thus generating a mixture of nine differently functionalized beads (A1B1–A3B3). The process is repeated (split, treat with C1–C3 and mix), and the final library (here, of 27 compounds) consists of resin beads, to each of which are attached multiple copies of a unique chemical (A1B1C1–A3B3C3). Screening of this library for some sort of biological or pharmacological activity is followed by isolation of the active bead – this is where the fun and games start, with identification of the unique chemical on the active bead!

Although it is often a straightforward matter to generate a library of many thousands of different compounds, the screening and structure identification steps present huge problems;[142–144] no one has put it better than Clark Still:[145]

> As for encoding in particular, it seems to complete one of the most powerful of combinatorial methodologies: split-and-pool synthesis and on-bead property screening. Thus, split-and-pool synthesis provides access to large libraries conveniently, on-bead screening allows the efficient selection of library members having a property of interest, and encoding provides a straightforward path to the structures of selected members. These three methods work exceptionally well together and provide a complete methodology for combinatorial exploration of chemical design problems.

Owing to the multifunctional nature of monosaccharides and to the general difficulties associated with anomeric stereocontrol, it was not surprising that it took until 1995 for Hindsgaul to announce the first library of carbohydrates (trisaccharides), prepared in solution without the aid of a polymer support.[146] This result was soon followed by a report from Boons, detailing the preparation of a library of trisaccharides with no attempt to control the anomeric stereoselectivity (this simple and expedient operation substantially increases the complexity of the library):[147]

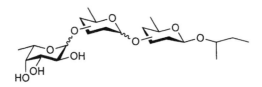

Polymer-supported syntheses were soon to follow, with one of the most notable syntheses involving the use of glycosyl sulfoxides and an ingenious screening/encoding system that allowed the identification and structure determination of an 'active' molecule.[148–150] Improvements in the construction of carbohydrate libraries, both on and off polymer supports,[151,152] continue to be announced.[153–161]

It is interesting to quote from a recent review (2007): 'the shift away from large combinatorial libraries has continued, with the emphasis now being on small, focused (100 to 3000) collections that contain much of the 'structural aspects' of natural products'.[162] The same review reports that only one de novo chemical entity has been discovered through combinatorial chemistry, namely sorafenib (Nexavar, from Bayer), an anti-tumour drug approved by the FDA in 2005.

References

1. Boons, G.-J. (1996). *Tetrahedron*, **52**, 1095.
2. Fügedi, P. (2006). In *The Organic Chemistry of Sugars* (D.E. Levy and P. Fügedi, eds) p. 181. CRC Press.
3. Demchenko, A.V. (2005). *Lett. Org. Chem.*, **2**, 580.
4. Pellissier, H. (2005). *Tetrahedron*, **61**, 2947.
5. Paulsen, H. (1982). *Angew. Chem. Int. Ed. Engl.*, **21**, 155.
6. Larsen, K., Worm-Leonhard, K., Olsen, P., Hoel, A. and Jensen, K.J. (2005). *Org. Biomol. Chem.*, **3**, 3966.
7. Takatani, M., Nakano, J., Arai, M.A., Ishiwata, A., Ohta, H. and Ito, Y. (2004). *Tetrahedron Lett.*, **45**, 3929.
8. Nicolaou, K.C., Caulfield, T.J. and Groneberg, R.D. (1991). *Pure Appl. Chem.*, **63**, 555.
9. Sliedregt, L.A.J.M., van der Marel, G.A. and van Boom, J.H. (1994). *Tetrahedron Lett.*, **35**, 4015.
10. Friesen, R.W. and Danishefsky, S.J. (1989). *J. Am. Chem. Soc.*, **111**, 6656.
11. Halcomb, R.L. and Danishefsky, S.J. (1989). *J. Am. Chem. Soc.*, **111**, 6661.
12. Williams, L.J., Garbaccio, R.M. and Danishefsky, S.J. (2000). In *Carbohydrates in Chemistry and Biology* (B. Ernst, G.W. Hart and P. Sinaÿ, eds) Vol. 1, p. 61. Wiley-VCH.
13. Fügedi, P., Birberg, W., Garegg, P.J. and Pilotti, A. (1987). *Carbohydr. Res.*, **164**, 297.
14. Mehta, S. and Pinto, B.M. (1993). *J. Org. Chem.*, **58**, 3269.

15. Kanie, O. (2000). In *Carbohydrates in Chemistry and Biology* (B. Ernst, G.W. Hart and P. Sinaÿ, eds) Vol. 1, p. 407. Wiley-VCH.
16. Bosse, F., Marcaurelle, L.A. and Seeberger, P.H. (2002). *J. Org. Chem.*, **67**, 6659.
17. Mootoo, D.R., Konradsson, P., Udodong, U. and Fraser-Reid, B. (1998). *J. Am. Chem. Soc.*, **110**, 5583.
18. Veeneman, G.H., van Leeuwen, S.H. and van Boom, J.H. (1990). *Tetrahedron Lett.*, **31**, 1331.
19. Boons, G.-J. and Zhu, T. (1997). *Synlett*, 809.
20. Zhu, T. and Boons, G.-J. (1998). *Tetrahedron Lett.*, **39**, 2187.
21. Zhu, T. and Boons, G.-J. (1998). *J. Chem. Soc., Perkin Trans. 1*, 857.
22. Yamada, H., Harada, T., Miyazaki, H. and Takahashi, T. (1994). *Tetrahedron Lett.*, **35**, 3979.
23. Yamada, H., Harada, T. and Takahashi, T. (1994). *J. Am. Chem. Soc.*, **116**, 7919.
24. Raghavan, S. and Kahne, D. (1993). *J. Am. Chem. Soc.*, **115**, 1580.
25. Ley, S.V. and Priepke, H.W.M. (1994). *Angew. Chem. Int. Ed. Engl.*, **33**, 2292.
26. Yu, B., Yang, Z. and Cao, H. (2005). *Curr. Org. Chem.*, **9**, 179.
27. Wang, Y., Ye, X.-S. and Zhang, L.-H. (2007). *Org. Biomol. Chem.*, **5**, 2189.
28. Huang, X., Huang, L., Wang, H. and Ye, X.-S. (2004). *Angew. Chem. Int. Ed.*, **43**, 5221.
29. Fridman, M., Solomon, D., Yogev, S. and Baasov, T. (2002). *Org. Lett.*, **4**, 281.
30. Hashihayata, T., Ikegai, K., Takeuchi, K., Jona, H. and Mukaiyama, T. (2003). *Bull. Chem. Soc. Jpn.*, **76**, 1829.
31. Codée, J.D.C., van den Bos, L.J., Litjens, R.E.J.N., Overkleeft, H.S., van Boom, J.H. and van der Marel, G.A. (2003). *Org. Lett.*, **5**, 1947.
32. Zhang, Z., Ollmann, I.R., Ye, X.-S., Wischnat, R., Baasov, T. and Wong, C.-H., *J. Am. Chem. Soc.*, 1999, **121**, 734.
33. Ye, X.-S. and Wong, C.-H. (2000). *J. Org. Chem.*, **65**, 2410.
34. Koeller, K.M. and Wong, C.-H. (2000). *Chem. Rev.*, **100**, 4465.
35. Mong, K.-K.T. and Wong, C.-H. (2002). *Angew. Chem. Int. Ed.*, **41**, 4087.
36. Hanson, S., Best, M., Bryan, M.C. and Wong, C.-H. (2004). *Trends Biochem. Sc.*, **29**, 656.
37. Lee, J.-C., Greenberg, W.A. and Wong, C.-H. (2006). *Nat. Protoc.*, **1**, 3143.
38. Miermont, A., Zeng, Y., Jing, Y., Ye, X.-s. and Huang, X. (2007). *J. Org. Chem.*, **72**, 8958.
39. Rana, S.S., Vig, R. and Matta, K.L. (1982–1983). *J. Carbohydr. Chem.*, **1**, 261.
40. Paulsen, H. (1990). *Angew. Chem. Int. Ed. Engl.*, **29**, 823.
41. Dan, A., Ito, Y. and Ogawa, T. (1995). *J. Org. Chem.*, **60**, 4680.
42. Demchenko, A. and Boons, G.-J. (1997). *Tetrahedron Lett.*, **38**, 1629.
43. Cato, D., Buskas, T. and Boons, G.-J. (2005). *J. Carbohydr. Chem.*, **24**, 503.
44. Nilsson, M. and Norberg, T. (1998). *J. Chem. Soc., Perkin Trans. 1*, 1699.
45. Allen, J.R. and Danishefsky, S.J. (2001). *ACS Symposium Series 796*, p. 299. Washington, DC: American Chemical Society.
46. Crich, D. and Dai, Z. (1998). *Tetrahedron Lett.*, **39**, 1681.
47. Weiler, S. and Schmidt, R.R. (1998). *Tetrahedron Lett.*, **39**, 2299.
48. Schmidt, R.R. (1998). *Pure Appl. Chem.*, **70**, 397.
49. Garegg, P.J., Konradsson, P., Lezdins, D., Oscarson, S., Ruda, K. and Öhberg, L. (1998). *Pure Appl. Chem.*, **70**, 293.
50. Baeschlin, D.K., Chaperon, A.R., Green, L.G., Hahn, M.G., Ince, S.J. and Ley, S.V. (2000). *Chem. Eur. J.*, **6**, 172.
51. Düffels, A. and Ley, S.V. (1999). *J. Chem. Soc., Perkin Trans. 1*, 375.
52. Lu, J. and Fraser-Reid, B. (2005). *Chem. Commun.*, 862.
53. Schratt, X. and Unverzagt, C. (2005). *Tetrahedron Lett.*, **46**, 691.
54. Matsuo, I., Kashiwagi, T., Totani, K. and Ito, Y. (2005). *Tetrahedron Lett.*, **46**, 4197.
55. Crich, D., Li, W. and Li, H. (2004). *J. Am. Chem. Soc.*, **126**, 15081.

56. López-Prados, J. and Martín-Lomas, M. (2005). *J. Carbohydr. Chem.*, **24**, 393.
57. Sundgren, A., Lahmann, M. and Oscarson, S. (2005). *J. Carbohydr. Chem.*, **24**, 379.
58. Gemma, E., Lahmann, M. and Oscarson, S. (2005). *Carbohydr. Res.*, **340**, 2558.
59. Dziadek, S., Brocke, C. and Kunz, H. (2004). *Chem. Eur. J.*, **10**, 4150.
60. Pozsgay, V., Coxon, B., Glaudemans, C.P.J., Schneerson, R. and Robbins, J.B. (2003). *Synlett*, 743.
61. Pozsgay, V., Ekborg, G. and Sampathkumar, S.-G. (2006). *Carbohydr. Res.*, **341**, 1408.
62. Attolino, E., Cumpstey, I. and Fairbanks, A.J. (2006). *Carbohydr. Res.*, **341**, 1609.
63. Zhang, J., Ma, Z. and Kong, F. (2003). *Carbohydr. Res.*, **338**, 2039.
64. Galoni, D.P. and Gin, D.Y. (2007). *Nature*, **446**, 1000.
65. van der Marel, G.A., van den Bos, L.J., Overkleeft, H.S., Litjens, R.E.J.N. and Codée, J.D.C. (2007). In *Frontiers in Modern Carbohydrate Chemistry* (A.V. Demchenko, ed.) ACS Symposium Series 960, p. 190. Washington, DC: American Chemical Society.
66. Zeng, Y. and Kong, F. (2003). *Carbohydr. Res.*, **338**, 2047.
67. Eller, S., Schuberth, R., Gundel, G., Seifert, J. and Unverzagt, C. (2007). *Angew. Chem. Int. Ed.*, **46**, 4173.
68. Joe, M., Bai, Y., Nacario, R.C. and Lowary, T.L. (2007). *J. Am. Chem. Soc.*, **129**, 9885.
69. Osborn, H.M.I. and Khan, T.H. (1999). *Tetrahedron*, **55**, 1807.
70. Brown, A.R., Hermkens, P.H.H., Ottenheijm, H.C.J. and Rees, D.C. (1998). *Synlett*, 817.
71. Danishefsky, S.J., McClure, K.F., Randolph, J.T. and Ruggeri, R.B. (1993). *Science*, **260**, 1307.
72. Seitz, O. (1998). *Angew. Chem. Int. Ed.*, **37**, 3109.
73. Früchtel, J.S. and Jung, G. (1996). *Angew. Chem. Int. Ed. Engl.*, **35**, 17.
74. Hermkens, P.H.H., Ottenheijm, H.C.J. and Rees, D. (1996). *Tetrahedron*, **52**, 4527.
75. Brown, R.C.D. (1998). *J. Chem. Soc., Perkin Trans. 1*, 3293.
76. Ito, Y. and Manabe, S. (1998). *Curr. Opin. Chem. Biol.*, **2**, 701.
77. Krepinsky, J.J. and Douglas, S.P. (2000). In *Carbohydrates in Chemistry and Biology* (B. Ernst, G.W. Hart and P. Sinaÿ, eds) Vol. 1, p. 239. Wiley-VCH.
78. Seeberger, P.H. (2002). *J. Carbohydr. Chem.*, **21**, 613.
79. Seeberger, P.H., ed. (2001). *Solid Support Oligosaccharide Synthesis and Combinatorial Carbohydrate Libraries*, 308 pp. Wiley-VCH.
80. Seeberger, P.H. and Haase, W.-C. (2000). *Chem. Rev.*, **100**, 4349.
81. Grathwohl, M., Drinnan, N., Broadhurst, M., West, M.L. and Meutermans, W. (2003). *Methods Enzymol.*, **369**, 248.
82. Diederichsen, U. and Wagner, T. (2003). In *Organic Synthesis Highlights V* (H.-G. Schmalz and T. Wirth, eds) p. 384. Wiley-VCH: Weinheim.
83. Schmidt, R.R., Jonke, S. and Liu, K.-g. (2007). In *Frontiers in Modern Carbohydrate Chemistry* (A.V. Demchenko, ed.) ACS Symposium Series 960, p. 209. American Chemical Society: Washington, DC.
84. Merrifield, R.B. (1985). *Angew. Chem. Int. Ed. Engl.*, **24**, 799.
85. Zhu, T. and Boons, G.-J. (2000). *Tetrahedron*: Asymm., **11**, 199.
86. Hünger, U., Ohnsmann, J. and Kunz, H. (2004). *Angew. Chem. Int. Ed.*, **43**, 1104.
87. Lam, S.N. and Gervay-Hague, J. (2002). *Carbohydr. Res.*, **337**, 1953.
88. Crich, D. and Smith, M. (2002). *J. Am. Chem. Soc.*, **124**, 8867.
89. Sherrington, D.C. (1998). *Chem. Commun.*, 2275.
90. Pittman, C.U., Jr. (2003). *Polymer News*, **28**, 183.
91. Wang, S.-S. (1973). *J. Am. Chem. Soc.*, **95**, 1328.
92. Hanessian, S. and Huynh, H.K. (1999). *Synlett*, 102.
93. Bayer, E. (1991). *Angew. Chem. Int. Ed. Engl.*, **30**, 113.
94. Adinolfi, M., Barone, G., De Napoli, L., Iadonisi, A. and Piccialli, G. (1996). *Tetrahedron Lett.*, **37**, 5007.

95. Heckel, A., Mross, E., Jung, K.-H., Rademann, J. and Schmidt, R.R. (1998). *Synlett*, 171.
96. Adinolfi, M., Barone, G., De Napoli, L., Iadonisi, A. and Piccialli, G. (1998). *Tetrahedron Lett.*, **39**, 1953.
97. Gravert, D.J. and Janda, K.D. (1997). *Chem. Rev.*, **97**, 489.
98. Krepinsky, J.J. (1996). In *Modern Methods in Carbohydrate Synthesis* (S.H. Khan and R.A. O'Neill, eds) p. 194. Harwood Academic.
99. Drinnan, N., West, M.L., Broadhurst, M., Kellam, B. and Toth, I., *Tetrahedron Lett.*, 2001, **42**, 1159.
100. Hodosi, G. and Krepinsky, J.J. (1996). *Synlett*, 159.
101. Roussel, F., Takhi, M. and Schmidt, R.R. (2001). *J. Org. Chem.*, **66**, 8540.
102. Nicolaou, K.C., Winssinger, N., Pastor, J. and DeRoose, F. (1997). *J. Am. Chem. Soc.*, **119**, 449.
103. Rodebaugh, R., Fraser-Reid, B. and Geysen, H.M. (1997). *Tetrahedron Lett.*, **38**, 7653.
104. Doi, T., Sugiki, M., Yamada, H., Takahashi, T. and Porco, J.A., Jr. (1999). *Tetrahedron Lett.*, **40**, 2141.
105. Weigelt, D. and Magnusson, G. (1998). *Tetrahedron Lett.*, **39**, 2839.
106. Timmer, M.S.M., Codée, J.D.C., Overkleeft, H.S., van Boom, J.H. and van der Marel, G.A. (2004). *Synlett*, 2155.
107. Mogemark, M., Gustafsson, L., Bengtsson, C., Elofsson, M. and Kihlberg, J. (2004). *Org. Lett.*, **6**, 4885.
108. Parlato, M.C., Kamat, M.N., Wang, H., Stine, K.J. and Demchenko, A.V., *J. Org. Chem.*, 2008, **73**, 1716.
109. Lubineau, A., Malleron, A. and Le Narvor, C. (2000). *Tetrahedron Lett.*, **41**, 8887.
110. Kantchev, E.A.B., Bader, S.J. and Parquette, J.R. (2005). *Tetrahedron*, **61**, 8329.
111. Wang, Z.-G., Douglas, S.P. and Krepinsky, J.J. (1996). *Tetrahedron Lett.*, **37**, 6985.
112. Verduyn, R., van der Klein, P.A.M., Douwes, M., van der Marel, G.A. and van Boom, J.H. (1993). *Recl. Trav. Chim. Pays-Bas*, **112**, 464.
113. Hunt, J.A. and Roush, W.R. (1996). *J. Am. Chem. Soc.*, **118**, 9998.
114. Rademann, J., Geyer, A. and Schmidt, R.R. (1998). *Angew. Chem. Int. Ed.*, **37**, 1241.
115. Yan, L., Taylor, C.M., Goodnow, R., Jr. and Kahne, D. (1994). *J. Am. Chem. Soc.*, **116**, 6953.
116. Nicolaou, K.C., Watanabe, N., Li, J., Pastor, J. and Winssinger, N. (1998). *Angew. Chem. Int. Ed.*, **37**, 1559.
117. Seeberger, P.H. and Danishefsky, S.J. (1998). *Acc. Chem. Res.*, **31**, 685.
118. Zheng, C., Seeberger, P.H. and Danishefsky, S.J. (1998). *Angew. Chem. Int. Ed.*, **37**, 786.
119. Zheng, C., Seeberger, P.H. and Danishefsky, S.J. (1998). *J. Org. Chem.*, **63**, 1126.
120. Ito, Y., Kanie, O. and Ogawa, T. (1996). *Angew. Chem. Int. Ed. Engl.*, **35**, 2510.
121. Ito, Y. and Ogawa, T. (1997). *J. Am. Chem. Soc.*, **119**, 5562.
122. Carrel, F.R. and Seeberger, P.H. (2008). *J. Org. Chem.*, **73**, 2058.
123. Ito, Y. and Manabe, S. (2002). *Chem. Eur. J.*, **8**, 3077.
124. Wu, J. and Guo, Z. (2006). *J. Org. Chem.*, **71**, 7067.
125. Seeberger, P.H. (2003). *Chem. Commun.*, 1115.
126. Plante, O.J., Palmacci, E.R. and Seeberger, P.H. (2003). *Methods Enzymol.*, **369**, 235.
127. Palmacci, E.R., Plante, O.J., Hewitt, M.C. and Seeberger, P.H. (2003). *Helv. Chim. Acta*, **86**, 3975.
128. Seeberger, P.H. and Werz, D.B. (2005). *Nat. Rev. Drug Discov.*, **4**, 751.
129. Werz, D.B. and Seeberger, P.H. (2005). *Chem. Eur. J.*, **11**, 3194.
130. Seeberger, P.H. and Werz, D.B. (2007). *Nature*, **446**, 1046.
131. Codée, J.D.C. and Seeberger, P.H. (2007). In *Frontiers in Modern Carbohydrate Chemistry* (A.V. Demchenko, ed.) ACS Symposium Series 960, p. 150. American Chemical Society: Washington, DC.
132. Castagner, B. and Seeberger, P.H. (2007). *Top. Curr. Chem.*, **278**, 289.

133. Seeberger, P.H. (2008). *Chem. Soc. Rev.*, **37**, 19.

134. Brown, R.D. and Newsam, J.M. (1998). *Chem. Ind.*, 785.

135. Terrett, N.K., Gardner, M., Gordon, D.W., Kobylecki, R.J. and Steele, J. (1995). *Tetrahedron*, **51**, 8135.

136. Lowe, G. (1995). *Chem. Soc. Rev.*, **24**, 309.

137. Thompson, L.A. and Ellman, J.A. (1996). *Chem. Rev.*, **96**, 555.

138. Balkenhohl, F., von dem Bussche-Hünnefeld, C., Lansky, A. and Zechel, C. (1996). *Angew. Chem. Int. Ed. Engl.*, **35**, 2288.

139. Tapolczay, D.J., Kobylecki, R.J., Payne, L.J. and Hall, B. (1998). *Chem. Ind.*, 772.

140. Sofia, M.J. (1996). *Drug Discov. Today*, **1**, 27.

141. Arya, P. and VNBS Sarma, B. (2006). In *The Organic Chemistry of Sugars* (D.E. Levy and P. Fügedi, eds) p. 729. CRC Press.

142. Asaro, M.F. and Wilson, R.B., Jr. (1998). *Chem. Ind.*, 777.

143. Nicolaou, K.C., Xiao, X.-Y., Parandoosh, Z., Senyei, A. and Nova, M.P. (1995). *Angew. Chem. Int. Ed. Engl.*, **34**, 2289.

144. Moran, E.J., Sarshar, S., Cargill, J.F., Shahbaz, M.M., Lio, A., Mjalli, A.M.M. and Armstrong, R.W. (1995). *J. Am. Chem. Soc.*, **117**, 10787.

145. Lebl, M. (1999). *J. Comb. Chem.*, **1**, 3.

146. Kanie, O., Barresi, F., Ding, Y., Labbe, J., Otter, A., Forsberg, L.S., Ernst, B. and Hindsgaul, O. (1995). *Angew. Chem. Int. Ed. Engl.*, **34**, 2720.

147. Boons, G.-J., Heskamp, B. and Hout, F. (1996). *Angew. Chem. Int. Ed. Engl.*, **35**, 2845.

148. Liang, R., Yan, L., Loebach, J., Ge, M., Uozumi, Y., Sekanina, K., Horan, N., Gildersleeve, J., Thompson, C., Smith, A., Biswas, K., Still, W.C. and Kahne, D. (1996). *Science*, **274**, 1520.

149. Borchardt, A. and Still, W.C. (1994). *J. Am. Chem. Soc.*, **116**, 373.

150. Nestler, H.P., Bartlett, P.A. and Still, W.C. (1994). *J. Org. Chem.*, **59**, 4723.

151. Kanie, O., Ohtsuka, I., Ako, T., Daikoku, S., Kanie, Y. and Kato, R. (2006). *Angew. Chem. Int. Ed.*, **45**, 3851

152. Ohtsuka, I., Ako, T., Kato, R., Daikoku, S., Koroghi, S., Kanemitsu, T. and Kanie, O. (2006). *Carbohydr. Res.*, **341**, 1476.

153. Johnson, M., Arles, C. and Boons, G.-J. (1998). *Tetrahedron Lett.*, **39**, 9801.

154. Izumi, M. and Ichikawa, Y. (1998). *Tetrahedron Lett.*, **39**, 2079.

155. Wong, C.-H., Ye, X.-S. and Zhang, Z. (1998). *J. Am. Chem. Soc.*, **120**, 7137.

156. Zhu, T. and Boons, G.-J. (1998). *Angew. Chem. Int. Ed.*, **37**, 1898.

157. Wunberg, T., Kallus, C., Opatz, T., Henke, S., Schmidt, W. and Kunz, H. (1998). *Angew. Chem. Int. Ed.*, **37**, 2503.

158. Kobayashi, S. (1999). *Chem. Soc. Rev.*, **28**, 1.

159. St. Hilaire, P.M., Lowary, T.L., Meldal, M. and Bock, K. (1998). *J. Am. Chem. Soc.*, **120**, 13312.

160. Boons, G.-J. (1999). *Carbohydrates in Europe*, Dec. 27, p. 28.

161. Yamago, S., Yamada, T., Ito, H., Hara, O., Mino, Y. and Yoshida, J.-i. (2005). *Chem. Eur. J.*, **11**, 6159.

162. Newman, D.J. and Cragg, G.M. (2007). *J. Nat. Prod.*, **70**, 461.

Chapter 6

Monosaccharide Metabolism

In the last five chapters we have looked at the structures, properties and reactivity of sugars, culminating in a discussion of methods available for the synthesis of the glycosidic linkage, and application of these methods to the synthesis of oligosaccharides. Now the book takes a different direction, ultimately leading us into the world of glycobiology. This (sixth) chapter provides an overview of the major pathways of monosaccharide metabolism.

In central metabolism, glucose is degraded to pyruvate, releasing energy and NADH (the glycolytic pathway). The pyruvate can enter into either aerobic metabolism to release additional energy (the citric acid cycle) or anaerobic metabolism to regenerate NAD^+ from NADH (homolactic or alcoholic fermentation). Alternatively, glucose enters the pentose phosphate pathway, generating the reducing power of cells in the form of NADPH, and 5-carbon precursors for nucleic acid biosynthesis. Glucose may also be synthesized from non-carbohydrate precursors (the glyoxylate cycle and gluconeogenesis). Glucose-6-phosphate is an important precursor for a range of other sugars and metabolites including fructose, mannose and inositols. Cells also produce large amounts of glycans and glycoconjugates containing a wide range of monosaccharides. These monosaccharides are biosynthesized in activated forms as sugar nucleoside (di)phosphates, which can act as glycosyl donors for glycosyltransferases, and as biosynthetic intermediates in the synthesis of other sugar nucleoside (di)phosphates and metabolites such as L-ascorbate. This chapter does not attempt to provide an encyclopaedic coverage of this topic, but focuses particularly on the more well-established vertebrate and bacterial pathways of carbohydrate metabolism. Interested readers are advised to consult any of a range of textbooks that provide authoritative coverage on many of these pathways.[1,2]

The Role of Charged Intermediates in Basic Metabolism

In 1958, in the essay entitled 'On the importance of being ionized', Davis noted that essentially all water-soluble, low molecular weight biosynthetic intermediates of metabolic reactions known at that time were predominantly completely

References start on page 249

ionized at neutral pH.[3] That is, these intermediates possessed either an acidic group with a pK_a value of less that 4 (e.g. carboxylate, phosphate, sulfonate) or a basic group with a pK_a value (for the conjugate acid) of greater than 10 (e.g. amine, amidine, guanidine, etc.). Davis suggested a possible explanation for this observation: charge modification of intermediates ensures more efficient retention of ionized compounds within the phospholipid membrane of the cell or its organelles. Exceptions to this generalization were poorly water soluble compounds such as lipids and steroids, polyhydroxylated compounds such as the higher sugars trehalose and starch, and the inositols. More generally, Davis noted that uncharged, water-soluble metabolites are found only as excretory and fermentation products, and occasionally as intermediates in purely degradative reactions (e.g. acetaldehyde in alcoholic fermentation). Nonetheless, while there are exceptions, Davis' generalization is useful in that it can assist in providing a rationale for the apparent complexity of many biochemical conversions and can help guide the understanding of the relative order of functional group transformations. For example, phosphorylation usually precedes decarboxylation, ensuring that no neutral intermediates are formed that may diffuse from the cell, and pathways that possess an amino or a carboxylate group throughout their entirety usually do not undergo phosphorylation, except to activate a leaving group.

Glucose-6-phosphate: a Central Molecule in Carbohydrate Metabolism

An entry point to understanding the pathways of carbohydrate metabolism is to consider the central position of glucose-6-phosphate. Glucose-6-phosphate lies near the start of two major pathways, glycolysis and the pentose phosphate pathway, and near the end of the β oxidation/glyoxylate/gluconeogenic pathways, which convert various non-carbohydrate precursors into glucose. Glucose-6-phosphate also acts as the key entry point into metabolism for dietary glucose, the glucose and glucose-1-phosphate released by the phosphorolysis/hydrolysis of glycogen, and galactose. Finally, glucose-6-phosphate is at the start of the biosynthetic pathways for most other sugars.

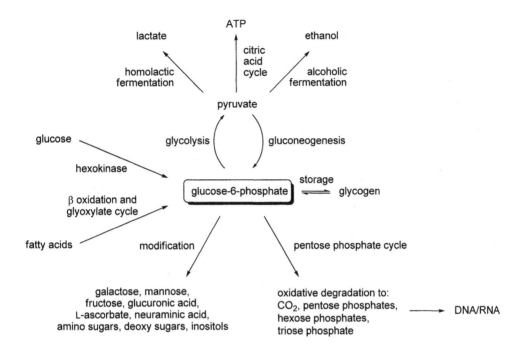

Glycolysis

Glucose and fructose are major sources of energy in higher organisms. Under normal conditions, blood glucose levels stay within the range of 4–8 mM (700–1400 mg l^{-1}).[a] Glucose arises in the blood through the direct consumption of glucose; the breakdown of higher polysaccharides in the mouth, duodenum or small intestine; to a small extent, by the breakdown of glycogen through the action of glycogen-debranching enzyme; or through its synthesis from non-carbohydrate sources through the pathway of gluconeogenesis. In the case of dietary consumption, glucose is actively transported from the intestinal lumen to the blood by sodium-glucose transporters (SGLTs).[1,5] SGLTs are also found in kidney tubules where they retrieve glucose from kidney filtrates. Once within the blood, glucose must be transported to cells. This is achieved by the action of energy-independent hexose transporters such as the family of glucose transporters (GLUTs).[5] After the entry of glucose into the cell, it is converted into glucose-6-phosphate by hexokinase, and is able to undergo glycolytic degradation. The

---

[a] We use a lowercase 'l' for litre (liter), following the rule that units that are not named after people are not capitalized. Interestingly, several fictional biographies have been invented purporting to provide details for a Claude Émile Jean-Baptiste Litre (who named his daughter Millicent!)[4] and for a Marco Guiseppe Litroni (who was speculated to have changed his name to Litre when he settled in France after a mafia-induced flight from his native Tuscany).

References start on page 249

process of glycolysis (also known as the Embden–Meyerhof–Parnas pathway)[b] converts glucose (a C_6 unit) into two molecules of pyruvate (C_3 units), and synthesizes two molecules each of ATP and NADH. While the major function of glycolysis is to generate energy in the form of ATP and NADH, a secondary function is to provide 3-carbon fragments that can be used in other metabolic pathways. Under aerobic conditions, pyruvate from glycolysis crosses into the mitochondria and enters the citric acid cycle, and NADH and $FADH_2$ are used to drive oxidative phosphorylation, regenerating NAD^+ and FAD^+ and leading to additional ATP. Alternatively, if insufficient oxygen is available for aerobic oxidation, NAD^+ must be regenerated by alternative means, and the cell will shift to anaerobic respiration and convert pyruvate to lactic acid in a process termed homolactic fermentation. In some organisms (e.g. yeast), alcoholic fermentation can take the place of homolactic fermentation, resulting in the formation of ethanol and carbon dioxide. As such, glycolysis provides the foundation for both aerobic and anaerobic metabolism.

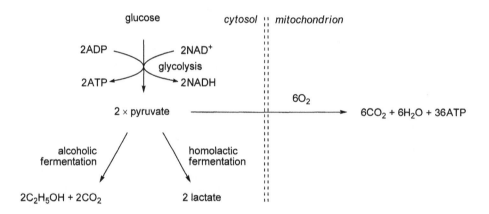

Glycolysis consists of the following steps:

1. *Phosphorylation of glucose and its conversion into glyceraldehyde-3-phosphate.* This occurs by the action of hexokinase, phosphoglucose isomerase (PGI), phosphofructokinase and aldolase. Aldolase affords glyceraldehyde-3-phosphate and dihydroxyacetone phosphate, which are interconverted by triose phosphate isomerase.
2. *Conversion of phosphorylated intermediates into 1,3-diphosphoglycerate (1,3-diPG) and phosphoenolpyruvate (PEP), compounds with high phosphate group transfer potential.* This occurs by the action of glyceraldehyde-3-phosphate

---

[b] Otto Fritz Meyerhof (1884–1951), M.D. (1909), University of Heidelberg, professorship at University of Pennysylvania (1940). Nobel Prize in Physiology or Medicine (1922).

dehydrogenase (GADPH; to 1,3-diPG), phosphoglycerate kinase and mutase, and enolase (to PEP). The action of GADPH reduces NAD$^+$ to NADH, which must be recycled back to NAD$^+$ for glycolysis to continue.

3. *Coupling of phosphate group transfer from 1,3-diPG and PEP to ATP synthesis.* This occurs by the action of phosphoglycerate kinase (for 1,3-diPG) and pyruvate kinase (for PEP).

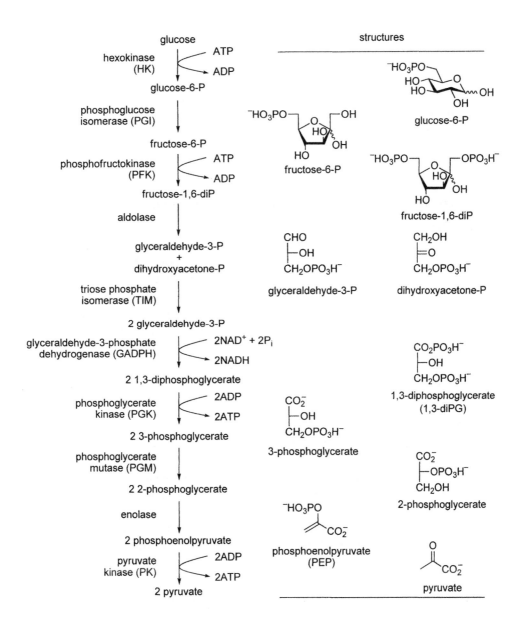

All enzymes in the glycolytic pathway have solved, three-dimensional structures and, despite the diversity of chemical steps that these enzymes perform, all possess a common α/β fold, with a core of β-sheet surrounded by α-helices.[6,7]

Fructose is a major dietary sugar and is also imported from the gut by an energy-independent transporter, GLUT-5.[1,5] Fructose is converted into fructose-1-phosphate in the liver and cleaved by fructose aldolase to dihydroxyacetone phosphate (which enters glycolysis) and glyceraldehyde. Glyceraldehyde is reduced to glycerol, and then phosphorylated and oxidized to dihydroxyacetone phosphate, when it is also able to enter the glycolytic pathway. Mannose can enter the glycolytic pathway through phosphorylation to mannose-6-phosphate and isomerization to fructose-6-phosphate by phosphomannose isomerase. Galactose can enter glycolysis through phosphorylation and epimerization reactions that convert it into glucose-1-phosphate, and then to glucose-6-phosphate through the action of phosphoglucomutase. These last two processes are discussed in more detail later in this chapter.

The Fate of Pyruvate in Primary Metabolism

Under aerobic conditions: For glycolysis to continue, NAD^+, which is present in cells in only limited quantities, must be recycled after its reduction to NADH (by GAPDH). Under aerobic conditions, NADH is oxidized to NAD^+ by the mitochondrial electron transport chain, and pyruvate enters the mitochondrion. Here, pyruvate is converted into acetyl-CoA by a three-component enzyme complex (pyruvate dehydrogenase, dihydrolipoyl transacetylase and dihydrolipoyl dehydrogenase), termed pyruvate dehydrogenase. The acetyl-CoA formed enters the citric acid cycle (also known as the tricarboxylic acid cycle or the Krebs cycle),[c] which results in the formation of numerous biosynthetic precursors and is responsible for a major portion of carbohydrate, fatty acid and amino acid oxidation, thereby generating the majority of ATP for the cell.

Under anaerobic conditions: If there is insufficient oxygen available to oxidize NADH, alternative processes are available for glycolysis to continue. Two processes that allow oxidation of NADH are *homolactic fermentation* and *alcoholic fermentation*. Homolactic fermentation occurs in muscles, particularly during vigorous activity when the demand for ATP is high and oxygen is depleted. Lactate dehydrogenase catalyses the oxidation of NADH by pyruvate and forms NAD^+ and (*S*)-lactate. Much of the lactate is exported from the muscle cell to the

---

[c] Hans Krebs (1900–1981), M.D. (1925), University of Hamburg, professorships at Sheffield (1945) and Oxford (1954). Nobel Prize in Physiology or Medicine (1953).

liver, where it is converted into glucose-6-phosphate by gluconeogenesis. In alcoholic fermentation, prevalent in yeasts, NAD^+ is regenerated from NADH by the action of two enzymes: pyruvate decarboxylase, which converts pyruvate into acetaldehyde, and alcohol dehydrogenase, which reduces acetaldehyde to ethanol, regenerating NAD^+.

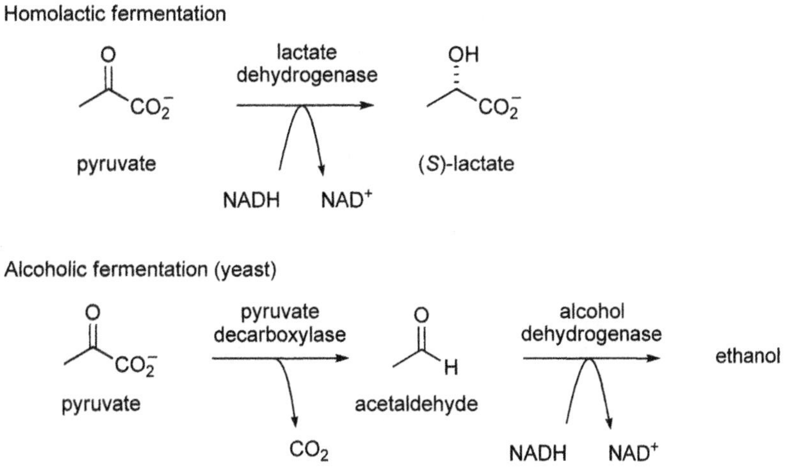

Homolactic fermentation

Alcoholic fermentation (yeast)

Gluconeogenesis

Gluconeogenesis is a process that converts various non-carbohydrate precursors into carbohydrates. Non-carbohydrate precursors for the gluconeogenic pathway include lactate, pyruvate, glycerol and the carbon skeletons of most amino acids (but not leucine or lysine). Fatty acids cannot act as precursors for glucose biosynthesis in animals as they are degraded to acetyl-CoA, and there is no pathway for the conversion of acetyl-CoA into oxaloacetate. However, plants do contain such a pathway, the glyoxylate cycle, which will be discussed later in this chapter.

In gluconeogenesis, each of the carbon precursors must first be converted into oxaloacetate, the starting material for the process. Gluconeogenesis in mammals occurs during fasting, starvation or intense exercise, and is an endergonic process, which requires ATP and GTP. The liver's capacity to store glucose as glycogen is only sufficient to supply the brain for about half a day when fasting. Isotopic labelling studies have shown that gluconeogenesis is responsible for 28% of total glucose production after 12 hrs of fasting, and for almost all glucose by 66 hrs.[8]

References start on page 249

Gluconeogenesis occurs predominantly in the liver and, to a small extent, in the kidney. Gluconeogenesis uses most of the enzymes of the glycolytic pathway, except for the steps catalysed by hexokinase, phosphofructokinase and pyruvate kinase, which have large negative free energy changes in the direction of glycolysis. These steps are subverted in gluconeogenesis as follows:

1. Pyruvate is converted to oxaloacetate, and then to PEP. This uses two enzymes: pyruvate carboxylase and PEP carboxykinase.
2. Hydrolytic reactions bypass phosphofructokinase and hexokinase. Fructose-1,6-diphosphatase converts fructose-1,6-diphosphate to fructose-6-phosphate. PGI of the glycolytic pathway allows interconversion of fructose-6-phosphate to glucose-6-phosphate, and glucose-6-phosphatase converts glucose-6-phosphate to glucose to complete the pathway.

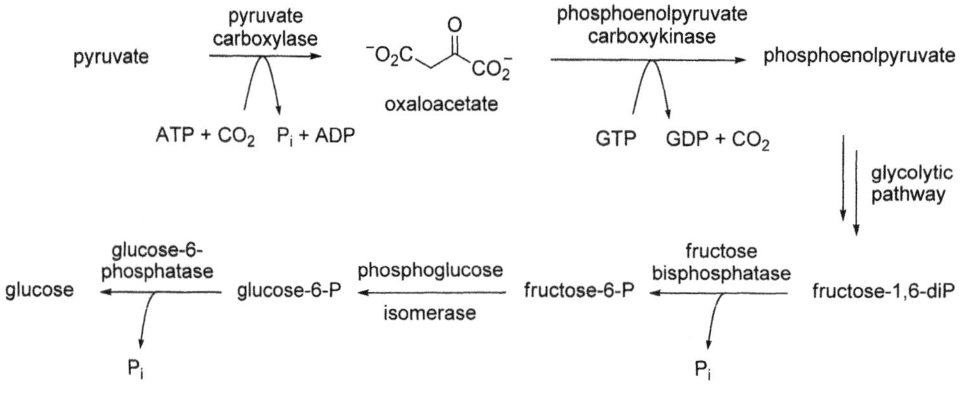

The Pentose Phosphate Pathway

Glycolysis and the citric acid cycle result in the formation of large amounts of ATP, which functions as the 'energy currency' of the cell. In these processes, NADH is used as a reducing agent, aiding in the conversion of the energy liberated from metabolite oxidation to the synthesis of ATP. In addition to the requirement of cells for large quantities of ATP, cells also require reducing power in the form of NADPH to drive biosynthetic pathways. The pentose phosphate pathway is the major source of NADPH, which also supplies 5-carbon precursors, especially ribose-5-phosphate, for the synthesis of nucleic acids.[9] The pentose phosphate pathway converts glucose-6-phosphate to ribose-5-phosphate (a precursor for nucleotide and nucleic acid biosynthesis), xylulose-5-phosphate or fructose-6-phosphate and glyceraldehyde-3-phosphate, depending on the needs of the cell. As such, the pentose phosphate pathway acts as a pool of intermediates under equilibrium, which can supply any of the intermediates to other biochemical pathways.[10,11] The pentose phosphate pathway can be considered to have three stages:

1. The oxidative branch of the pentose phosphate pathway, which converts glucose-6-phosphate to ribulose-5-phosphate and generates NADPH from $NADP^+$. This branch includes the steps catalysed by glucose-6-phosphate dehydrogenase, 6-phosphogluconolactonase and 6-phosphogluconate dehydrogenase.

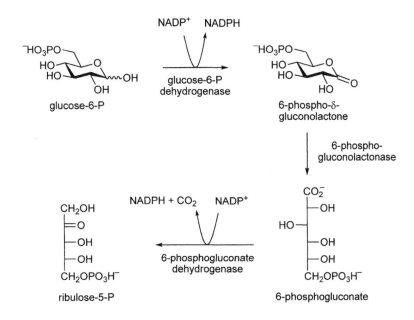

2. Isomerization and epimerization reactions that convert ribulose-5-phosphate to ribose-5-phosphate or xylulose-5-phosphate. The steps are catalysed by ribulose-5-phosphate isomerase or ribulose-5-phosphate epimerase, respectively.
3. A series of carbon–carbon bond cleavage and formation reactions that ultimately transform xylulose-5-phosphate and ribose-5-phosphate to fructose-6-phosphate and glyceraldehyde-3-phosphate. The reactions are catalysed by transketolase and transaldolase.

References start on page 249

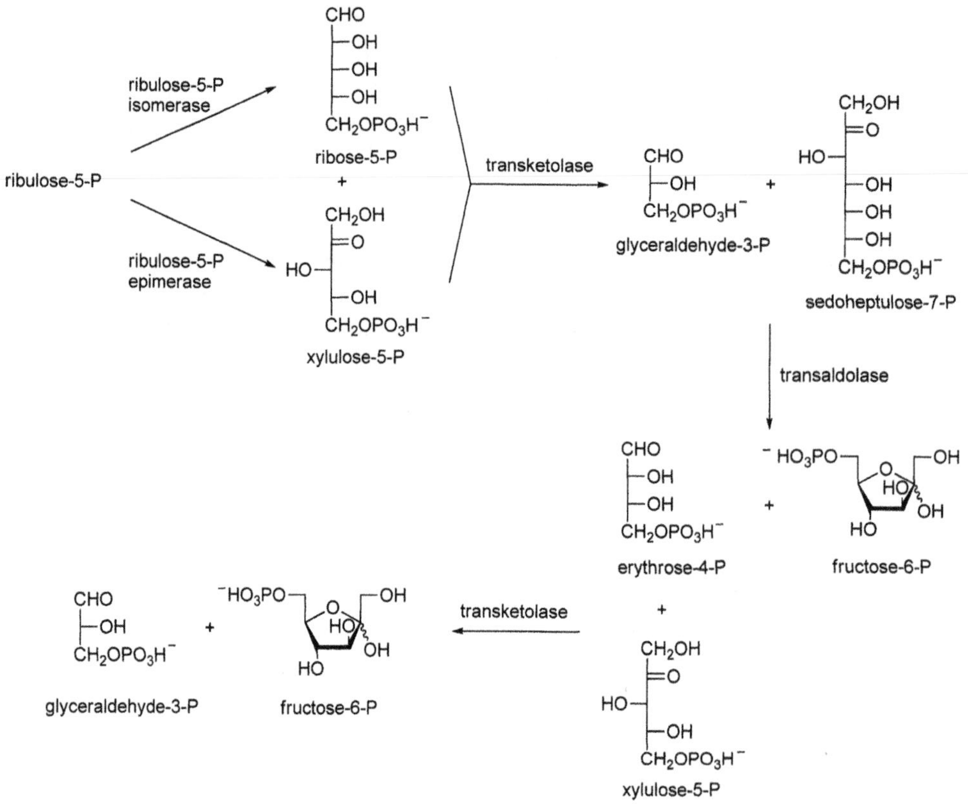

CHO
—OH
—OH
—OH
CH$_2$OPO$_3$H$^-$
ribose-5-P

ribulose-5-P isomerase

ribulose-5-P

ribulose-5-P epimerase

CH$_2$OH
=O
HO—
—OH
CH$_2$OPO$_3$H$^-$
xylulose-5-P

transketolase

CHO
—OH
CH$_2$OPO$_3$H$^-$
glyceraldehyde-3-P

CH$_2$OH
=O
HO—
—OH
—OH
—OH
CH$_2$OPO$_3$H$^-$
sedoheptulose-7-P

transaldolase

CHO
—OH
—OH
CH$_2$OPO$_3$H$^-$
erythrose-4-P

$^-$HO$_3$PO— —OH
HO
HO —OH
fructose-6-P

transketolase

CHO
—OH
CH$_2$OPO$_3$H$^-$
glyceraldehyde-3-P

$^-$HO$_3$PO— —OH
HO
HO —OH
fructose-6-P

CH$_2$OH
=O
HO—
—OH
CH$_2$OPO$_3$H$^-$
xylulose-5-P

The latter two stages are commonly referred to as the non-oxidative branch of the pentose phosphate pathway.

The Glyoxylate Cycle

Plants, fungi and certain bacteria, but not animals, possess the ability to convert acetyl-CoA (from either β oxidation of fatty acids or pyruvate from glycolysis) to succinate, which is then converted to oxaloacetate, via malate. In plants, the enzymes that mediate this process are located within the glyoxysome, a specialized plant cell organelle. The glyoxylate cycle consists of five reactions:

1. Citrate synthase condenses oxaloacetate and acetyl-CoA to form citrate.
2. Citrate is isomerized to isocitrate, most likely by cytosolic aconitase.
3. Isocitrate lyase cleaves isocitrate to succinate and glyoxylate.
4. Malate synthase condenses glyoxylate with another molecule of acetyl-CoA to give malate.
5. Malate dehydrogenase catalyses the oxidation of malate to oxaloacetate using NAD$^+$.

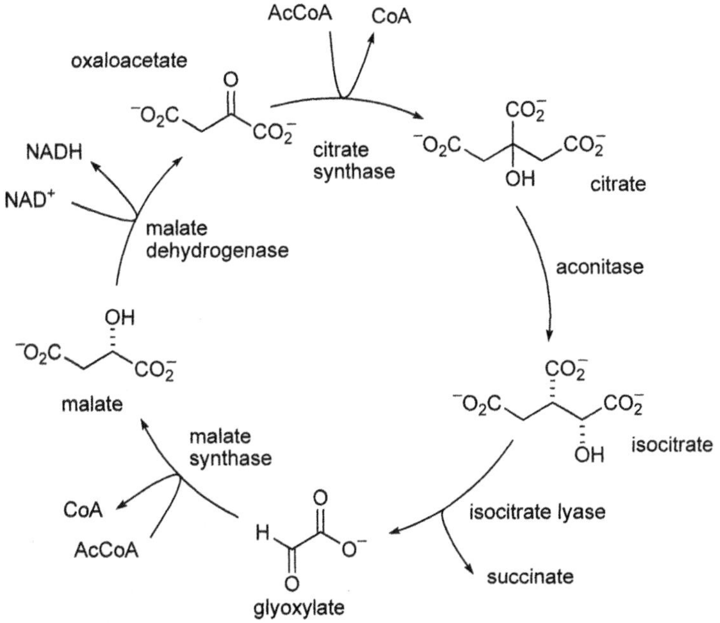

The glyoxylate cycle allows organisms to convert triacylglycerols to glucose, via acetyl-CoA. Thus, acetyl-CoA enters the glyoxylate cycle and is converted into oxaloacetate, which enters the gluconeogenesis pathway. The glyoxylate cycle is used in seeds, and by certain pathogenic microorganisms, particularly residents of the macrophage,[12] including *M. tuberculosis*,[13] *Candida albicans*[14] and *Leishmania* sp.[15]

Biosynthesis of Sugar Nucleoside Diphosphates

Sugar nucleoside (di)phosphates are central intermediates in the biosynthesis of glycoconjugates (through the action of glycosyltransferases) and other metabolites. Sugar nucleoside (di)phosphates arise from two sources: de novo synthesis from primary metabolites including glucose-6-phosphate or fructose-6-phosphate, or through salvage pathways from dietary or breakdown products. This section will cover only de novo synthesis from primary metabolites, but the interested reader is directed to several reviews that provide an entry point to the literature on salvage pathways.[1,16,17]

Nucleotidylyltransferases: Nucleotidylyltransferases catalyse the condensation of (deoxy)nucleoside triphosphates (NTPs or dNTPs) and sugar-1-phosphates to afford sugar (deoxy)nucleoside mono- or diphosphates [sugar-1-(d)NMPs or sugar-1-(d)NDPs]. The prototypical enzyme is UDP-glucose pyrophosphorylase, which catalyses the condensation of glucose-1-phosphate and UTP to give UDP-glucose and pyrophosphate.[18] This reaction has a free energy change of close to zero, and the presence of the enzyme pyrophosphatase can be used to shift the equilibrium to favour product, by hydrolysing the released pyrophosphate.[19]

$$\text{sugar-1-P} \xrightarrow[\text{NTP or dNTP}]{\text{nucleotidylyltransferase}} \begin{array}{c}\text{sugar-1-(d)NDP}\\\text{or}\\\text{sugar-1-NMP}\end{array}$$

Nucleotidylyltransferases play an important role in the biosynthesis of all sugar nucleoside (di)phosphates and, more recently, they have been investigated as catalysts for preparative synthesis. As will be seen in Chapter 8, sugar nucleoside (di)phosphates are of special interest for use as glycosyl donors for sugar nucleoside (di)phosphate-dependent glycosyltransferases. Nucleotidylyltransferases have been exploited in the synthesis of such glycosyl donors from NTPs and sugar-1-phosphates.[19–21] In particular, it has been found that many nucleotidylyltransferases possess substantial substrate flexibility, allowing the use of alternative NTPs and sugars. For example, E. coli α-glucopyranose-1-phosphate thymidylyltransferase (E_p) can convert a wide range of hexopyranose-1-phosphates to UDP and dTDP sugars,[22] which were then used to synthesize 'glycorandomized' analogues of the antibiotics vancomycin and teicoplanin using the natural vancomycin pathway glycosyltransferases.[23] Based on the crystal structure of E_p, introduction of specific mutations into this enzyme resulted in an expansion of its substrate specificity, allowing the production of an even wider set of sugar nucleoside diphosphates.[24]

Biosynthesis of UDP-glucose, UDP-galactose and galactose: UDP-glucose is a biosynthetic precursor for a range of sugar nucleotides including UDP-galactose, UDP-glucuronic acid, UDP-xylose and UDP-arabinose.[25] In vertebrates, UDP-glucose is converted into glucuronic acid, and then to L-ascorbic acid. UDP-glucose is formed in two steps from glucose-6-phosphate by the action of phosphoglucomutase, which forms glucose-1-phosphate, and UDP-glucose pyrophosphorylase (vide infra). Phosphoglucomutase transfers an enzyme-bound phosphoryl group to glucose-6-phosphate to generate glucose-1,6-diphosphate, which then rephosphorylates the enzyme and releases glucose-1-phosphate.[26,27] Glucose-1,6-diphosphate occasionally dissociates from the enzyme in the midst of catalysis and, as the enzyme must be phosphorylated to be active, this results in inactivation of the enzyme. The presence of small amounts of the cofactor glucose-1,6-diphosphate enables the enzyme to rephosphorylate and ensures full activity. A stable high-energy phosphorane intermediate corresponding to the transfer of a phosphoryl group from the anomeric oxygen to an enzymic carboxylate was claimed to be observed by X-ray crystallography;[28] an alternative interpretation of this structure revealed it to be a complex of MgF_3^-, derived from buffer salts, and the enzyme.[29]

Galactose comprises half of the milk sugar lactose and is liberated by the action of intestinal lactase. Galactose therefore comprises a major dietary sugar of dairy products but cannot enter the glycolytic pathway directly. In addition, the formation of UDP-galactose as a substrate for galactosyltransferases requires activation of galactose or, alternatively, in the absence of sufficient dietary quantities, its synthesis from glucose. Galactose and glucose are interconverted in a three-enzyme sequence known as the Leloir pathway,[d] which comprises galactokinase, galactose-1-phosphate uridylyltransferase and UDP-galactose 4-epimerase.[30,31] The Leloir pathway also provides UDP-galactose for the synthesis of glycoconjugates. A fourth auxiliary enzyme, galactose mutarotase, assists in this pathway, interconverting β-galactose and α-galactose, thereby providing the substrate α-galactose for galactokinase. Galactokinase catalyses the union of α-galactose and ATP to afford galactose-1-phosphate. The next step is the reversible transfer of a UMP moiety from UDP-glucose to galactose-1-phosphate, catalysed by galactose-1-phosphate uridylyltransferase. This enzyme affords glucose-1-phosphate and UDP-galactose in a reaction proposed to proceed through a covalently bound uridylyl-enzyme intermediate. An X-ray crystal structure of this

---

[d] Luis L. Leloir (1906–1987), M.D. (1933), University of Buenos Aires, professorship at University of Buenos Aires. Nobel Prize in Chemistry (1970). The study of this pathway by Leloir and co-workers constituted the original discovery of the first nucleotide sugar, UDP-glucose.

References start on page 249

intermediate has been determined.[32] The final step of the Leloir pathway regenerates UDP-glucose from UDP-galactose in a reaction catalysed by UDP-galactose 4-epimerase. UDP-galactose 4-epimerase achieves this transformation using an NAD$^+$ cofactor, which acts in a purely catalytic manner, operating in a redox cycle involving oxidation to the 4-keto sugar nucleoside diphosphate and reduction from the alternate face. The second step of the Leloir pathway therefore converts galactose-1-phosphate to UDP-galactose, the substrate for the final step that is catalysed by the 4-epimerase. The uridylyl group is transferred from UDP-glucose, generating glucose-1-phosphate. The 4-epimerase is frequently termed UDP-glucose 4-epimerase, and the enzyme from mammals is also able to catalyse the interconversion of UDP-N-acetylglucosamine and UDP-N-acetylgalactosamine.[33] 4-Epimerases specific for UDP-glucose/UDP-galactose and UDP-N-acetylglucosamine/UDP-N-acetylgalactosamine are also known.[34]

In humans, defects in the enzymes of the Leloir pathway lead to a variety of conditions termed the galactosemias. Mutations have been found in galactokinase, galactose-1-phosphate uridylyltransferase and UDP-galactose 4-epimerase, leading to galactosemias II, I and III, respectively.[31] Clinical manifestations range from mild to severe and include intellectual retardation, liver dysfunction and cataracts. These diseases result in galactose toxicity from accumulation of this sugar, which may be alleviated by limiting consumption of dairy products and other sources of galactose.

Biosynthesis of UDP-glucuronic acid and UDP-xylose: UDP-glucuronic acid serves as the glycosyl donor for the synthesis of glycosaminoglycans such as heparin/heparan sulfate, hyaluronan and chondroitin. UDP-glucuronic acid also acts as the substrate for UDP-glucuronyltransferases that catalyse the formation of glucuronide conjugates with endogenous and xenobiotic substances including bilirubin and various drugs, aiding in their excretion in Phase II metabolism. Finally, UDP-glucuronic acid is hydrolysed by UDP-glucuronyltransferases to glucuronic acid, which can act as a precursor for L-ascorbic acid biosynthesis in animals (vide infra). UDP-glucuronic acid is synthesized directly from UDP-glucose by UDP-glucose dehydrogenase.[35] This enzyme catalyses the oxidation of the 6-hydroxyl using NAD^+ as a cofactor. The proposed mechanism for the enzyme proceeds through the initial oxidation of UDP-glucose to the corresponding aldehyde, then through a second step involving the formation of a thiohemiacetal with an active site cysteine, and oxidation to a covalent thioester intermediate, which is hydrolysed to afford UDP-glucuronic acid.[36]

In higher animals, xylose is relatively rare but is used to initiate proteoglycan biosynthesis through an O-link of xylose to serine and is found in some proteins including factors VII, IX and Notch-1.[1] In plants, xylose is found in a wide range of glycoconjugates including xyloglucan and xylan.[37] UDP-xylose is formed from UDP-glucuronic acid by the action of UDP-glucuronic acid decarboxylase, an NAD^+-dependent enzyme that is proposed to operate through a 4-keto sugar intermediate.[25,38,39]

Biosynthesis of GDP-mannose: GDP-mannose is a source of mannosyl units found in glycoconjugates including glycoproteins, dolichol phosphomannose and glycolipids, and also acts as an important biochemical feedstock for the synthesis of deoxy sugars, including GDP-fucose. In plants GDP-mannose is a precursor for L-ascorbic acid synthesis. GDP-mannose is synthesized from fructose-6-phosphate by the action of three enzymes: phosphomannose isomerase, which converts fructose-6-phosphate to mannose-6-phosphate; phosphomannomutase, which converts mannose-6-phosphate to mannose-1-phosphate; and mannose-1-phosphate guanylyltransferase, which converts mannose-1-phosphate to GDP-mannose. Phosphomannose isomerases (sometimes referred to as mannose-6-phosphate isomerase and mannosephosphate isomerase) are believed to operate through a mechanism involving an ene-diol intermediate. Those from *Saccharomyces cerevisiae* and *Candida albicans* are zinc-dependent metalloenzymes that are highly specific for the conversion of fructose-6-phosphate to mannose-6-phosphate.[40] By contrast, dual-function PGI/ phosphomannose isomerase enzymes are found in archea that convert fructose-6-phosphate to glucose-6-phosphate and mannose-6-phosphate at equal rates.[41] Phosphomannomutase, like phosphoglucomutase, utilizes mannose-1,6-diphosphate as a cofactor, which acts as a donor to phosphorylate an active-site aspartate residue.[42,43] The catalytic cycle involves conversion of mannose-6-phosphate into the cofactor using the active-site phosphoryl-aspartate group, and rephosphorylation of the enzyme by the cofactor, forming mannose-1-phosphate.

Defects in GDP-mannose biosynthesis lead to abnormal protein glycosylation, known collectively as congenital disorders of glycosylation (CDG). The most common CDG is CDG type 1a, also known as classical Jaeken syndrome, and is caused by mutations in the gene encoding human α-phosphomannomutase.[42] Mutations in phosphomannose isomerase lead to another CDG: CDG type 1b.[44]

Biosynthesis of UDP-*N*-acetylglucosamine and UDP-*N*-acetylgalactosamine: Biosynthesis of the sugar nucleotide UDP-*N*-acetylglucosamine commences from fructose-6-phosphate, which is available from the glycolytic or gluconeogenic pathways.[45] Fructose-6-phosphate is converted into glucosamine-6-phosphate by the action of glucosamine-6-phosphate synthase.[46] This enzyme is a bifunctional glutaminase/synthetase that possesses two active sites.[47] The first site is located in the glutaminase domain of the enzyme and utilizes the *N*-terminal cysteine residue as a nucleophile to catalyse the hydrolysis of L-glutamine to L-glutamic acid, liberating ammonia. The ammonia is used by the second active site, located in the synthetase domain, and is condensed with fructose-6-phosphate, forming an intermediate imine, which is isomerized to afford glucosamine-6-phosphate. Remarkably, exogenous ammonia cannot substitute for ammonia derived from hydrolysis of L-glutamine, and it is suggested that ammonia released by hydrolysis of glutamine at the glutaminase active site is channelled through the protein directly to the synthetase active site.[47] After the formation of glucosamine-6-phosphate, the pathways used by eukaryotes and prokaryotes differ.

In eukaryotes, glucosamine-6-phosphate is converted into N-acetylglucosamine-6-phosphate by the action of glucosamine-6-phosphate acetyltransferase, using acetyl-CoA as the acetyl donor.[48] Next, phosphoacetylglucosamine mutase catalyses the isomerization of N-acetylglucosamine-6-phosphate to N-acetylglucosamine-1-phosphate.[49] Like phosphoglucomutase and phosphomannomutase, this enzyme operates through a two-step mechanism that utilizes a diphosphorylated cofactor, in this case N-acetylglucosamine-1,6-diphosphate.[50] In the first step, N-acetylglucosamine-6-phosphate is converted into the cofactor, and in the second step the cofactor is converted into GlcNAc-1-phosphate. The final step in UDP-N-acetylglucosamine biosynthesis is the conversion of N-acetylglucosamine-1-phosphate to UDP-N-acetylglucosamine, catalysed by the nucleotidylyltransferase UDP-N-acetylglucosamine pyrophosphorylase.[45,51]

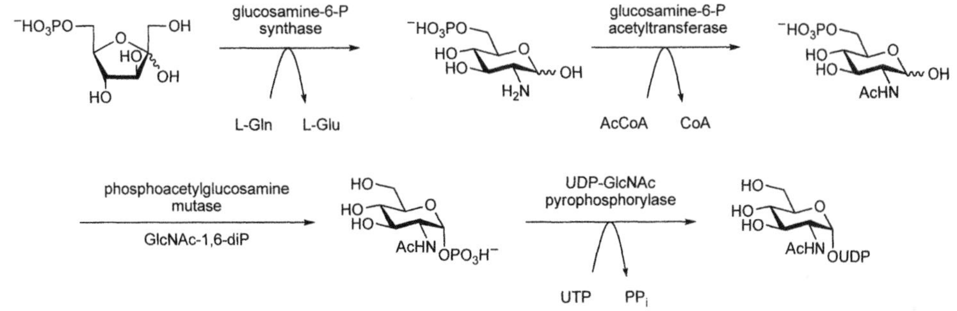

In prokaryotes, glucosamine-6-phosphate is converted into glucosamine-1-phosphate through the action of phosphoglucosamine mutase (GlmM). E. coli phosphoglucosamine mutase must be phosphorylated to be active; this occurs by autophosphorylation using ATP.[52] Finally, N-acetylation (using acetyl-CoA) and uridylylation (using UTP) are carried out by the bifunctional enzyme GlmU.[51]

UDP-N-acetylgalactosamine is formed from UDP-N-acetylglucosamine in one step by the action of a 4-epimerase. The enzyme in mammals is identical to UDP-galactose 4-epimerase,[33] but certain bacteria possess enzymes that have a strong preference for UDP-N-acetylglucosamine over UDP-glucose.[34]

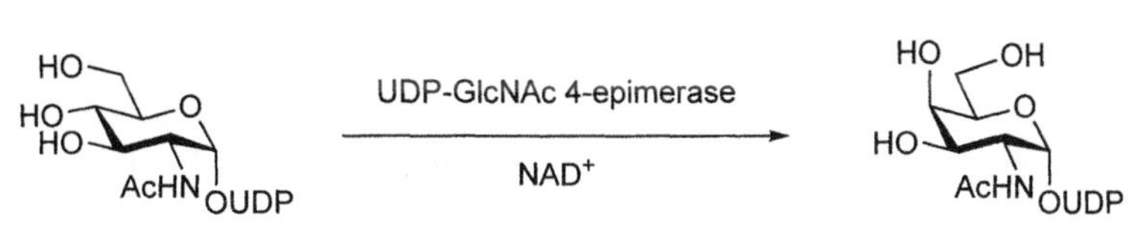

References start on page 249

Biosynthesis of UDP-*N*-acetylmuramic acid: Two monosaccharide glycosyl donors are needed for the biosynthesis of the peptidoglycan component of the bacterial cell wall: UDP-*N*-acetylglucosamine and UDP-*N*-acetylmuramic acid. UDP-*N*-acetylmuramic acid is formed in two steps catalysed by the enzymes UDP-*N*-acetylglucosamine enolpyruvyl transferase (MurA) and UDP-*N*-acetylenolpyruvylglucosamine reductase (MurB).[53,54] MurA catalyses the transfer of an enolpyruvyl group from PEP to UDP-*N*-acetylglucosamine to generate an enol ether (at C3). The enzyme mechanism is proposed to proceed through an addition–elimination process involving a sugar-bound tetrahedral intermediate.[55] MurA is inhibited by the clinically used antibiotic fosfomycin, which alkylates an active-site cysteine residue, thereby inactivating the enzyme.[55] MurB is a flavoenzyme that contains a single molecule of FAD, which transfers hydride from NADPH to the substrate, affording UDP-*N*-acetylmuramic acid.[56]

Biosynthesis of GDP-fucose: L-Fucose is a component of glycoconjugates and glycoproteins, where it is usually found as a terminal sugar. L-Fucose is also found in the so-called fucans, polymers of L-fucose that are frequently heavily sulfated. GDP-fucose is derived from GDP-mannose through the action of two enzymes.[57] The first enzyme, GDP-mannose 4,6-dehydratase, catalyses the dehydration of GDP-mannose.[58] In this reaction the 6-position is reduced and the 4-position is oxidized, affording GDP-6-deoxy-4-keto-mannose. GDP-mannose 4,6-dehydratase utilizes an NADP$^+$ cofactor, and evidence has been obtained showing that hydride is transferred from the 4-position to the 6-position of the sugar in the course of the reaction.[46] The second step is catalysed by a dual-function GDP-4-keto-6-deoxy-mannose 3,5-epimerase/4-reductase. This remarkable NADPH-dependent enzyme catalyses the epimerization of the 3- and 5-positions, and the reduction of the 4-keto group.[59,60]

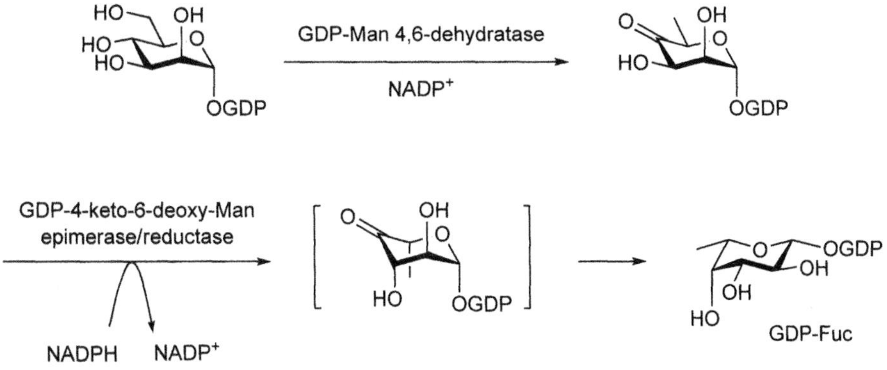

Biosynthesis of furanosyl nucleoside diphosphates: UDP-galactofuranose and UDP-arabinofuranose:

Hexoses are most commonly found in pyranose forms in glycoconjugates; however, galactofuranose is found in a variety of glycoconjugates in bacteria,[61] protozoa,[62] algae,[63] and fungi,[61] and arabinofuranose is found in plants and bacteria. For example, the cell wall of mycobacteria contains a galactan chain comprised of β-Galf-(1,5)-β-Galf-(1,6)- residues that are capped with branched arabinofuranosyl-containing oligosaccharides,[64] and arabinofuranosyl residues are found in pectins and arabinogalactan proteins, commonplace in the cell wall and extracellular materials of plants.[65]

The biosynthesis of UDP-galactofuranose has been most intensively studied. It is achieved in a single step from UDP-galactopyranose by the action of UDP-galactopyranose mutase.[61] The position of this equilibrium strongly favours the pyranose form, with a K_{eq} value of 0.057 at 37°C.[66] The enzyme possesses an FAD cofactor and is able to perform catalysis only when the FAD is reduced. Several mechanisms have been proposed,[67] and experimental evidence for a glycosyl–FAD intermediate has been obtained.[68,69]

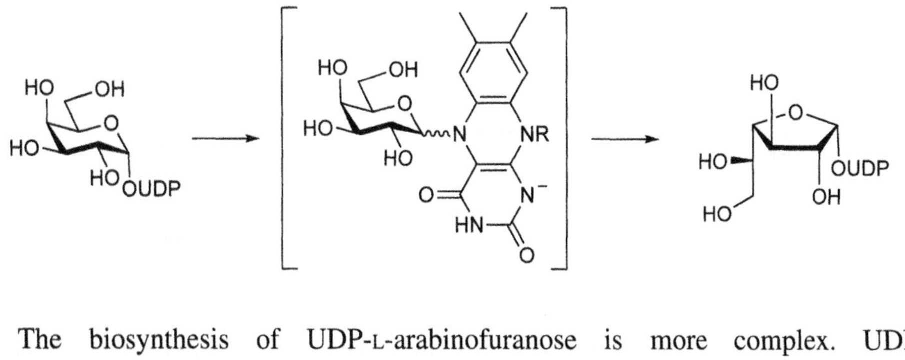

The biosynthesis of UDP-L-arabinofuranose is more complex. UDP-L-arabinopyranose is formed from UDP-xylose by a UDP-xylose 4-epimerase, or through a salvage pathway from arabinose kinase acting on L-arabinose to afford Ara$p$-1-phosphate, and the reaction of this phosphate with UTP catalysed by a UDP-L-arabinopyranose-1-phosphate pyrophosphorylase.[70] A UDP-L-arabinopyranose mutase has been isolated that can catalyse the isomerization of the pyranose form to UDP-L-arabinofuranose.[71]

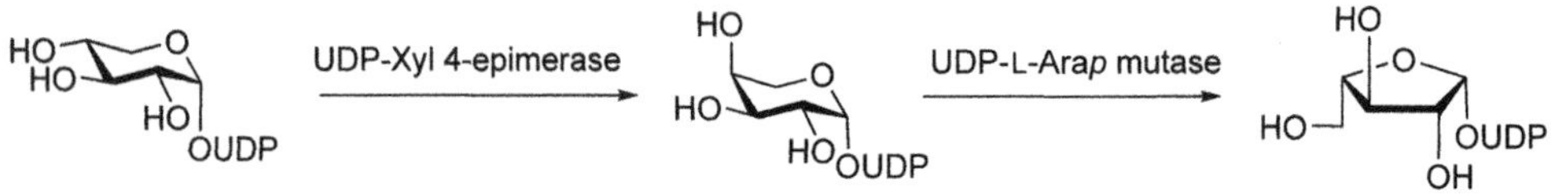

Biosynthesis of Sialic Acids and CMP-Sialic Acids

Sialic acids[e] are 9-carbon α-keto acids and include N-acetylneuraminic acid (Neu5Ac), 2-keto-3-deoxy-D-*glycero*-D-*galacto*-nonulosonic acid (Kdn), 5,7-diamino-3,5,7,9-tetradeoxy-D-*glycero*-D-*galacto*-nonulosonic acid (legionaminic acid) and 5,7-diamino-3,5,7,9-tetradeoxy-L-*glycero*-L-*manno*-nonulosonic acid (pseudaminic acid), the last three being considerably rarer than Neu5Ac.[72] The 8-carbon α-keto acid, 2-keto-3-deoxy-D-*manno*-octulosonic acid (Kdo), is not a sialic acid but bears a striking resemblance to them.

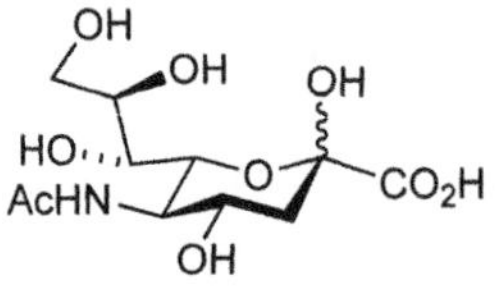

N-acetylneuraminic acid
(Neu5Ac)

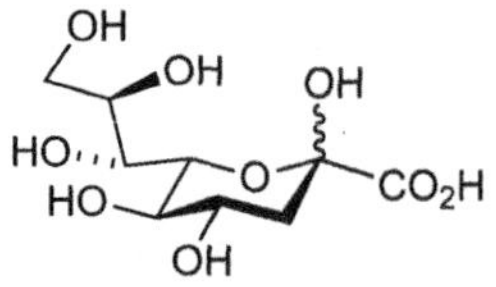

2-keto-3-deoxy-D-*glycero*-D-*galacto*-nonulosonic acid
(Kdn)

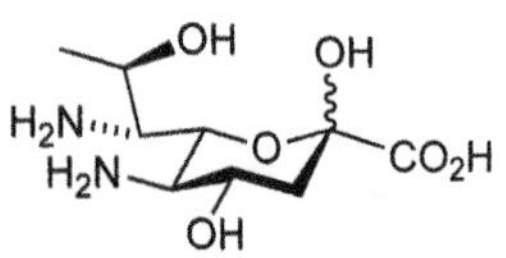

legionaminic acid

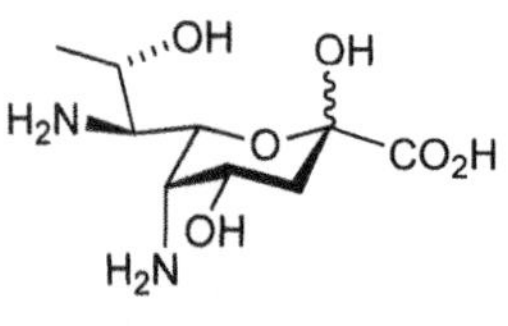

pseudaminic acid

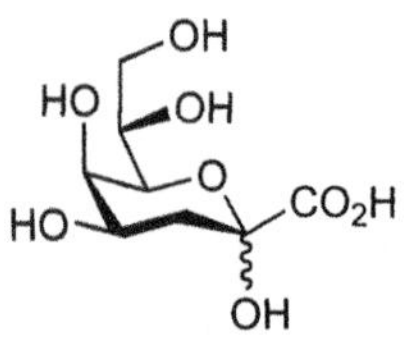

2-keto-3-deoxy-D-*manno*-octulosonic acid
(Kdo)

---

[e] The names neuraminic acid and sialic acid derive from their original discovery in brain glycolipids (neuro + amine + acid) and salivary submaxillary mucins (Greek: sialos = saliva).[72]

In vertebrates, the first two steps in the biosynthesis of *N*-acetylneuraminic acid are performed by the bifunctional enzyme UDP-*N*-acetylglucosamine 2-epimerase/ManNAc kinase.[73] The epimerase domain catalyses the elimination of UDP to give 2-acetamidoglucal, which is hydrated to afford *N*-acetylmannosamine.[74,75] The phosphorylase domain catalyses the phosphorylation of the released *N*-acetylmannosamine to afford *N*-acetylmannosamine-6-phosphate. Several alternative pathways for *N*-acetylmannosamine biosynthesis exist in bacteria, including the 2-epimerization of *N*-acetylglucosamine-6-phosphate to *N*-acetylmannosamine-6-phosphate by the action of *N*-acetylglucosamine-6-phosphate 2-epimerase.[76] Next, Neu5Ac-9-phosphate synthase catalyses the condensation of *N*-acetylmannosamine-6-phosphate and PEP to afford Neu5Ac-9-phosphate, which is dephosphorylated to afford Neu5Ac.[72] Again, bacteria can perform this transformation by an alternative route, catalysing the condensation of *N*-acetylmannosamine with PEP to afford Neu5Ac directly, using the enzyme Neu5Ac synthetase.[75] Finally, CMP-Neu5Ac synthetase catalyses the formation of CMP-Neu5Ac from Neu5Ac and CTP. In vertebrates, CMP-Neu5Ac synthetase is located in the nucleus, the only such nucleoside mono- or diphosphate synthetase located in this compartment.[77]

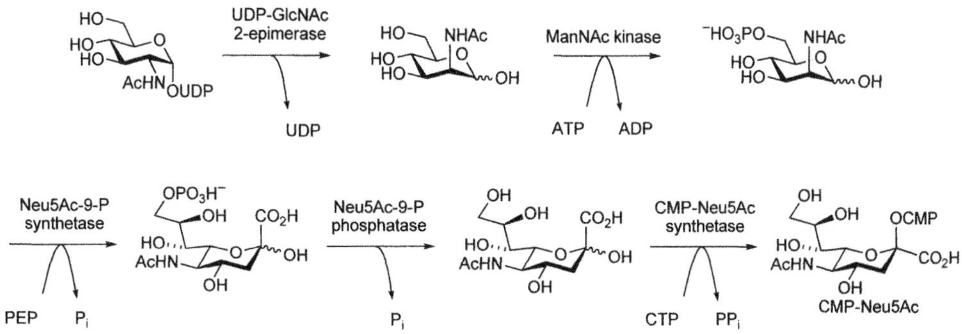

Kdo is an 8-carbon α-keto acid found in the bacterial cell wall.[78] The biosynthesis of Kdo bears considerable similarity to that of Neu5Ac and involves three steps from ribulose-5-phosphate through the action of arabinose-5-phosphate isomerase, Kdo-8-phosphate synthetase and Kdo-8-phosphate phosphatase.[79] Kdo is predominantly found in cell surface capsular polysaccharides, and for this, requires conversion to a glycosyl donor, CMP-Kdo, through the action of Kdo-CMP synthetase.[80]

References start on page 249

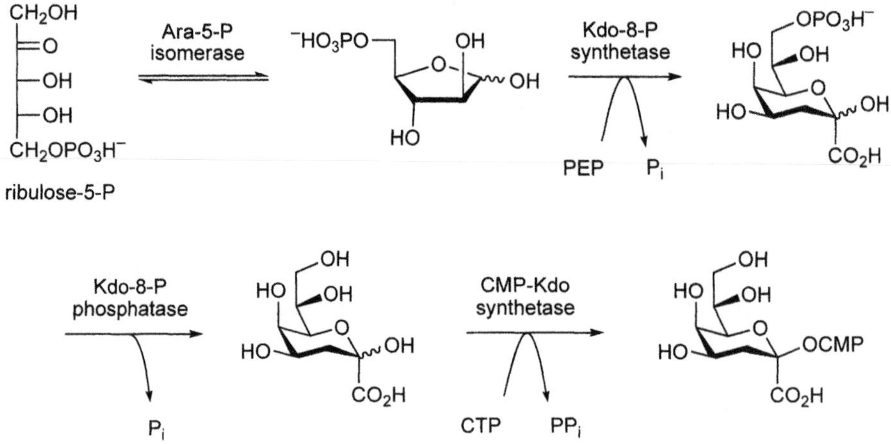

Biosynthesis of *myo*-Inositol

Inositols are cyclohexane hexaols. *myo*-Inositol is the most common of the inositols and is present in signalling molecules (inositol phosphates and phosphatidylinositol phosphates),[81,82] and as a structural component (as phosphatidylinositol) of a wide range of cell surface–associated glycoconjugates. These glycoconjugates include glycosylphosphatidylinositol (GPI) anchors of cell surface proteins and lipids,[83] and bacterial glycolipids such as the lipoarabinomannans of mycobacteria.[84] Many plant seeds and tissues contain large amounts of *myo*-inositol hexaphosphate, also known as phytic acid, which acts as a metabolic store of energy for the conversion to ATP, as well as an antioxidant.[85] *myo*-Inositol is formed in two steps from glucose-6-phosphate by the action of L-*myo*-inositol-1-phosphate synthase and inositol-1-phosphate phosphatase.[86] L-*myo*-Inositol-1-phosphate synthase is an NADH-dependent enzyme, and the catalytic mechanism is proposed to involve oxidation of C5 of glucose-6-phosphate, aldol condensation of the acyclic ketone to afford the inosose and reduction of the inosose to L-*myo*-inositol-1-phosphate (identical to D-*myo*-inositol-3-phosphate).[87]

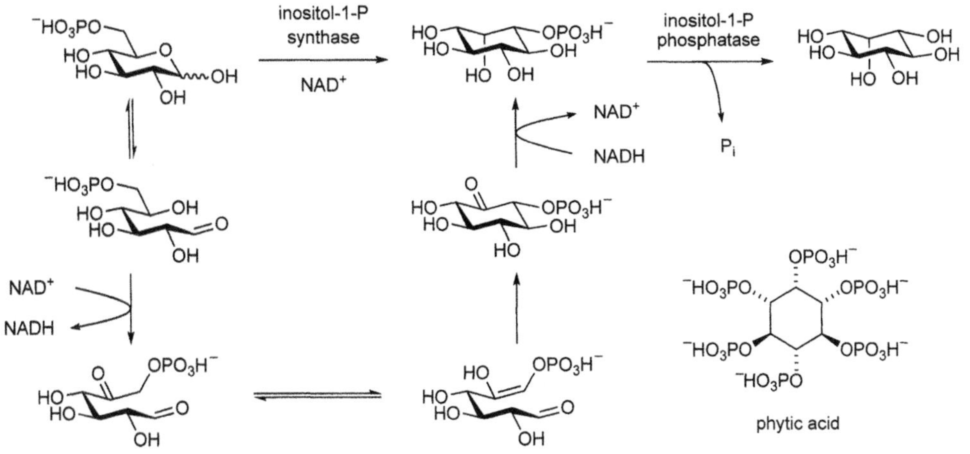

Biosynthesis of L-Ascorbic Acid[f]

L-Ascorbic acid is an important reducing agent and a cofactor for many enzymatic processes, including some glycoside hydrolases[88] and copper(I)-dependent mono-xygenases and iron(II)-dependent dioxygenases.[89] In addition, L-ascorbic acid is a biosynthetic precursor for metabolites including oxalic acid and tartaric acids in certain plants.[90] While most vertebrates can biosynthesize L-ascorbic acid, some mammals including man, guinea pigs and bats have lost this ability, and require it from their diet, hence giving it the status of a vitamin, vitamin C.[91] Fish, reptiles and amphibians produce L-ascorbic acid in the kidney, whereas mammals produce it in the liver. L-Ascorbic acid biosynthesis in animals occurs from UDP-glucuronic acid, which is hydrolysed by UDP-glucuronic acid transferases to afford D-glucuronic acid.[91] Next, an NADPH-dependent glucuronate reductase (also known as aldehyde reductase) reduces the anomeric position, resulting in the formation of L-gulonic acid. Aldonolactonase converts L-gulonic acid to L-gulono-1,4-lactone and, finally, this lactone is oxidized to L-ascorbic acid by the action of L-gulono-1,4-lactone dehydrogenase (GLO) with the production of hydrogen peroxide. The immediate product of GLO is the 2-keto lactone, which

---

[f] One-half of the Nobel Prize in Chemistry (1937) 'for his investigations on carbohydrates and vitamin C' was awarded to Walter Norman Haworth (1883–1950), Ph.D. (1910) under Otto Wallach at the University of Gottingen, D.Sc. (1911) under W.H. Perkin Jr. at the University of Manchester, professorship at Birmingham (1925).

References start on page 249

tautomerizes to L-ascorbic acid. Defects in the GLO gene are responsible for the inability of humans to synthesize L-ascorbic acid.[91]

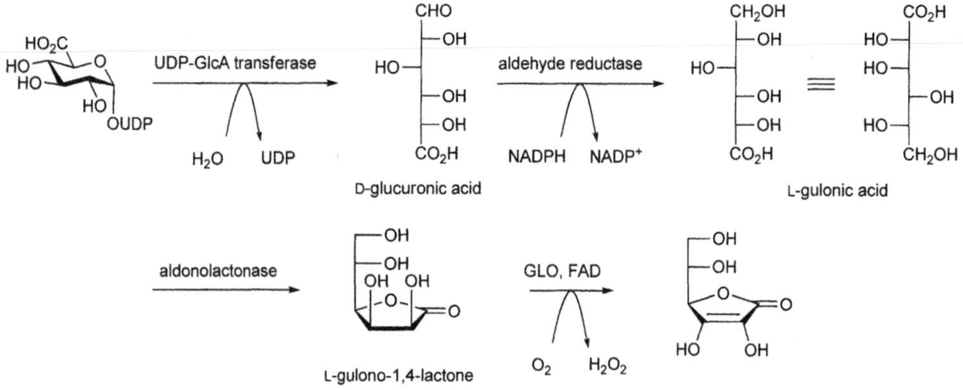

D-glucuronic acid

L-gulonic acid

L-gulono-1,4-lactone

L-Ascorbic acid biosynthesis apparently occurs in all plants, and no deficient mutants have ever been isolated, suggesting that it is essential for these organisms. Evidence for several pathways to L-ascorbic acid in plants has been obtained.[70,90,92] The major route is termed the Smirnoff–Wheeler pathway and proceeds from GDP-mannose to GDP-L-galactose through the action of GDP-mannose epimerase. GDP-L-galactose phosphorylase converts GDP-L-galactose to L-galactose-1-phosphate,[93,94] and this anomeric phosphate is hydrolysed by the action of L-galactose-1-phosphatase to L-galactose. L-Galactose is oxidized to L-galactono-1,4-lactone by L-galactose dehydrogenase, and finally this lactone is oxidized to L-ascorbic acid by L-galactonolactone-1,4-dehydrogenase.[95] Other pathways to L-ascorbic acid from *myo*-inositol and D-galacturonic acid are possible, and these may interlink with the Smirnoff–Wheeler pathway.[96]

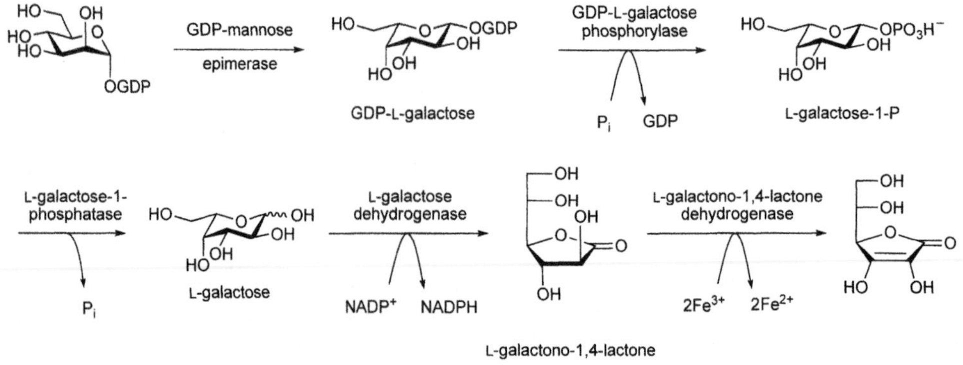

GDP-L-galactose

L-galactose-1-P

L-galactose

L-galactono-1,4-lactone

References

1. Varki, A., Cummings, R., Esko, J., Hart, G. and Marth, J., eds (1999). *Essentials of Glycobiology*. New York: Cold Spring Harbor.
2. Stamford, N.P.J. (2000). In *Glycoscience: Chemistry and Chemical Biology* (B.O. Fraser-Reid, K. Tatsuta, J. Thiem, eds) Vol. 2, p. 1215. Springer-Verlag: Berlin.
3. Davis, B.D. (1958). *Arch. Biochem. Biophys.*, **78**, 497.
4. Reese, K.M. (1980). *Chem. Eng. News*, 14 January, p. 64.
5. Brown, G.K. (2000). *J. Inherit. Metab. Dis.*, **23**, 237.
6. Erlandsen, H., Abola, E.E. and Stevens, R.C. (2000). *Curr. Opin. Struct. Biol.*, **10**, 719.
7. Sternberg, M.J., Cohen, F.E., Taylor, W.R. and Feldmann, R.J. (1981). *Phil. Trans. R. Soc. Lond. B*, **293**, 177.
8. Consoli, A., Kennedy, F., Miles, J. and Gerich, J. (1987). *J. Clin. Invest.*, **80**, 1303.
9. Sprenger, G.A. (1995). *Arch. Microbiol.*, **164**, 324.
10. Kruger, N.J. and von Schaewen, A. (2003). *Curr. Opin. Plant Biol.*, **6**, 236.
11. Tozzi, M.G., Camici, M., Mascia, L., Sgarrella, F. and Ipata, P.L. (2006). *FEBS J.*, **273**, 1089.
12. Lorenz, M.C. and Fink, G.R. (2002). *Eukaryot. Cell*, **1**, 657.
13. McKinney, J.D., Höner zu Bentrup, K., Muñoz-Elias, E.J., Miczak, A., Chen, B., Chan, W.-T., Swenson, D., Sacchettini, J.C., Jacobs, W.R., Jr. and Russell, D. (2000). *Nature*, **406**, 735.
14. Lorenz, M.C. and Fink, G.R. (2001). *Nature*, **412**, 83.
15. Simon, M.W., Martin, E. and Mukkada, A.J. (1978). *J. Bacteriol.*, **135**, 895.
16. Goon, S. and Bertozzi, C.R. (2001). In *Glycochemistry. Principles, Synthesis, and Applications* (P.G. Wang and C.R. Bertozzi, eds), p. 641. Marcel Dekker, Inc.: New York, USA.
17. Yarema, K.J. and Bertozzi, C.R. (2001). *Genome Biol.*, **2**, 1.
18. Roeben, A., Plitzko, J.M., Korner, R., Bottcher, U. M., Siegers, K., Hayer-Hartl, M. and Bracher, A. (2006). *J. Mol. Biol.*, **364**, 551.
19. Timmons, S.C., Mosher, R.H., Knowles, S.A. and Jakeman, D.L. (2007). *Org. Lett.*, **9**, 857.
20. Langenhan, J.M., Griffith, B.R. and Thorson, J.S. (2005). *J. Nat. Prod.*, **68**, 1696.
21. Griffith, B.R., Langenhan, J.M. and Thorson, J.S. (2005). *Curr. Opin. Biotechnol.*, **16**, 622.
22. Jiang, J., Biggins, J.B. and Thorson, J.S. (2000). *J. Am. Chem. Soc.*, **122**, 6803.
23. Losey, H.C., Jiang, J., Biggins, J.B., Oberthur, M., Ye, X.Y., Dong, S.D., Kahne, D., Thorson, J.S. and Walsh, C.T. (2002). *Chem. Biol.*, **9**, 1305.
24. Barton, W.A., Biggins, J.B., Jiang, J., Thorson, J.S. and Nikolov, D.B. (2002). *Proc. Natl. Acad. Sci. USA*, **99**, 13397.
25. Seifert, G.J. (2004). *Curr. Opin. Plant Biol.*, **7**, 277.
26. Dai, J.B., Liu, Y., Ray, W.J., Jr. and Konno, M. (1992). *J. Biol. Chem.*, **267**, 6322.
27. Zhang, G., Dai, J. and Wang, L. (2005). *Biochemistry*, **44**, 9404.
28. Lahiri, S.D., Zhang, G., Dunaway-Mariano, D. and Allen, K.N. (2003). *Science*, **299**, 2067.
29. Baxter, N.J., Olguin, L.F., Goličnik, M., Feng, G., Hounslow, A.M., Bermel, W., Blackburn, G.M., Hollfelder, F., Waltho, J.P. and Williams, N.H. (2006). *Proc. Natl. Acad. Sci. USA*, **103**, 14732.
30. Frey, P.A. (1996). *FASEB J.*, **10**, 461.
31. Holden, H.M., Rayment, I. and Thoden, J.B. (2003). *J. Biol. Chem.*, **278**, 43885.
32. Wedekind, J.E., Frey, P.A. and Rayment, I. (1996). *Biochemistry*, **35**, 11560.
33. Piller, F., Hanlon, M.H. and Hill, R.L. (1983). *J. Biol. Chem.*, **258**, 10774.
34. Demendi, M., Ishiyama, N., Lam, J.S., Berghuis, A.M. and Creuzenet, C. (2005). *Biochem. J.*, **389**, 173.
35. Spicer, A.P., Kaback, L.A., Smith, T.J. and Seldin, M.F. (1998). *J. Biol. Chem.*, **273**, 25117.
36. Campbell, R.E., Mosimann, S.C., van De Rijn, I., Tanner, M.E. and Strynadka, N.C. (2000). *Biochemistry*, **39**, 7012.
37. Harper, A.D. and Bar-Peled, M. (2002). *Plant Physiol.*, **130**, 2188.

38. Moriarity, J.L., Hurt, K.J., Resnick, A.C., Storm, P.B., Laroy, W., Schnaar, R.L. and Snyder, S.H. (2002). *J. Biol. Chem.*, **277**, 16968.
39. Wilson, I.B. (2004). *Cell. Mol. Life Sci.*, **61**, 794.
40. Cleasby, A., Wonacott, A., Skarzynski, T., Hubbard, R.E., Davies, G.J., Proudfoot, A.E., Bernard, A.R., Payton, M.A. and Wells, T.N. (1996). *Nat. Struct. Biol.*, **3**, 470.
41. Swan, M.K., Hansen, T., Schönheit, P. and Davies, C. (2004). *J. Biol. Chem.*, **279**, 39838.
42. Silvaggi, N.R., Zhang, C., Lu, Z., Dai, J., Dunaway-Mariano, D. and Allen, K.N. (2006). *J. Biol. Chem.*, **281**, 14918.
43. Kedzierski, L., Malby, R.L., Smith, B.J., Perugini, M.A., Hodder, A.N., Ilg, T., Colman, P.M. and Handman, E. (2006). *J. Mol. Biol.*, **363**, 215.
44. Schollen, E., Dorland, L., de Koning, T.J., Van Diggelen, O.P., Huijmans, J.G., Marquardt, T., Babovic-Vuksanovic, D., Patterson, M., Imtiaz, F., Winchester, B., Adamowicz, M., Pronicka, E., Freeze, H. and Matthijs, G. (2000). *Hum. Mutat.*, **16**, 247.
45. Milewski, S., Gabriel, I. and Olchowy, J. (2006). *Yeast*, **23**, 1.
46. He, X., Agnihotri, G. and Liu, H.-w. (2000). *Chem. Rev.*, **100**, 4615.
47. Teplyakov, A., Leriche, C., Obmolova, G., Badet, B. and Badet-Denisot, M.A. *Nat. Prod. Rep.*, **19**, 60.
48. Peneff, C., Mengin-Lecreulx, D. and Bourne, Y. (2001). *J. Biol. Chem.*, **276**, 16328.
49. Nishitani, Y., Maruyama, D., Nonaka, T., Kita, A., Fukami, T.A., Mio, T., Yamada-Okabe, H., Yamada-Okabe, T. and Miki, K. (2006). *J. Biol. Chem.*, **281**, 19740.
50. Cheng, P.W. and Carlson, D.M. (1979). *J. Biol. Chem.*, **254**, 8353.
51. Brown, K., Pompeo, F., Dixon, S., Mengin-Lecreulx, D., Cambillau, C. and Bourne, Y. (1999). *EMBO J.*, **18**, 4096.
52. Jolly, L., Pompeo, F., van Heijenoort, J., Fassy, F. and Mengin-Lecreulx, D. (2000). *J. Bacteriol.*, **182**, 1280.
53. El Zoeiby, A., Sanschagrin, F. and Levesque, R.C. (2003). *Mol. Microbiol.*, **47**, 1.
54. Bugg, T.D.H. and Walsh, C.T. (1992). *Nat. Prod. Rep.*, **9**, 199.
55. Skarzynski, T., Mistry, A., Wonacott, A., Hutchinson, S.E., Kelly, V.A. and Duncan, K. (1996). *Structure*, **4**, 1465.
56. Benson, T.E., Walsh, C.T. and Hogle, J.M. (1996). *Structure*, **4**, 47.
57. Becker, D.J. and Lowe, J.B. (2003). *Glycobiology*, **13**, 41R.
58. Sullivan, F.X., Kumar, R., Kriz, R., Stahl, M., Xu, G.Y., Rouse, J., Chang, X.J., Boodhoo, A., Potvin, B. and Cumming, D.A. (1998). *J. Biol. Chem.*, **273**, 8193.
59. Rizzi, M., Tonetti, M., Vigevani, P., Sturla, L., Bisso, A., Flora, A.D., Bordo, D. and Bolognesi, M. (1998). *Structure*, **6**, 1453.
60. Somers, W.S., Stahl, M.L. and Sullivan, F.X. (1998). *Structure*, **6**, 1601.
61. Sanders, D.A., Staines, A.G., McMahon, S.A., McNeil, M.R., Whitfield, C. and Naismith, J.H. (2001). *Nat. Struct. Biol.*, **8**, 858.
62. Kleczka, B., Lamerz, A.C., van Zandbergen, G., Wenzel, A., Gerardy-Schahn, R., Wiese, M. and Routier, F.H. *J. Biol. Chem.*, **282**, 10498.
63. Igarashi, T., Satake, M. and Yasumoto, T. (1999). *J. Am. Chem. Soc.*, **121**, 8499.
64. Crick, D.C., Mahapatra, S. and Brennan, P.J. (2001). *Glycobiology*, **11**, 107R.
65. Carpita, N. and McCann, M. (2000). In *Biochemistry and Molecular Biology of Plants* (B.B. Buchanan, W. Gruissem, and R.L. Jones, eds) American Society of Plant Physiologists: Rockville, MD.
66. Zhang, Q. and Liu, H.-w. (2000). *J. Am. Chem. Soc.*, **122**, 9065.
67. Itoh, K., Huang, Z. and Liu, H.-w. (2007). *Org. Lett.*, **9**, 879.
68. Soltero-Higgin, M., Carlson, E.E., Gruber, T.D. and Kiessling, L.L. (2004). *Nat. Struct. Mol. Biol.*, **11**, 539.
69. Huang, Z., Zhang, Q. and Liu, H.-w. (2003). *Bioorg. Chem.*, **31**, 494.
70. Reiter, W.D. and Vanzin, G.F. (2001). *Plant Mol. Biol.*, **47**, 95.

71. Konishi, T., Takeda, T., Miyazaki, Y., Ohnishi-Kameyama, M., Hayashi, T., O'Neill, M.A. and Ishii, T. (2007). *Glycobiology*, **17**, 345.

72. Angata, T. and Varki, A. (2002). *Chem. Rev.*, **102**, 439.

73. Stäsche, R., Hinderlich, S., Weise, C., Effertz, K., Lucka, L., Moormann, P. and Reutter, W. (1997). *J. Biol. Chem.*, **272**, 24319.

74. Chou, W.K., Hinderlich, S., Reutter, W. and Tanner, M.E. (2003). *J. Am. Chem. Soc.*, **125**, 2455.

75. Tanner, M.E. (2005). *Bioorg. Chem.*, **33**, 216.

76. Plumbridge, J. and Vimr, E. (1999). *J. Bacteriol.*, **181**, 47.

77. Münster-Kühnel, A.K., Tiralongo, J., Krapp, S., Weinhold, B., Ritz-Sedlacek, V., Jacob, U., Gerardy-Schahn, R. (2004). *Glycobiology*, **14**, 43R.

78. Unger, F.M. (1981). *Adv. Carbohydr. Chem. Biochem.*, **38**, 323.

79. Meredith, T.C. and Woodard, R.W. (2006). *Biochem. J.*, **395**, 427.

80. Jelakovic, S. and Schulz, G.E. (2001). *J. Mol. Biol.*, **312**, 143.

81. Hilbi, H. (2006). *Cell. Microbiol.*, **8**, 1697.

82. Michell, R.H., Perera, N.M. and Dove, S.K. (2003). *Biochem. Soc. Trans.*, **31**, 11.

83. Ferguson, M.A. (1999). *J. Cell. Sci.*, **112**, 2799.

84. Nigou, J., Gilleron, M. and Puzo, G. (2003). *Biochimie*, **85**, 153.

85. Raboy, V. (2003). *Phytochemistry*, **64**, 1033.

86. Majumder, A.L., Johnson, M.D. and Henry, S.A. (1997). *Biochim. Biophys. Acta*, **1348**, 245.

87. Stein, A.J. and Geiger, J.H. (2002). *J. Biol. Chem.*, **277**, 9484.

88. Burmeister, W.P., Cottaz, S., Rollin, P., Vasella, A. and Henrissat, B. (2000). *J. Biol. Chem.*, **275**, 39385.

89. Michal, G., ed. (1999). *Biochemical Pathways. An Atlas of Biochemistry and Molecular Biology.* New York, USA: John Wiley and Sons.

90. Debolt, S., Melino, V. and Ford, C.M. (2007). *Ann. Bot.*, **99**, 3.

91. Linster, C.L. and Van Schaftingen, E. (2007). *FEBS J.*, **274**, 1.

92. Valpuesta, V. and Botella, M.A. (2004). *Trends Plant Sci.*, **9**, 573.

93. Linster, C.L., Gomez, T.A., Christensen, K.C., Adler, L.N., Young, B.D., Brenner, C. and Clarke, S.G. (2007). *J. Biol. Chem.*, **282**, 18879.

94. Laing, W.A., Wright, M.A., Cooney, J. and Bulley, S.M. (2007). *Proc. Natl. Acad. Sci. USA*, **104**, 9534.

95. Wheeler, G.L., Jones, M.A. and Smirnoff, N. (1998). *Nature*, **393**, 365.

96. Major, L.L., Wolucka, B.A. and Naismith, J.H. (2005). *J. Am. Chem. Soc.*, **127**, 18309.

Chapter 7

Enzymatic Cleavage of Glycosides: Mechanism, Inhibition and Synthetic Applications

Nature has contrived a multitude of ingenious methods to cleave glycosides. The enzymes that undertake these duties include glycoside hydrolases, transglycosylases, lyases and phosphorylases. This chapter will give an overview of each of these enzymes, their mechanism, reagents/inhibitors for their study and applications in synthesis, with a particular focus on the most abundant, the glycoside hydrolases.

Glycoside Hydrolases

Glycoside hydrolases (or glycosidases) are enzymes that catalyse the hydrolysis of glycosidic bonds in all manner of glycosides, glycans and glycoconjugates. As a class they are arguably the best studied of the carbohydrate-processing enzymes. The cleavage of a glycosidic bond (catalysed by a glycoside hydrolase) occurs through C–O scission of the *exo*-cyclic acetal bond:

Measurements of the rates of spontaneous hydrolysis of the glycosidic bond reveal it to be an extraordinarily slow reaction, with a half-life for hydrolysis of a glycosidic bond in cellulose estimated to be 4.7 million years.[1] Glycoside hydrolases are remarkably proficient catalysts and accelerate the spontaneous hydrolysis reaction by rates of up to 10^{17}, making these one of the most powerful known enzyme catalysts.[2]

Mechanistically, scission of the glycosidic bond can occur with either retention or inversion of configuration, and these two outcomes provide a critical distinction between two sets of glycoside hydrolases.

References start on page 280

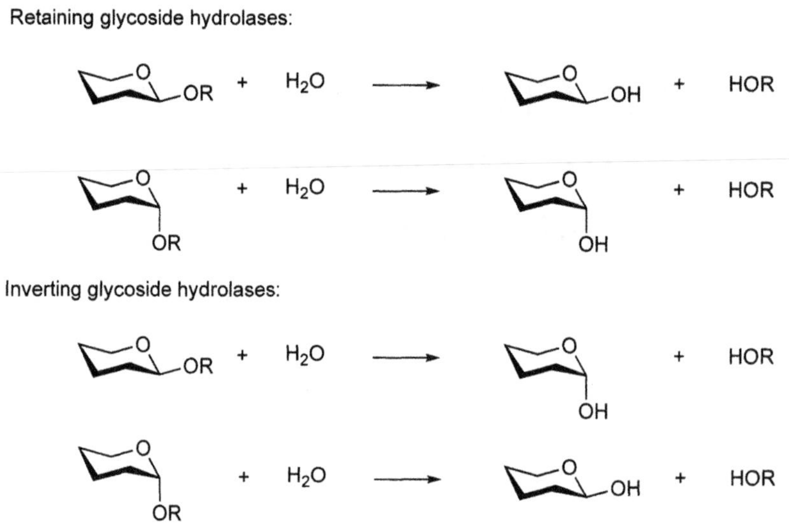

Retaining glycoside hydrolases:

Inverting glycoside hydrolases:

While over the years, many conflicting mechanisms for catalysis by glycoside hydrolases have been proposed,[3–5] there is now convincing evidence for detailed mechanisms that apply to the major classes of inverting and retaining glycoside hydrolases.[2,6] Remarkably, and almost without exception, these mechanisms conform to the general predictions outlined by Koshland[a] more than half a century ago.[8] The mechanistically discrete retaining and inverting glycoside hydrolases will be discussed in detail in the following sections.

Glycoside hydrolases may also be usefully classified into *endo-* or *exo-*glycoside hydrolases. This classification is made on the basis of whether a glycoside hydrolase cleaves an internal glycosidic linkage in a chain (*endo*) or at the end of a chain (*exo*; most commonly the non-reducing end, but not always[9]). Whether an enzyme is *exo-* or *endo-*acting has no relationship to the mechanism utilized – thus cellulases, for example, can be *endo-* or *exo-*acting and retaining or inverting β-glucanases.

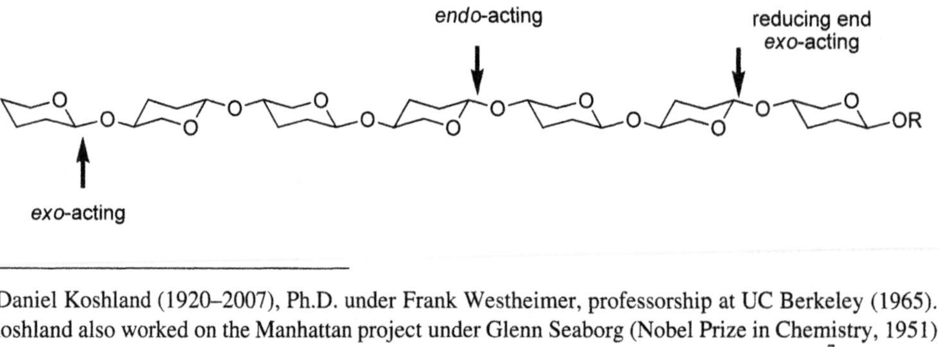

[a] Daniel Koshland (1920–2007), Ph.D. under Frank Westheimer, professorship at UC Berkeley (1965). Koshland also worked on the Manhattan project under Glenn Seaborg (Nobel Prize in Chemistry, 1951) and is credited with devising the 'induced fit' model of substrate recognition by enzymes.[7]

Retaining and Inverting Mechanisms

In the first major class of glycoside hydrolases to be discussed, the retaining enzymes, hydrolysis of a glycosidic bond occurs with overall retention of configuration through nucleophile-assisted catalysis. The nature of the nucleophile can vary greatly, being either an enzymic nucleophile or a nucleophile present in the substrate. Koshland originally proposed that hydrolysis with retention of configuration must occur in two steps, each involving inversion of anomeric stereochemistry, through the involvement of a nucleophilic group forming an intermediate in the first step, which in the second step is hydrolysed with an overall net retention of configuration.[8] Since the original proposals of Koshland, many studies have been performed that have cast light on the diverse identities of the groups involved in executing this reaction.

The second major class of glycoside hydrolases are inverting enzymes. Inverting glycoside hydrolases utilize a one-step mechanism in which two strategically positioned residues, both carboxylic acids, act as general acid and base, respectively, and simultaneously act to protonate the glycosidic oxygen and deprotonate a nucleophilic water molecule, resulting in a stereo-inversion upon hydrolysis.[10]

Sinnott has proposed an extension to the classification of inverting and retaining glycoside hydrolases based on whether the enzymes hydrolyse axial (a) or equatorial (e) bonds to give rise to axial or equatorial products. Thus, an enzyme that catalyses the cleavage of an equatorial glycoside with retention of configuration to give an equatorial product is classified as an e → e enzyme.[6]

Sequence-based classification of glycoside hydrolases: Owing to direct gene-sequencing and indirect genome-sequencing efforts, a tremendous number of sequences is known for glycoside hydrolases. Much effort has been expended on trying to make sense of this treasure trove of potentially useful information. Owing in great part to the authoritative and seminal contributions of Henrissat and colleagues, sequence comparison methods have allowed the generation of a series of sequence-based Glycoside Hydrolase (GH) families.[11–14] There currently exists more than 110 GH families, but the number is not static, and updates frequently occur as new sequence and biochemical data come to hand. The updated classification database is available over the World Wide Web through the *c*arbohydrate *a*ctive *e*nzyme (CAZy) server.[15] The glycoside hydrolase classification scheme has proved invaluable as a means of correlating large amounts of biochemical and structural information with sequence data and has allowed mechanistic and functional assignments for arbitrary glycoside hydrolase genes that fall within each sequence-based family.[16,17] A particularly useful feature of the sequence-based classification is that it almost always results in classification of glycoside hydrolases with identical enzymatic mechanisms (retaining or inverting) into the same family. Moreover, if specific functions can be assigned to individual amino acids in the active site (catalytic nucleophile, acid/base), then, owing to sequence similarities inherent in

References start on page 280

the classification, it is almost always found that conserved equivalent residues are present in all members of the sequence-based family, and it is therefore likely that identical catalytic roles are undertaken by these analogous residues.

The sequence-based classification can also be used in a predictive manner to suggest possible substrates for previously unstudied glycoside hydrolases. Thus, certain activities that have been demonstrated for a specific family member are possible candidate activities for other sequence-related family members. As with any in silico functional annotation, such predictions should always be made with care and remain subject to experimental confirmation.

Mechanism of inverting glycoside hydrolases: Inverting glycoside hydrolases are arguably the simplest mechanistic class of enzymes that cleave the glycosidic linkage. These enzymes act through a one-step mechanism that involves strategically positioned active-site base and acid residues.[6] A single molecule of water is positioned such that deprotonation by the base residue occurs simultaneously to formation of a new bond to the anomeric centre, and cleavage of the bond to the aglycon, which occurs with general acid assistance from the acid residue. Secondary kinetic isotope effects reveal that the transition state for hydrolysis has significant oxocarbenium ion character;[6] there is little evidence that a discrete oxocarbenium ion intermediate has any meaningful existence.[6] The roles of acid and base are typically played by either glutamate or aspartate residues.

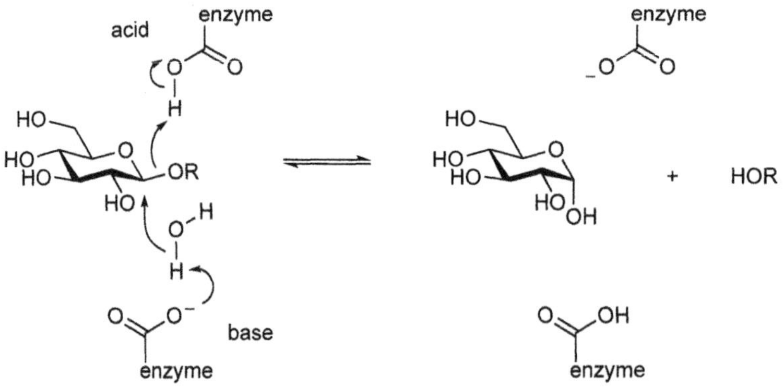

Mechanism of retaining glycoside hydrolases that use carboxylic acids as nucleophiles: The prototypical example of a glycoside hydrolase is provided by the enzyme Hen Egg White Lysozyme (HEWL). HEWL catalyses the cleavage of β-1,4-glycosidic linkages between N-acetylmuramic acid and N-acetylglucosamine in peptidoglycan and was the first enzyme to have its three-dimensional structure elucidated by X-ray crystallography, by Phillips[b] and co-workers.[18] Based on these and other results, HEWL was the first glycoside hydrolase to have its mechanism studied in detail. However, such are the challenges associated with working with HEWL, in particular, the difficulty of obtaining chemically homogeneous substrates, that many mechanistic features, which are now ascribed to lysozyme, were originally determined for other related glycoside hydrolases, before ultimately being demonstrated for this enzyme.

HEWL possesses two carboxylic acid residues that are strategically positioned about the glycosidic linkage and catalyse the cleavage of the glycosidic linkage in two steps. These two residues perform quite different functions, one acting as an acid/base residue and the other as an enzymic nucleophile. In the first step (glycosylation), the acid/base residue Glu35 acts as a general acid and protonates the glycosidic linkage. At the same time, the deprotonated carboxylate Asp52 performs a nucleophilic attack on the anomeric carbon, resulting in the displacement of the aglycon, and the formation of a glycosyl enzyme. In the second step (deglycosylation), the now deprotonated acid/base Glu35 acts as a general base to deprotonate a suitably positioned water molecule, which performs a nucleophilic substitution on the glycosyl enzyme, releasing the hydrolysed product and regenerating the active-site nucleophile. In the case of HEWL, the identities of the acid/base and nucleophile have been assigned through detailed kinetic and mass spectrometric studies with acid/base mutant enzymes, and through trapping experiments with 2-deoxy-2-fluoro sugars and determination of the corresponding X-ray structure of the covalent adduct.[19]

glycosyl-enzyme intermediate

The mechanism utilized by HEWL outlined above is common to most retaining glycoside hydrolases, and typically, the acid/base and nucleophilic residues are carboxylic acid side chains of glutamic and/or aspartic acids. Detailed insight into the structure of the transition states leading to and from the glycosyl-enzyme intermediate

---

[b] David Chilton Phillips (1924–1999), Ph.D. (1951) at the University College, Cardiff, professorship at Oxford (1968).

References start on page 280

has been obtained for a good many retaining glycoside hydrolases. An excellent illustrative case is that of the retaining β-glucosidase Abg from *Agrobacterium*.[2] A Brønsted plot of log k_{cat} versus leaving group pK_a value for a series of aryl β-glucosides afforded a concave downward relationship, indicating a change in rate-determining step as the aglycone leaving group ability increases.[20] This was interpreted to indicate rate-limiting glycosylation for poor leaving groups and rate-limiting deglycosylation for good leaving groups. Deuterium α-secondary kinetic isotope effects were then measured for substrates that were glycosylation or deglycosylation rate limiting. In each case the kinetic isotope effects revealed $k_H/k_D > 1$, indicating that the transition states for both steps possess oxocarbenium ion character.[20]

Mechanism of retaining glycoside hydrolases that use tyrosine as a catalytic nucleophile: While most retaining glycoside hydrolases utilize carboxylate groups as acid/base and nucleophile, there are some exceptions. Retaining neuraminidases (sialidases) have been shown to utilize tyrosine as a catalytic nucleophile rather than a carboxylate nucleophile.[21] One rationale for this replacement is that the anomeric centre in these sugars bears an anionic carboxylate residue, and nucleophilic attack by an anionic nucleophile is therefore disfavoured. Consequently, a neutral nucleophile such as a tyrosine (with a suitably positioned general base that can tune the acidity of this tyrosine)[21] may be more effective. The glycosyl-enzyme intermediate formed is a glycosyl tyrosine residue, which is an aryl glycoside and is, in effect, an activated intermediate that can be readily hydrolysed in the second step of the reaction.

sialyl-enzyme intermediate

Mechanism of retaining glycoside hydrolases that use substrate-assisted catalysis: While most glycoside hydrolases utilize an enzymic nucleophile (e.g. aspartate, glutamate or tyrosine), certain retaining glycoside hydrolases utilize substrate-assisted catalysis. Mainly these are enzymes that possess a 2-acetamido group, such as chitinases,[22] hyaluronidases[23] and *N*-acetylhexosaminidases.[24,25] In these enzymes the enzymic nucleophile is replaced with a substrate-borne nucleophilic residue capable of anchimeric assistance. Thus, in the first step of the reaction the 2-acetamido group acts as a nucleophile, forming an oxazoline (or oxazolinium ion) intermediate. In the second step an enzymic carboxylate acts as a general base and assists the nucleophilic attack of a water molecule to hydrolyse the intermediate.

oxazolinium intermediate

Not all retaining glycoside hydrolases that catalyse the hydrolysis of substrates bearing neighbouring groups do so with substrate-assisted catalysis. It has been demonstrated that Family 3 glycoside hydrolases, some of which are retaining hexosaminidases, and the Family 22 enzymes discussed earlier (which include HEWL), utilize an enzymic nucleophile rather than a 2-acetamido group.[19,26,27]

Unusual Enzymes that Catalyse Glycoside Cleavage

Some enzymes catalyse the cleavage of the glycosidic linkage through rather unusual mechanisms similar to or sometimes quite distinct from those described above.

Myrosinases are retaining glycoside hydrolases that cleave glucosinolates, naturally occurring sulfated thioglycosides. Glucosinolates are plant defence compounds, which are activated by grazing insects. Through the act of mastication, the spatially separated glucosinolate substrate and myrosinase glycoside hydrolase are combined.[28] Following hydrolysis of the thioglycoside, a range of thiocyanate and nitrile products is formed, which are toxic to the insect. Myrosinases are unusual retaining glycoside hydrolases in that they lack an enzymic acid/base residue, with the usual glutamic acid residue being replaced by glutamine. These enzymes have been shown to exhibit a striking dependence of catalytic activity on the coenzyme, L-ascorbic acid.[29] Through studies with 2-deoxy-2-fluoro sugars, myrosinases have been demonstrated to accumulate a glycosyl-enzyme intermediate, showing that the sole active-site carboxylate acts as a nucleophile and not as an acid/base.[30,31] Additionally, a structure of a ternary complex formed with an aza sugar inhibitor and L-ascorbate revealed binding of the latter in the aglycon binding site, and it was thus suggested that L-ascorbate fulfils the role of a catalytic base.[32]

References start on page 280

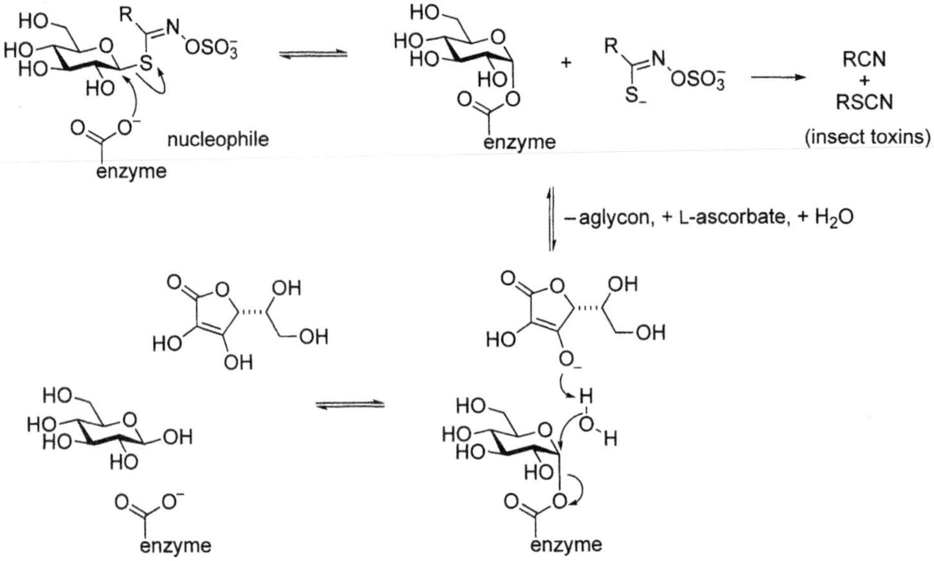

Many bacteria possess enzymes termed lyases, which cleave glycosidic linkages through an elimination mechanism.[33] There are two main sets of lyases: those that act on carbohydrate substrates bearing an acidic proton adjacent to the leaving group and those that do not have an obviously acidic proton that is nonetheless lost in the elimination process. The first set of lyases is responsible for the degradation of glycosaminoglycans and pectin. These enzymes facilitate a relatively simple elimination process and result in the formation of a product containing an alkene. Studies of chondroitin AC lyase from *Flavobacterium heparinum* using a nitro sugar anion with slow reprotonation kinetics identified a conserved active-site tyrosine as the catalytic base.[34]

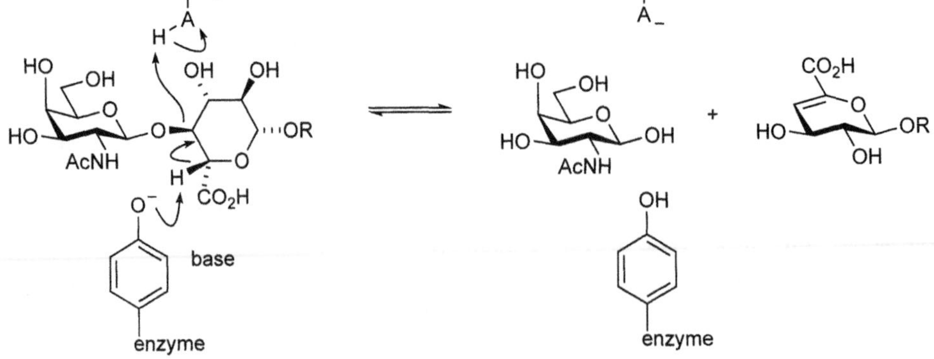

The second set of lyases acts on substrates that are not obviously acidic.[33] An example of such enzymes are the α-glucan lyases that act on α-glucans and result in cleavage to afford 1,5-anhydrofructose.[35] Detailed mechanistic studies have shown that the α-1,4-glucan lyase of *Gracilariopsis* performs the cleavage reaction through a β-elimination pathway involving a covalent glycosyl-enzyme intermediate. Thus, nucleophilic attack by an enzymic carboxylate forms a glycosyl enzyme, and the glycosyl enzyme then undergoes an elimination reaction, affording a transient 2-hydroxy glucal. Finally, the 2-hydroxy glucal tautomerizes to 1,5-anhydrofructose. Strong evidence was obtained for this mechanism through kinetic isotope effect studies,[36] by inhibition with acarbose and 1-deoxynojirimycin (DNJ; known inhibitors of retaining α-glucanases),[36] and by detection of the covalently labelled glycosyl enzyme formed upon reaction of the *Gracilariopsis* α-1,4-glucan lyase with the mechanism-based inactivator, 5-fluoro-β-L-idopyranosyl fluoride.[37] Interestingly, the α-1,4-glucan lyases are classified into family GH31, which also contains retaining α-glucosidases that proceed through a typical retaining mechanism involving a similar glycosyl-enzyme intermediate.[38]

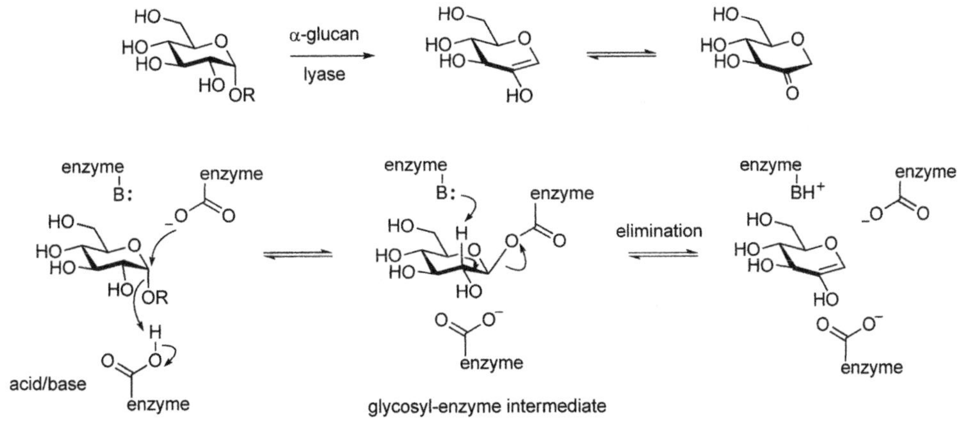

glycosyl-enzyme intermediate

Phosphorylases are enzymes that catalyse the phosphorolysis of the glycosidic bond. They are usually named using a combination of the 'substrate name' and 'phosphorylase'.[39] Phosphorolysis of a glycosidic bond can occur with retention or inversion of configuration and always occurs in an *exo*-fashion, leading to the formation of a sugar-1-phosphate.[39] The energy content of the sugar-1-phosphate product means that the cleavage reaction is in a practical sense reversible and, in nature, these enzymes may be used for either synthesis or cleavage of the glycosidic bond. As such there is a relatively fine distinction among sugar phosphorylases, glycoside hydrolases and classical sugar nucleoside (di)phosphate–dependent glycosyltransferases. In the last case, the synthetic reaction is normally, but not always, irreversible because of the much higher energy of a sugar nucleoside (di)phosphate.

The classical example of a phosphorylase is glycogen phosphorylase.[40] This 'workhorse' enzyme catalyses the cleavage of individual glucosyl residues from glycogen (up to five residues from a branchpoint), forming sequentially deglyco-sylated glycogen and glucose-1-phosphate. Glycogen phosphorylase has a complex mechanism that is not fully understood and requires pyridoxal phosphate (PLP) as a cofactor.[10,41] Glycogen phosphorylase is classified into the same sequence-related glycosyltransferase family as starch phosphorylases (GT35), which also require a PLP cofactor.[41] Most sugar phosphorylases act on gluco-sides, and many cleave simple disaccharides such as sucrose, trehalose, cellobiose and maltose, leading to glucose-1-phosphate and the other component sugar (glucose or fructose).[39] Other sugar phosphorylases are known that act on chitobiose, laminaribiose, 1,3-β-glucan and nucleosides.[10,39] Sequence and structural analysis of sugar phosphorylases reveal that many have sequences and structures (and likely mechanisms) similar to glycosyltransferases, whereas others have sequences and structures that more closely resemble glycoside hydrolases (GH13, GH65).[42]

Transglycosidases

Hydrolysis of a glycoside by a glycoside hydrolase can be considered an enzyme-catalysed transfer of a glycosyl residue from its aglycon to water. Several enzymes closely related to glycoside hydrolases use essentially the same mechanism to catalyse the transfer of glycosyl residues to acceptors other than water. In doing so they are acting as transglycosidases (also known as transglycosylases). It is important to note the difference between transglycosidases and glycosyltrans-ferases (to be covered in the next chapter): the former use glycosides as substrates, whereas the latter use sugar nucleotides (or related species) as sub-strates. Transglycosidases operate through a retaining mechanism that proceeds through a glycosyl-enzyme intermediate and that is the same as used by retaining glycoside hydrolases. Such mechanisms have been demonstrated for various transglycosidases including xyloglucan *endo*-transglycosidases (which remodel xylo-glucan in the plant cell wall),[43,44] cyclodextrin glucosyltransferases (which convert starch into cyclodextrins),[10,45–47] trans-sialidases[48] and the transglycosidase activity of glycogen-debranching enzyme.[10,49]

By analogy with glycoside hydrolases, a suitable enzymic nucleophile (aspartate, glutamate or tyrosine) performs a nucleophilic attack on the glycoside substrate, generating a glycosyl-enzyme intermediate. While in glycoside hydrolases the intermediate glycosyl enzyme reacts with water, in transglycosidases this intermediate is unusually stable to hydrolysis and is instead transferred to an alternative nucleophile, such as the hydroxyl group of another sugar residue.

cyclodextrin glucosyltransferase:

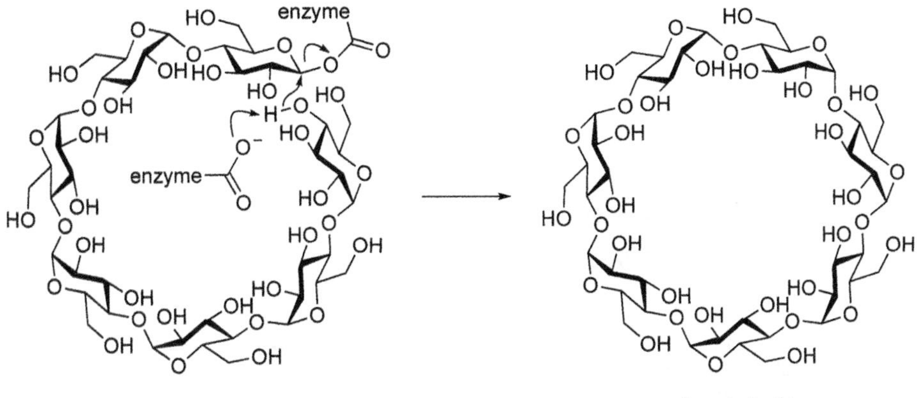

β-cyclodextrin

Presumably, specificity for transfer to another carbohydrate rather than to water is achieved by active-site features that utilize favourable bonding to the acceptor carbohydrate to stabilize the transition state for transglycosidation relative to hydrolysis. As discussed below, transglycosidases can be used in the synthesis of the glycoside bond using kinetically controlled transglycosidation with activated glycosyl donors.

Structure-based Studies of Glycoside Hydrolases

A staggering amount of effort has been invested into the structural analysis of glycoside hydrolases. New structures for previously unsolved families are always keenly sought and, to date, representatives for more than 60 of the glycoside hydrolase families have been determined. These structures have revealed a multitude of three-dimensional folds that range from complete β-strand to almost complete α-helical structures.[50] As expected for sequence-based classifications, the overall fold of a glycoside hydrolase is conserved within a family. Structural information can provide direct confirmation of the enzymatic mechanism used by a glycoside hydrolase and provide structural insight into these mechanisms. For example, comparison of structures for a considerable number of glycoside hydrolases reveals that the typical spacing of the acid/base and nucleophile residues of retaining glycosides is of the order of 5.5 Å, whereas that between acid/base residues of inverting hydrolases is almost 10 Å.[51] This observation is consistent with that expected for the two mechanisms: for retaining glycoside hydrolases, the enzymic nucleophile must be positioned close to the anomeric centre to allow for the formation of a covalent bond in the glycosyl-enzyme intermediate. By contrast, in inverting glycoside hydrolases, there must be sufficient space between the basic and acidic residues to accommodate the nucleophilic water molecule and the glycoside.

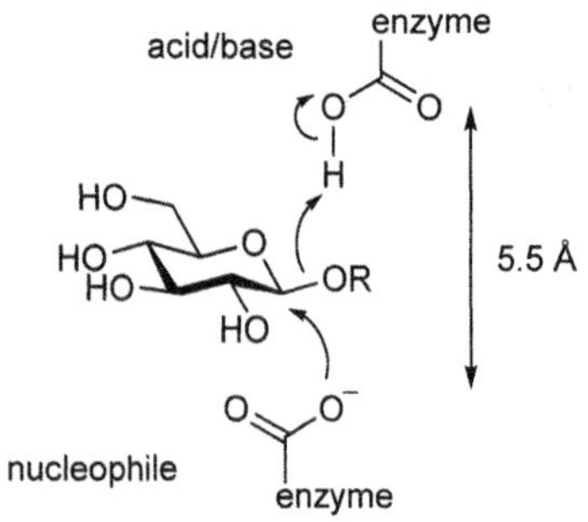

retaining glycoside hydrolase 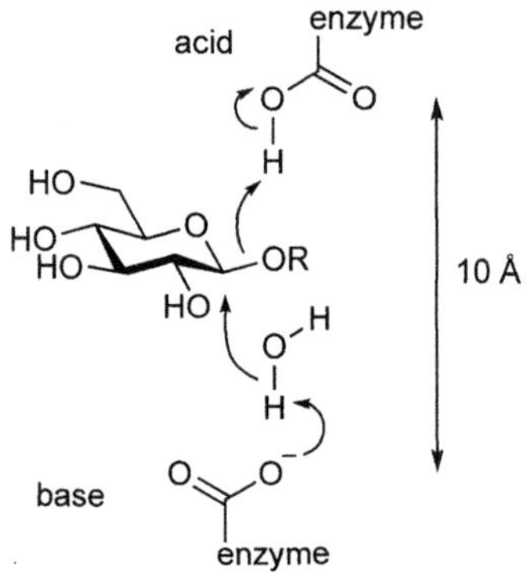   inverting glycoside hydrolase

Further sophistication in the glycoside hydrolase mechanism has been elaborated from the examination of three-dimensional structures of glycoside hydrolases. Comparison of large numbers of three-dimensional structures with bound ligands reveals that there are only two orientations in which the acid residues are found, both providing protonation laterally with respect to the plane of the pyranose ring.[52] Those with the acid/base providing protonation from the front of the sugar have been termed anti-protonators, and those with an acid/base situated to protonate from the rear of the sugar are termed syn-protonators ('anti' and 'syn' refer to the orientation of the acid residue relative to the C1–O5 bond).[52,53] Such details have been exploited in inhibitors that have been found to be specific for anti-protonators over syn-protonators – the corresponding inhibitors selective for syn- over anti-protonators have not yet been described.[54]

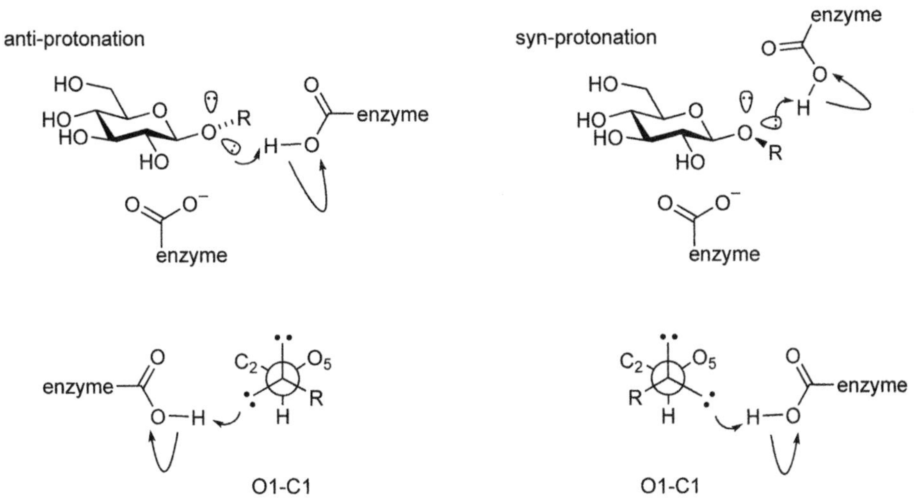

Reagents and Tools for the Study of Glycoside Hydrolases

The study of glycoside hydrolases has been ably assisted by the development of a range of chemical reagents and tools. These reagents include molecules capable of selectively reacting with specific amino acid residues and thereby assisting in defining their catalytic roles through sequencing and X-ray crystallography, affinity labelling reagents for selectively labelling glycoside hydrolases in complex mixtures, and inhibitors that can be used to control the activity of the target glycoside hydrolase. These molecules can assist in defining the function of glycoside hydrolases in complex biological systems, as well as acting as therapeutic agents.

A range of affinity labelling reagents has been developed over the years for the identification of specific active-site residues in glycoside hydrolases.[55,56] Mainly these consist of carbohydrates that have been modified to install various electrophilic groups capable of reacting with specific nucleophilic or acid/base residues in the enzyme active site. An important class of affinity labelling reagents includes the 2-deoxy-2-fluoro and 5-fluoro sugars.

As discussed earlier, the standard reaction mechanism for a retaining glycoside hydrolase proceeds through an enzyme-linked covalent intermediate. While there are various strategies for trapping this intermediate, by far the most reliable approach was pioneered by Withers and co-workers.[2] In this approach, one or more fluorine atoms are introduced into the 2-[57] or 5-position[58] to reduce the rate of the enzyme-catalysed hydrolysis reaction by destabilizing the oxocarbenium ion-like transition states leading to and from the glycosyl-enzyme intermediate. Since such substitution slows both the glycosylation and the deglycosylation step, a good leaving group,

References start on page 280

typically fluoride, 2,4-dinitrophenolate or 2,4,6-trinitrophenolate ion, is also incorporated to ensure that the glycosylation step remains faster than deglycosylation, thereby resulting in accumulation of the glycosyl-enzyme intermediate.

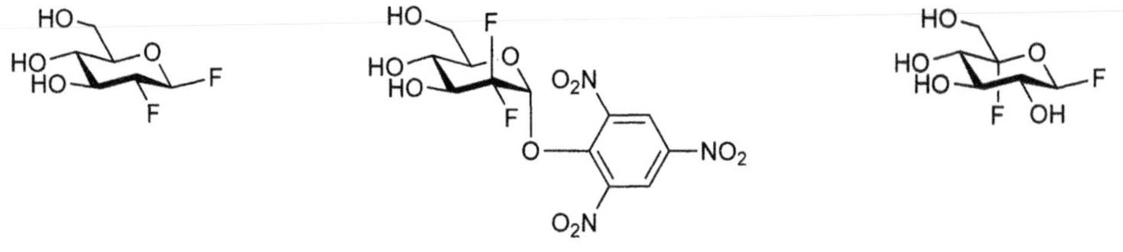

2-deoxy-2-fluoro   2-deoxy-2,2-difluoro   5-fluoro

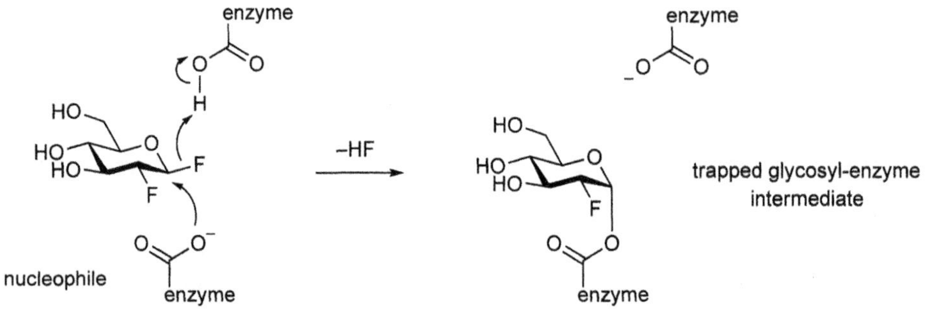

trapped glycosyl-enzyme intermediate

Such 2-deoxy-2-fluoro, 2-deoxy-2,2-difluoro and 5-fluoro sugars act as mechanism-based, time-dependent inhibitors of retaining glycoside hydrolases and have proven extremely useful in the study of the glycosyl-enzyme intermediate. For example, these reagents can be used to identify the enzymic nucleophile of a glycoside hydrolase through mass spectrometric sequencing methods.[56,59] Just as importantly, the trapped glycosyl-enzyme intermediate can be sufficiently stable to allow X-ray crystallographic analysis of the resulting covalent complex, providing a 'snap-shot' of an important species along the reaction coordinate of glycoside hydrolysis.[60,61] Finally, such covalent adducts can be sufficiently stable to allow affinity chromatography and purification of the covalently labelled peptide, which has been used in activity-based protein profiling and cloning strategies.[62,63]

Various epoxide-containing carbohydrates act as time-dependent covalent inactivators of glycoside hydrolases.[55] The earliest of these were the epoxyalkyl glycosides.[64] These have been prepared as glucosides, xylosides, cellobiosides, hexosaminides, and so on, and are broadly active against most retaining glycosidases. Investigations into the effect of chain length on activity have revealed that different enzymic residues can be alkylated, including the nucleophile or acid/base residues of retaining glycosidases.[65] The presence of the epoxide introduces a new stereocentre into the aglycon, and studies of individual diastereomers have shown differences in rates of inactivation of the target enzymes. Thus, (3S)-epoxybutyl β-cellobioside inactivates the (1,3;1,4)-β-glucan 4-glucanohydrolase from *Bacillus subtilis* more so than the (3R)-isomer, whereas the reverse is true for the corresponding enzyme from barley.[66]

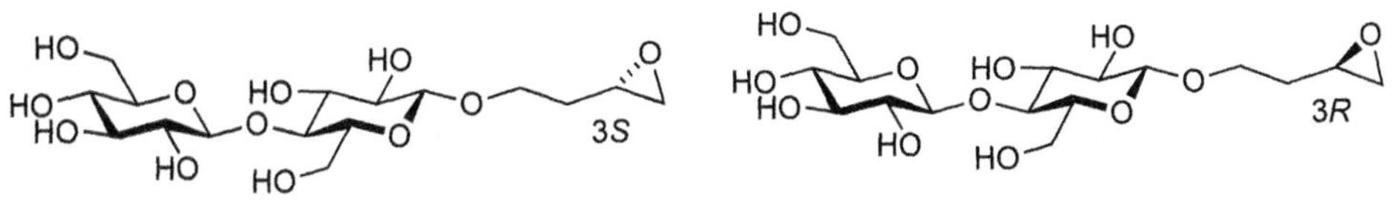

Several bicyclic inactivators of glycosidases have been found as natural or chemically synthesized products, including cyclophellitol,[67] various conduritol epoxides[68,69] and conduritol aziridine.[70] These compounds are inactivators of many glycoside hydrolases and have been shown to modify their target enzymes covalently. A labelling study of *E. coli* β-galactosidase with conduritol C epoxide revealed labelling of a residue that was proposed to be the catalytic nucleophile.[71] Later studies with a 2-deoxy-2-fluoro sugar inactivator revealed the original assignment to be incorrect.[72] The related conduritol aziridine has been demonstrated to act as a time-dependent inactivator of a glycoside hydrolase.[70]

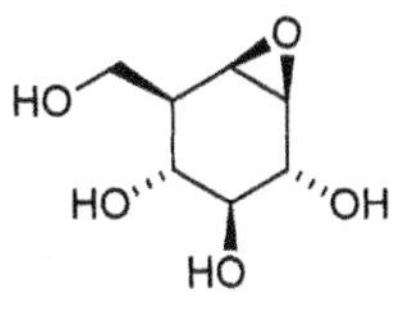

(+)-cyclophellitol 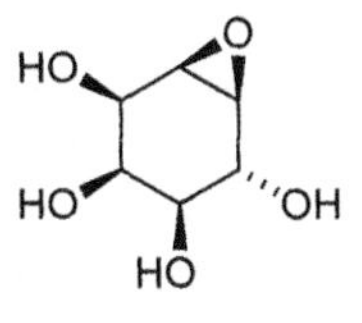   conduritol C epoxide 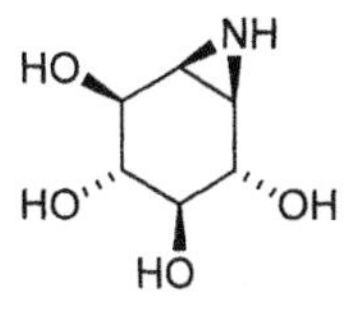   conduritol aziridine

α-Halocarbonyl compounds are effective electrophiles, with the right combination of reactivity towards enzymic nucleophiles and stability in aqueous solution. Incorporation of α-halocarbonyl groups into carbohydrates affords affinity labels that are time-dependent inactivators of glycoside hydrolases. The most commonly studied α-halocarbonyl derivatives are *N*-bromoacetyl glycosides, but bromomethyl ketones have also been used.[73,74] These reagents typically exhibit good specificity for labelling of the enzymic acid/base residue of glycoside hydrolases.

References start on page 280

However, examples of non-specific labelling[75] and labelling of residues not involved in catalysis[76] are known.

Non-covalent Glycoside Hydrolase Inhibitors

Competitive inhibitors of glycoside hydrolases have a rich history.[77–79] The interest in developing such compounds has ranged from the hope that they can act as therapeutic agents for the control and treatment of disease, to the simple delight of generating molecules that can stop the powerful catalytic action of glycoside hydrolases. The major glycoside hydrolase inhibitors are nitrogen-containing 'sugar-shaped' heterocycles. Many of these compounds have been found in nature, and others have been designed based on considerations of enzyme mechanisms. We have already discussed many of these compounds in Chapter 3, and there are several comprehensive reviews on nitrogen-containing heterocycle inhibitors of glycoside hydrolases.[52,80–84]

The progenitor of the nitrogen-containing heterocycles is DNJ. Its synthesis by Paulsen and co-workers predated its discovery as a natural product;[79,85,86] it was later discovered in a range of natural sources, including what may be the richest source discovered to date, mulberry latex.[87] DNJ is a powerful inhibitor of many glucoside hydrolases, having greater activity towards inverting glucoside hydrolases rather than retaining ones. Structural variants of DNJ have also been uncovered as natural products, including the epimer 1-deoxymannojirimycin and various dideoxy derivatives.[88] DNJ is a powerful inhibitor of intestinal sucrase, and the related derivative Miglitol has been developed as a treatment for non-insulin-dependent (type II) diabetes.[89] N-Butyl-DNJ has been studied for its ability to treat lysosomal storage disorders and inhibit HIV infection.[83,90] Much has been made of the similarity of the protonated forms of aza sugars such as DNJ and the oxocarbenium-ion-like transition state of glycoside hydrolysis.[84,91]

1-deoxynojirimycin (DNJ)   Miglitol   N-butyldeoxynojirimycin

A range of sugar-shaped heterocycles related to DNJ has been discovered in natural sources. For example, swainsonine was discovered in *Swainsonia canescens*, a plant native to Western Australia; the digestion of swainsonine by livestock leads to symptoms that have the appearance of a neurological condition termed mannosidiasis.[92] These and other studies have shown swainsonine to be a powerful inhibitor of the N-linked glycan-processing enzyme α-mannosidase II,[93] and the pharmacological inhibition of this enzyme matches the effect of genetic mutation of the enzyme. A related compound, castanospermine, is a powerful inhibitor of many glycoside hydrolases.[94] Castanospermine was isolated from the seeds of *Castanospermine australae*, a tree found in the rainforests of Queensland, Australia,[95] and commonly planted as a delightful flowering specimen tree in botanical gardens around the world.

swainsonine   castanospermine

Isofagomine, a structural isomer of fagomine (1,2-dideoxynojirimycin), was originally synthesized by Lundt, Bols and co-workers[96] but and has never been found in nature. Isofagomine and analogues are powerful inhibitors of retaining glycoside hydrolases.[82] The closely related molecule siastatin B and analogues have been isolated from various *Streptomyces* strains and are inhibitors of sialidases, *N*-acetylhexosaminidases and glucuronidases.[82] As suggested for DNJ, there is a similarity between the protonated form of isofagomine and the oxocarbenium-ion-like transition state of glycoside hydrolysis.[84,91] An isofagomine derivative in complex with the *endo*-cellulase Cel5A from *Bacillus agaradhaerans* has been studied by high-resolution X-ray crystallography, which has provided evidence that the isofagomine was bound in the protonated form.[97] Some isofagomine-inspired compounds, termed the immucillins (see Chapter 12), have been developed as inhibitors of 5′-methylthioadenosine/*S*-adenosylhomocysteine nucleosidase (a nucleoside hydrolase involved in purine and methionine salvage reactions)[98] and 5′-methylthioadenosine phosphorylase (a nucleoside phosphorylase involved in catabolism of methylthioadenosine).[99] 5′-(4-Chlorophenylthio)-DADMe-immucillin-A binds the former enzyme from *E. coli* with a dissociation constant (K_i) of 47 femtomolar, making it one of the most powerful, non-covalent inhibitors reported for any enzyme and a promising candidate for therapeutic development.[98]

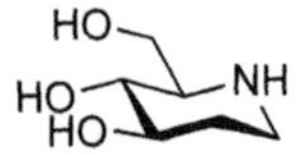

fagomine 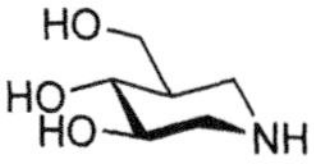   isofagomine 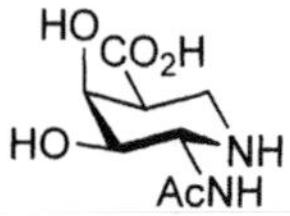   siastatin B

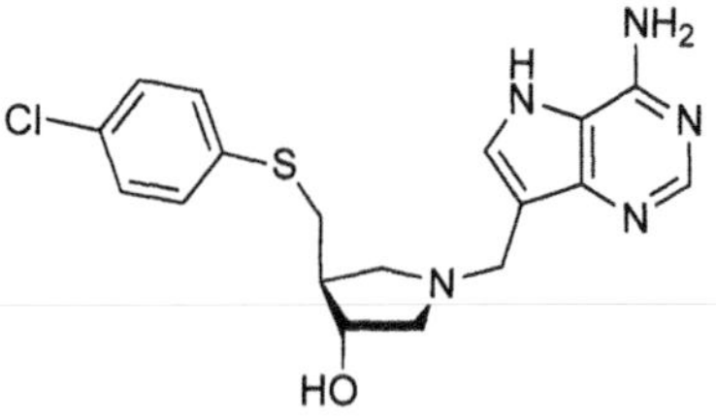

4-ClPhT-DADMe-ImmA

A range of sugar heterocycles with sp^2 hybridized 'anomeric' centres has been developed, largely through the efforts of the groups of Ganem and Vasella.[52,77] These compounds typically bear a 'glycosidic' nitrogen that has a lone pair of electrons oriented in the plane of the sugar ring, away from the sugar. It has been shown that these heterocycles are good inhibitors of anti-protonating glycoside hydrolases, viz. enzymes with their acid/base residue located in the plane of the sugar ring and positioned anti to the C1–O5 bond. These inhibitors have been shown to interact with anti-protonating glycoside hydrolases through the formation of a strong hydrogen bond with the acid/base residue.[100–102] While this series of inhibitors was largely the product of rational design, at least one example, nagstatin, has been discovered as a natural product.[103–105]

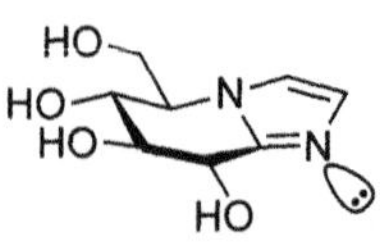

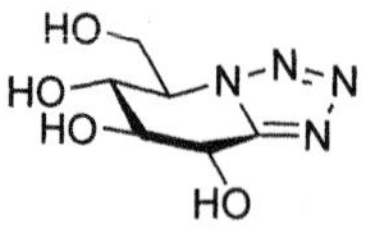

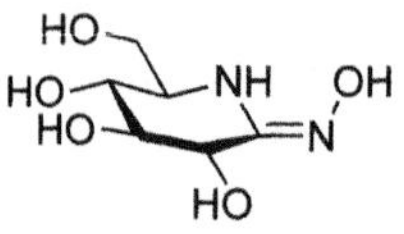

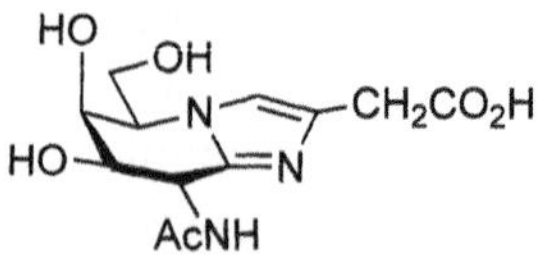

nagstatin

Another group of interesting glycoside hydrolase inhibitors is those that contain a cyclitol ring. The most significant of these compounds is acarbose, which was originally isolated from an *Actinoplanes* sp. strain.[106] Acarbose contains an aminocyclitol, which is essential for its activity; it is an extraordinarily powerful inhibitor of intestinal α-amylase and has been approved for the treatment of type I and type II diabetes.[89] There is a family of closely related five-membered ring cyclitol natural products that are found as inhibitors of many glycoside hydrolases.[107] For example, trehazolin is an inhibitor of trehalases, and mannostatin is effective against mannosidases.[107]

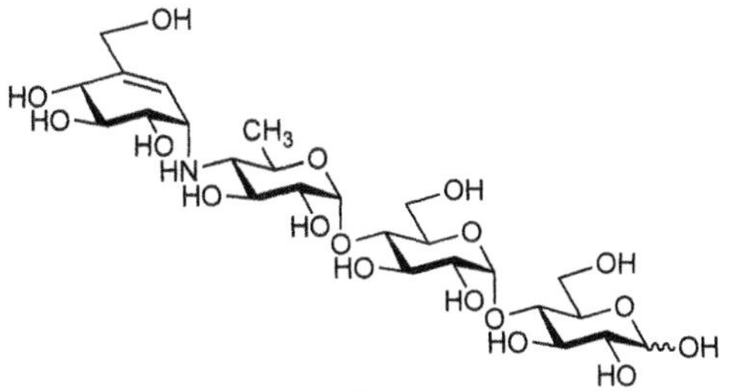

acarbose

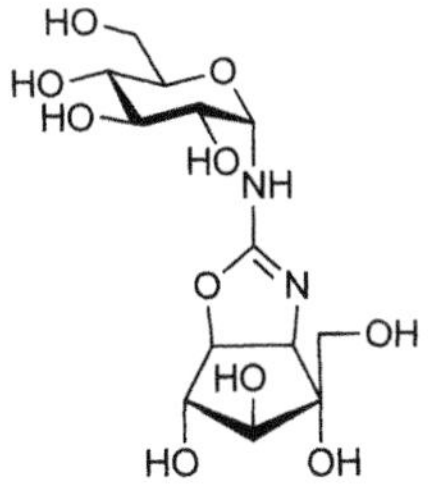

trehazolin

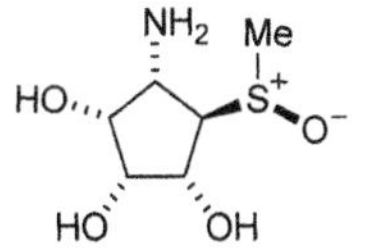

mannostatin

Most of the above-mentioned molecules are inhibitors of retaining or inverting glycoside hydrolases that utilize enzymic acid/base or nucleophile residues. Interactions of the inhibitors and these residues can be used to rationalize their inhibitory activity. For enzymes that use neighbouring group participation, mainly hexosaminidases, several powerful inhibitors have been discovered or designed. Allosamidin, an unusual N-acetylallosamine-containing carba-trisaccharide, was originally isolated from a *Streptomyces* sp. strain and is a powerful inhibitor of chitinases that use substrate-assisted catalysis.[108] The related compound, NAG-thiazoline, was originally synthesized by Knapp and co-workers and is an excellent inhibitor of retaining hexosaminidases that utilize substrate-assisted catalysis.[24,25,109]

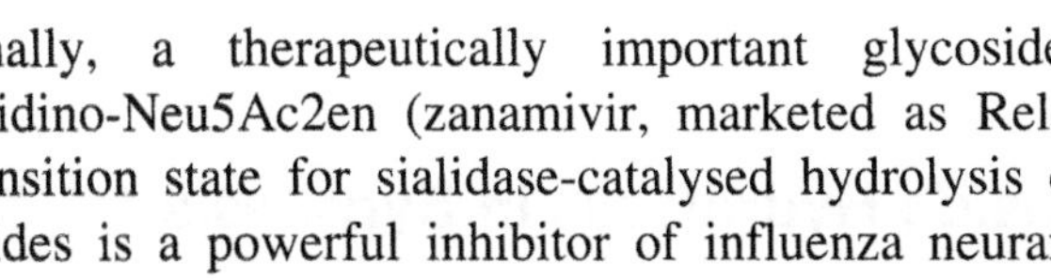

allosamidin

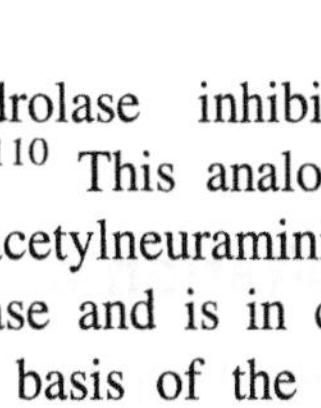

NAG-thiazoline

Finally, a therapeutically important glycoside hydrolase inhibitor is 4-guanidino-Neu5Ac2en (zanamivir, marketed as Relenza).[110] This analogue of the transition state for sialidase-catalysed hydrolysis of N-acetylneuraminic acid glycosides is a powerful inhibitor of influenza neuraminidase and is in clinical use.[111] Dimers of zanamivir have been developed on the basis of the known trimeric arrangement of neuraminidase on the influenza virus surface and have been shown to be superior to the monomeric form of the drug in terms of potency and duration of action.[112,113] The related carbocyclic analogue oseltamivir (marketed as Tamiflu), the design of which was based on zanamivir, is also widely used for the treatment or prevention of influenza as, in contrast to zanamivir, it is orally available.[114–116]

References start on page 280

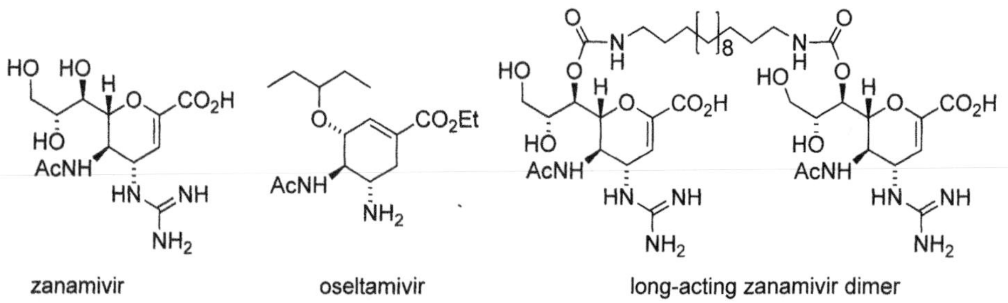

zanamivir

oseltamivir

long-acting zanamivir dimer

Oligosaccharides in which one or more glycosidic oxygens have been replaced by some other atom or group (e.g. sulfur, CH_2) have the potential to act as non-hydrolysable substrate analogues. The most significant of these analogues are those in which oxygen has been replaced by sulfur. Such thiooligosaccharides typically act as competitive inhibitors of glycoside hydrolases and are useful tools for studying the initially formed Michaelis complex of substrate with a glycoside hydrolase.[117,118] Thiooligosaccharides are usually relatively poor inhibitors, likely as a consequence of resembling substrate, rather than the much more tightly bound transition state. Nonetheless, binding can be sufficiently tight for structural determination of enzyme complexes.[119,120] The X-ray structure of *Fusarium oxysporum* EGI in complex with an S-linked methyl α-cellopentaoside revealed substrate distortion from the ground-state $^4C_1$ structure to a conformation with a pseudo-axially oriented aglycon.[121]

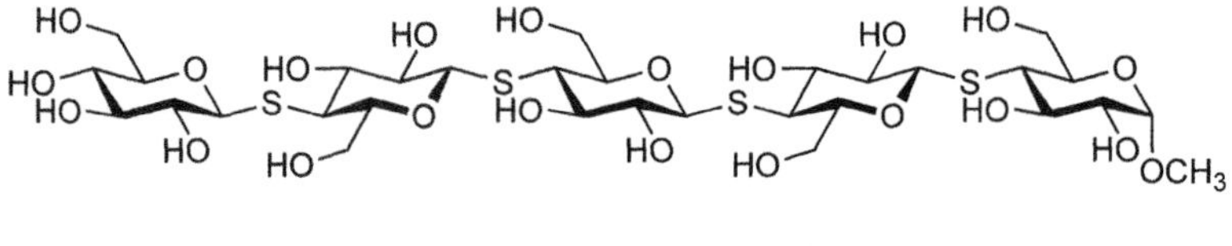

Exploitation of Glycoside Hydrolases in Synthesis[122–128]

Glycoside hydrolases usually operate in the direction of hydrolysis of substrate glycosides. However, under suitable conditions, these enzymes can be made to operate in the reverse direction, i.e. towards the synthesis of glycosides. Alternatively, if these enzymes perform catalysis by way of a reactive intermediate, this intermediate can be intercepted by a suitable nucleophile, allowing the synthesis of glycosides from 'activated' glycosyl donors, which may themselves be glycosides.

Thermodynamic control (reversed hydrolysis):[124] Glycoside hydrolases are enzyme catalysts that accelerate the forward and reverse reactions of glycoside hydrolysis. Thermodynamic control (also called equilibrium control or reversed hydrolysis) refers to the use of a glycoside hydrolase as a synthetic enzyme where the conditions of the enzymic reaction are perturbed so that the position of equilibrium is shifted to favour glycoside products. All glycoside hydrolases can be used in this fashion, regardless of the mechanism of the enzyme under operation.

Consideration of the overall reaction equation for glycoside hydrolysis gives some clues as to how the position of the equilibrium may be shifted to favour products: large amounts of the alcohol acceptor, reduction of the effective concentration of water by the addition of co-solvents, removal of the product glycoside as it is formed and varying the reaction temperature (the hydrolysis reaction is an exothermic process). While thermodynamic control is notable for its simplicity and the moderate-to-good yields of products that can be obtained (15–65%),[129] owing to difficulty in purification of products and in achieving high yields, the method is not widely used.

Kinetic control (transglycosidation):[122–125] Kinetic control of glycoside hydrolase reactions refers to situations where perturbations in reaction rate of hydrolysis/transglycosidation are used to increase yields of transglycosidation products. Kinetic control can be used only for enzymes that operate by way of an activated intermediate (i.e. only retaining glycoside hydrolases). The general approach is to utilize an activated glycosyl donor, e.g. aryl glycosides or glycosyl fluorides, which are usually excellent substrates for retaining glycoside hydrolases. The excellent anomeric leaving group in these substrates results in a rapid formation of a glycosyl enzyme. This glycosyl enzyme can suffer two fates: either (effectively irreversible) reaction with water (hydrolysis) or reversible reaction with a suitable alcohol nucleophile, leading to transglycosidation.

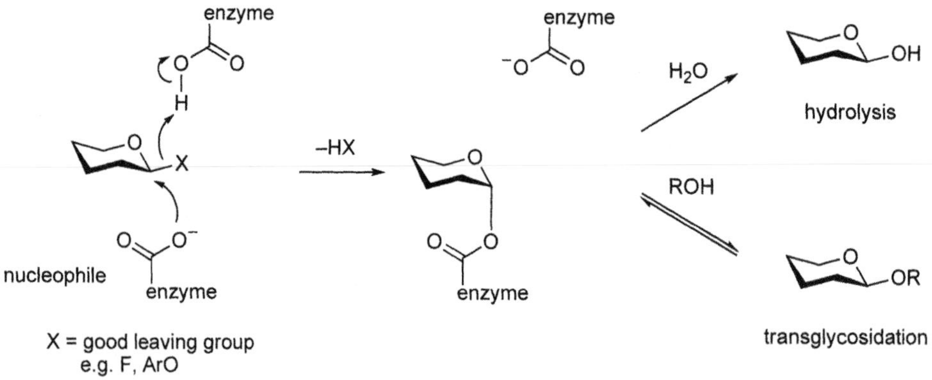

X = good leaving group
e.g. F, ArO

Further consideration of the outcomes of this reaction reveals that as the product formed is a glycoside with the same anomeric stereochemistry as that preferred by the enzyme, it can also function as a substrate. However, as this product glycoside bears a poorer anomeric leaving group than the activated donor, it will usually be a poorer substrate for the enzyme, and careful timing of the reaction will allow it to accumulate to useful levels. Use of an excellent leaving group that is much more reactive than in the product, reduction of the effective concentration of water and removal of product as it is formed will generally ensure a good yield of transglycosylated product (typically up to 50%).[130,131]

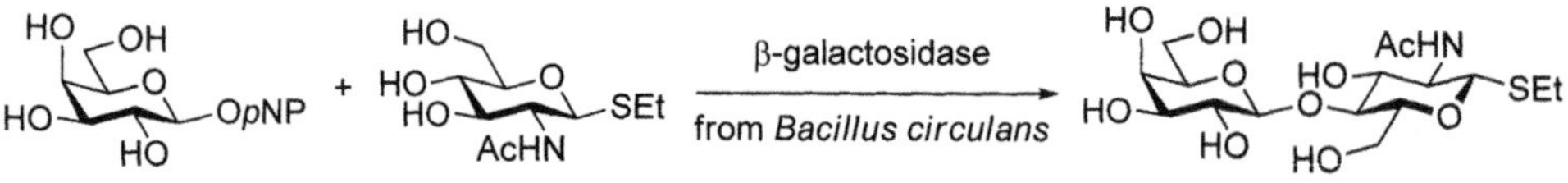

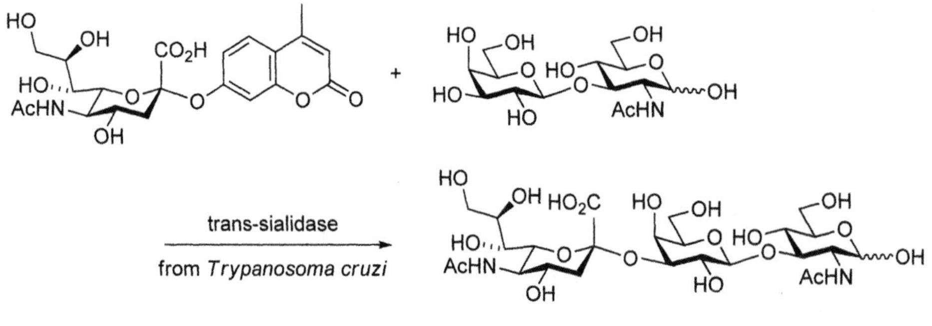

Kinetically controlled glycosidations have been used for the synthesis of a wide variety of products ranging from relatively simple glycosides to more complex glycoconjugates including glycoproteins.[131,132]

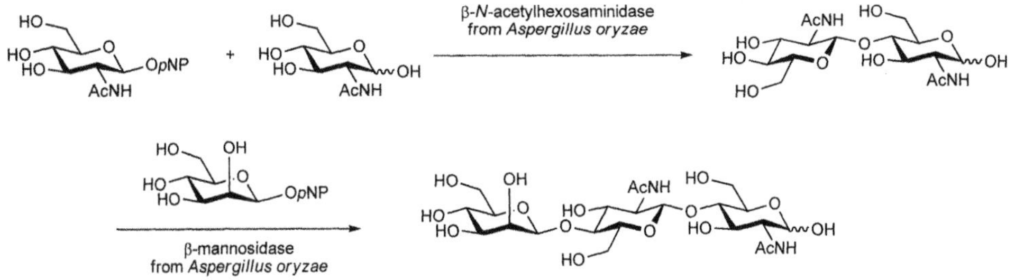

While most kinetically controlled transglycosidations are performed using *exo*-glycoside hydrolases, many *endo*-glycoside hydrolases exhibit powerful transglycosidation abilities. Use of these enzymes as transglycosidation catalysts is complicated as the product can act as a substrate for further transglycosidation, leading to the formation of polymers. An elegant solution to this problem has been advanced by Shoda and co-workers, where a lactosyl donor is used as a substrate for an *endo*-β-glucanase.[133–135] The incorrect stereochemistry at C4 of the glycosyl donor is tolerated by the enzyme for the transfer reaction, but the newly formed product cannot act as a glycosyl acceptor, preventing hyperglycosylation. The product is isolated and then trimmed in a second reaction using an *exo*-acting β-galactosidase to give the net transfer of a single β-glucosyl unit.[136]

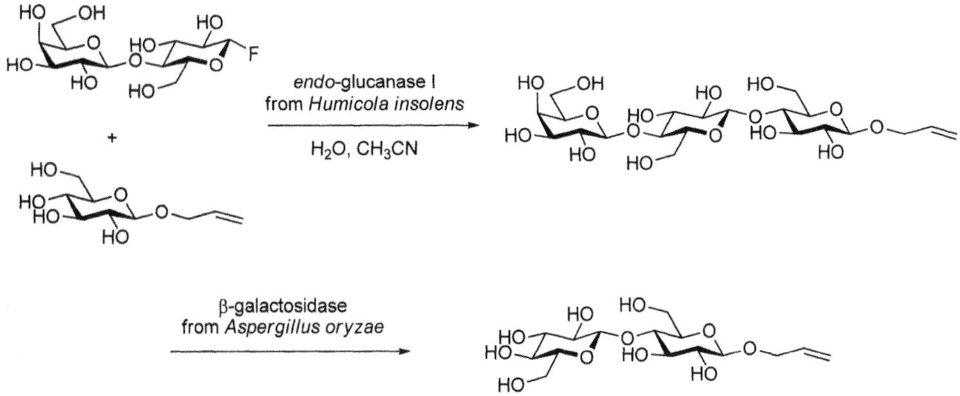

Kinetically controlled glycosidations typically use one of the two main sets of glycosyl donors: aryl glycosides or glycosyl fluorides. Both possess good leaving groups and accelerate the glycosylation step of the enzymatic reaction compared to alkyl glycosides. Aryl glycosides and glycosyl fluorides have individual merits, and the final choice for a particular application may reflect only personal preference. Many aryl glycosides are commercially available, and alteration of substituents on the aryl ring can tune the reactivity of the glycosyl donor. On the contrary, many aryl glycosides possess relatively poor solubility in water and do not allow high concentrations to be achieved. Some enzymes do not tolerate aryl glycosides as substrates. Glycosyl fluorides usually have to be chemically synthesized and require the use of potentially hazardous fluorination reagents; some glycosyl fluorides have poor stability. On the contrary, glycosyl fluorides possess excellent reactivity and can be prepared on very large scale.[137]

Glycosynthases: Mutant Glycosidases for Glycoside Synthesis[138–140]

While thermodynamically and kinetically controlled transglycosidation reactions can give rise to usable amounts of product(s), their application is limited as a result of incomplete reactions and the concomitant difficulties associated with isolation of the product and, in the case of kinetically controlled reactions, the need for careful monitoring of reactions. Genetic engineering techniques may be used to produce glycoside hydrolase mutants that overcome many of the problems associated with transglycosidation methods. One major approach is the use of site-directed mutagenesis to remove the catalytic nucleophile from a retaining glycoside hydrolase and replace it with a non-nucleophilic residue such as glycine,[141] serine[142] or alanine.[143] Such an enzyme is a severely wounded catalyst and can no longer form the intermediate glycosyl enzyme. If such mutants are supplied with a glycosyl fluoride of the 'wrong' anomeric configuration, the glycosyl fluoride can mimic the glycosyl enzyme and act as a glycosyl donor towards an acceptor sugar, generating a glycoside with inverted anomeric stereochemistry.[143,144] Such disabled nucleophile mutants are termed glycosynthases. One benefit of the use of glycosynthases is that the product of the reaction, a glycoside, cannot be degraded by the enzyme (as it lacks the enzymic nucleophile). Thus, unlike kinetically controlled transglycosylation reactions, there is no requirement for careful monitoring of the reaction, and product(s) can accumulate to potentially quantitative yields.

Glycosynthases have been developed for the synthesis of many different glycosidic bonds, including β-glucoside, β-galactoside, β-mannoside and β-xyloside and, most recently, α-glucoside and β-glucuronide.[140,145]

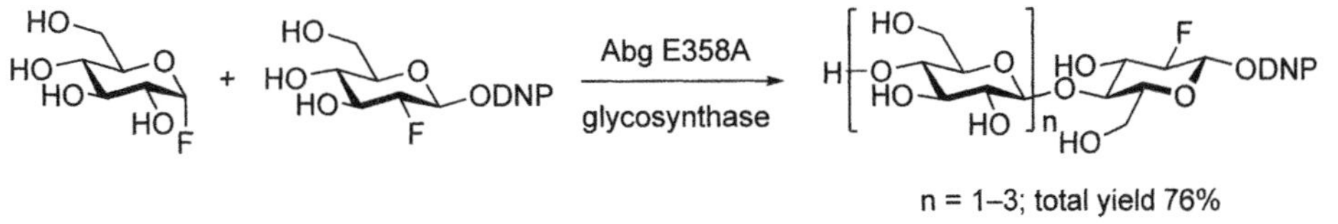

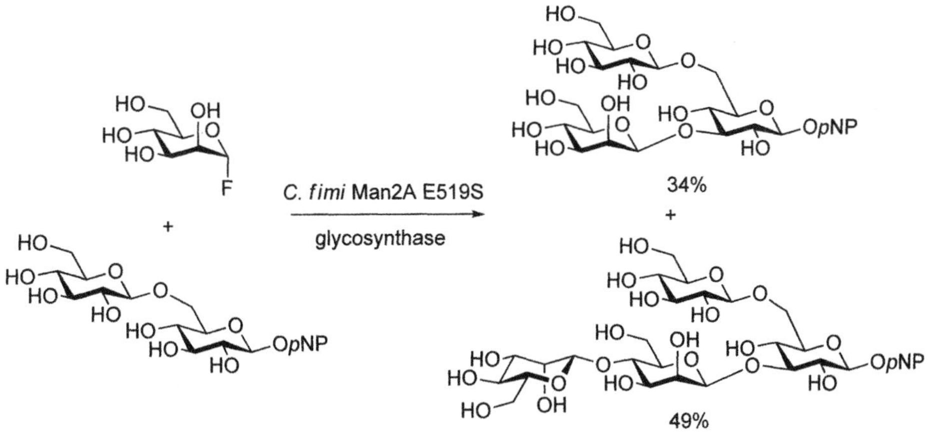

n = 1–3; total yield 76%

Glycosynthases have also been developed from *endo*-glycanases. When using disaccharide donors, these enzymes can act as powerful polymerization catalysts. Treatment of α-cellobiosyl fluoride with Cel7B E197A results in the formation of low molecular weight cellulose.[146] *endo*-Glucanases can tolerate functionalized glycosyl donors, and so protecting groups can be included to prevent polymerization. A tetrahydropyranyl group is an effective protecting group for such reactions.[147]

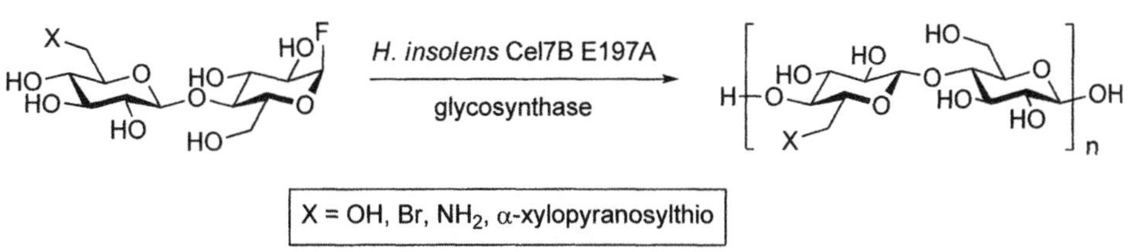

X = OH, Br, NH$_2$, α-xylopyranosylthio

References start on page 280

Glycosynthases have been utilized for the synthesis of glycoproteins[148] and glycolipids[149] and for glycoside synthesis on solid support.[150] Much effort is being expended on extending the glycosynthase approach through a combination of rational and combinatorial screening methods.[141,151,152]

Thioglycoligases: Mutant Glycosidases for Thioglycoside Synthesis[138,140]

Thioglycoligases are retaining glycoside hydrolases that have been mutated to remove the acid/base catalyst residue.[153] These mutant enzymes can cleave highly activated substrates such as 2,4-dinitrophenyl glycosides or glycosyl fluorides (which do not require acid catalysis) to form a glycosyl enzyme. Without a functional basic residue, deglycosylation is slowed but can be accelerated by the use of a more powerful nucleophile, such as a thiol, thereby forming a thioglycoside. Thioglycoligases have been developed that are effective catalysts for the synthesis of S-linked α- and β-glucosides, β-mannosides and α-xylosides.[140]

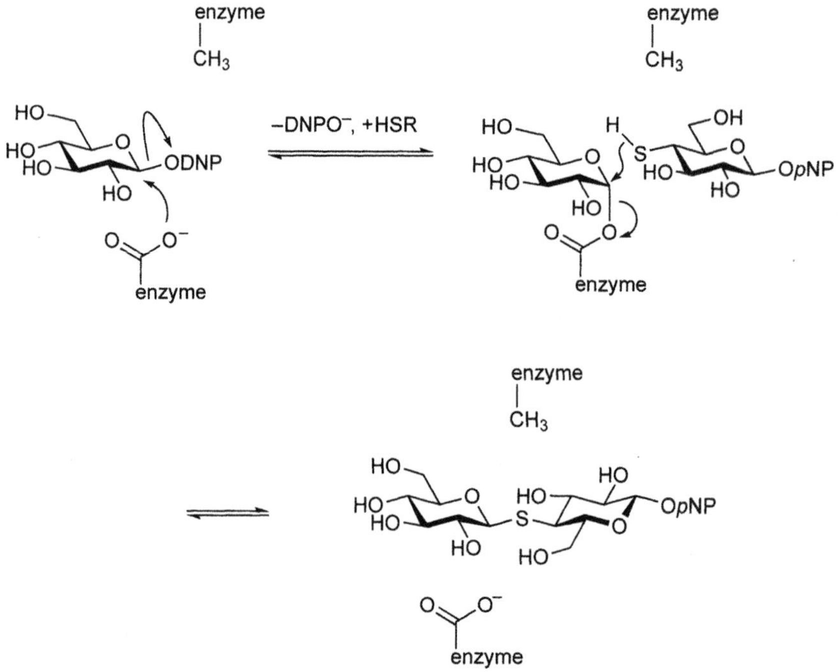

In an astute application of the thioglycoligase methodology, a thioglycoligase generated from the acid/base mutant (D482A) of the α-xylosidase YicI from *E. coli*

was used to synthesize the S-linked 4-nitrophenyl xylopyranosyl-α-1, 4-glucopyranoside.[154] This compound, in turn, was found to be a moderate inhibitor of the parent enzyme (K_i 2 μM) and was used as a non-hydrolysable substrate analogue to determine the X-ray structure of the pseudo-Michaelis complex with the wild-type parent enzyme.[154]

Hehre Resynthesis/Hydrolysis Mechanism

Inverting glycoside hydrolases cannot be used for kinetically controlled transglycosylation reactions using activated donors.[6] A simple explanation for this is to consider the principle of microscopic reversibility, which states that because catalysts promote both the forward and reverse reactions, the mechanisms of the forward and reverse reactions must be identical. Thus, if an inverting glycoside hydrolase can utilize one anomer of substrate to generate a product with inverted anomeric stereochemistry, the product of inverted stereochemistry must also be a substrate for the enzyme. This is clearly not the case.

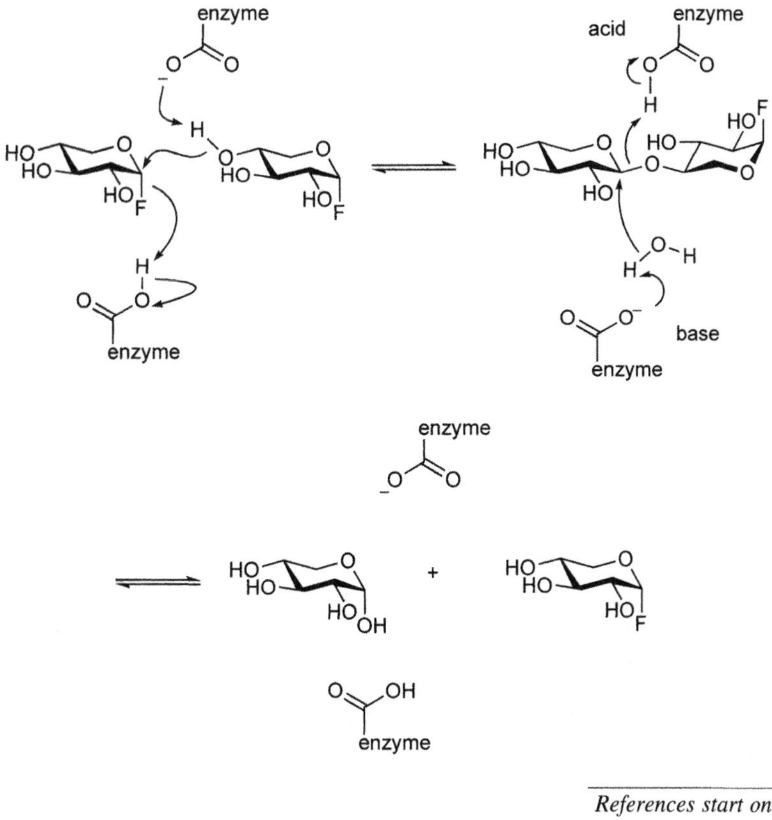

There is, however, an alternative mechanism that enables transglycosylation reactions of inverting glycoside hydrolases. Some inverting glycoside hydrolases, when presented with an activated sugar of the 'wrong' anomeric configuration, can catalyse a transglycosylation reaction. Typically, the only feasible substrates for such reactions are glycosyl fluorides, which have the right combination of a small anomeric leaving group, high reactivity and water stability. Through such a mechanism, the wrong glycosyl fluoride is transglycosylated onto a suitable alcohol acceptor, generating a glycoside product. As this product is now of the correct stereochemistry for the glycoside hydrolase, it acts as an excellent substrate and is typically rapidly hydrolysed. This overall reaction therefore leads to hydrolysis of wrong glycosyl fluorides, by way of a transglycosidation intermediate, and is termed the Hehre resynthesis/hydrolysis mechanism.[155]

While this process is capable of producing transglycosylation products, these are rarely produced in sufficient amounts to be synthetically useful.[137]

References

1. Wolfenden, R., Lu, X. and Young, G. (1998). *J. Am. Chem. Soc.*, **120**, 6814.
2. Zechel, D.L. and Withers, S.G. (2000). *Acc. Chem. Res.*, **33**, 11.
3. Phillips, D.C. (1967). *Proc. Natl. Acad. Sci. USA*, **57**, 484.
4. Sinnott, M.L. (1993). *Bioorg. Chem.*, **21**, 34.
5. Franck, R.W. (1992). *Bioorg. Chem.*, **20**, 77.
6. Sinnott, M.L. (1990). *Chem. Rev.*, **90**, 1171.
7. Alberts, B. (2007). *Nature*, **448**, 882.
8. Koshland, D.E., Jr. (1953). *Biol. Rev. Cambridge Philos. Soc.*, **28**, 416.
9. Honda, Y. and Kitaoka, M. (2004). *J. Biol. Chem.*, **279**, 55097.
10. Zechel, D.L. and Withers, S.G. (1999). In *Comprehensive Natural Products Chemistry* (C.D. Poulter, ed.) Vol. 5, p. 279, Pergamon: Amsterdam.
11. Henrissat, B. and Davies, G. (1997). *Curr. Opin. Struct. Biol.*, **7**, 637.
12. Henrissat, B. and Bairoch, A. (1996). *Biochem. J.*, **316**, 695.
13. Henrissat, B. and Bairoch, A. (1993). *Biochem. J.*, **293**, 781.
14. Henrissat, B. (1991). *Biochem. J.*, **280**, 309.
15. Coutinho, P.M. and Henrissat, B. *Carbohydrate-Active Enzymes Server at URL: http://www.cazy.org* accessed 19/6/08.
16. Davies, G.J., Gloster, T.M. and Henrissat, B. (2005). *Curr. Opin. Struct. Biol.*, **15**, 637.
17. Stam, M.R., Blanc, E., Coutinho, P.M. and Henrissat, B. (2005). *Carbohydr. Res.*, **340**, 2728.
18. Blake, C.C.F., Koenig, D.F., Mair, G.A., North, A.C.T., Phillips, D.C. and Sarma, V.R. (1965). *Nature*, **206**, 757.
19. Vocadlo, D.J., Davies, G.J., Laine, R. and Withers, S.G. (2001). *Nature*, **412**, 835.
20. Kempton, J.B. and Withers, S.G. (1992). *Biochemistry*, **31**, 9961.
21. Watts, A.G., Oppezzo, P., Withers, S.G., Alzari, P.M. and Buschiazzo, A. (2006). *J. Biol. Chem.*, **281**, 4149.
22. Tews, I., Terwisscha van Scheltinga, A.C., Perrakis, A., Wilson, K.S. and Dijkstra, B.W. (1997). *J. Am. Chem. Soc.*, **119**, 7954.

23. Markovic-Housley, Z., Miglierini, G., Soldatova, L., Rizkallah, P.J., Muller, U. and Schirmer, T. (2000). *Structure*, **8**, 1025.

24. Mark, B.L., Vocadlo, D.J., Knapp, S., Triggs-Raine, B.L., Withers, S.G., James, M.N.G. (2001). *J. Biol. Chem.*, **276**, 10330.

25. Knapp, S., Vocadlo, D., Gao, Z., Kirk, B., Lou, J. and Withers, S.G. (1996). *J. Am. Chem. Soc.*, **118**, 6804.

26. Mayer, C., Vocadlo, D.J., Mah, M., Rupitz, K., Stoll, D. and Warren, R.A. (2006). Withers, S.G. *FEBS J.*, **273**, 2929.

27. Vocadlo, D.J., Mayer, C., He, S. and Withers, S.G. (1999). *Biochemistry*, **39**, 117.

28. Wittstock, U. and Halkier, B.A. (2002). *Trends Plant Sci.*, **7**, 263.

29. Ettlinger, M.G., Dateo, G.P., Jr., Harrison, B.W., Mabry, T.J. and Thompson, C.P. (1961). *Proc. Natl. Acad. Sci. USA*, **47**, 1875.

30. Burmeister, W.P., Cottaz, S., Driguez, H., Iori, R., Palmieri, S. and Henrissat, B. (1997). *Structure*, **5**, 663.

31. Cottaz, S., Henrissat, B. and Driguez, H. (1996). *Biochemistry*, **35**, 15256.

32. Burmeister, W.P., Cottaz, S., Rollin, P., Vasella, A. and Henrissat, B. (2000). *J. Biol. Chem.*, **275**, 39385.

33. Yip, V.L. and Withers, S.G. (2006). *Curr. Opin. Chem. Biol.*, **10**, 147.

34. Rye, C.S., Matte, A., Cygler, M. and Withers, S.G. (2006). *ChemBioChem*, **7**, 631.

35. Kametani, S., Shiga, Y. and Akanuma, H. (1996). *Eur. J. Biochem.*, **242**, 832.

36. Lee, S.S., Yu, S. and Withers, S.G. (2003). *Biochemistry*, **42**, 13081.

37. Lee, S.S., Yu, S. and Withers, S.G. (2002). *J. Am. Chem. Soc.*, **124**, 4948.

38. Lovering, A.L., Lee, S.S., Kim, Y.W., Withers, S.G. and Strynadka, N.C. (2005). *J. Biol. Chem.*, **280**, 2105.

39. Kitaoka, M. and Hayashi, K. (2002). *Trends Glycosci. Glycotechnol.*, **14**, 35.

40. Johnson, L.N. (1992). *FASEB J.*, **6**, 2274.

41. Geremia, S., Campagnolo, M., Schinzel, R. and Johnson, L.N. (2002). *J. Mol. Biol.*, **322**, 413.

42. Hidaka, M., Honda, Y., Kitaoka, M. Nirasawa, S., Hayashi, K., Wakagi, T., Shoun, H. and Fushinobu, S. (2004). *Structure*, **12**, 937.

43. Campbell, P. and Braam, J. (1999). *Trends Plant Sci.*, **4**, 361.

44. Johansson, P., Brumer, H., 3rd, Baumann, M.J., Kallas, A.M., Henriksson, H., Denman, S.E., Teeri, T.T. and Jones, T.A. (2004). *Plant Cell*, **16**, 874.

45. Davies, G.J. and Wilson, K.S. (1999). *Nat. Struct. Biol.*, **6**, 406.

46. Mosi, R., He, S., Uitdehaag, J., Dijkstra, B.W. and Withers, S.G. (1997). *Biochemistry*, **36**, 9927.

47. Uitdehaag, J.C., Mosi, R., Kalk, K.H., Veen, B. A.v.d., Dijkhuizen, L., Withers, S.G. and Dijkstra, B.W. (1999). *Nat. Struct. Biol.*, **6**, 432.

48. Watts, A.G., Damager, I., Amaya, M.L., Buschiazzo, A., Alzari, P., Frasch, A.C. and Withers, S.G. (2003). *J. Am. Chem. Soc.*, **125**, 7532.

49. Liu, W., Madsen, N.B., Braun, C. and Withers, S.G. (1991). *Biochemistry*, **30**, 1419.

50. Davies, G. and Henrissat, B. (1995). *Structure*, **3**, 853.

51. McCarter, J.D. and Withers, S.G. (1994). *Curr. Opin. Struct. Biol.*, **4**, 885.

52. Heightman, T.D. and Vasella, A.T. (1999). *Angew. Chem. Int. Ed.*, **38**, 750.

53. Nerinckx, W., Desmet, T., Piens, K. and Claeyssens, M. (2005). *FEBS Lett.*, **579**, 302.

54. Vasella, A., Davies, G.J. and Bohm, M. (2002). *Curr. Opin. Chem. Biol.*, **6**, 619.

55. Legler, G. (1990). *Adv. Carb. Chem. Biochem.*, **48**, 319.

56. Withers, S.G. and Aebersold, R. (1995). *Protein Sci.*, **4**, 361.

57. Withers, S.G., Street, I.P., Bird, P. and Dolphin, D.H. (1987). *J. Am. Chem. Soc.*, **109**, 7530.

58. McCarter, J.D. and Withers, S.G. (1996). *J. Am. Chem. Soc.*, **118**, 241.

59. Miao, S., McCarter, J.D., Grace, M.E., Grabowski, G.A., Aebersold, R. and Withers, S.G. (1994). *J. Biol. Chem.*, **269**, 10975.

60. Davies, G.J., Mackenzie, L., Varrot, A., Dauter, M., Brzozowski, A.M., Schulein, M. and Withers, S.G. (1998). *Biochemistry*, **37**, 11707.
61. White, A., Tull, D., Johns, K., Withers, S.G. and Rose, D.R. (1996). *Nat. Struct. Biol.*, **3**, 149.
62. Hekmat, O., Kim, Y.W., Williams, S.J., He, S.M. and Withers, S.G. (2005). *J. Biol. Chem.*, **280**, 35126.
63. Stubbs, K.A., Scaffidi, A., Debowski, A.W., Mark, B.L., Stick, R.V. and Vocadlo, D.J. (2008). *J. Am. Chem. Soc.*, **130**, 327.
64. Moult, J., Eshdat, Y. and Sharon, N. (1973). *J. Mol. Biol.*, **75**, 1.
65. Havukainen, R., Törrönen, A., Laitinen, T. and Rouvinen, J. (1996). *Biochemistry*, **35**, 9617.
66. Høj, P.B., Rodriguez, E.B., Iser, J.R., Stick, R.V. and Stone, B.A. (1991). *J. Biol. Chem.*, **266**, 11628.
67. Atsumi, S., Umezawa, K., Iinuma, H., Naganawa, H., Nakamura, H., Iitaka, Y. and Takeuchi, T. (1990). *J. Antibiot.*, **43**, 49.
68. Legler, G. and Bause, E. (1973). *Carbohydr. Res.*, **28**, 45.
69. Akiyama, T., Shima, H., Ohnari, M., Okazaki, T. and Ozaki, S. (1993). *Bull. Chem. Soc. Jpn.*, **66**, 3760.
70. Caron, G. and Withers, S.G. (1989). *Biochem. Biophys. Res. Commun.*, **163**, 495.
71. Herrchen, M. and Legler, G. (1984). *Eur. J. Biochem.*, **138**, 527.
72. Gebler, J.C., Aebersold, R. and Withers, S.G. (1992). *J. Biol. Chem.*, **267**, 11126.
73. Howard, S. and Withers, S.G. (1998). *Biochemistry*, **37**, 3858.
74. Howard, S. and Withers, S.G. (1998). *J. Am. Chem. Soc.*, **120**, 10326.
75. Black, T.S., Kiss, L., Tull, D. and Withers, S.G. (1993). *Carbohydr. Res.*, **250**, 195.
76. Naider, F., Bohak, Z. and Yariv, J. (1972). *Biochemistry*, **11**, 3202.
77. Ganem, B. (1996). *Acc. Chem. Res.*, **29**, 340.
78. Fellows, L.E. (1987). *Chem. Brit.*, **23**, 842.
79. Paulsen, H. (1991). In *Iminosugars as Glycosidase Inhibitors: Nojirimycin and Beyond* (A.E. Stütz, ed.) Wiley-VCH: Weinheim, p. 1.
80. Stütz, A.E. ed. (1999). *Iminosugars as Glycosidase Inhibitors: Nojirimycin and Beyond*. Wiley-VCH: Weinheim.
81. Hughes, A.B. and Rudge, A.J. (1994). *Nat. Prod. Rep.*, **11**, 135.
82. Lillelund, V.H., Jensen, H.H., Liang, X. and Bols, M. (2002). *Chem. Rev.*, **102**, 515.
83. Watson, A.A., Fleet, G.W., Asano, N., Molyneux, R.J. and Nash, R.J. (2001). *Phytochemistry*, **56**, 265.
84. Legler, G. (1999). In *Iminosugars as Glycosidase Inhibitors: Nojirimycin and Beyond* (A.E. Stütz, ed.) Wiley-VCH: Weinheim, p. 31.
85. Paulsen, H. (1966). *Angew. Chem. Int. Ed. Engl.*, **5**, 495.
86. Paulsen, H., Sangster, I. and Heyns, K. (1967). *Chem. Ber.*, **100**, 802.
87. Konno, K., Ono, H., Nakamura, M., Tateishi, K., Hirayama, C., Tamura, Y., Hattori, M., Koyama, A. and Kohno, K. (2006). *Proc. Natl. Acad. Sci. USA*, **103**, 1337.
88. Asano, N., Nash, R.J., Molyneux, R.J. and Fleet, G.W.J. (2000). *Tetrahedron: Asymm.*, **11**, 1645.
89. Lebovitz, H.E. (1998). *Diabetes Rev.*, **6**, 132.
90. Sawkar, A.R., D'Haeze, W. and Kelly, J.W. (2006). *Cell. Mol. Life Sci.*, **63**, 1179.
91. Withers, S.G., Namchuk, M. and Mosi, R. (1999). In *Iminosugars as Glycosidase Inhibitors: Nojirimycin and Beyond* (A.E. Stütz, ed.) p. 188. Wiley-VCH: Weinheim.
92. Colegate, S.M., Dorling, P.R. and Huxtable, C.R. (1979). *Aust. J. Chem.*, **32**, 2257.
93. van den Elsen, J.M.H., Kuntz, D.A. and Rose, D.R. (2001). *EMBO J.*, **20**, 3008.
94. Cutfield, S.M., Davies, G.J., Murshudov, G., Anderson, B.F., Moody, P.C., Sullivan, P.A. and Cutfield, J.F. (1999). *J. Mol. Biol.*, **294**, 771.
95. Hohenschutz, L.D., Bell, E.A., Jewess, P.J., Leworthy, D.P., Pryce, R.J., Arnold, E. and Clardy, J. (1981). *Phytochemistry*, **20**, 811.

96. Jespersen, T.M., Dong, W., Sierks, M.R., Skrydstrup, T., Lundt, I. and Bols, M. (1994). *Angew. Chem. Int. Ed. Engl.*, **33**, 1778.

97. Varrot, A., Tarling, C.A., Macdonald, J.M., Stick, R.V., Zechel, D.L., Withers, S.G. and Davies, G.J. (2003). *J. Am. Chem. Soc.*, **125**, 7496.

98. Singh, V., Evans, G.B., Lenz, D.H., Mason, J.M., Clinch, K., Mee, S., Painter, G.F., Tyler, P.C., Furneaux, R.H., Lee, J.E., Howell, P.L. and Schramm, V.L. (2005). *J. Biol. Chem.*, **280**, 18265.

99. Evans, G.B., Furneaux, R.H., Lenz, D.H., Painter, G.F., Schramm, V.L., Singh, V. and Tyler, P.C. (2005). *J. Med. Chem.*, **48**, 4679.

100. Varrot, A., Schülein, M., Pipelier, M., Vasella, A. and Davies, G.J. (1999). *J. Am. Chem. Soc.*, **121**, 2621.

101. Notenboom, V., Williams, S.J., Hoos, R., Withers, S.G. and Rose, D.R. (2000). *Biochemistry*, **39**, 11553.

102. Heightman, T.D., Locatelli, M. and Vasella, A. (1996). *Helv. Chim. Acta*, **79**, 2190.

103. Aoyagi, T., Suda, H., Uotani, K., Kojima, F., Aoyama, T., Horiguchi, K., Hamada, M. and Takeuchi, T. (1992). *J. Antibiot.*, **45**, 1404.

104. Tatsuta, K. and Miura, S. (1995). *Tetrahedron Lett.*, **36**, 6712.

105. Tatsuta, K., Miura, S., Ohta, S. and Gunji, H. (1995). *J. Antibiot.*, **48**, 286.

106. Truscheit, E., Frommer, W., Junge, B., Muller, L., Schmidt, D.D. and Wingender, W. (1981). *Angew. Chem. Int. Ed. Engl.*, **20**, 744.

107. Berecibar, A., Grandjean, C. and Siriwardena, A. (1999). *Chem. Rev.*, **99**, 779.

108. Terwisscha van Scheltinga, A.C., Armand, S., Kalk, K.H., Isogai, A., Henrissat, B. and Dijkstra, B.W. (1995). *Biochemistry*, **34**, 15619.

109. Macauley, M.S., Whitworth, G.E., Debowski, A.W., Chin, D. and Vocadlo, D.J. (2005). *J. Biol. Chem.*, **280**, 25313.

110. von Itzstein, M., Wu, W.-Y., Kok, G.B., Pegg, M.S., Dyason, J.C., Jin, B., Phan, T.V., Smythe, M.L., White, H.F., Oliver, S.W., Colman, P.M., Varghese, J.N., Ryan, D.M., Woods, J.M., Bethell, R.C., Hotham, V.J., Cameron, J.M. and Penn, C.R. (1993). *Nature*, **363**, 418.

111. von Itzstein, M., Wua, W.-Y. and Jina, B. (1994). *Carbohydr. Res.*, **259**, 301.

112. Macdonald, S.J., Cameron, R., Demaine, D.A., Fenton, R.J., Foster, G., Gower, D., Hamblin, J.N., Hamilton, S., Hart, G.J., Hill, A.P., Inglis, G.G., Jin, B., Jones, H.T., McConnell, D.B., McKimm-Breschkin, J., Mills, G., Nguyen, V., Owens, I.J., Parry, N., Shanahan, S.E., Smith, D., Watson, K.G., Wu, W.Y. and Tucker, S.P. (2005). *J. Med. Chem.*, **48**, 2964.

113. Macdonald, S.J., Watson, K.G., Cameron, R., Chalmers, D.K., Demaine, D.A., Fenton, R.J., Gower, D., Hamblin, J.N., Hamilton, S., Hart, G.J., Inglis, G.G., Jin, B., Jones, H.T., McConnell, D.B., Mason, A.M., Nguyen, V., Owens, I.J., Parry, N., Reece, P.A., Shanahan, S.E., Smith, D. and Wu, W.-Y., Tucker, S.P. (2004). *Antimicrob. Agents Chemother.*, **48**, 4542.

114. Kim, C.U., Lew, W., Williams, M.A., Liu, H., Zhang, L., Swaminathan, S., Bischofberger, N., Chen, M.S., Mendel, D.B., Tai, C.Y., Laver, W.G. and Stevens, R.C. (1997). *J. Am. Chem. Soc.*, **119**, 681.

115. Yeung, Y.Y., Hong, S. and Corey, E.J. (2006). *J. Am. Chem. Soc.*, **128**, 6310.

116. Karpf, M. and Trussardi, R. (2001). *J. Org. Chem.*, **66**, 2044.

117. Driguez, H. (2001). *ChemBioChem*, **2**, 311.

118. Driguez, H. (1997). *Topics Curr. Chem.*, **187**, 85.

119. Parsiegla, G., Reverbel-Leroy, C., Tardif, C., Belaich, J.P., Driguez, H. and Haser, R. (2000). *Biochemistry*, **39**, 11238.

120. Zou, J., Kleywegt, G.J., Stahlberg, J., Driguez, H., Nerinckx, W., Claeyssens, M., Koivula, A., Teeri, T.T. and Jones, T.A. (1999). *Structure*, **7**, 1035.

121. Sulzenbacher, G., Driguez, H., Henrissat, B., Schulein, M. and Davies, G.J. (1996). *Biochemistry*, **35**, 15280.

122. Trincone, A. and Giordano, A. (2006). *Curr. Org. Chem.*, **10**, 1163.
123. Palcic, M.M. (1999). *Curr. Opin. Biotechnol.*, **10**, 616.
124. Crout, D.H.G. and Vic, G. (1998). *Curr. Opinion Chem. Biol.*, **2**, 98.
125. Vocadlo, D.J., Withers, S.G. (2000). In *Carbohydrates in Chemistry and Biology* (B. Ernst, G.W. Hart and P. Sinaÿ, eds) Vol. I, 2, p. 723. Wiley-VCH: Weinheim.
126. Shoda, S.-i. (2001). In *Glycoscience: Chemistry and Chemical Biology* (B.O. Fraser-Reid, K. Tatsuta, and J. Thiem, eds) Vol 2, p. 1465. Springer-Verlag: Berlin.
127. Kren, V. and Thiem, J. (1997). *Chem. Soc. Rev.*, **26**, 463.
128. Wong, C.-H., Halcomb, R.L., Ichikawa, Y. and Kajimoto, T. (1995). *Angew. Chem. Int. Ed. Engl.*, **34**, 412.
129. Vic, G. and Crout, D.H.G. (1995). *Carbohydr. Res.*, **279**, 315.
130. Vic, G., Tran, C.H., Scigelova, M. and Crout, D.H.G. (1997). *Chem. Commun.*, 169.
131. Vetere, A., Miletich, M., Bosco, M. and Paoletti, S. (2000). *Eur. J. Biochem.*, **267**, 942.
132. Kröger, L. and Thiem, J. (2005). *J. Carbohydr. Chem.*, **24**, 717.
133. Kobayashi, S., Kawasaki, T., Obata, K. and Shoda, S.-i. (1993). *Chem. Lett.*, 685.
134. Shoda, S.-i., Kawasaki, T., Obata, K. and Kobayashi, S. (1993). *Carbohydr. Res.*, **249**, 127.
135. Shoda, S.-i., Obata, K., Karthaus, O. and Kobayashi, S. (1993). *J. Chem. Soc., Chem. Commun.*, 1402.
136. Fairweather, J.K., Stick, R.V., Tilbrook, D.M.G. and Driguez, H. (1999). *Tetrahedron*, **55**, 3695.
137. Williams, S.J. and Withers, S.G. (2000). *Carbohydr. Res.*, **327**, 27.
138. Perugino, G., Trincone, A., Rossi, M. and Moracci, M. (2004). *Trends Biotechnol.*, **22**, 31.
139. Williams, S.J. and Withers, S.G. (2002). *Aust. J. Chem.*, **55**, 3.
140. Hancock, S.M., Vaughan, M.D. and Withers, S.G. (2006). *Curr. Opin. Chem. Biol.*, **10**, 509.
141. Mayer, C., Jakeman, D.L., Mah, M., Karjala, G., Gal, L., Warren, R.A. and Withers, S.G. (2001). *Chem. Biol.*, **8**, 437.
142. Mayer, C., Zechel, D.L., Reid, S.P., Warren, R.A. and Withers, S.G. (2000). *FEBS Lett.*, **466**, 40.
143. Mackenzie, L.F., Wang, Q.P., Warren, R.A.J. and Withers, S.G. (1998). *J. Am. Chem. Soc.*, **120**, 5583.
144. Malet, C. and Planas, A. (1998). *FEBS Lett.*, **440**, 208.
145. Wilkinson, S.M., Liew, C.W., Mackay, J.P., Salleh, H.M., Withers, S.G. and McLeod, M.D. (2008). *Org. Lett.*, **10**, 1585.
146. Fort, S., Boyer, V., Greffe, L., Davies, G.J., Moroz, O., Christiansen, L., Schulein, M., Cottaz, S. and Driguez, H. (2000). *J. Am. Chem. Soc.*, **122**, 5429.
147. Fort, S., Christiansen, L., Schulein, M., Cottaz, S. and Driguez, H. (2000). *Isr. J. Chem.*, **40**, 217.
148. Mullegger, J., Chen, H.M., Warren, R.A. and Withers, S.G. (2006). *Angew. Chem. Int. Ed.*, **45**, 2585.
149. Vaughan, M.D., Johnson, K., DeFrees, S., Tang, X., Warren, R.A. and Withers, S.G. (2006). *J. Am. Chem. Soc.*, **128**, 6300.
150. Tolborg, J.F., Petersen, L., Jensen, K.J., Mayer, C., Jakeman, D.L., Warren, R.A. and Withers, S.G. (2002). *J. Org. Chem.*, **67**, 4143.
151. Kim, Y.W., Lee, S.S., Warren, R.A.J. and Withers, S.G. (2004). *J. Biol. Chem.*, **279**, 42787.
152. Lin, H., Tao, H. and Cornish, V.W. (2004). *J. Am. Chem. Soc.*, **126**, 15051.
153. Jahn, M., Marles, J., Warren, R.A. and Withers, S.G. (2003). *Angew. Chem. Int. Ed.*, **42**, 352.
154. Kim, Y.W., Lovering, A.L., Chen, H., Kantner, T., McIntosh, L.P., Strynadka, N.C. and Withers, S.G. (2006). *J. Am. Chem. Soc.*, **128**, 2202.
155. Hehre, E.J., Brewer, C.F. and Genghof, D.S. (1979). *J. Biol. Chem.*, **254**, 5942.

Chapter 8

Glycosyltransferases

Nature's principal agents for the construction of the glycosidic bond are the glycosyltransferases (GTs). These enzymes use sugar phosphates as donors for glycosyl transfer to nucleophilic acceptors. This chapter will give an overview of GTs with a focus on mechanism and structure, reagents/inhibitors for their study and their application in synthesis.

Classification and Mechanism

There are three main sets of glycosyl donors for GTs: monosaccharide (di)phosphonucleosides, mono- or oligosaccharide phospholipids, and sugar-1-phosphates or sugar-1-pyrophosphates.

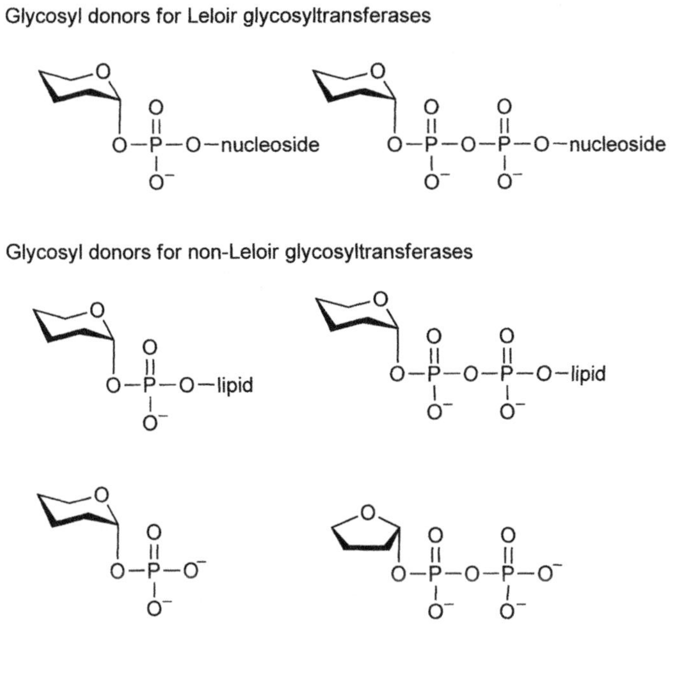

Glycosyl donors for Leloir glycosyltransferases

Glycosyl donors for non-Leloir glycosyltransferases

Classification: Sugar (di)phosphonucleosides were discovered during the pioneering work of Leloir and co-workers, and as a result sugar (di)phosphonucleoside-dependent GTs are now frequently referred to as Leloir-type enzymes.[1] GTs that do not utilize sugar (di)phosphonucleosides are referred to as non-Leloir-type enzymes. Non-Leloir-type GTs largely utilize mono- or oligosaccharide phospholipids containing phosphate or pyrophosphate moieties linked to dolichol (an isoprenoid), undecaprenol or polyprenol lipids. In addition, there is a small group of non-Leloir-type GTs that utilize sugar-1-phosphates or sugar-1-pyrophosphates as glycosyl donors, e.g. glycogen phosphorylase, which catalyses the reversible transfer of a glucosyl unit from glucose-1-phosphate to the growing glycogen molecule (see Chapter 7), and phosphoribosyltransferases, which transfer 5′-phosphoribosylpyrophosphate to nucleophiles such as water and nucleobases (see Chapter 12).[2,3]

Mechanism: GT-catalysed glycosyl transfer can occur with either retention or inversion of configuration at the anomeric centre of the glycosyl donor. This distinction, by analogy with glycoside hydrolases, allows classification of GTs as either retaining or inverting enzymes.

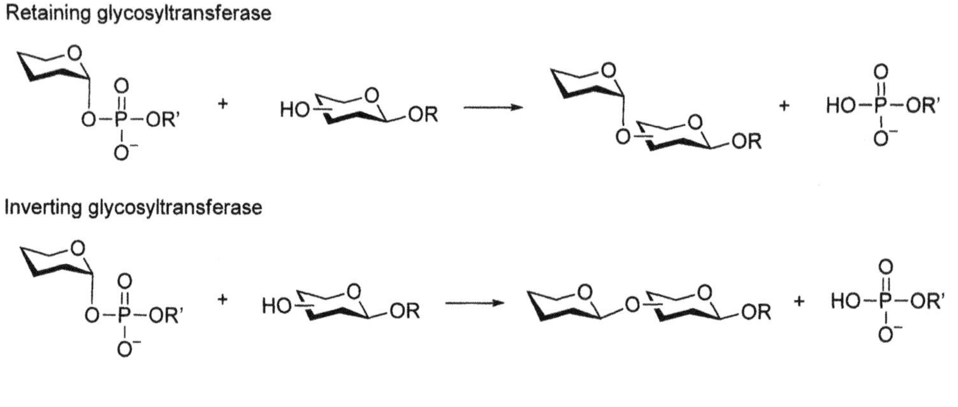

Retaining glycosyltransferase

Inverting glycosyltransferase

Structural and kinetic data for inverting GTs support a mechanism that proceeds by direct displacement through an oxocarbenium-ion-like transition state, with general base catalysis by a carboxylate[4] or histidine residue.[5] For example, the β-glucosyltransferase GtfB, which transfers a glucosyl unit from UDP-glucose to the vancomycin aglycon, has been proposed to utilize Asp332 as catalytic base; mutation of this residue to alanine reduced catalytic proficiency by 250-fold.[6]

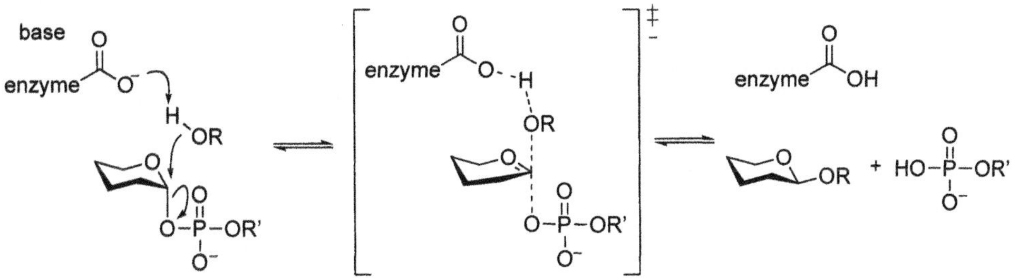

For retaining GTs, the situation is considerably more complex. Two possibilities have been suggested for glycosyl transfer with retention of configuration: a two-step mechanism involving sequential inversions by an enzyme-borne nucleophile similar to that proposed for retaining glycoside hydrolases, or an $S_N i$ mechanism involving non-concerted loss of the leaving group and front-side attack by the acceptor moiety.[7]

Two-step mechanism

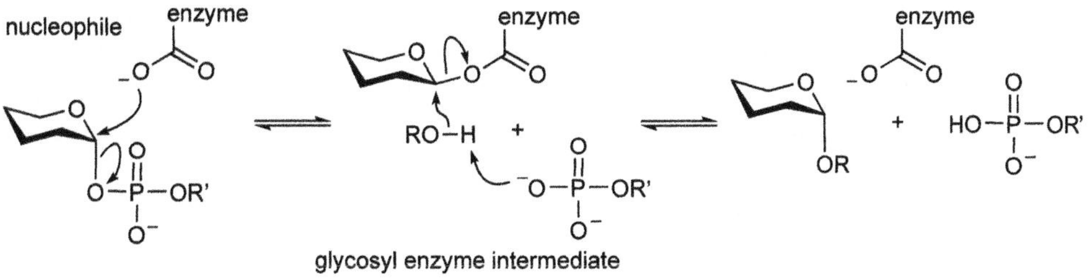

glycosyl enzyme intermediate

$S_N i$ mechanism

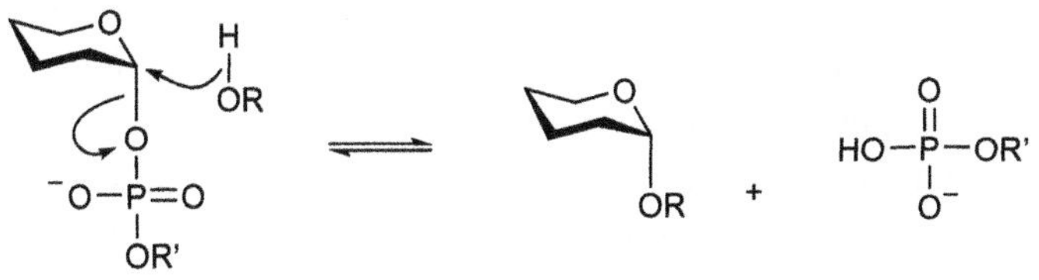

In either case the transition states are expected to bear significant oxocarbenium-ion-like character and, to date, no convincing evidence has been reported that can clearly distinguish between the two possibilities. Confounding the situation, crystal structures of retaining GTs are ambiguous in identifying possible candidate nucleophiles, with some lacking any suitable nucleophile situated close enough to the glycosyl donor to be able to form a covalent intermediate. Analysis of some structures has identified possible candidates, which mutagenesis studies have supported;[8] however, in other cases, mutagenesis of the candidate residue to a non-nucleophilic amino acid has not resulted in the expected, dramatic loss in catalytic prowess.[7,9] Attempts to trap covalent intermediates on wild-type GTs have proven unsuccessful;[10] in one case, mutation of the proposed aspartate nucleophile to a non-nuclophilic asparagine residue resulted in the formation of a covalently labelled enzyme bearing a glycosyl moiety on yet another carboxylate residue.[9] The technique of chemical rescue, whereby a small exogenous nucleophile is used to reactivate an inactive enzyme mutant, was applied to an inactive retaining galactosyltransferase mutant lacking the proposed glutamate nucleophile and led to a more than 100-fold increase in activity and the formation of β-D-galactosyl azide.[11] This result was taken to support a double-displacement mechanism but does not definitively exclude the $S_N i$ mechanism.

Detailed kinetic analyses have been performed on many retaining and inverting GTs. Generally, ordered kinetic mechanisms have been seen with either the glycosyl donor or the glycosyl acceptor binding to the enzyme first, followed by a catalytic step, and then either the (di)phosphonucleoside or glycosylated product departing from the active site first.[12] For example, the UDP-xylose-dependent core protein β-xylosyltransferase operates by way of an ordered single displacement with UDP-xylose as the leading substrate and the xylosylated peptide as the first released product.[13] By contrast, OleD, a UDP-glucose-dependent β-glucosyltransferase from *Streptomyces antibioticus*, operates through an ordered mechanism where the acceptor, the macrolide lankamycin, binds before UDP-glucose, and UDP is released prior to β-glucosylated lankamycin.[14]

Many GTs utilize divalent metal ions such as Mn^{2+} and Mg^{2+} as cofactors. These metals are found mainly in GTs that are diphosphonucleoside dependent, and an analysis by X-ray crystallography has shown that the metal ion is coordinated to an oxygen of each of the two phosphate groups, as well as to side-chain carboxyl groups of the protein. Much has been made of the so-called DXD amino acid sequence as an identifier of GTs, where the aspartate residues of this sequence are presumed to comprise the metal-binding residues of the active site;[15] in this respect, it is to be noted that no part of the DXD motif is invariant among GTs and that this motif is present in more than 50% of all protein sequences examined.[16] More-over, it is emphasized that many GTs are metal-ion independent, and thus do not bind metals at the active site.[4]

Leloir-type GTs utilize sugar (di)phosphonucleoside donors, and many hundreds of such glycosyl donors have been identified. Fortunately, in mammals the situation is relatively simple, where only nine sugar (di)phosphonucleoside donors are used – six uridine diphosphates: UDP-glucose, UDP-galactose, UDP-xylose, UDP-glucuronic acid, UDP-*N*-acetylglucosamine and UDP-*N*-acetylgalactosamine; two guanosine diphosphates: GDP-mannose and GDP-fucose; and one cytidine monophosphate: CMP-*N*-acetylneuraminic acid (CMP-Neu5Ac).

UDP-Glc

UDP-Gal

UDP-Xyl

UDP-GlcNAc

UDP-GalNAc

UDP-GlcA

GDP-Fuc

GDP-Man

CMP-Neu5Ac

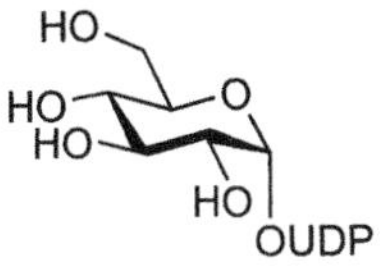

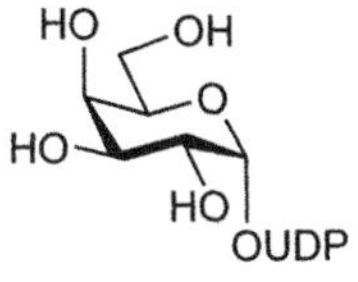

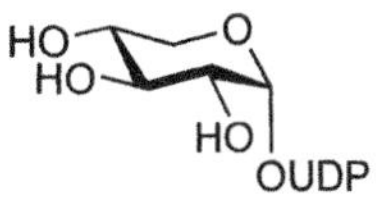

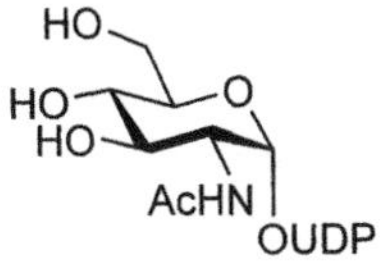

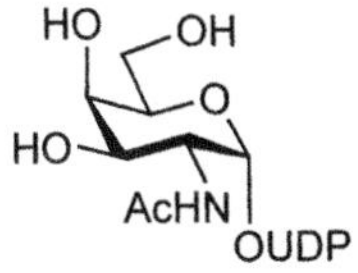

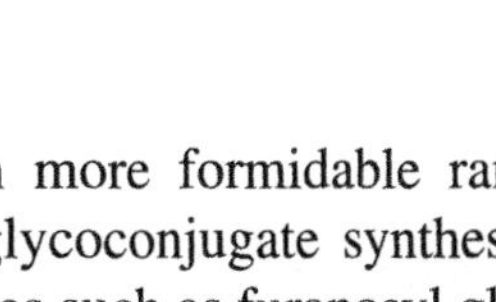

Across nature's profusion of life forms, a much more formidable range of sugar (di)phosphonucleoside donors is employed in glycoconjugate synthesis and include the above, plus hundreds of more unusual species such as furanosyl glycosyl donors (e.g. UDP-galactofuranose), various deoxy, amino and methylated sugar donors (e.g. GDP-rhamnose, dTDP-desosamine and TDP-3-*O*-methyl-rhamnose) and hept- and octulosyl donors (e.g. CMP-Kdo[17] and ADP-L-*glycero*-β-D-*manno*-heptopyranose).[18]

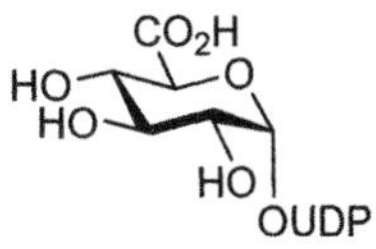

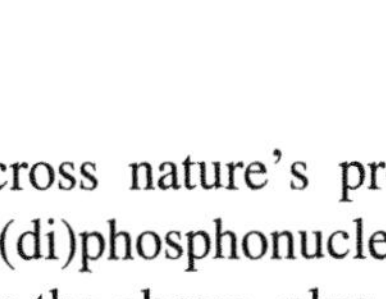

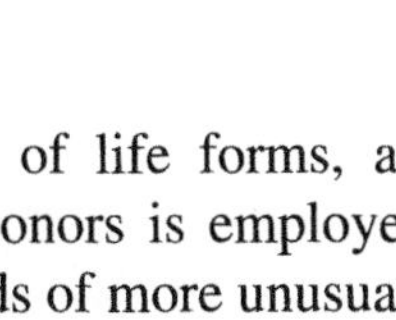

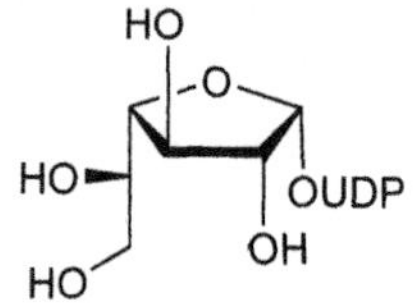

UDP-Gal*f*

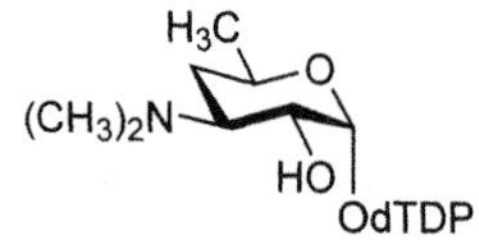

dTDP-desosamine

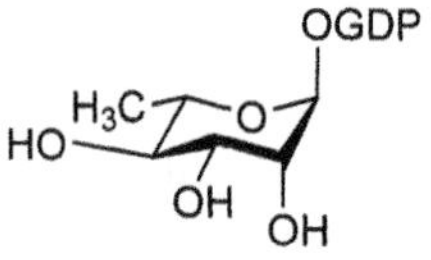

GDP-Rha

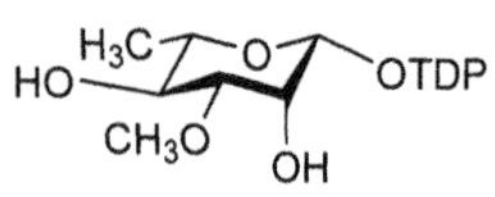

TDP-3-O-methyl-Rha

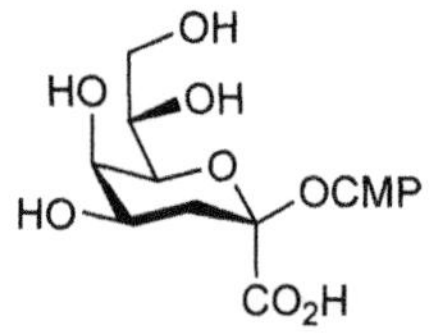

CMP-Kdo

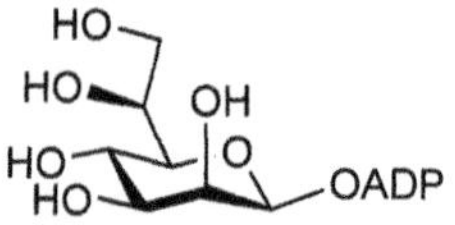

ADP-L-*glycero*-β-D-*manno*-heptopyranose

Many non-Leloir GTs utilize sugar phospholipids as glycosyl donors.[19] Such glycosyl donors are usually membrane associated, and the corresponding GTs are typically membrane proteins. The sugar phospholipid donors can be divided into monosaccharide and oligosaccharide donors, and both are used by lower and higher organisms as glycosyl donors for the synthesis of complex glycans. For example, many bacteria utilize monosaccharide donors such as polyprenolphosphomannopyranose[20] and decaprenolphosphoarabinofuranose,[21] including *Actinomycetes* such as mycobacteria. Mammals utilize lipid-linked glycosyl donors such as dolicholphosphomannose,[21] which is a glycosyl donor involved in the synthesis of dolicholphosphooligosaccharides, culminating in the synthesis of the N-linked glycoprotein core, dolichol-PP-GlcNAc$_2$Man$_9$Glc$_3$ (see Chapter 11). This lipid-linked oligosaccharide is itself a glycosyl donor for transfer by oligosaccharyltransferase to asparagine residues in nascent proteins in the lumen of the rough endoplasmic reticulum.[22]

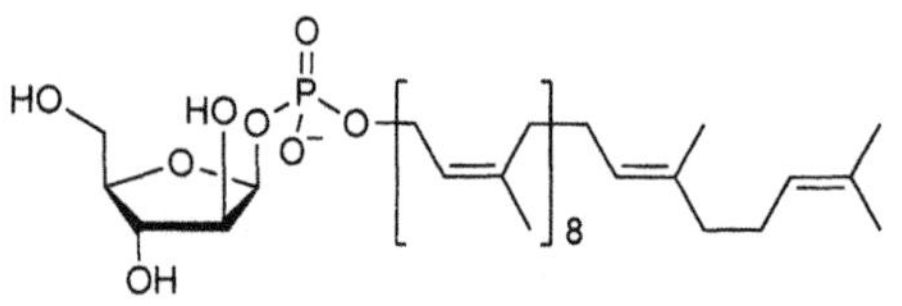

decaprenolphosphoarabinofuranose

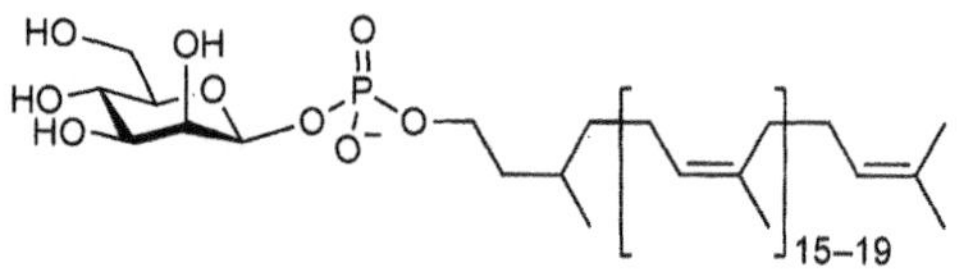

dolicholphosphomannose

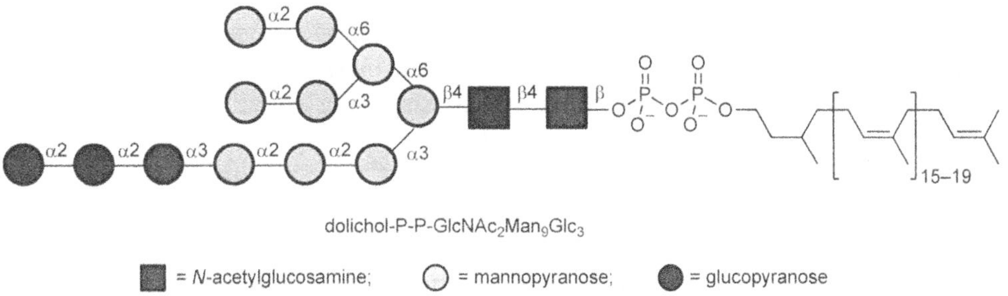

dolichol-P-P-GlcNAc$_2$Man$_9$Glc$_3$

■ = *N*-acetylglucosamine;   ◯ = mannopyranose;   ● = glucopyranose

In bacteria, the repeating MurNAc–GlcNAc backbone of peptidoglycan is synthesized from a phospholipid donor, lipid II, through the action of GTs termed peptidoglycan transglycosylases.[23,24]

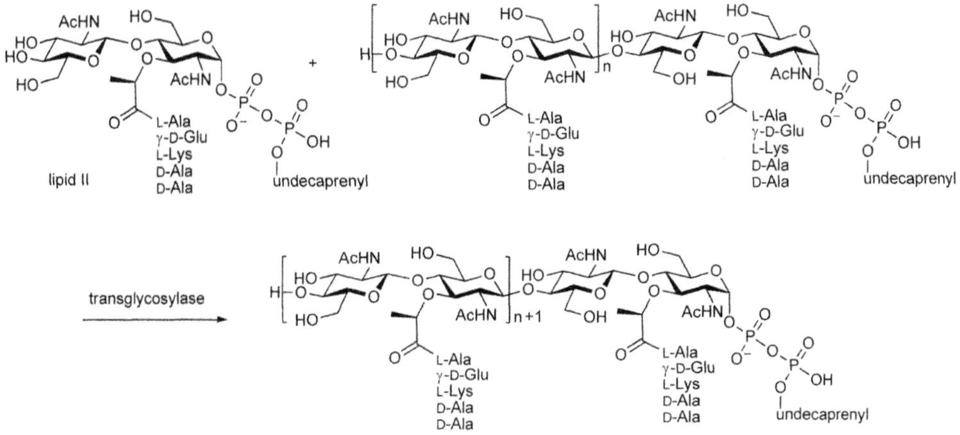

Glycosyltransferases and the 'One-enzyme One-linkage' Hypothesis

In contrast to glycoside hydrolases, which in many cases can catalyse the hydrolysis of a wide range of linkages between a variety of glycon and aglycons, GTs typically possess much greater specificity for transfer of glycosyl

References start on page 316

moieties to well-defined acceptors, forming a single type of glycosidic linkage in what has been termed the 'one-enzyme one-linkage' hypothesis, originally proposed by Hagopian and Eylar.[25] However, an increasing number of GTs has been discovered that fail to obey this hypothesis, using either the same glycosyl donor to make more than one type of glycosidic linkage or different glycosyl donors to make each distinct glycosidic linkage.[26] For example, the mycobacterial β-galactofuranosyltransferases glfT1 and glfT2, which catalyse the synthesis of the galactofuranose glycan core of cell wall arabinogalactan, are bifunctional and can synthesize both β-1,5 and β-1,6-Galf linkages.[27,28]

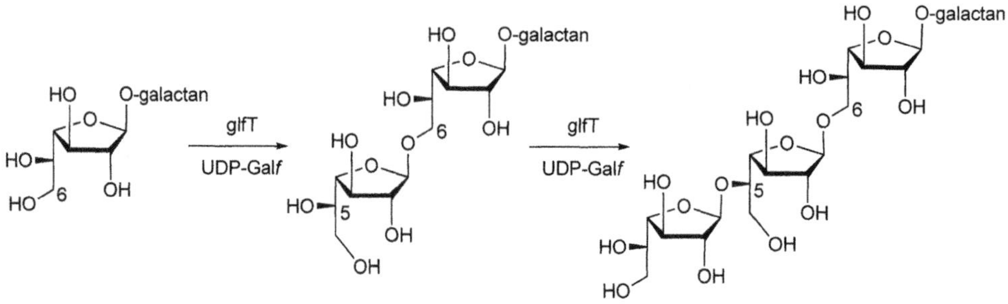

Similarly, β-GalT1 is a mammalian enzyme that catalyses the transfer of galactose from UDP-galactose to terminal *N*-acetyl-β-glucosaminyl residues, forming poly-1,4-β-*N*-acetyllactosamine core structures found in glycoproteins and glycosphingolipids.[29] Upon binding of the protein α-lactalbumin to β-GalT1, the resultant complex (termed lactose synthase) alters its specificity to favour glucose as an acceptor (with a decrease in the K_m value for glucose from 1 M to 1 mM), resulting in the formation of lactose.[30,31]

Sequence-based Classification and Structure

Despite a great amount of effort, the understanding of the chemical mechanism of GT action lags behind that of glycoside hydrolases. Sequence-based classification methods have proven powerful allies in the correlation of biochemical data, mechanism and structure, and in the prediction of structure and function for newly sequenced GTs. GTs have been classified into a series of sequence-based families – at the time of writing, more than 90 families have been assigned.[16,32]

As for glycoside hydrolases, the updated classification database is available over the World Wide Web through the *c*arbohydrate *a*ctive enzyme (CAZy) server.[33] The sequence-based classification scheme reliably differentiates retaining glycoside hydrolases from inverting ones but has been less effective at predicting mechanistic outcomes without a biochemically characterized representative being present in the family.[34]

Structural analysis of GTs has accelerated markedly in recent years. Prior to 1999 only one GT had been characterized by X-ray crystallography – since then representative structures for more than 20 GT families have been solved.[35,36] In contrast to the remarkable diversity of structures that has been observed for glycoside hydrolases, only a few topological folds have been seen for GTs. The majority of GTs fall into just two groups,[37] termed GT-A and GT-B,[35,38] each of which include members that are retaining and inverting GTs; thus, no stereochemical predictions on reaction outcome can be made based only on folds. Two additional folds have recently been identified: that exemplified by the sialyltransferase *Cst*II from *Campylobacter jejuni*[4,5] and that of the peptidoglycan GT of *Staphylococcus aureus* PBP2.[24,39] GT-A fold members can be both metal-ion dependent or independent but, to date, only metal-ion-independent GTs have been found in GT-B.

Reversibility of Glycosyl Transfer by Glycosyltransferases

Sugar (di)phosphonucleosides, sugar phospholipids and sugar-1-phosphates, the glycosyl donors for GTs, all possess excellent anomeric leaving groups. As a result, it is widely believed that in practice, GTs act only in the direction of glycosyl transfer. In fact, many GTs have been shown to act in a practically reversible manner. For example, sucrose synthase, which catalyses the reaction of fructose and UDP-glucose to give sucrose, has an equilibrium constant (K_{eq}) of 1.8–6.7 for the forward reaction.[40,41] The reversibility of this system allows the ready preparation of UDP-glucose from sucrose and UDP, which has been applied in multi-enzyme reaction cycles to produce this valuable substrate for use by other enzymes.[42] The GTs CalG1 and CalG4 involved in the biosynthesis of the antibiotic calicheamycin, and GtfD and GtfE involved in vancomycin biosynthesis, have been demonstrated to be reversible in their action, allowing preparation of complex sugar nucleoside (di)phosphates from the glycosylated natural product.[43,44] In addition, taking advantage of the reversibility of the glycosyl transfer reactions, the aglycon could be exchanged between natural

References start on page 316

products, allowing transfer of sugar moieties from the vancomycin aglycon to the calicheamycin aglycon.

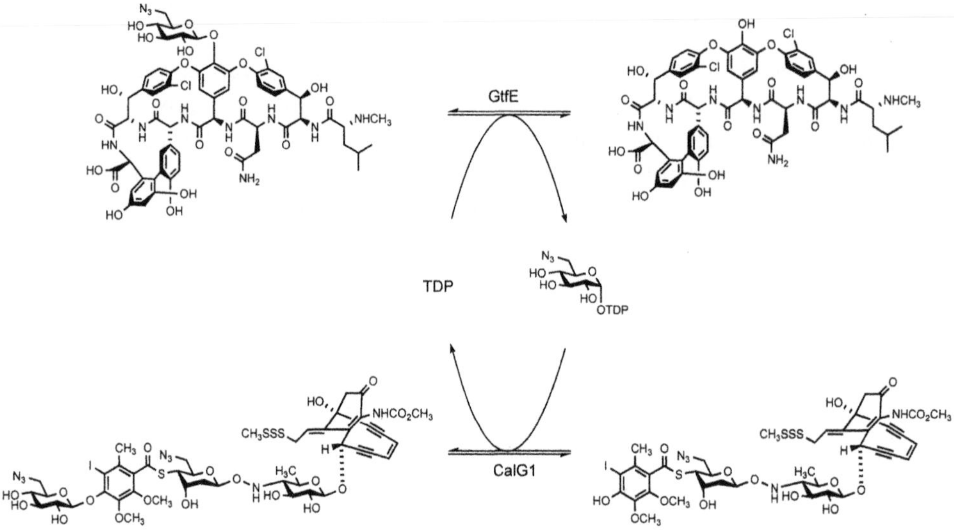

Inhibitors of Glycosyltransferases

Despite the sustained efforts of the scientific community, relatively few effective competitive inhibitors of glycosyl transferases have been developed and little progress has been made in defining general strategies for GT inhibition. However, several strategies that use small molecules as agents to effect inhibition have been developed, including the use of alternative primers, chain terminators and modified substrates. On the contrary, nature has provided many examples of effective inhibitors of GTs as natural products.

'Direct' inhibition of glycosyltransferases:[45] Numerous carbohydrate-based natural products are effective inhibitors of GTs.[46] Such compounds possess aspects of the complex, three-component transition state of the enzymatic reaction comprising the sugar of the glycosyl donor, the nucleotide or lipid phosphate leaving group and the acceptor. The most striking example of this class is moenomycin, an inhibitor of peptidoglycan GTs involved in transfer of lipid II to the growing peptidoglycan chain.[23,24,47–50]

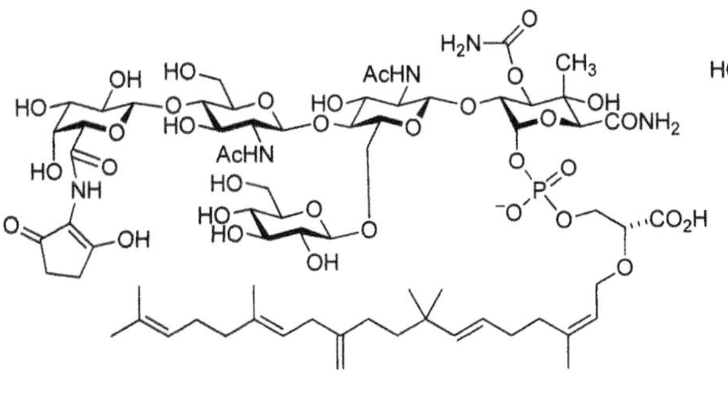

moenomycin A

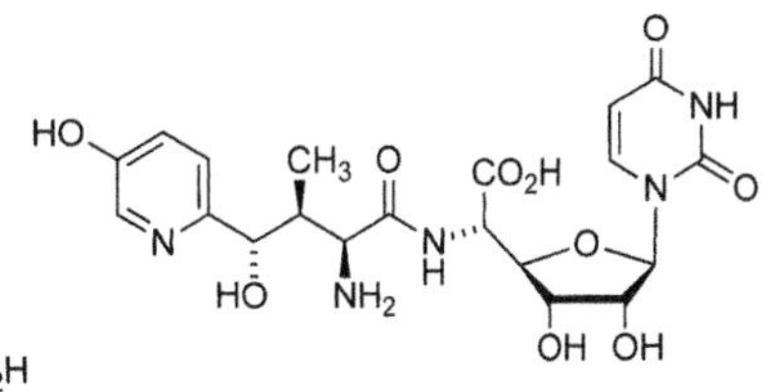

nikkomycin Z

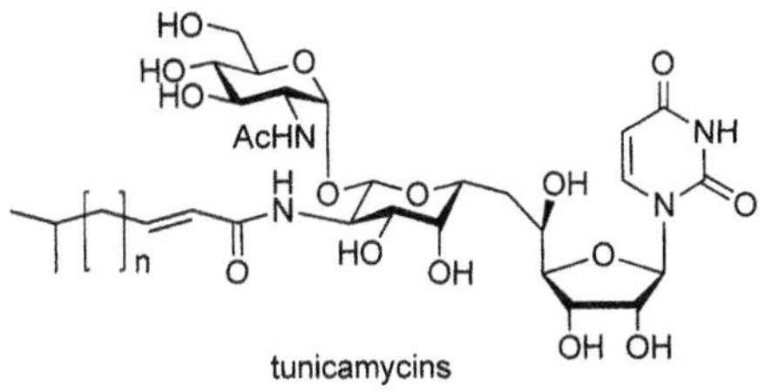

tunicamycins

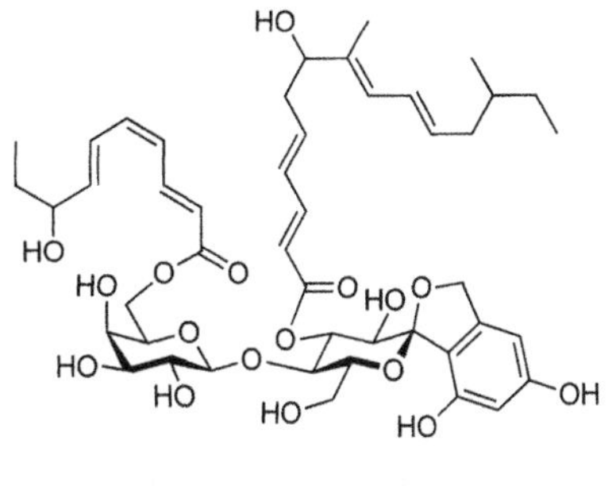

papulacandin B

The natural products of the nikkomycin family[51] are inhibitors of chitin synthase,[52] which catalyses the condensation of UDP-N-acetylglucosamine and the growing chitin chain, and the papulacandin family of glycolipids are inhibitors of fungal glucan synthases.[53] A related family of compounds is the tunicamycins, first isolated from *Streptomyces lysosuperificus*. Tunicamycin is not a direct inhibitor of a GT but is an inhibitor of an enzyme that synthesizes a GT donor substrate, N-acetylglucosamine phosphorotransferase, which catalyses the transfer of N-acetylglucosamine-1-phosphate from UDP-N-acetylglucosamine to dolichol phosphate. Inhibition of N-acetylglucosamine phosphorotransferase by tunicamycin thereby decreases the amount of dolichol-PP-N-acetylglucosamine available for synthesis of N-linked glycoproteins; as a result, tunicamycin is a widely used reagent for the study of N-linked glycosylation.[54]

Various cyclic peptides are effective inhibitors of GTs. Caspofungin (Cancidas) is a semi-synthetic cyclic peptide belonging to the echinocandin family of natural products and is synthesized from fermentation products of *Glarea lozoyensis*.[53,55] Caspofungin is a newly approved drug for the treatment of fungal infections.[56] It is a powerful inhibitor of fungal 1,3-β-glucan synthases, which catalyse the polymerization of UDP-glucose.

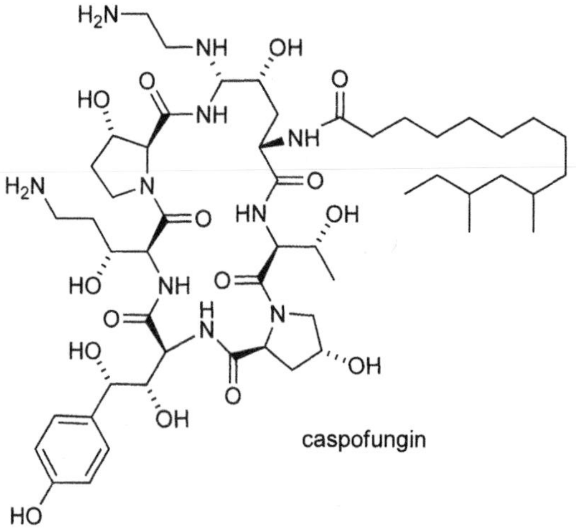

caspofungin

A cyclic peptidyl compound was developed by Imperiali and co-workers as an inhibitor of oligosaccharyl transferase, an enzyme that catalyses the *en bloc* transfer of dolichol-PP-GlcNAc$_2$Man$_9$Glc$_3$ to asparagine within a polypeptide chain in the endoplasmic reticulum to form N-linked glycoproteins.[57] The design of this compound incorporated features of the known consensus sequence for oligosaccharyltransferase, the preference of this enzyme for a specific peptide conformation (achieved by cyclization) and the known, albeit weak, inhibition of oligosaccharyltransferase by the rare amino acid, γ-aminobutyrine, when incorporated into tripeptides. Attachment of a membrane-permeable import peptide ensures that the inhibitor crosses the intracellular membrane of the endoplasmic reticulum.[58]

NH-Val-Thr-Ala-Ala-Val-Leu-Leu-Pro-Val-Leu-Leu-Ala-Ala-Pro-CO$_2^-$

Efforts to generate inhibitors of GTs through mimicry of the proposed enzymic transition state have had only limited success. Such mimicry has been attempted through linking together groups that resemble features of the glycosyl donor and acceptor. In most cases such inhibitors have potencies in the micromolar range, similar to the affinity towards most sugar (di)phosphonucleoside substrates. For example, Wong and co-workers reported the synthesis of an aza sugar, homofuconojirimycin, conjugated to LacNAc as an inhibitor of human α-1,3-fucosyltransferase V (FucT V), which catalyses the transfer of a fucosyl moiety from GDP-fucose to a LacNAc acceptor.[59] However, this conjugate was only a modest inhibitor of FucT V (IC$_{50}$ value of 2.4 mM). Remarkably, in the presence of GDP, a 77-fold improvement in the inhibitory potency of the conjugate (IC$_{50}$ value of 31 μM) was observed, suggesting the formation of a complex with GDP that mimics the transition state.

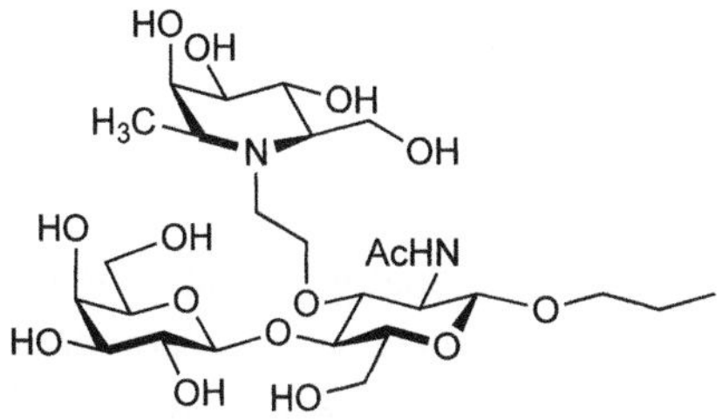

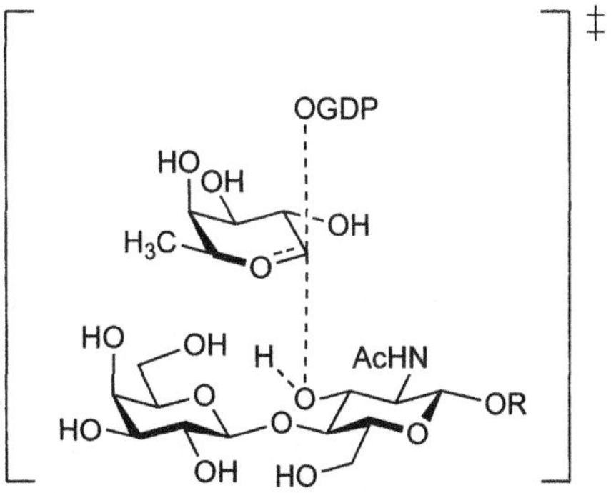

Perhaps, one of the most effective transition-state analogue inhibitors for a GT was developed by Schmidt and co-workers for α-2,6-sialyltransferase from rat liver.[60] α-2,6-Sialyltransferase catalyses the transfer of sialyl residues from CMP-Neu5Ac to terminal N-acetyllactosamine residues and has been suggested to operate through an exploded, S$_N$1-type transition state. A series of phosphonates was generated, the most potent of which exhibited a K_i value of 40 nM towards this enzyme.

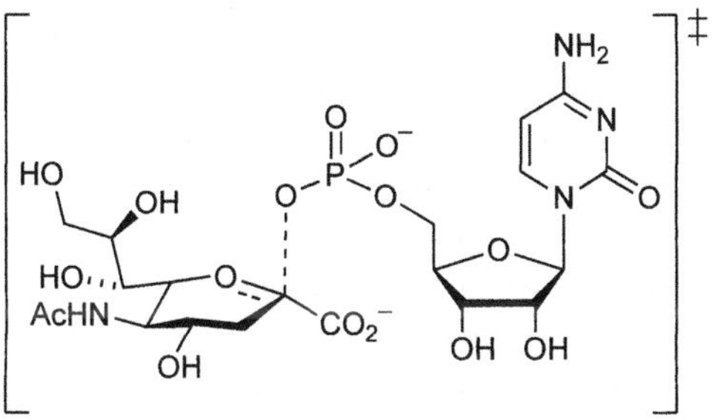

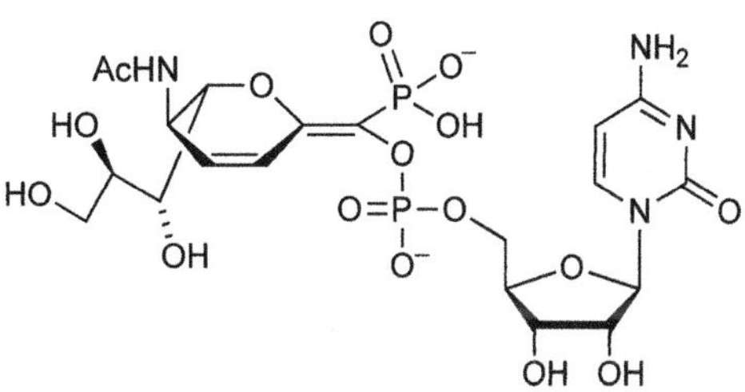

Hindsgaul and co-workers have described the design of a bisubstrate analogue comprised of a conjugate of the acceptor sugar and the diphosphonucleoside portion of the donor. The conjugate was an inhibitor (K_i value of 2.3 μM) of porcine α-1,2-fucosyltransferase, which transfers a fucosyl moiety from GDP-fucose to β-galactosides.[61]

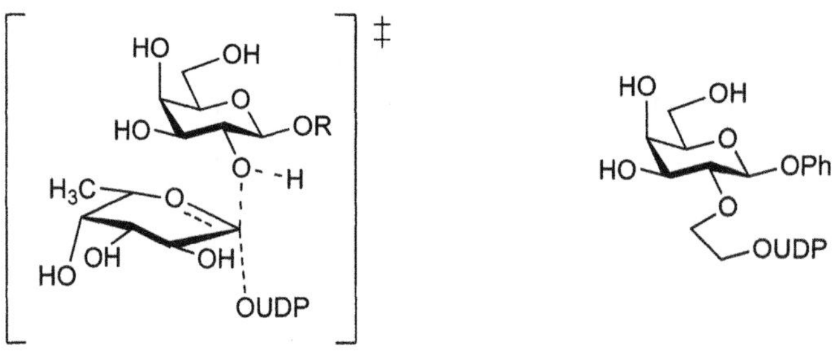

Hashimoto and co-workers have reported a bisubstrate analogue that is again a conjugate of a sugar diphosphonucleoside and the glycosyl acceptor. This compound was found to be a reasonably potent inhibitor (K_i value of 1.35 μM relative to the acceptor, and K_i value of 3.3 μM relative to the donor, UDP-galactose) of β-1,4-galactosyltransferase from bovine milk.[62]

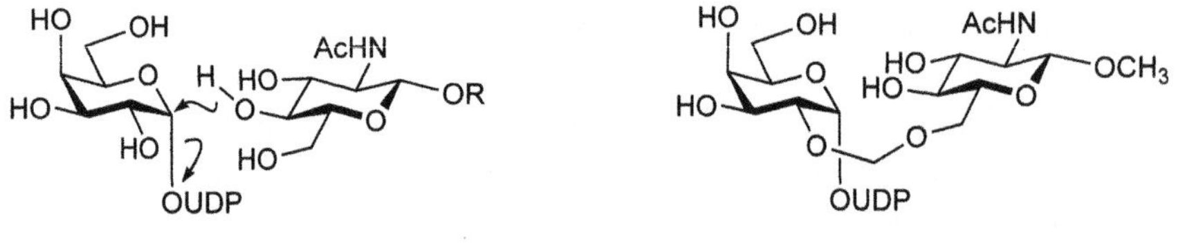

Lowary and Hindsgaul have reported the synthesis of a powerful inhibitor of blood group A GT obtained by simple modification of the acceptor substrate.[63] Blood group A GT catalyses the synthesis of the blood group A determinant, α-GalNAc-(1,3)-[α-Fuc-(1,2)-]-β-Gal-OR, by transfer of an N-acetylgalactosaminyl residue from UDP-N-acetylgalactosamine to the H determinant, α-Fuc-(1,2)-β-Gal-OR. α-Fuc-(1,2)-β-Gal-O-octyl is an effective artificial substrate for this GT, and modification of the acceptor hydroxyl to an amino group results in a powerful inhibitor (K_i value of 200 nM) of this GT. This compound was subsequently shown to be effective in inhibiting blood group A determinant synthesis in cell culture.[64]

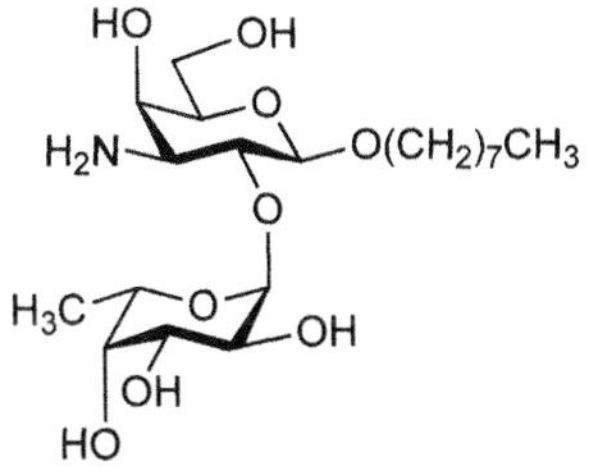

More often than not, the design of effective inhibitors for GTs has yielded limited success. A novel strategy for the discovery of an inhibitor of human α-1,3-fucosyltransferase, an enzyme involved in the biosynthesis of sialyl Lewis[x] and sialyl Lewis[a], was described in the laboratories of Sharpless and Wong.[65,a] Here, combinatorial modification of GDP, which has been shown to contribute to the greater part of donor-substrate binding energy, resulted in a library of triazole-linked GDP analogues. Screening of this library of compounds led to the identification of a biphenyl GDP adduct, which is an effective inhibitor (K_i value of 62 nM) of the target fucosyltransferase. Similarly, Bertozzi and co-workers reported the synthesis of a targeted combinatorial library of uridine derivatives that yielded a uridine-based inhibitor (K_i value of 7.8 µM) of polypeptide UDP-N-acetylgalactosamine transferase, a GT that initiates the synthesis of O-linked mucins through the synthesis of an α-GalNAc linkage to serine/threonine.[67,68]

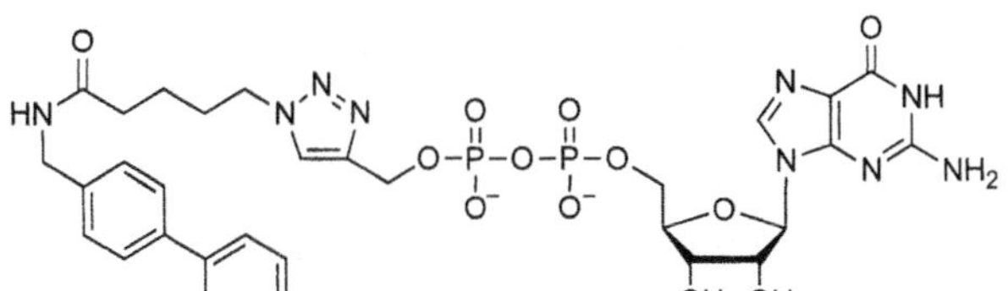

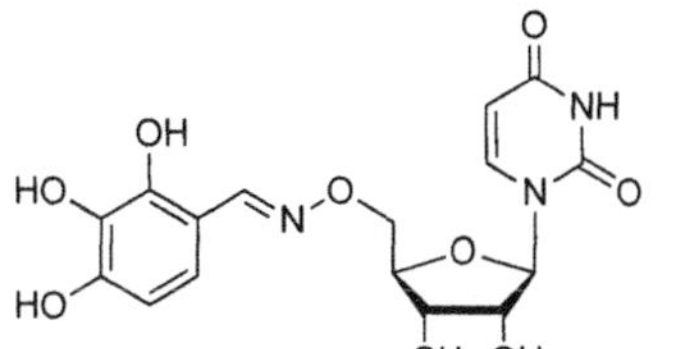

Therapeutically-useful glycosyltransferase inhibitors: While inhibition of GTs might be forecast to provide effective therapeutic solutions to a variety of disease states, few drugs along these lines have hitherto been developed. Aside from Caspofungin (described earlier), ethambutol is a frontline drug used for the treatment of tuberculosis and is an inhibitor of the decaprenolphosphoarabinofuranose-dependent arabinosyltransferases EmbA and

---

[a] The term Lewis is derived from the family name of Mrs. H.D.G. Lewis, whose new-born baby was found to suffer from jaundice caused by a hemolytic disease.[66]

References start on page 316

EmbB.[69,70] *N*-Butyldeoxynojirimycin (miglustat, Zavesca)[71] is an inhibitor of glucosylceramide synthase and is used for the so-called substrate-reduction therapy in the treatment of the lysosomal storage disorder, type I Gaucher disease (see Chapter 11).[72] Lufenuron is used for the control of fleas in animals and acts to inhibit chitin synthesis.[73] While the exact mode of action of Lufenuron is unclear, it has been proposed to act as a chitin synthase inhibitor.[74,75]

ethambutol   *N*-butyldeoxynojirimycin   Lufenuron

Several strategies have been elaborated for the development of non-reactive analogues of sugar donors. The transition states for glycosyl transfer for inverting or retaining GTs involve substantial oxocarbenium-ion-like character. Consequently, the installation of an electron-withdrawing fluorine atom at C2 of the glycosyl donor should act to destabilize such a transition state, and the corresponding 2-deoxy-2-fluoro glycosyl donors should act as non-hydrolysable substrate analogues. Several of these sugars have been prepared and shown to act as moderate-to-good inhibitors of GTs. For example, UDP-2F-glucose is an inhibitor (K_i value of 190 μM) of the *N*-acetylglucosaminyltransferase GlnT from rabbit,[76] UDP-2F-galactose is an inhibitor (K_i value of 149 μM) of β-1,4-galactosyltransferase from bovine milk[77] and CMP-3F-sialic acid is an inhibitor of the sialyltransferase CstII from *Campylobacter jejuni* (K_i value of 657 μM).[5,78] In a similar vein, introduction of an electron-withdrawing fluorine atom into the 5-position of a glycosyl donor also results in an incompetent donor; UDP-5F-*N*-acetylglucosamine is an inhibitor for the GT chitobiosyl-phospholipid synthase.[79]

A second approach to the development of non-hydrolysable analogues of sugar donors has involved the removal of the glycosidic oxygen or its replacement with another group or an atom such as sulfur or carbon. For example, phosphonates are hydrolytically stable analogues of phosphates, and a variety of phosphonate analogues of sugar donors has been synthesized and assessed as inhibitors of GTs.[80,81] The *C*-glycoside phosphonate analogue of UDP-*N*-acetylglucosamine is an effective inhibitor (K_i value of 28 μM) of the *N*-acetylglucosaminyltransferase GlnT from rabbit.[76] A *C*-glycoside analogue of decaprenolphosphoarabinose was synthesized, and while not directly assessed as an inhibitor of the corresponding GT, was found to be a powerful inhibitor of the growth of *Mycobacterium tuberculosis*.[82] Thioglycoside-based analogues of UDP-sugars have been prepared, but their efficacy as GT inhibitors has not been reported.[83]

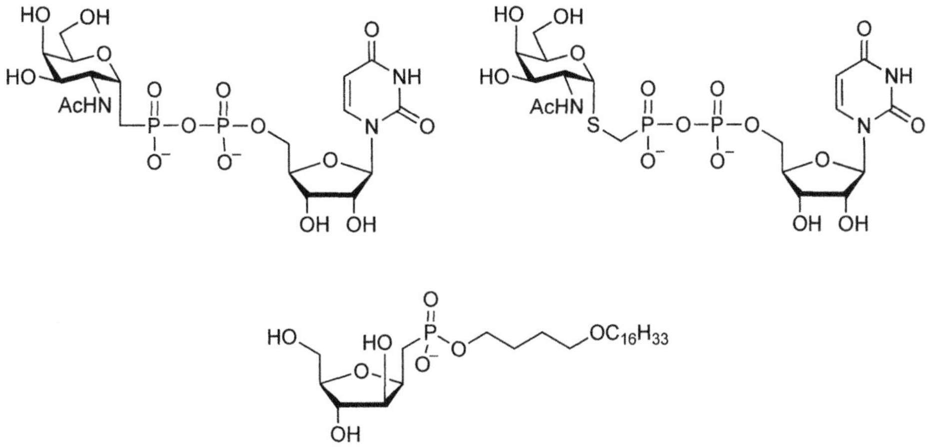

'Indirect' inhibition of glycosyltransferases by metabolic interference: With a lack of success in delineating general approaches to the design of effective GT inhibitors, alternative approaches have been developed that provide the functional equivalent of GT inhibition and that allow the study of the effect of GT blockade in biological settings. Two major approaches to this end have been developed: the use of *glycoside primers*, unnatural analogues of the acceptor that compete with the natural acceptor, and the use of *metabolic chain terminators*, unnatural analogues of the donor that are incorporated into the glycoconjugate product but that, owing to structural modifications, are unable to act as acceptors for the transfer of additional carbohydrate moieties.

Glycoside primers[54,67] are alternative acceptor substrates that compete with the natural substrate for the active site of the GT, thereby diverting the action of the GT to the synthesis of unnatural glycoconjugates and reducing the production of the natural glycoconjugates, thus providing functional inhibition.[54] The use of glycoside primers has proven to be a relatively general approach for the functional inhibition of GTs. Such inhibitors are also referred to as competitive

References start on page 316

substrate-based primers. The first glycoside primer inhibitor was discovered by Okayama and co-workers, who reported that 4-nitrophenyl β-D-xylopyranoside was able to inhibit glycosaminoglycan synthesis on proteoglycans when administered to embryonic chicken cartilage.[84] Inhibition was achieved through the added xyloside priming the biosynthesis of the glycosaminoglycans through competition with endogenous β-D-xylosylated core proteins. A naphthol aglycon has been shown to provide excellent uptake of aldopentosides into animal cells, and 2-naphthyl β-D-xylopyranoside is a widely used reagent for studying the functional inhibition of glycans bearing a β-D-xylopyranoside core.[85] A variety of glycoside primers for the inhibition of a range of glycosylation processes has now been developed, including various β-D-glucosides (as inhibitors of N-acetylglucosamine β-1,4-galactosyltransferase),[86] β-D-galactosides (as inhibitors of glycosaminoglycan biosynthesis),[87] N-acetyl-β-D-glucosaminides (as inhibitors of polylactosamine biosynthesis),[88] N-acetyl-α-D-galactosaminides (e.g. benzyl N-acetyl-α-D-galactosaminide as an inhibitor of mucin-type O-linked glycan biosynthesis)[67,89] and various di- and trisaccharides (e.g. 2-naphthylmethyl N-acetyl-β-lactosaminide as an inhibitor of blood group Lewis^x synthesis).[85]

For studies in cells, conversion of glycoside primers to their peracetylated derivatives usually improves uptake, with the acetyl groups being removed by promiscuous esterases within the cell. While such primers have proven to be important research tools, the indirect nature of their action can limit the conclusions that can be drawn from their use. Issues that may arise in their use include the following: (i) incomplete inhibition may leave residual levels of glycoconjugates intact, (ii) glycoconjugates synthesized on the primer may have activities that complicate the analysis of biological phenomena, (iii) inhibition occurs at the level of entire classes of glycoconjugates that share a common core structure (that of the primer), and therefore specific targeting to a single glycoconjugate is not always possible and (iv) enhanced rates of glycoconjugate synthesis on the primer may deplete sugar (di)phosphonucleoside pools, leading to reduced formation of other, unrelated glycoconjugates.

An alternative approach to inhibition of GTs by metabolic interference is to utilize modified donor substrates that are altered so as to prevent the glycoconjugate formed by glycosyltransfer from acting as an acceptor for subsequent transfer of additional sugar residues, in effect acting as glycan chain terminators.

One way in which chain termination can be achieved is by using a deoxyge-nated glycosyl donor that lacks the hydroxyl group required for the resultant glycoconjugate to act as an acceptor in a subsequent glycosylation, resulting in functional inhibition of the latter GT.[90] One of the earliest examples of this approach was the supplementation of liver cells with '2-deoxy-D-galactose', which acted to provide functional inhibition of 2-O-fucosylation.[91] 2-Deoxy-D-galactose was incorporated into glycoproteins, which were found to contain lowered levels of 2-O-fucosylation, a result attributed to the missing 2-hydroxyl. Interestingly, higher levels of α-1,3- and α-1,4-fucosylation were noted, a result that can be attributed to the elevation of intracellular pools of GDP-fucose resulting in unnatural fucosylation patterns.[92] Bertozzi and co-workers have presented an alternative approach to chain termination with the finding that N-butanoyl-D-mannosamine inhibits formation of polysialic acid in neurons and tumour cells.[93] N-Butanoyl-D-mannosamine is metabolized to the corresponding CMP-sialic acid as the cellular machinery can tolerate modestly sized N-acyl substituents on D-mannosamine. While the sialyltransferase of interest was able to transfer CMP-N-butanoylsialic acid onto the acceptor, neural cell adhesion molecule, after transfer the resultant glycoconjugate did not act as an acceptor for additional transfer events, resulting in inhibition of poly-α-2,8-sialic acid expression. Inhibition of α-2,8-sialyltransferase by N-butanoyl-D-mannosamine, which still bears the hydroxyl group required to act as acceptor once incorporated into a glycoconjugate, likely results from a lack of tolerance by the sialyltransferase of the sterically demanding N-butanoyl group when found in the acceptor.

Chemical Modification of Glycoconjugates Using Metabolic Pathway Promiscuity

Studies with unnatural sugars have revealed that many biosynthetic pathways leading to the synthesis of glycoconjugates can tolerate suitably chosen unna-tural substituents on the sugars and that these substituents are incorporated unchanged into the final glycoconjugate. For example, Reutter and co-workers showed that unnatural D-mannosamine derivatives, in which the natural N-acetyl

References start on page 316

group was replaced by *N*-propanoyl, *N*-butanoyl or *N*-pentanoyl, are converted into the corresponding *N*-acylsialoses and incorporated into sialic acid–containing glycoconjugates in cell culture and in rats.[94,95] An elegant extension of this observation is to incorporate metabolically silent functional groups that, following incorporation into glycoconjugates, can react selectively with reagents to enable the detection or isolation of the chemically labelled glycoconjugate. This approach has been termed 'metabolic labelling of glycans'. Of particular importance to this approach is the identification of so-called bio-orthogonal conjugations, reactions that result in the ligation of two species with a chemoselectivity that allows them to be performed in a biological context.[96,97] Several functional groups have been utilized to achieve metabolic labelling of glycans and are well tolerated by entire metabolic pathways leading to the glycoconjugates, including azido,[98] thiol,[99] keto[100] and alkynyl groups.[101] Of this set of functional groups, the azido group is the premier choice, being distinguished by its lack of reactivity in cellular systems, its rarity in biological settings and its exquisite specificity for clean, high-yielding reactions with suitable reactive probes. For example, *N*-azidoacetyl-D-mannosamine can be converted into azide-modified cell-surface sialic acid residues by way of the respective biosynthesis pathway.[98,102] The azido group is not altered as it passes through the biosynthetic pathway but, once in place, can react with azide-specific reagents such as bifunctional phosphines bearing electrophilic traps (the Staudinger ligation), terminal alkynes in the presence of copper(I) catalysis (the CuAAC reaction) and strain-promoted Huisgen [3 + 2] cycloaddition with cycloalkynes.[103–106] By such ligation reactions, fluorescent dyes and epitope tags, including biotin and antigenic peptides, can be conjugated to facilitate detection and enable enrichment/isolation by immuno-precipitation.

A variety of different carbohydrates are excellent starting materials for incorporation into glycans, including the following: various *N*-acyl-D-mannosamines[98–100] and *N*-acylsialic acids[107] for incorporation as sialic acid derivatives into glycans; various analogues of *N*-acetylgalactosamine[108,109] leading to O-linked mucin-type glycans; *N*-acyl-D-glucosamines for incorporation as O-GlcNAc modifications;[108] and fucose derivatives for incorporation into glycoproteins.[110]

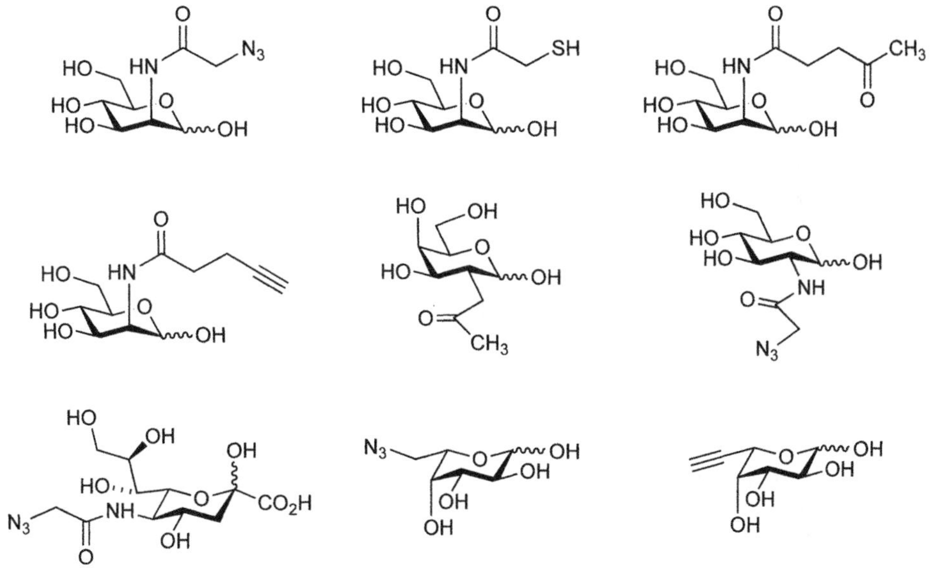

Metabolic labelling has even been performed in living animals to modify cell metabolites in their native cellular environment. Peracetylated *N*-azidoacetyl-D-mannosamine was metabolized by living mice into cell surface sialic acids, which were labelled in vivo by epitope labelling with an antigenic peptide using the Staudinger ligation.[111] Isolated splenocytes were harvested from sacrificed mice, and modified cells were detected after labelling with a fluorescently labelled anti-epitope antibody using flow cytometry.

Use of Glycosyltransferases in Synthesis[112]

GTs have been widely used in synthesis, providing a direct and protecting-group-free method for the synthesis of complex glycoconjugates. However, the application of GTs for large-scale synthesis is more technically demanding owing to the difficulty of acquiring and purifying large amounts of many GTs and their glycosyl donors, the

References start on page 316

significant inhibition often seen with the released nucleoside mono- or diphosphates, and the difficulty of purifying the resultant glycoconjugates. As a result a variety of elegant approaches has been developed to side-step these problems, including the use of enzymatic systems capable of synthesizing sugar (di)phosphonucleosides in situ, enzymatic synthesis on solid supports and approaches that use whole living cells as biochemical reactors.

Enzymatic synthesis using glycosyltransferases and sugar (di)phosphonucleoside donors:[113] Many GTs are commercially available and are in common use for the synthesis of milligram quantities of glycoconjugates. β-1,4-Galactosyltransferase, which catalyses the condensation of UDP-galactose with the 4-hydroxyl of terminal N-acetyl-β-D-glucosaminides, is in widespread use and is commercially available in significant quantities. This enzyme is an excellent catalyst for the direct synthesis of N-acetyllactosamine analogues, as the enzyme can tolerate a wide range of modifications on the acceptor and the donor. An elegant example employed β-1,4-galactosyltransferase to prepare ^{13}C-enriched linear oligo-N-acetyllactosamines, molecules that would be especially challenging to prepare through standard chemical approaches.[114]

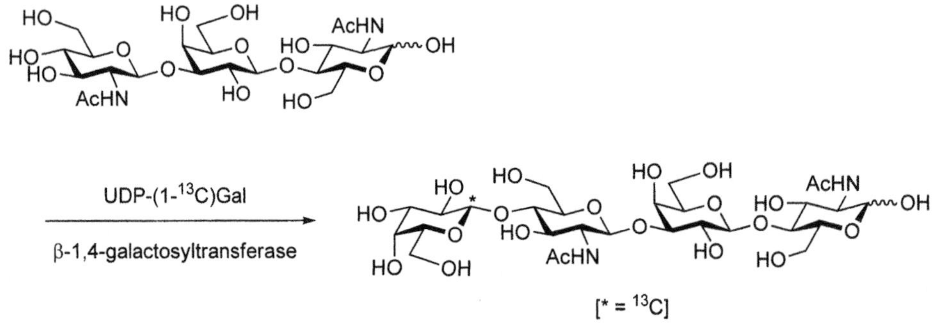

GTs represent particularly valuable agents for chemical synthesis, enabling the late-stage introduction of difficult glycosidic linkages. The commercial availability of various sialyltransferases and fucosyltransferases has meant that these catalysts are commonly the method of choice for direct installation of sialyl and fucosyl groups. Wong and co-workers have reported a spectacular enzymatic synthesis of a dimeric sialyl Lewis^x by the sequential application of an α-2,3-sialyltransferase and an α-1,3-fucosyltransferase.[115] Here, the nonasaccharide substrate was obtained by enzymatic degradation of hen egg yolk.

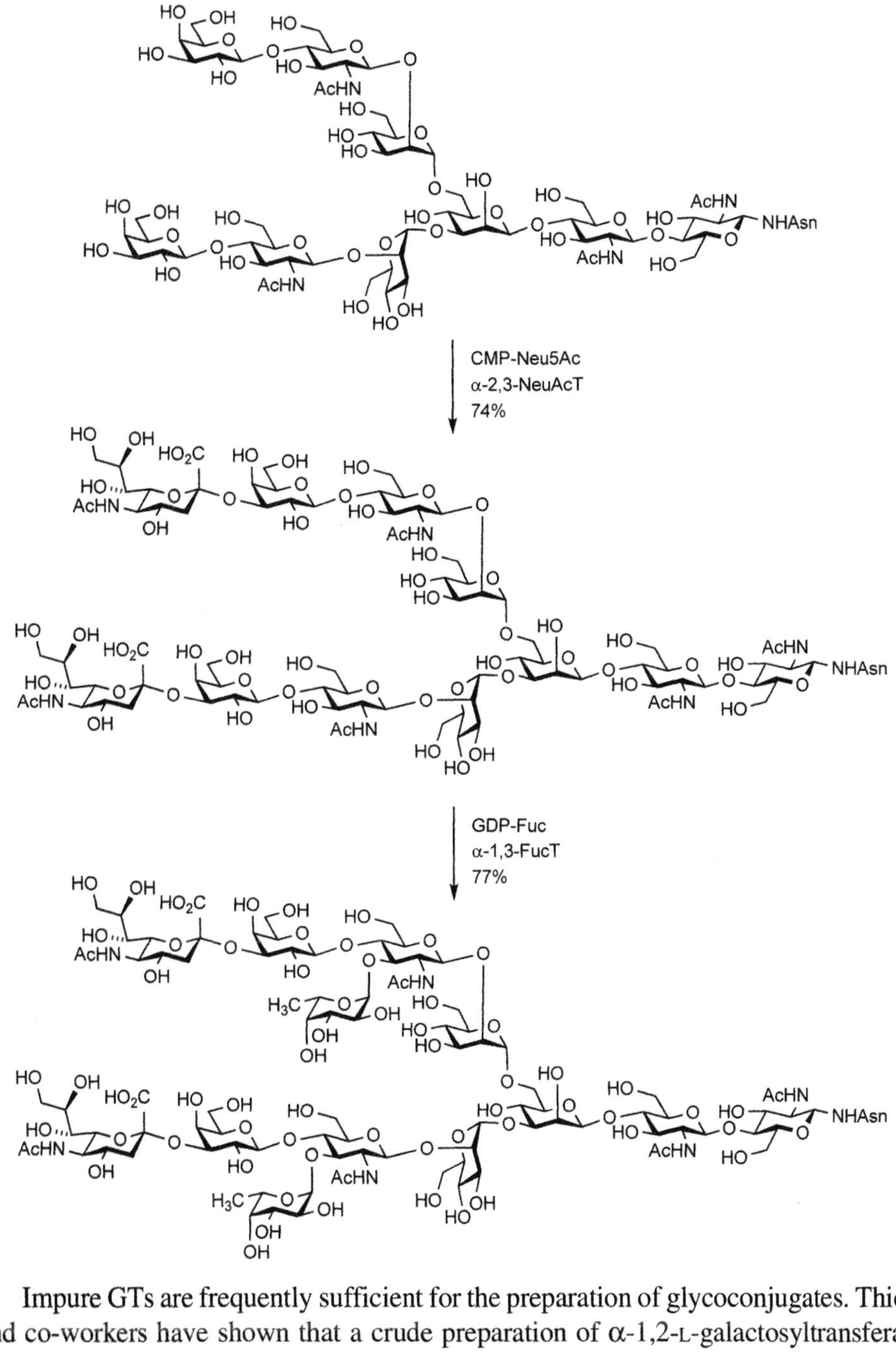
CMP-Neu5Ac
α-2,3-NeuAcT
74%
GDP-Fuc
α-1,3-FucT
77%
NHAsn

Impure GTs are frequently sufficient for the preparation of glycoconjugates. Thiem and co-workers have shown that a crude preparation of α-1,2-L-galactosyltransferase from the albumen gland of the vineyard snail *Helix pomatia* can be used to catalyse the transfer of an L-fucosyl moiety from GDP-fucose to various galactoside

References start on page 316

acceptors.[116] One snail provides sufficient enzyme to fucosylate around 7 mg of acceptor.

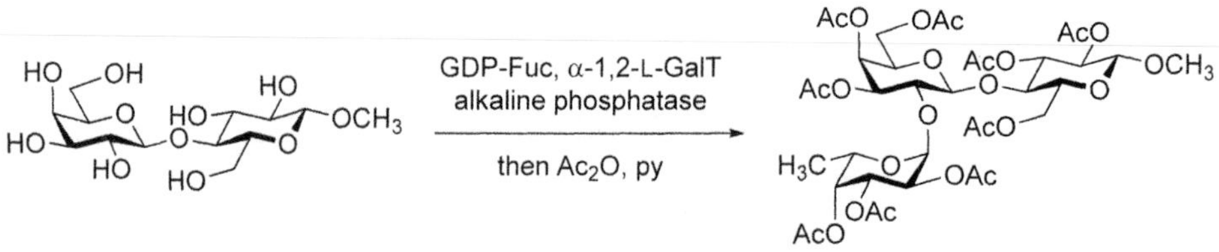

GTs can be used for glycoside synthesis on a solid phase. Solid-phase synthesis offers an advantage for sequential glycosylations in that individual enzymic products can be purified at each step by simple filtration and washing. An impressive application of sequential glycosylations on a solid phase was reported by Blixt and Norberg for the synthesis of sialyl Lewis[x] on a Sepharose support using three different GTs.[117]

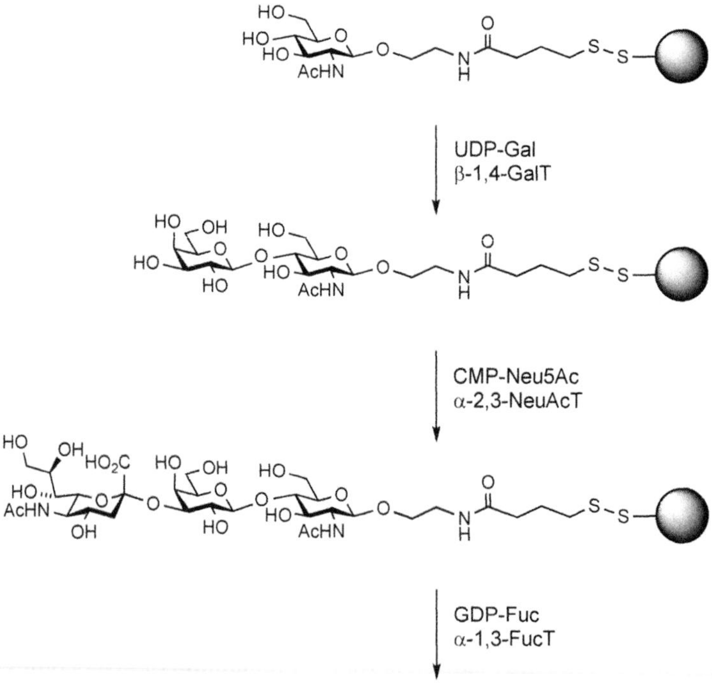

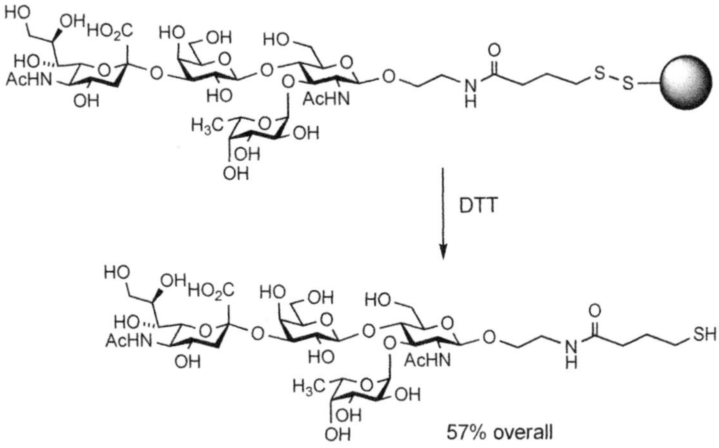

For synthesis with GTs on a large scale, significant problems need to be overcome, including the following: securing adequate supplies of enzymes and sugar (di)phosphonucleoside donors; preventing reduction of reaction rates caused by enzyme inhibition by the (di)phosphonucleoside by-product; and the purification of product from the large quantities of buffer salts and by-products. Ratcliffe, Palcic and co-workers have reported the 62 g scale synthesis of a *Clostridium difficile* trisaccharide from a lactoside using an α-galactosyltransferase.[118] Here, adequate supply of the GT was achieved through recombinant expression in *E. coli*, and the high cost of the donor UDP-galactose was overcome through the use of the much less expensive donor UDP-glucose in concert with recombinant UDP-galactose epimerase, an enzyme that catalyses the formation of UDP-galactose from UDP-glucose. Inhibition of the GT by the by-product UDP was prevented by the inclusion of alkaline phosphatase, which degrades UDP to uridine and phosphate. Finally, purification of the product was achieved by taking advantage of a hydrophobic methoxycarbonyloctyl aglycon, which allowed a simple purification by reversed-phase chromatography on C_{18} silica gel.

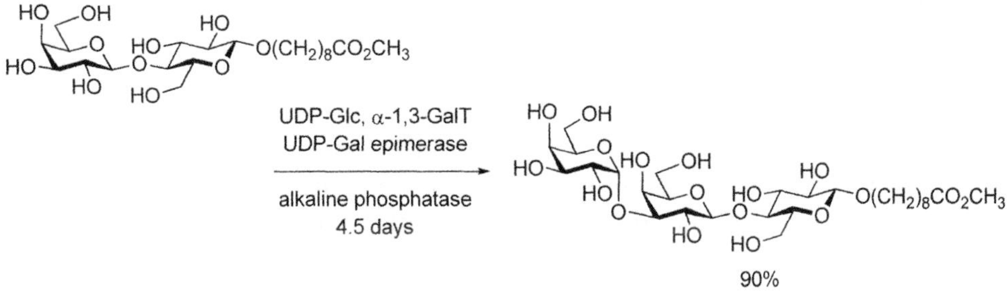

Bundle and co-workers have shown that GTs can catalyse the synthesis of thioglycosides from sugar donors and thiol acceptors.[119] In the case of bovine α-1,3-galactosyltransferase, UDP-galactose was an effective glycosyl donor towards octyl 3'-deoxy-3'-thio-β-lactoside, resulting in the formation of a doubly galactosylated tetrasaccharide product, with transfer of the first galactosyl residue to the thiol resulting in the formation of a thioglycoside linkage. These workers exploited the resistance of an S-glycoside over an O-glycoside to cleavage by a glycosidase to remove selectively the terminal O-galactoside using α-galactosidase from green coffee beans.

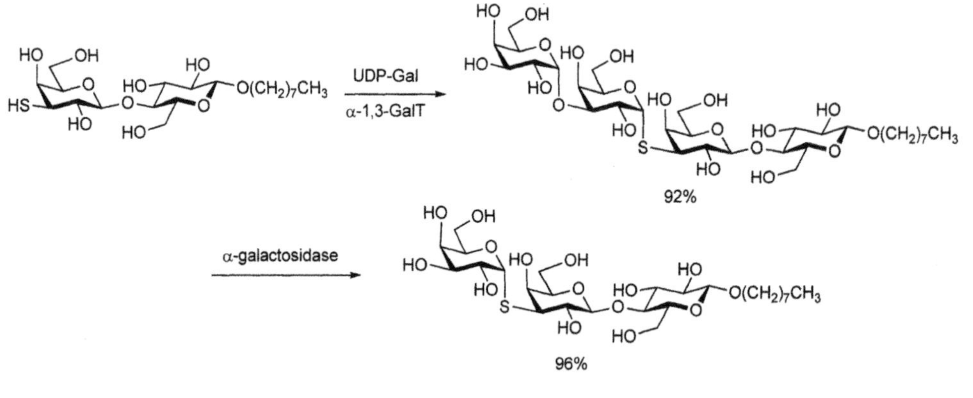

Withers and co-workers have reported a strategy for expanding the substrate specificity of a GT through a process called 'substrate engineering'.[120] In this approach, a readily removed functional group is attached to the acceptor, altering the binding orientation and presenting alternative hydroxyl groups to the enzyme that differ from the normal orientation of the substrate, thereby allowing the rational control of substrate regioselectivity. For example, the α-1,4-galactosyltransferase LgtC normally transfers a galactosyl moiety from UDP-galactose to lactose, with a second-order rate constant, $k_{cat}/K_m = 240 \, \text{min}^{-1} \, \text{mM}^{-1}$. Unsurprisingly, this enzyme can also utilize simple galactosides such as octyl β-D-galactopyranoside ($k_{cat}/K_m = 3.8 \, \text{min}^{-1} \, \text{mM}^{-1}$) as substrates, but with reduced proficiency. The introduction of a 6-O-benzoyl group into mannose provides an acceptor that mimics a β-D-galactoside, and which can act as a glycosyl acceptor ($k_{cat}/K_m = 5.0 \, \text{min}^{-1} \, \text{mM}^{-1}$), resulting in the formation of an α-Gal-(1,2)-Man linkage in a fashion readily predicted by direct structural comparison with the natural substrate.

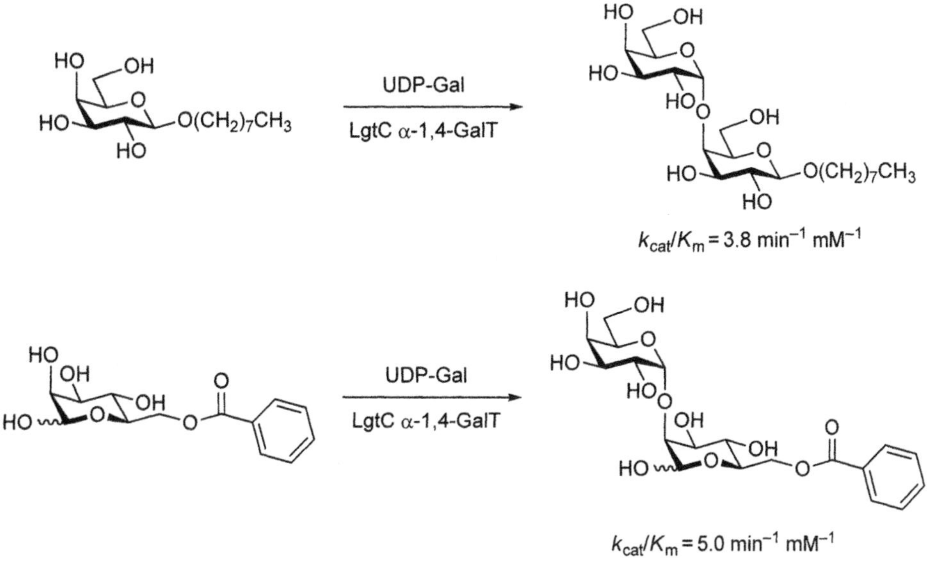

$$k_{cat}/K_m = 3.8 \text{ min}^{-1} \text{ mM}^{-1}$$

$$k_{cat}/K_m = 5.0 \text{ min}^{-1} \text{ mM}^{-1}$$

The concept of altering the specificity of enzymatic glycosyl transfer by modifying the structure of the acceptor alcohol so as to structurally mimic the 'natural' acceptor was first applied by Stick and co-workers to alter the regioselectivity of a glycosynthase[121] and may be applicable to other enzymes that modify carbohydrates.

Multienzyme systems including sugar (di)phosphonucleoside generation and recycling: Owing to the expense of obtaining large quantities of sugar (di)phosphonucleoside donors, several so-called (di)phosphonucleoside recycling systems have been developed that can both synthesize the sugar (di)phosphonucleoside and recycle the released (di)phosphonucleoside.[122,123] The recycling of the sugar (di)phosphonucleoside allows simple access to the required amounts of sugar donor in situ and at the same time reduces the concentration of the (di)phosphonucleoside by-products, which frequently inhibit GT activity. The earliest of the recycling systems was reported by Whitesides and co-workers, who utilized immobilized β-1,4-galactosyltransferase to synthesize N-acetyllactosamine on a >10 g scale from UDP-galactose and N-acetylglucosamine.[124] UDP-galactose was synthesized in three steps from glucose-6-phosphate through the sequential actions of phosphoglucomutase, UDP-glucose pyrophosphorylase and UDP-galactose 4′-epimerase. Released UDP was converted to UTP (needed for the formation of UDP-glucose) by reaction with phosphoenolpyruvate catalysed by pyruvate kinase.

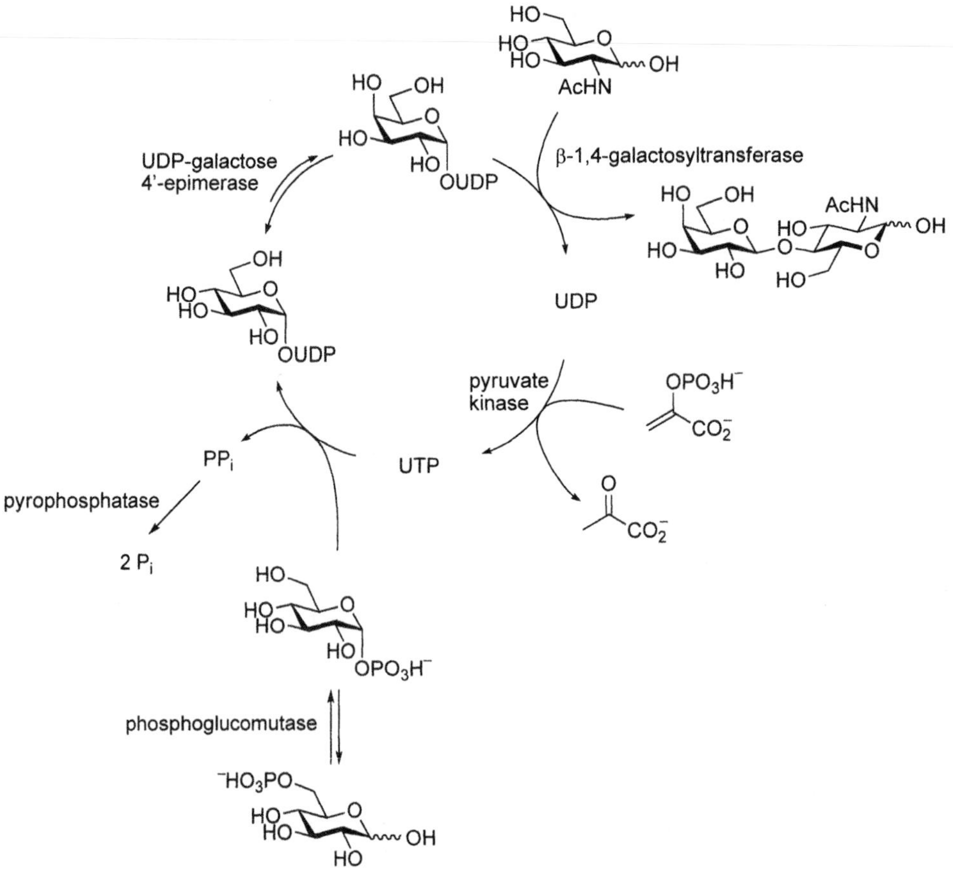

As mentioned earlier, sucrose synthase reversibly catalyses the formation of sucrose from UDP-glucose and fructose, and this enzyme can be used to generate UDP-glucose from sucrose inexpensively. Zervosan and Elling have applied sucrose synthase in a three-enzyme system comprised of sucrose synthase, UDP-galactose 4′-epimerase and β-1,4-galactosyltransferase, allowing the synthesis of N-acetyllactosamine from sucrose and N-acetylglucosamine.[42] Notably, this simpler system does not result in the formation of large amounts of pyrophosphate or phosphate, which can act to inhibit some enzymes in this and related systems.

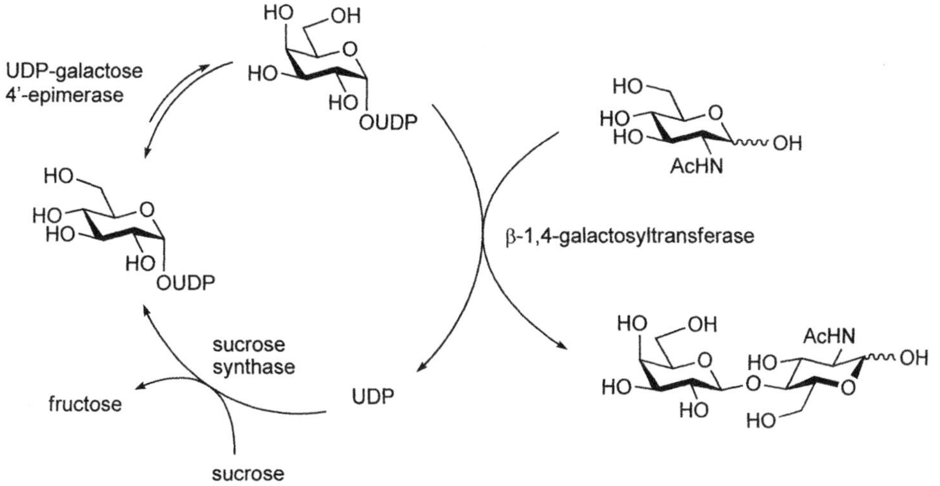

Given the difficulty of obtaining large quantities of sugar (di)phospho-nucleoside donors, it is noteworthy that LgtC from *Neisseria meningiditis* can utilize galactosyl fluoride as a glycosyl donor for transfer to lactose.[125] Whether other GTs can utilize simple glycosyl fluorides as glycosyl donors has not been widely investigated.

Enzyme systems that involve the immobilization of GTs either alone, or in combination with other enzymes that allow synthesis of sugar (di)phosphonucleo-sides and recycling of (di)phosphonucleoside by-products back to sugar (di)pho-sphonucleosides, have been described by several groups. Such immobilized enzyme systems allow enzyme recycling, reducing the cost of enzymatic synthesis. Wang and colleagues have developed the so-called superbeads, which are a matrix support on which a complete suite of enzymes required for sugar (di)phosphonu-cleoside regeneration is immobilized.[126,127] For example, individually recombi-nantly expressed and His$_6$-tagged galactokinase (GalK), galactose-1-phosphate uridyltransferase (GalT), UDP-glucose pyrophosphorylase (GalU) and pyruvate kinase (PykF) were combined and immobilized onto Ni^{2+}-NTA resin to afford superbeads capable of converting D-galactose into UDP-galactose. When *E. coli* lysate containing recombinant bovine α-1,3-galactosyltransferase was added to the above enzyme mixture and the resultant blend co-immobilized on Ni^{2+}-NTA resin, the resultant superbeads could be used to synthesize the benzyl glycoside of the α-Gal trisaccharide, involved in xenotransplant rejection (see Chapter 11), from benzyl β-lactoside.[128] The immobilized enzyme could be reused for synthesis and afforded 72%, 69% and 66% yields for three repeat reactions over a 3-week period.

References start on page 316

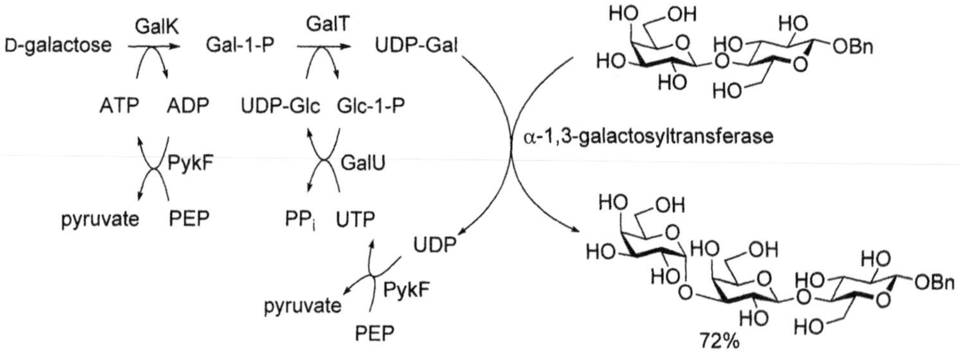

Synthesis using glycosyltransferases in engineered whole cell systems:

GTs have proven to be particularly efficacious for synthesis of glycoconjugates when expressed recombinantly in a host cell strain, and the resulting whole cells are used as catalysts for glycoconjugate synthesis.[129] In this approach, recombinant cells express the GTs, which are not purified but used either in fermentation culture or as isolated cells, with the cells acting as a 'living factory' for production of glycoconjugates. If the host cells are able to synthesize sufficient amounts of the required sugar (di)phosphonucleoside donors, it may prove sufficient to express recombinantly only the GTs of interest. For example, α-Man-(1,2)-α-Man-CbzThrOCH$_3$ could be synthesized by $E.\ coli$ expressing the α-1,2-mannosyltransferase of $Saccharomyces\ cerevisiae$.[130]

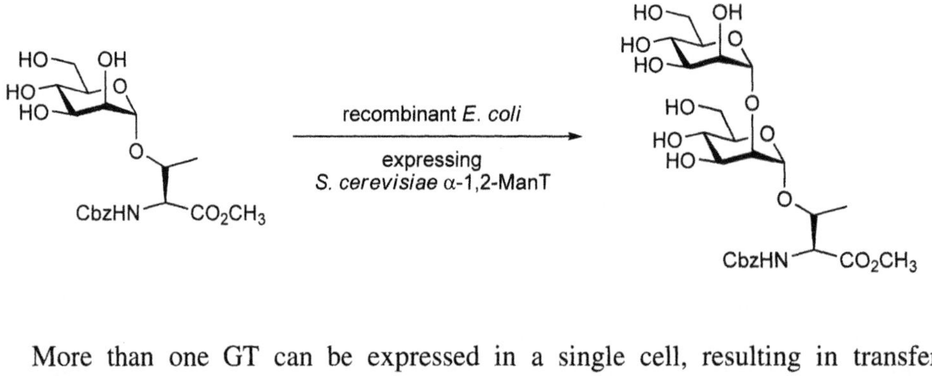

More than one GT can be expressed in a single cell, resulting in transfer of multiple sugar moieties. Bettler and co-workers have reported the recombinant co-expression of two GTs in $E.\ coli$, chitin pentaose synthase NodC from $Azorhizobium$ and β-1,4-galactosyltransferase LgtB from $Neisseria\ meningitidis$, which together were able to synthesize >1 g l^{-1} of the hexasaccharide β-Gal-(1,4)-[β-GlcNAc-(1,4)-]$_4$-GlcNAc when grown in high-density cell culture.[131]

Low endogenous levels of the required sugar (di)phosphonucleosides may limit the quantities of products formed from the recombinant GT(s) of interest. Better results are obtained if the entire pathways for sugar (di)phosphonucleoside synthesis and (di)phosphonucleoside recycling as well as the GTs of interest are transferred into the host strain. Wang and colleagues have referred to systems developed for the synthesis of sugar (di)phosphonucleosides and the corresponding GTs using recombinantly expressed enzymes within one cell strain as 'super-bugs'.[127,132] For example, a single plasmid bearing genes for sucrose synthase, UDP-galactose 4'-epimerase and α-1,4-galactosyltransferase (LgtC) was transformed into *E. coli*, and the resultant cells were harvested, made permeable by freeze thawing, and then employed as a catalyst for synthesis, using lactose and sucrose as feedstocks, resulting in the accumulation of α-Gal-(1,4)-β-Gal-(1,4)-Glc (globotriose) to 22 g l^{-1}.[133]

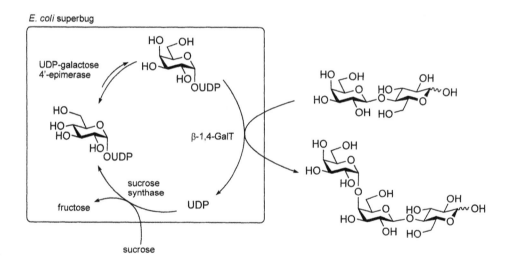

Enzymes for the production of sugar (di)phosphonucleosides and (di)phosphonucleoside recycling can also be recombinantly expressed in separate cell lines that are co-cultured with those expressing the GT of interest, with sugar (di)phosphonucleosides diffusing between cells as required. The use of co-cultures of multiple bacterial strains that individually undertake aspects of an entire cycle, with transfer of reaction intermediates between strains, has been referred to as 'bacterial coupling'. Cell lines have been reported that can synthesize UDP-galactose or CMP-NeuAc that accumulate in culture media at levels of 44 g l^{-1} and 17 g l^{-1},[134,135] respectively, and these sugar (di)phosphonucleosides can be taken up by cell lines possessing functional GTs for the synthesis of glycoconjugates. In one of the most remarkable achievements in the area, Ozaki

References start on page 316

and colleagues have reported the development of a multiplasmid, three-strain system for the synthesis of α-Gal-(1,4)-β-Gal-(1,4)-Glc (globotriose).[134] The bacterium *Corynebacterium ammoniagenes* was engineered to convert orotic acid into UTP. The resultant UTP is taken up by an *E. coli* strain (NM522/pNT25/pNT32) that has been engineered to produce UDP-galactose from galactose using fructose as energy source. This strain in turn releases UDP-galactose where it is taken up by a second *E. coli* strain (NM522/pGT5) that expresses an α-1,4-galactosyltransferase from *Neisseria gonorrhoeae*, which utilizes UDP-galactose and lactose to produce globotriose. This coupled system results in accumulation of the trisaccharide product to a remarkable 188 g l^{-1}!

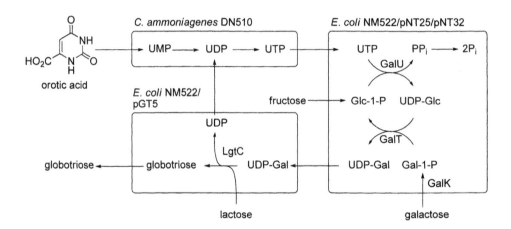

References

1. Leloir, L.F. (1971). *Science*, **172**, 1299.
2. Smith, J.L. (1995). *Curr. Opin. Struct. Biol.*, **5**, 752.
3. Focia, P.J., Craig, S.P., 3rd and Eakin, A.E. (1998). *Biochemistry*, **37**, 17120.
4. Pak, J.E., Arnoux, P., Zhou, S., Sivarajah, P., Satkunarajah, M., Xing, X. and Rini, J.M. (2006) *J. Biol. Chem.*, **281**, 26693.
5. Chiu, C.P., Watts, A.G., Lairson, L.L., Gilbert, M., Lim, D., Wakarchuk, W.W., Withers, S.G. and Strynadka, N.C. *Nat. Struct. Mol. Biol.*, **11**, 163.
6. Mulichak, A.M., Losey, H.C., Walsh, C.T. and Garavito, R.M. (2001). *Structure*, **9**, 547.
7. Pedersen, L.C., Dong, J., Taniguchi, F., Kitagawa, H., Krahn, J.M., Pedersen, L.G., Sugahara, K. and Negishi, M. (2003). *J. Biol. Chem.*, **278**, 14420.
8. Patenaude, S.I., Seto, N.O., Borisova, S.N., Szpacenko, A., Marcus, S.L. and Palcic, M.M. and Evans, S.V. (2002). *Nat. Struct. Biol.*, **9**, 685.
9. Lairson, L.L., Chiu, C.P., Ly, H.D., He, S., Wakarchuk, W.W., Strynadka, N.C. and Withers, S.G. (2004). *J. Biol. Chem.*, **279**, 28339.

10. Ly, H.D., Lougheed, B., Wakarchuk, W.W. and Withers, S.G. (2002). *Biochemistry*, **41**, 5075.

11. Monegal, A. and Planas, A. (2006). *J. Am. Chem. Soc.*, **128**, 16030.

12. Chen, L., Men, H., Ha, S., Ye, X.-Y., Brunner, L., Hu, Y. and Walker, S. (2002). *Biochemistry*, **41**, 6824.

13. Kearns, A.E., Campbell, S.C., Westley, J. and Schwartz, N.B. (1991). *Biochemistry*, **30**, 7477.

14. Quirós, L.M., Carbajo, R.J., Braña, A.F. and Salas, J.A. (2000). *J. Biol. Chem.*, **275**, 11713.

15. Breton, C., Bettler, E., Joziasse, D.H., Geremia, R.A. and Imberty, A. (1998). *J. Biochem.*, **123**, 1000.

16. Coutinho, P.M., Deleury, E., Davies, G.J. and Henrissat, B. (2003). *J. Mol. Biol.*, **328**, 307.

17. Lin, C.-H., Murray, B.W., Ollmann, I.R. and Wong, C.-H. (1997). *Biochemistry*, **36**, 780.

18. Zamyatina, A., Gronow, S., Oertelt, C., Puchberger, M., Brade, H. and Kosma, P. (2000). *Angew. Chem. Int. Ed.*, **39**, 4150.

19. Waechter, C.J. and Lennarz, W.J. (1976). *Annu. Rev. Biochem.*, **45**, 95.

20. Besra, G.S., Morehouse, C.B., Rittner, C.M., Waechter, C.J. and Brennan, P.J. (1997). *J. Biol. Chem.*, **272**, 18460.

21. Scherman, M.S., Kalbe-Bournonville, L., Bush, D., Xin, Y., Deng, L. and McNeil, M. (1996). *J. Biol. Chem.*, **271**, 29652.

22. Silberstein, S. and Gilmore, R. (1996). *FASEB J.*, **10**, 849.

23. van Heijenoort, J. (2001). *Glycobiology*, **11**, 25R.

24. Lovering, A.L., de Castro, L.H., Lim, D. and Strynadka, N.C. (2007). *Science*, **315**, 1402.

25. Hagopian, A. and Eylar, E.H. (1968). *Arch. Biochem. Biophys.*, **128**, 422.

26. Rose, N.L., Completo, G.C., Lin, S.J., McNeil, M., Palcic, M.M. and Lowary, T.L. (2006). *J. Am. Chem. Soc.*, **128**, 6721.

27. Kremer, L., Dover, L.G., Morehouse, C., Hitchin, P., Everett, M., Morris, H.R., Dell, A., Brennan, P.J., McNeil, M.R., Flaherty, C., Duncan, K. and Besra, G.S. (2001). *J. Biol. Chem.*, **276**, 26430.

28. Beláňová, M., Dianišková, P., Brennan, P.J., Completo, G.C., Rose, N.L., Lowary, T.L. and Mikušová, K. (2008). *J. Bacteriol.*, **190**, 1141.

29. Gastinel, L.N., Cambillau, C. and Bourne, Y. (1999). *EMBO J.*, **18**, 3546.

30. Brodbeck, U., Denton, W.L., Tanahashi, N. and Ebner, K.E. (1967). *J. Biol. Chem.*, **242**, 1391.

31. Ramakrishnan, B. and Qasba, P.K. (2001). *J. Mol. Biol.*, **310**, 205.

32. Campbell, J.A., Davies, G.J., Bulone, V. and Henrissat, B. (1997). *Biochem. J.*, **326**, 929.

33. Coutinho, P.M. and Henrissat, B. *Carbohydrate-Active Enzymes Server at URL: http://www.cazy.org* accessed 19/6/08.

34. Ullman, C.G. and Perkins, S.J. (1997). *Biochem. J.*, **326**, 929.

35. Davies, G.J., Gloster, T.M. and Henrissat, B. (2005). *Curr. Opin. Struct. Biol.*, **15**, 637.

36. Breton, C., Šnajdrová, L., Jeanneau, C., Koca, J. and Imberty, A. (2006). *Glycobiology*, **16**, 29R.

37. Ünligil, U.M. and Rini, J.M. (2000). *Curr. Opin. Struct. Biol.*, **10**, 510.

38. Hu, Y. and Walker, S. (2002). *Chem. Biol.*, **9**, 1287.

39. Yuan, Y., Barrett, D., Zhang, Y., Kahne, D., Sliz, P. and Walker, S. (2007). *Proc. Natl. Acad. Sci. USA*, **104**, 5348.

40. Geigenberger, P. and Stitt, M. (1993). *Planta*, **189**, 329.

41. Cardini, C.E., Leloir, L.F. and Chiriboga, J. (1955). *J. Biol. Chem.*, **214**, 149.

42. Zervosan, A. and Elling, L. (1996). *J. Am. Chem. Soc.*, **118**, 1836.

43. Zhang, C., Griffith, B.R., Fu, Q., Albermann, C., Fu, X., Lee, I. K., Li, L. and Thorson, J.S. (2006). *Science*, **313**, 1291.

44. Melançon, C.E., III, Thibodeaux, C.J. and Liu, H.-w. (2006). *ACS Chem. Biol.*, **1**, 499.

45. Qian, X. and Palcic, M.M. (2000). In *Carbohydrates in Chemistry and Biology* (B. Ernst, G.W. Hart, P. Sinaÿ, eds.) Wiley-VCH: Weinheim, Vol. 3, 2, p. 293.

46. Knapp, S. (1995). *Chem. Rev.*, **95**, 1859.

47. Ostash, B. and Walker, S. (2005). *Curr. Opin. Chem. Biol.*, **9**, 459.
48. Halliday, J., McKeveney, D., Muldoon, C., Rajaratnam, P. and Meutermans, W. (2006). *Biochem. Pharmacol.*, **71**, 957.
49. Taylor, J.G., Li, X., Oberthur, M., Zhu, W. and Kahne, D.E. (2006). *J. Am. Chem. Soc.*, **128**, 15084.
50. Welzel, P. (2005). *Chem. Rev.*, **105**, 4610.
51. Dähn, U., Hagenmaier, H., Höhne, H., König, W.A., Wolf, G. and Zähner, H. (1976). *Arch. Microbiol.*, **107**, 143.
52. Cabib, E. (1991). *Antimicrob. Agents Chemother.*, **35**, 170.
53. Debono, M. and Gordee, R.S. (1994). *Annu. Rev. Microbiol.*, **48**, 471.
54. Varki, A., Cummings, R., Esko, J., Hart, G. and Marth, J., eds. (1999). *Essentials of Glycobiology*. New York: Cold Spring Harbor.
55. Journet, M., Cai, D., DiMichele, L.M., Hughes, D.L., Larsen, R.D., Verhoeven, T.R. and Reider, P.J. (1999). *J. Org. Chem.*, **64**, 2411.
56. Letscher-Bru, V. and Herbrecht, R. (2003). *J. Antimicrob. Chemother.*, **51**, 513.
57. Hendrickson, T.L., Spencer, J.R., Kato, M. and Imperiali, B. (1996). *J. Am. Chem. Soc.*, **118**, 7636.
58. Eason, P.D. and Imperiali, B. (1999). *Biochemistry*, **38**, 5430.
59. Qiao, L., Murray, B.W., Shimazaki, M., Schultz, J.E. and Wong, C.-H. (1996). *J. Am. Chem. Soc.*, **118**, 7653.
60. Müller, B., Schaub, C. and Schmidt, R.R. (1998). *Angew. Chem. Int. Ed.*, **37**, 2893.
61. Palcic, M.M., Heerze, L.D., Srivastava, O.P. and Hindsgaul, O. (1989). *J. Biol. Chem.*, **264**, 17174.
62. Hashimoto, H., Endo, T. and Kajihara, Y. (1997). *J. Org. Chem.*, **62**, 1914.
63. Lowary, T.L. and Hindsgaul, O. (1994). *Carbohydr. Res.*, **251**, 33.
64. Laferté, S., Chan, N.W., Sujino, K., Lowary, T.L. and Palcic, M.M. (2000). *Eur. J. Biochem.*, **267**, 4840.
65. Lee, L.V., Mitchell, M.L., Huang, S.J., Fokin, V.V., Sharpless, K.B. and Wong, C.H. (2003). *J. Am. Chem. Soc.*, **125**, 9588.
66. Mourant, A.E. (1946). *Nature*, **4007**, 237.
67. Hang, H.C. and Bertozzi, C.R. (2005). *Bioorg. Med. Chem.*, **13**, 5021.
68. Hang, H.C., Yu, C., Ten Hagen, K.G., Tian, E., Winans, K.A., Tabak, L.A. and Bertozzi, C.R. (2004). *Chem. Biol.*, **11**, 337.
69. Belanger, A.E., Besra, G.S., Ford, M.E., Mikusova, K., Belisle, J.T., Brennan, P.J. and Inamine, J.M. (1996). *Proc. Natl. Acad. Sci. USA*, **93**, 11919.
70. Lee, R.E., Protopopova, M., Crooks, E., Slayden, R.A., Terrot, M. and Barry, C.E., 3rd *J. Comb. Chem.*, **5**, 172.
71. Wennekes, T., Berg, R.J., Donker, W., Marel, G.A., Strijland, A., Aerts, J.M. and Overkleeft, H.S. (2007). *J. Org. Chem.*, **72**, 1088.
72. Elstein, D., Hollak, C., Aerts, J.M.F.G., van Weely, S., Maas, M., Cox, T.M., Lachmann, R.H., Hrebicek, M., Platt, F.M., Butters, T.D., Dwek, R.A. and Zimran, A. (2004). *J. Inherit. Metab. Dis.*, **27**, 757.
73. Meinke, P.T. (2001). *J. Med. Chem.*, **44**, 641.
74. Hajjar, N.P. and Casida, J.E. (1978). *Science*, **200**, 1499.
75. Wilson, T.G. and Cryan, J.R. (1997). *J. Exp. Zool.*, **278**, 37.
76. Gordon, R.D., Sivarajah, P., Satkunarajah, M., Ma, D., Tarling, C.A., Vizitiu, D., Withers, S.G. and Rini, J.M. (2006). *J. Mol. Biol.*, **360**, 67.
77. Hayashi, T., Murray, B.W., Wang, R. and Wong, C.-H. (1997). *Bioorg. Med. Chem.*, **5**, 497.
78. Watts, A.G. and Withers, S.G. (2004). *Can. J. Chem.*, **82**, 1581.
79. Hartman, M.C., Jiang, S., Rush, J.S., Waechter, C.J. and Coward, J.K. (2007). *Biochemistry*, **46**, 11630.

80. Zou, W. (2005). *Curr. Top. Med. Chem.*, **5**, 1363.
81. Qian, X. and Palcic, M.M. (2000) In *Carbohydrates in Chemistry and Biology* (B. Ernst, G.W. Hart, P. Sinaÿ, eds.) Wiley-VCH: Weinheim, Vol. 2, 3, p. 293.
82. Centrone, C.A. and Lowary, T.L. (2002). *J. Org. Chem.*, **67**, 8862.
83. Zhu, X., Stolz, F. and Schmidt, R.R. (2004). *J. Org. Chem.*, **69**, 7367.
84. Okayama, M., Kimata, K. and Suzuki, S. (1973). *J. Biochem.*, **74**, 1069.
85. Sarkar, A.K., Fritz, T.A., Taylor, W.H. and Esko, J.D. (1995). *Proc. Natl. Acad. Sci. USA*, **92**, 3323.
86. Portner, A., Etchison, J.R., Sampath, D. and Freeze, H.H. (1996). *Glycobiology*, **6**, 7.
87. Robinson, H.C., Brett, M.J., Tralaggan, P.J., Lowther, D.A. and Okayama, M. (1975). *Biochem. J.*, **148**, 25.
88. Neville, D.C., Field, R.A. and Ferguson, M.A. (1995). *Biochem. J.*, **307**, 791.
89. Kuan, S.F., Byrd, J.C., Basbaum, C. and Kim, Y.S. (1989). *J. Biol. Chem.*, **264**, 19271.
90. Danac, R., Ball, L., Gurr, S.J., Muller, T. and Fairbanks, A.J. (2007). *ChemBioChem*, **8**, 1241.
91. Buchsel, R., Hassels-Vischer, B., Tauber, R. and Reutter, W. (1980). *Eur. J. Biochem.*, **111**, 445.
92. Geilen, C.C., Kannicht, C., Orthen, B., Heidrich, C., Paul, C., Grunow, D., Nuck, R. and Reutter, W. (1992). *Arch. Biochem. Biophys.*, **296**, 108.
93. Mahal, L.K., Charter, N.W., Angata, K., Fukuda, M., Koshland, D.E., Jr. and Bertozzi, C.R. (2001). *Science*, **294**, 380.
94. Kayser, H., Zeitler, R., Kannicht, C., Grunow, D., Nuck, R. and Reutter, W. (1992). *J. Biol. Chem.*, **267**, 16934.
95. Keppler, O.T., Stehling, P., Herrmann, M., Kayser, H., Grunow, D., Reutter, W. and Pawlita, M. (1995). *J. Biol. Chem.*, **270**, 1308.
96. Laughlin, S.T., Agard, N.J., Baskin, J.M., Carrico, I.S., Chang, P.V., Ganguli, A.S., Hangauer, M.J., Lo, A., Prescher, J.A. and Bertozzi, C.R. (2006). *Methods Enzymol.*, **415**, 230.
97. Saxon, E. and Bertozzi, C.R. (2001). *Annu. Rev. Cell Dev. Biol.*, **17**, 1.
98. Saxon, E. and Bertozzi, C.R. (2000). *Science*, **287**, 2007.
99. Sampathkumar, S.G., Li, A.V., Jones, M.B., Sun, Z. and Yarema, K.J. (2006). *Nat. Chem. Biol.*, **2**, 149.
100. Mahal, L.K., Yarema, K.J. and Bertozzi, C.R. (1997). *Science*, **276**, 1125.
101. Hsu, T.L., Hanson, S.R., Kishikawa, K., Wang, S.K., Sawa, M. and Wong, C.H. (2007). *Proc. Natl. Acad. Sci. USA*, **104**, 2614.
102. Saxon, E., Luchansky, S.J., Hang, H.C., Yu, C., Lee, S.C. and Bertozzi, C.R. (2002). *J. Am. Chem. Soc.*, **124**, 14893.
103. Agard, N.J., Baskin, J.M., Prescher, J.A., Lo, A. and Bertozzi, C.R. (2006). *ACS Chem. Biol.*, **1**, 644.
104. Baskin, J.M. and Bertozzi, C.R. (2007). *QSAR Comb. Sci.*, **26**, 1211.
105. Baskin, J.M., Prescher, J.A., Laughlin, S.T., Agard, N.J., Chang, P.V., Miller, I.A., Lo, A., Codelli, J.A. and Bertozzi, C.R. (2007). *Proc. Natl. Acad. Sci. USA*, **104**, 16793.
106. Ning, X., Guo, J., Wolfert, M.A. and Boons, G.-J. (2008). *Angew. Chem. Int. Ed.*, **47**, 2253.
107. Goon, S., Schilling, B., Tullius, M.V., Gibson, B.W. and Bertozzi, C.R. (2003). *Proc. Natl. Acad. Sci. USA*, **100**, 3089.
108. Vocadlo, D.J., Hang, H.C., Kim, E.J., Hanover, J.A. and Bertozzi, C.R. (2003). *Proc. Natl. Acad. Sci. USA*, **100**, 9116.
109. Hang, H.C. and Bertozzi, C.R. (2001). *J. Am. Chem. Soc.*, **123**, 1242.
110. Rabuka, D., Hubbard, S.C., Laughlin, S.T., Argade, S.P. and Bertozzi, C.R. (2006). *J. Am. Chem. Soc.*, **128**, 12078.
111. Prescher, J.A., Dube, D.H. and Bertozzi, C.R. (2004). *Nature*, **430**, 873.
112. Koeller, K.M. and Wong, C.H. (2000). *Chem. Rev.*, **100**, 4465.

113. Qian, X., Sujino, K., Palcic, M.M. and Ratcliffe, R.M. (2002). *J. Carbohydr. Chem.*, **21**, 911.
114. Di Virgilio, S., Glushka, J., Moremen, K. and Pierce, M. (1999). *Glycobiology*, **9**, 353.
115. Lin, C.H., Shimazaki, M., Wong, C.H., Koketsu, M., Juneja, L.R. and Kim, M. (1995). *Bioorg. Med. Chem.*, **3**, 1625.
116. Scheppokat, A.M., Bretting, H. and Thiem, J. (2003). *Carbohydr. Res.*, **338**, 2083.
117. Blixt, O. and Norberg, T. (1998). *J. Org. Chem.*, **63**, 2705.
118. Ratcliffe, R.M., Kamath, V.P., Yeske, R.E., Gregson, J.M., Fang, Y.R. and Palcic, M.M. (2004). *Synthesis*, 2293.
119. Rich, J.R., Szpacenko, A., Palcic, M.M. and Bundle, D.R. (2004). *Angew. Chem. Int. Ed.*, **43**, 613.
120. Lairson, L.L., Watts, A.G., Wakarchuk, W.W. and Withers, S.G. (2006). *Nat. Chem. Biol.*, **2**, 724.
121. Stick, R.V., Stubbs, K.A. and Watts, A.G. (2004). *Aust. J. Chem.*, **57**, 779.
122. Bülter, T. and Elling, L. (1999). *Glycoconj. J.*, **16**, 147.
123. Koeller, K.M. and Wong, C.-H. (2000) In *Carbohydrates in Chemistry and Biology* (B. Ernst, G. Hart, P. Sinaÿ, eds) Wiley-VCH: Weinheim, Vol. 3, 2, p. 663.
124. Wong, C.-H., Haynie, S.L. and Whitesides, G.M. (1982). *J. Org. Chem.*, **47**, 5416.
125. Lougheed, B., Ly, H.D., Wakarchuk, W.W. and Withers, S.G. (1999). *J. Biol. Chem.*, **274**, 37717.
126. Nahalka, J., Liu, Z., Chen, X. and Wang, P.G. (2003). *Chem. Eur. J.*, **9**, 372.
127. Zhang, J., Chen, X., Shao, J., Liu, Z., Kowal, P., Lu, Y. and Wang, P.G. (2003). *Methods Enzymol.*, **362**, 106.
128. Chen, X., Fang, J., Zhang, J., Liu, Z., Shao, J., Kowal, P., Andreana, P. and Wang, P.G. (2001). *J. Am. Chem. Soc.*, **123**, 2081.
129. Endo, T. and Koizumi, S. (2000). *Curr. Opin. Struct. Biol.*, **10**, 536.
130. Herrmann, G.F., Wang, P., Shen, G.-J. and Wong, C.-H. (1994). *Angew. Chem. Int. Ed.*, **33**, 1241.
131. Bettler, E., Samain, E., Chazalet, V., Bosso, C., Heyraud, A., Joziasse, D.H., Wakarchuk, W.W., Imberty, A. and Geremia, A.R. (1999). *Glycoconj. J.*, **16**, 205.
132. Chen, X., Liu, Z., Zhang, J., Zhang, W., Kowal, P. and Wang, P.G. (2002). *ChemBioChem*, **3**, 47.
133. Chen, X., Zhang, J., Kowal, P., Liu, Z., Andreana, P.R., Lu, Y. and Wang, P. (2001). G. *J. Am. Chem. Soc.*, **123**, 8866.
134. Koizumi, S., Endo, T., Tabata, K. and Ozaki, A. (1998). *Nat. Biotechnol.*, **16**, 847.
135. Endo, T., Koizumi, S., Tabata, K. and Ozaki, A. (2000). *Appl. Microbiol. Biotechnol.*, **53**, 257.

Chapter 9

Disaccharides, Oligosaccharides and Polysaccharides

Monosaccharides may be linked together in an almost limitless number of ways to form disaccharides, oligosaccharides and polysaccharides. There are over 134 million ways to link the eight D-hexopyranoses together into a linear hexasaccharide and, if different ring forms (e.g. furanose), L-sugars, deoxy sugars, aminodeoxy sugars, chain branching and assorted modifying groups are allowed, the number of potential structures increases combinatorially. That said there is a much smaller set of structures commonly found in nature. This chapter will cover the common di- and trisaccharides and oligomers and polymers derived from (mostly) single carbohydrate monomers. Later chapters will cover more complex oligo- and polysaccharides comprised of more than one sugar monomer and bearing various functional group adornments. To begin we provide a list of the commonly accepted names for a range of disaccharides (Table 1); many of these compounds are not found in appreciable amounts in nature but are derived by controlled hydrolysis of more complex polysaccharides.

Cellulose and Cellobiose

The molecular constitution of cellulose was first described by Payen in 1838.[a] Cellulose is a β-1,4-linked polymer of D-glucose and is one of the most abundant biopolymers on earth. It is the major structural component of green plants in which it is found in primary and secondary cell walls, and it is estimated that 10^{15} kg of cellulose are synthesized annually. Cellulose is the major component of plant secondary walls and thus the main component of plant fibres. In the cell wall, the cellulose of plant fibres is embedded in a matrix of other polysaccharides, forming a biocomposite. Aside from cellulose, other components of the cell wall include

---

[a] Anselme Payen (1795–1871), professorships at the Ecole Centrale des Arts et Manufactures, Châtenay-Malabry, and Conservatoire National des Arts et Métiers, France. Discoverer of the enzyme diastase that catalyzes the conversion of starch into maltose. The practice of adding the suffix '-ase' to denote an enzyme, and '-ose' to denote a sugar is attributed to Payen. Ironically, the term 'cellulose' was not originally used by Payen but was applied by a panel of the French Academy convened to evaluate Payen's work in 1839.[1]

References start on page 339

Table 1 Trivial names of common disaccharides

allolactose	β-D-Galp-(1,6)-D-Glc
cellobiose	β-D-Glcp-(1,4)-D-Glc
chitobiose	β-D-GlcpNAc-(1,4)-D-GlcNAc
gentiobiose	β-D-Glcp-(1,6)-D-Glc
isomaltose	α-D-Glcp-(1,6)-D-Glc
isomaltulose	α-D-Glcp-(1,6)-D-Fru
kojibiose	α-D-Glcp-(1,2)-D-Glc
lactose	β-D-Galp-(1,4)-D-Glc
lactulose	β-D-Galp-(1,4)-D-Fru
laminaribiose	β-D-Glcp-(1,3)-D-Glc
leucrose	α-D-Glcp-(1,5)-D-Fru
maltose	α-D-Glcp-(1,4)-D-Glc
maltulose	α-D-Glcp-(1,4)-D-Fru
mannobiose	β-D-Manp-(1,4)-D-Man
melibiose	α-D-Galp-(1,6)-D-Glc
nigerose	α-D-Glcp-(1,3)-D-Glc
planteobiose	α-D-Galp-(1,6)-D-Fru
sophorose	β-D-Glcp-(1,2)-D-Glc
sucrose	α-D-Glcp-(1,2)-β-D-Fruf
trehalose	α-D-Glcp-(1,1)-α-D-Glcp
trehalulose	α-D-Glcp-(1,1)-D-Fru
turanose	α-D-Glcp-(1,3)-D-Fru
xylobiose	β-D-Xylp-(1,4)-D-Xyl

hemicelluloses (xyloglucans in dicotyledons, glucuronoxylans in hardwoods, gluco-mannan and arabinoxylans in softwoods and arabinoxylans in grasses), pectins (galacturonate-rich branched polysaccharides), and lignins (complex phenolic polymers of substituted cinnamyl alcohols).[2] Cellulose forms long inelastic micro-fibrils that wrap around the cell and rigidify the cell wall, providing the turgor that allows plants to stand upright.[3] While green plants are the major producers of cellulose, tunicates and certain bacteria (e.g. *Acetobacter xylinium*, reclassified as *Gluconacetobacter xylinus*) have developed the ability to synthesize cellulose.

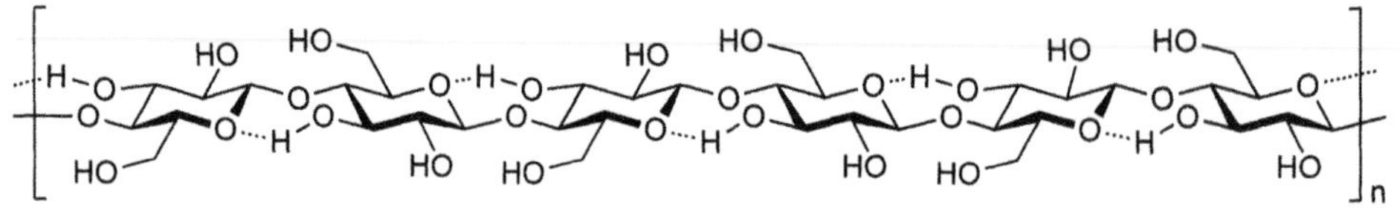

Cellulose structure has been the subject of much investigation, and a complete crystallographic description remains elusive. There is evidence from powder diffraction studies of cellulose for the presence of a hydrogen bond between the endocyclic oxygen of one sugar and the 3-OH of its neighbour. This hydrogen bond results from a structure where every second sugar residue is 'flipped' with respect to the first, forming a ribbon-like chain. Each of these chains pack together in the crystal in either a parallel or an anti-parallel sense, forming cellulose I or II, respectively.[4] Cellulose I is the natural form found in plant fibres, and cellulose II is formed upon recrystallization of cellulose.

Cellulose biosynthesis in plants is orchestrated by glycosyltransferases termed cellulase synthases, using UDP-glucose as a substrate, itself derived from sucrose. Cellulose synthase enzymes are believed to be arranged in sixfold symmetric rosettes, each comprised of six catalytic subunits, in the plasma membrane of the plant cell.[5] Consistent with this biosynthetic hypothesis, individual cellulose fibres are proposed to consist of 36 parallel β-glucan chains.

Cellulose is the major constituent of paper and is therefore used heavily by the pulp and paper industry. In the textile industry, cellulose is the major component of fabrics made from cotton, linen and other plant fibres. In recent years there has been great interest in moving away from chemical-intensive traditional bleaching processes for fibre modification, towards enzyme-based procedures. Glycoside hydrolases, termed xylanases, are useful for fibre modification and assist in the removal of lignins from crude kraft[b] pulps by degrading unwanted hemicellulose components.[6,7]

Cellulose is particularly challenging to degrade, and organisms capable of doing so require many enzymes to achieve the task. Enzyme complexes, termed cellulosomes, have been found in cellulolytic fungi and bacteria.[8] These complexes contain *endo*- and *exo*-cellulases, and β-glucosidases, as well as carbohydrate-binding proteins that are commonly fused to a glycoside hydrolase. Vertebrates are unable to degrade cellulose, and the digestive tracts of herbivores contain symbiotic cellulolytic bacteria. Similarly, termites achieve the degradation of cellulose either through gut microflora or through the practice of fungiculture.[9] In fungiculture, termites apply pre-digested plant material to a fungal colony, and older cultured material is removed and consumed by the termite.

Cellulolytic enzymes may be of potential use in cellulose degradation for bioethanol production. However, to date, limited success has been achieved in developing commercial processes capable of degrading cellulose economically, with the bulk of bioethanol being produced from more easily fermented precursors containing starch, such as grains. Cellulases have been developed as additives for laundry detergents.[10] During wear, microfibres rise from the surface of cotton thread and affect the feel and brightness of cotton fabrics. Alkali-stable cellulases have been

---

[b] Kraft. *noun* (German), strength or power.

References start on page 339

developed that degrade these microfibres, restoring the feel and brightening the appearance of the garment.

Cellulose is a feedstock for the preparation of a wide range of materials. The nitric acid ester of cellulose, cellulose nitrate or 'nitrocellulose', is the oldest cellulose derivative and is prepared by treatment of cellulose with nitric acid, sulfuric acid and water.[11] It has excellent film-forming properties and is used in lacquers and thermoplastics. A combination of nitrocellulose and camphor provides celluloid, the first synthetic thermoplastic material. Nitrocellulose is combustible and explosive, and was used in gun cotton and various military devices. However, its shock sensitivity led to a range of industrial disasters in plants and mines. Nobel[c] discovered a more stable combination of cellulose nitrate, termed blasting gelatine or gelignite, comprised of 'nitroglycerine', sodium nitrate and wood pulp, and also a smokeless military explosive termed ballistite, composed of cellulose nitrate and nitroglycerine blended with a stabilizer.[11] Cellulose acetate is formed by acetylation of cellulose with a mixture of acetic acid, acetic anhydride and sulfuric acid.[11] Cellulose acetate is used as an electrical insulating film, as a lacquer, in adhesives and in the production of cellulose acetate fibres, which are used in the textile industry. Cellulose may be converted to rayon fibres or cellophane sheets by the viscose process, which involves dissolution of cellulose in alkali and carbon disulfide, and reprecipitation using acid.[11]

Carboxymethylcellulose is an ether derivative of cellulose prepared by reaction of cellulose and chloroacetic acid in alkali.[12] It is used in the food industry as a thickener and stabilizer, in the mining industry in drilling muds, and in the paper and textile industry to alter surface characteristics of fibres. Controlled acetolysis of cellulose with acetic anhydride and sulfuric acid leads to α-cellobiose octaacetate.[13]

Starch, Amylopectin, Amylose and Maltose

Starch is the major source of dietary energy for the world's human population. Plants produce a mixture of α-glucans that are deposited in the cytoplasm as insoluble, semicrystalline granules and that act as an energy storage reserve. Starch granules are composed of a linear polymer, termed amylose, of several thousand α-1,4-linked D-glucose residues, and a branched polymer, termed amylopectin, consisting of mainly α-1,4-linked D-glucose residues, with an α-1,6-linked branch every 24–30 residues.[14] Amylopectin molecules are much larger than amylose; each amylopectin molecule contains up to 10^6 D-glucose units. Amylose adopts a helical conformation

---

[c] Alfred Nobel (1833–1896), Swedish inventor and industrialist noted for the inventions of dynamite, gelignite and ballistite, and the creation of the Nobel Foundation by bequest of his will.

with six D-glucose residues per turn, an outer diameter of 13 Å and an inner diameter of 5 Å. This inner hydrophobic cavity can occlude lipophilic molecules such as butanol;[14] the starch–iodine complex provides a sensitive means for the detection of iodine and results from the binding of polyiodide ion (likely to be I_5^-) within this cavity, thereby generating a characteristic blue colour.[15]

amylose

amylopectin branchpoint
fragment

The biosynthesis of amylose and amylopectin in plants requires a glycosyltransferase (starch synthase), a transglycosidase (branching enzyme) and a glycoside hydrolase (debranching enzyme).[16] Starch synthase condenses the glycosyl donor ADP-glucose to give α-1,4-linked linear chains of D-glucose residues. ADP-glucose itself is synthesized from glucose-1-phosphate and ATP by the action of ADP-glucose pyrophosphorylase.[16] Starch-branching enzymes cleave internal α-1,4-linkages within the long chains and transfer the released chains to the 6-position of a terminal sugar residue, resulting in the formation of branch points within amylopectin. Debranching enzymes are involved in amylopectin biosynthesis and cleave α-1,6-glucose linkages, removing branches that are inappropriately positioned within amylopectin and that affect its ability to pack into a dense, semicrystalline structure.

Starch is widely used in foods for controlling texture and structure, and for making gels. Maltose syrups and maltodextrins are produced from starch by partial hydrolysis using acid or enzymes and consist of mixtures of D-glucose, maltose and maltose oligosaccharides. They are used as a fermentation substrate in brewing, for improving 'mouth feel' of foods, and for altering the physical properties of foodstuffs through their gelling and crystallization prevention properties.[17] Starch may be isomerized to form high-fructose corn syrups, which possess sweetness similar to invert syrup derived from sucrose (see Epilogue). Such syrups can be used to maintain the texture of baked goods during storage and to prevent crystallization in frozen foods such as ice cream.[18] Aside from its use as a dietary component, starch

References start on page 339

finds various industrial applications as a textile sizing agent, in adhesives, in the pharmaceutical industry as an excipient and in cosmetics. Finally, a growing use of starch is as a renewable feedstock providing the so-called biofuels, which are an alternative to fuels derived from fossil reserves. The most significant biofuel is bioethanol, which in the United States is in large part derived from fermentation of starch from corn.[19] In the United States alone, 15 megalitres of bioethanol are produced annually.[20] Despite much interest, no process for conversion of cellulosic material into ethanol is yet in large-scale use.

Glycogen

Glycogen is a branched polymer of D-glucose that serves as a store of energy and carbon in vertebrates, and is found largely in the liver and skeletal muscle.[15] It possesses considerable structural similarity to amylopectin but is more highly branched, and is sometimes referred to as animal starch. Glycogen particles contain 10^7–10^8 D-glucose molecules per unit, and varying amounts of associated proteins. The basic polymerization occurs through α-1,4-linkages, with branch points introduced through α-1,6-linkages. After feeding in mammals, elevated blood glucose stimulates secretion of insulin, promoting glucose uptake into tissues and production of glycogen. When blood glucose levels fall, the hormone glucagon stimulates the breakdown of glycogen in the liver.

Glycogen particles contain at their core a dimeric protein termed glycogenin. Glycogenin is a glycosyltransferase that initiates glycogen biosynthesis and acts as a primer for the glycogen molecule.[21] Glycogenin catalyses the transfer of D-glucose from UDP-glucose to a tyrosine residue (Tyr194) on glycogenin to form an α-glucosyl tyrosine. Additional catalytic cycles lead to elongation of the α-glucosyl tyrosine, building an α-1,4-linked D-glucose chain of about 10 residues.[22] There is evidence suggesting that one molecule of glycogenin within the dimer glucosylates the other partner,[23] although an analysis of an X-ray crystal structure of glycogenin suggested that self-glucosylation may occur for the earliest glucosylation events.[21] Following the initiation of glycogen biosynthesis by glycogenin, an α-glucosyltransferase, glycogen synthase, elongates the short chains formed on glycogenin, leading to extended α-1,4-linked chains. A transglycosidase, glycogen-branching enzyme, introduces α-1,6-linked branches into the glycogen structure, through transfer of the terminal seven glucosyl residues from an α-1,4-linked chain to O6 of a glucosyl residue.[24,25] This reaction introduces a branch point, and both chains can then be extended by glycogen synthase. Introduction of branches increases the number of terminal D-glucose residues and affects the rate of release of D-glucose during glycogen metabolism. Additionally, introduction of branches into glycogen increases its solubility.

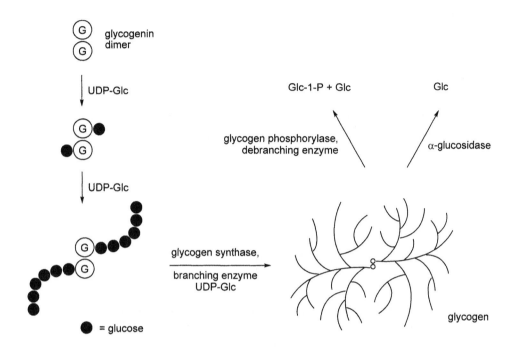

Glycogen degradation occurs through the action of three enzymes: glycogen phosphorylase, debranching enzyme and acid α-glucosidase.[22] Glycogen phosphorylase cleaves terminal α-1,4-glucosidic linkages by phosphorolysis, releasing glucose-1-phosphate; the enzyme stalls four residues from an α-1, 6-branch point, leading to the so-called limit dextran. The branch point blockage of limit dextran is removed by the action of debranching enzyme, a protein that contains two catalytic subunits.[26] One subunit is a transglycosidase that cleaves maltotriose from O6 of the remaining D-glucose residue and transfers it to the main α-1,4-linked chain, in a reaction that occurs with retention of configuration. The second subunit is an α-glucosidase that hydrolyses the remaining α-1,6-linked glucoside unit, releasing D-glucose with inversion of configuration. The final enzyme required for glycogen degradation is the lysosomal enzyme acid α-glucosidase, which catalyses the hydrolysis of α-1,4- and α-1,6-linkages in glycogen. Defects in acid α-glucosidase activity lead to the inherited condition type II glycogen storage disease (Pompe disease) in which glycogen accumulates in the lysosome.

Cyclodextrins

Cyclodextrins are α-1,4-linked cyclic oligosaccharides produced from starch by the action of cyclodextrin glucanotransferase, a transglycosidase that cleaves an α-1,4-linkage in amylose and forms the cycle (see Chapter 7).[27] The most commonly studied members are α-, β- and γ-cyclodextrins, which are comprised of six, seven and eight D-glucose residues, respectively. Cyclodextrins are toroidal with a hydrophobic interior and a hydrophilic exterior. They are able to form inclusion complexes with hydrophobic molecules; the odour freshener Febreze (Proctor and Gamble) contains cyclodextrins, which are claimed to entrap odour-producing lipophilic molecules. Cyclodextrin and cyclodextrin derivatives are used in the pharmaceutical industry for drug formulation; Sporanox (Janssen) is an intravenous formulation of the antifungal itraconazole with hydroxypropyl-β-cyclodextrin for treatment of fungal infections.

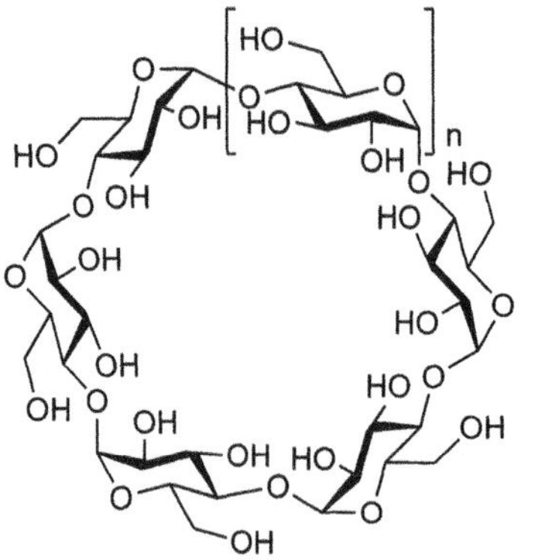

n = 1, α-cyclodextrin
n = 2, β-cyclodextrin
n = 3, γ-cyclodextrin

Sucrose, Sucrose Analogues and Sucrose Oligosaccharides

Sucrose is a non-reducing disaccharide abundant in photosynthetic organisms including plants. In higher plants, sucrose is a major product of photosynthesis and plays roles in growth and development. Sucrose is a major human macronutrient and is widely used in foods as a sweetening agent, stimulating consumption even in the absence of hunger through its pleasant taste. Sucralose is a halogenated sucrose derivative that is used as an artificial sweetener (Splenda) and possesses sweetness 600 times that of sucrose.[28] Hough and Phadnis discovered its sweetening properties by accident when they were planning to test chlorinated sugars as chemical insecticides. Hough asked Phadnis to test the powder but Phadnis thought that Hough asked him to *taste* it, and so he did![29] A non-reducing disaccharide closely related to sucrose, mannosylfructose, has been discovered as an osmolyte in *Agrobacterium tumefaciens*.[30]

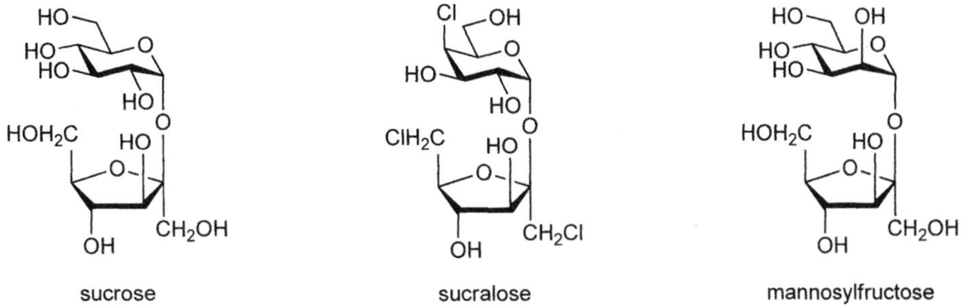

sucrose   sucralose   mannosylfructose

The biosynthesis of sucrose occurs in the cytosol of plant cells in two steps: sucrose-6-phosphate synthase condenses UDP-glucose and fructose-6-phosphate to give sucrose-6-phosphate, and sucrose-6-phosphate phosphatase cleaves the phosphate affording sucrose.[31] The first of these two reactions is reversible, but the phosphatase step results in an essentially irreversible process.

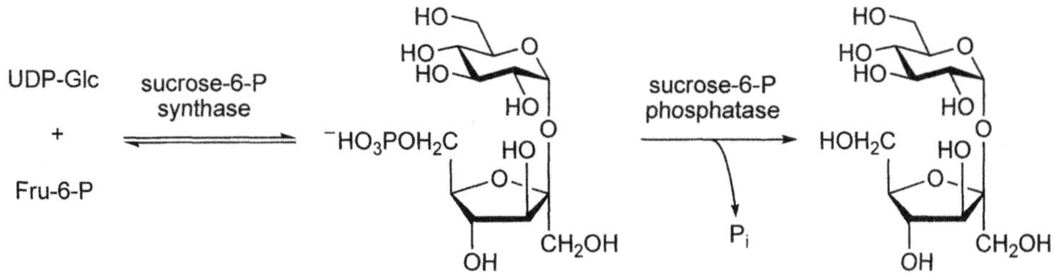

In higher plants, sucrose suffers two major fates: hydrolysis by fructofuranosidases (commonly called invertases, see below) to afford D-glucose and D-fructose, or reversible cleavage by sucrose synthase, which catalyses the reversible phosphorolysis of the glycosidic bond by UDP, affording UDP-glucose and D-fructose. Sucrose may also act as a primer for the synthesis of oligo- and polysaccharides based on a sucrose core (vide infra). Sucrose is rapidly cleaved by the gastrointestinal enzyme sucrase to D-glucose and D-fructose, which enter the glycolysis pathway (see Chapter 6). Sucrase produced by yeast is termed invertase, the name deriving from the observation that upon hydrolysis of sucrose to a mixture of D-glucose and D-fructose, an inversion of the sign of the optical rotation of the solution occurs. Invertase is used widely in industry to produce invert sugar, a 1:1 mixture of D-glucose and D-fructose with a sweetness slightly less than that of sucrose.[32]

A range of oligosaccharides containing a sucrose core is found in plants including the following: the so-called raffinose oligosaccharides including raffinose, stachyose, verbascose, and ajugose; lactosucrose; and melezitose, a glucosylated sucrose derivative. Lactosucrose, stachyose and raffinose are believed to be largely undigestible and may cause flatulence in humans.[33] They are used as low-calorie, non-cariogenic sugar substitutes in a range of foods.[34,35] Additionally, lactosucrose may have effects in promoting the growth of beneficial bacteria in the gut of humans and canine and feline pets.[36] The raffinose oligosaccharides are of almost ubiquitous occurrence in plant seeds and are biosynthesized from sucrose and UDP-galactose, by the intermediacy of various galactosylinositols, including galactinol.[37] These oligosaccharides, along with the fructans discussed later in this chapter, represent rare examples of oligosaccharides that are synthesized without the direct involvement of sugar 1-phosphate donors.[37]

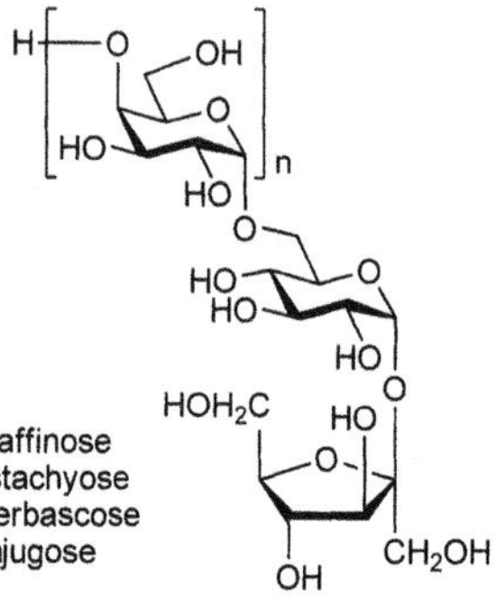

n = 1 raffinose
n = 2 stachyose
n = 3 verbascose
n = 4 ajugose

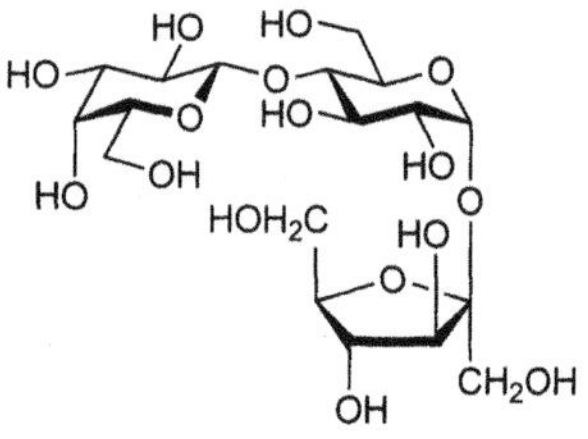

lactosucrose

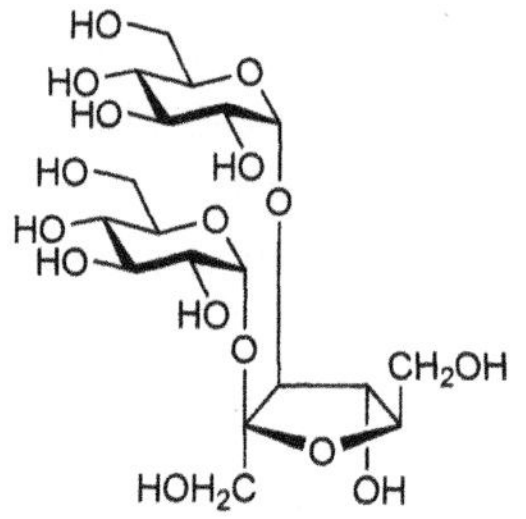

melezitose

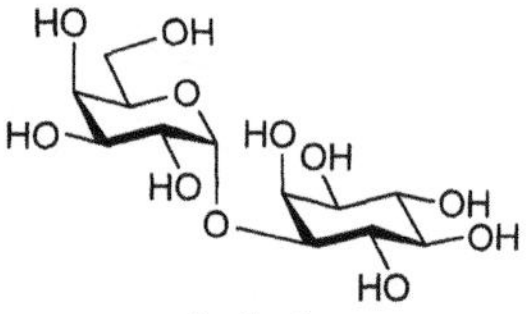

galactinol

Sucrose is a precursor for the production of a variety of disaccharides used in the food industry including isomaltulose (Palatinose) by rearrangement with the enzyme trehalulose synthase,[38] and isomalt (Palatinit), a mixture of two diastereoisomeric disaccharides derived by reduction of isomaltulose (see Chapter 3).[39]

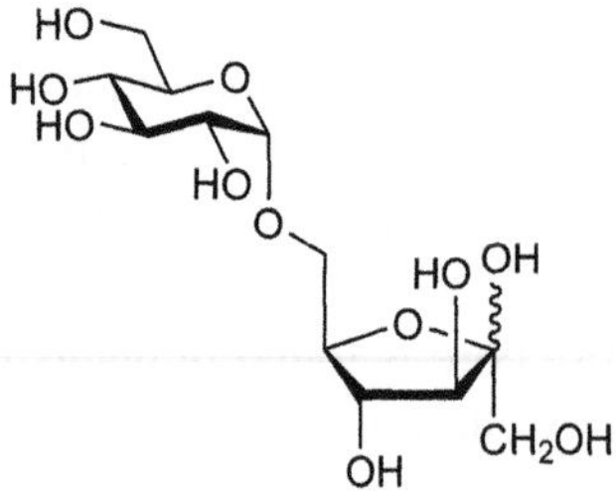

isomaltulose

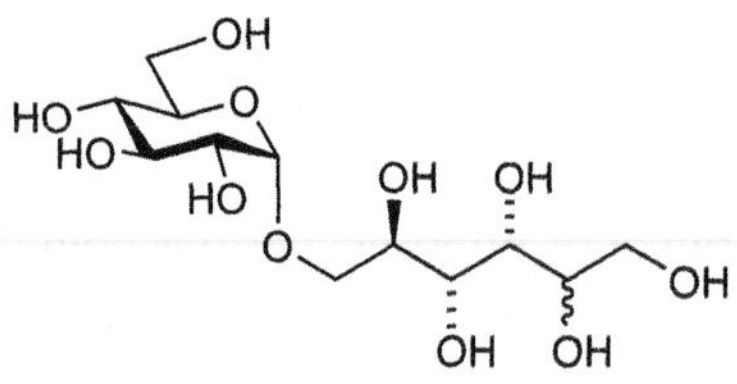

isomalt

Lactose and Milk Oligosaccharides

Milk contains a range of free oligosaccharides; the content varies between and within species. For all lactating mammals, lactose is the preponderant milk sugar and is found at $55-70\,\text{g}\,\text{l}^{-1}$ in human milk and $40-50\,\text{g}\,\text{l}^{-1}$ in cow's milk. Cow's milk contains very little other free sugars, but human milk contains a range of oligosaccharides in a total concentration of $5-8\,\text{g}\,\text{l}^{-1}$.[40] These oligosaccharides mostly contain lactose at their reducing terminus and are further modified with D-galactose, D-glucose, D-fucose and N-acetylneuraminic acid residues; lacto-N-tetraose $(0.5-1.5\,\text{g}\,\text{l}^{-1})$, lacto-N-fucopentaose I $(1.2-1.7\,\text{g}\,\text{l}^{-1})$, and Neu5Ac-lactose $(0.3-0.5\,\text{g}\,\text{l}^{-1})$ are among the most abundant. There is evidence that milk oligosaccharides have roles in defence against pathogenic microorganisms through prevention of adhesion processes in the early stages of infection and, in this regard, it is notable that the structures of milk oligosaccharides resemble those found on mucosal membranes as glycolipids and glycoproteins. Some variation in the structure of milk oligosaccharides is found within humans; for example only 77% of Caucasians possess the α-1,2-fucosyltransferase required for formation of the α-1,2-fucosyl linkage. Thus, only these females secrete lacto-N-fucopentaose, and others secrete α-1,2-fucosyl containing oligosaccharides. Other milk oligosaccharide structures are synthesized by-products of the Lewis genes, which are also involved in blood group oligosaccharide synthesis; such structures therefore occur only in carriers of the specific genes.

lactose

lacto-N-tetraose

Neu5Ac-lactose

lacto-N-fucopentaose

Lactose is biosynthesized by the glycosyltransferase lactose synthase from UDP-galactose and D-glucose. Lactose synthase is a two-protein complex containing a catalytic protein, β-GalT1, and an auxiliary protein, α-lactalbumin

References start on page 339

(which is produced in the mammary gland). In the absence of α-lactalbumin, the galactosyltransferase β-GalT1 has a preference for the transfer of galactosyl residues to terminal N-acetyl-β-glucosaminyl residues, thereby forming poly-1,4-β-N-acetyllactosamine core structures found in glycoproteins and glycosphingolipids (see Chapter 11).[41,42]

Lactose is digested through the action of the glycoside hydrolase lactase and is essential during infancy when milk is the main source of nutrition. Most mammals and most humans undergo a natural decline in levels of lactase following weaning in what is termed lactose non-persistance or hypolactasia.[43,44] In Asian and black groups, levels decline during early childhood, whereas in most Caucasians the decline occurs in later childhood or adolescence. Hypolactasia leads to incomplete digestion of lactose, resulting in a rise in osmotic load in the colon after milk product consumption, causing diarrhoea. In some groups, particularly northern Europeans, lactase activity persists into adulthood, allowing digestion of lactose as adults.[43] Lactose intolerance may also be caused by colonization of the gastrointestinal tract by lactose-metabolizing bacteria leading to gas production and aberrant flatulence.[44] The symptoms of lactose intolerance can be managed by reducing consumption of lactose-containing foods.

The bacterium *Escherichia coli* can metabolize lactose to D-glucose and D-galactose. The genetic system that controls this function, the *lac* operon, was described by Jacob and Monod in 1961.[d,45,46] The *lac* operon, a series of three genes (*lacZ*, *lacY* and *lacA*), a promoter and terminator region, and an operator, transports and hydrolytically cleaves lactose. In the absence of lactose as a food source, lactose permease LacY and the β-galactosidase LacZ are constitutively expressed at only low levels, as the *lac* repressor protein LacI, which binds to a region of the *lac* operon termed the *lac* operator, prevents their transcription by RNA polymerase. When lactose is supplied, the constitutively expressed LacZ β-galactosidase catalyses its hydrolysis and its transglycosidation, producing D-glucose and D-galactose, and a small amount of a lactose isomer, allolactose. Allolactose binds tightly to LacI, preventing the *lac* repressor binding to the *lac* operator and leading to high-level expression of lactose-metabolizing enzymes. This system of genetic regulation was one of the first to be described in any organism and forms the basis of an expression system used for recombinant production of proteins. In this system the gene of interest is cloned downstream of the *lac* promoter. Addition of the non-hydrolysable allolactose analogue isopropyl 1-thio-β-D-galactopyranoside (IPTG) leads to induction of protein expression by preventing the *lac* repressor LacI from binding to the *lac* operator.

---

[d] François Jacob (1920–), Ph.D. (1947) at the Faculty of Paris, D.Sc. (1951) at the Sorbonne, professorship at the Collège de France (1964); and Jacques Monod (1910–1976), Ph.D. (1941) at the Faculty of Paris, professorship at the Collège de France (1967). Nobel Prizes in Physiology or Medicine (1965).

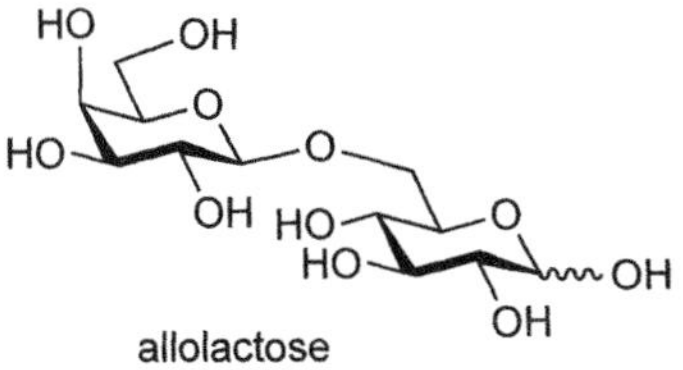

allolactose

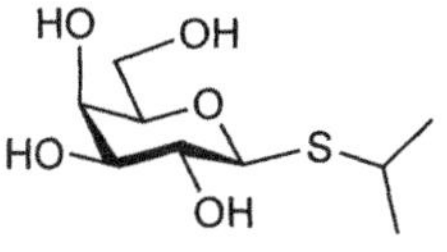

Fructans

Fructans are fructose polymers of four to several hundred residues that possess a sucrose core. Fructans are found widely in plants and appear to act as reserve carbohydrates, and also improve cold and drought tolerance.[47] Some bacteria and fungi can produce fructans. There are three main types of fructans: inulins, levans (or phleins) and gramminans. Linear inulin consists of β-1,2-linked fructose residues attached to a sucrose core and is found in members of the Asterales family (e.g. chicory). Inulin neo-series possesses two β-1,2-linked fructose chains attached to the sucrose core and is found in members of the Liliacae family (e.g. onions). Levan fructans are found in grasses and consist of a β-2,6-linked fructose chain attached to sucrose. Mixed fructans consisting of β-2,6-linked fructose residues with β-1,2-branches are termed gramminans and are also found in grasses.

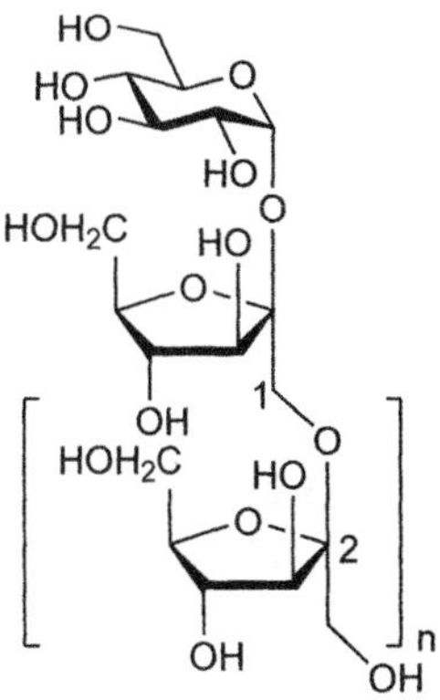

inulin

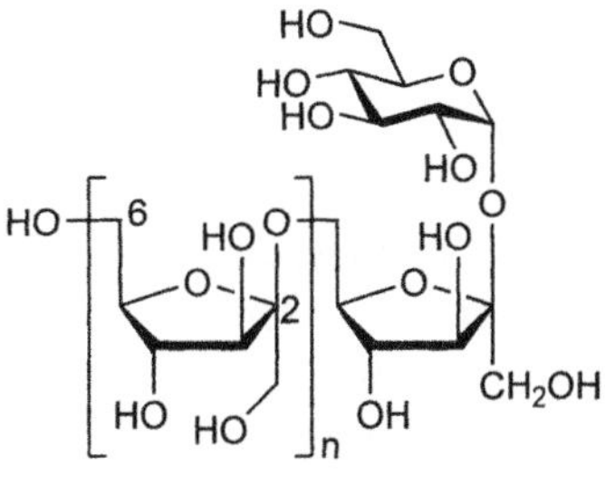

levan (phlein)

Fructans are used widely in the food industry. Chicory inulin is the main fructan food additive and forms an emulsion with water that possesses a similar texture to fat and is used in yoghurts and ice-cream as a fat substitute.[48] Fructans are classified as soluble fibres and are available as a food supplement. Fructans are touted to have the capability of promoting the growth of beneficial bacteria within the gastrointestinal tract and may provide relief from inflammatory bowel disease. 1-Kestose (isoketose) and nystose are inulin oligosaccharides comprised of one or two β-1,2-linked fructose residues, respectively, attached to sucrose. They are obtained from sucrose using a fructosyltransferase and are marketed as low-calorie, non-cariogenic sweeteners.[49] Fructan biosynthesis takes place in the vacuole of the plant cell and uses 1-kestose as a primer.[50] Unusually, for a polysaccharide biosynthesis pathway, fructan biosynthesis uses sucrose as a fructosyl donor, rather than a sugar phosphate donor.

Chitin and Chitosan

Chitin is a β-1,4-linked polymer of N-acetyl-D-glucosamine and is the major structural component of the exoskeletons of insects and crustaceans such as crabs and shrimps, and the cell wall of many fungi (edible mushrooms contain approximately 20% chitin by dry weight).[51] At least 10^{13} kg of chitin is synthesized each year, and it is estimated to be the most abundant nitrogenous compound in nature.[52] Chitin usually occurs in microfibrils that are embedded within a matrix of proteins, waxes, carotenoids, calcium carbonate and lipoproteins.[53] Chitin isolated from various sources possesses differing degrees of N-acetylation – rarely are more than 90% of amino groups found N-acetylated.[52]

Chitin is synthesized from UDP-N-acetylglucosamine by the action of glycosyltransferases termed chitin synthases.[54] In insects, UDP-N-acetylglucosamine is formed from trehalose, the major sugar of insect hemolymph (the insect equivalent of blood).[55] Chitin is degraded by glycoside hydrolases termed chitinases. Chitinaceous organisms typically possess both chitin synthases and chitinases, which work in partnership to remodel chitinous structures. For example, the chitinaceous exoskeleton of insects and crustaceans is synthesized by chitin synthases but has only limited abilities to expand and cannot keep pace with body growth. Such organisms must therefore periodically replace their old cuticle with a newer, loose one during molting. Molting is assisted by release of a molting fluid that separates epidermal cells from the old exoskeleton. The molting fluid contains proteases and chitinases, allowing partial digestion and recovery of exoskeleton components for recycling.[55] Chitinases are also used by plants as a defence against fungal pathogen attack,[56] and many bacteria produce chitinases and N-acetylglucosaminidases, presumably to liberate N-acetylglucosamine as a nutrient source. Chitinolytic bacteria have been used in agriculture as biocontrol agents to protect against fungus attack.[56] The use of chitinolytic enzymes has been proposed to convert shell-fish waste into N-acetyl-D-glucosamine and D-glucosamine; however, the major approach to the degradation of such wastes is hydrolysis with cold, dilute sulfuric acid to give N-acetyl-D-glucosamine.[53]

Chitin is widely used in industry as non-toxic and antibactericidal wound dressings, as a pharmaceutical excipient, and for the production of biodegradable packaging materials. Additionally, chitin is a feedstock for the production of D-glucosamine and chitosan.

Chitosan is a β-1,4-linked polymer of D-glucosamine; it occurs naturally or can be made from chitin by alkaline N-deacetylation.[53] Alternatively, enzymes termed chitin deacetylases produced by a range of fungi and bacteria can be used to hydrolyse the N-acetyl group of chitin.[57] Unlike chitin, chitosan is water soluble, and the amino groups of the polymer can be modified to introduce substituents onto the chitosan chain. Chitosan and its derivatives have a variety of uses including the following: on wound dressings as a hemostatic agent for rapid clotting of blood; in filtration devices for the removal of particulates and various ions, and as a flocculant (Flonac); as food additives with anticholesterolemic activity; in agriculture as a seed- or fruit coating, a fertilizer and a fungicide; and as an antibactericidal agent.[53,57,58]

Trehalose and Trehalose Oligosaccharides

Trehalose (α-D-glucopyranosyl α-D-glucopyranoside) is a non-reducing disaccharide in which the two D-glucose residues are linked through the anomeric positions to one another. Trehalose is widespread in bacteria, fungi, yeast, insects and plants, but is absent from vertebrates.[59] Trehalose is the major sugar in the blood equivalent of insects, the hemolymph, and is rapidly consumed during highly energy intensive activities such as flight. Relative to many other sugars, trehalose possesses a powerful ability to stabilize membranes and proteins against dehydration.[60] Some organisms can survive even when greater than 99% of their water content is removed, and this ability is in many cases linked to the production of large quantities of trehalose (and in some cases sucrose and oligosaccharides).[61] For example, upon slow dehydration the nematode *Aphelenchus avenae* converts up to 20% of its dry weight into trehalose, and its survival is correlated with production of this sugar. In addition, trehalose can protect cells against heat and oxidative stress.[59] The ability of trehalose to act as a protectant for lipid bilayers appears to relate to its unique solid and solution structure, in which it exists in a so-called clam-shaped conformation with the two sugar rings opposed to one another; trehalose is also able to form a non-crystalline glass upon dehydration.[60]

There are five known trehalose biosynthesis routes, of which three are more common.[62] The first was described in yeast[63] and involves the condensation of UDP-glucose and glucose-6-phosphate by the glycosyltransferase trehalose-6-phosphate synthase (OtsA)[64] to afford trehalose-6-phosphate, and the hydrolysis of the phosphoric acid ester by trehalose-6-phosphate phosphatase (OtsB). The second route was first described in *Pimelobacter* sp.[65] and involves the rearrangement of maltose to trehalose catalysed by the transglycosidase, trehalose synthase (TreS). The third route to trehalose was first reported in Rhizobia,[66] *Arthrobacter* sp.[67] and *Sulfolobus acidocaldarius*[68] and involves the isomerization of the terminal residue of maltooligosaccharides at the reducing end to maltooligosaccharyltrehalose by the action of maltooligosaccharyltrehalose synthase (TreY), and hydrolysis of this product by maltooligosaccharyltrehalose trehalohydrolase (TreZ) to yield trehalose. While most organisms possess only one of these pathways, mycobacteria[69,70] and corynebacteria[71] possess all three.

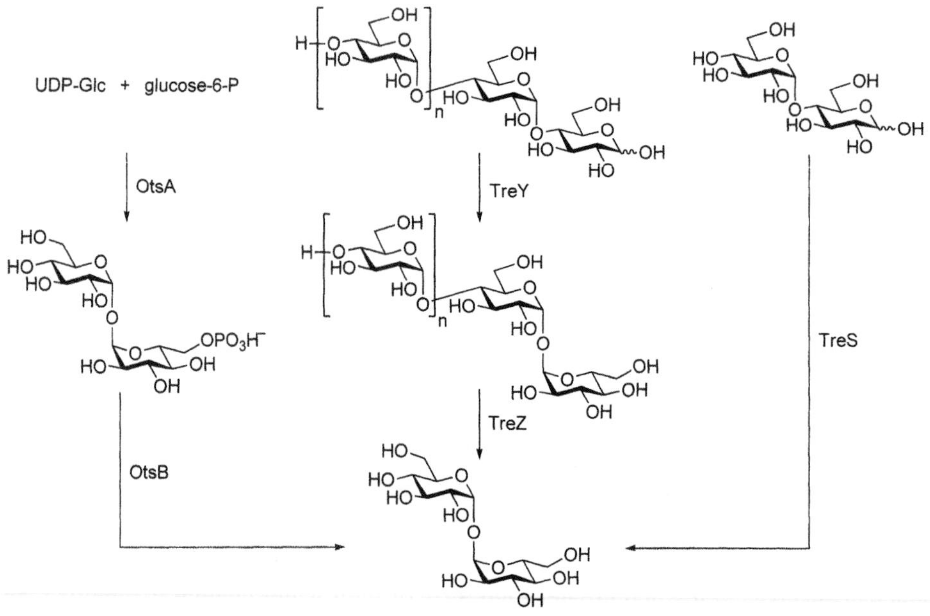

Trehalose is hydrolysed by glycoside hydrolases termed trehalases. While vertebrates do not biosynthesize trehalose, trehalase is found in the intestinal villae membranes where it can hydrolyse ingested trehalose. Like other intestinal hydrolases, trehalase is a glycosylphosphatidylinositol-anchored protein.[72] Individuals with a defect in intestinal trehalase suffer diarrhoea when foods with high trehalose content are ingested.

While trehalose is a significant sugar in its own right, it is also found at the core of a range of oligosaccharides. Bemisiose is a trehalose-containing trisaccharide originally isolated from the honeydew of the whitefly *Bemisia argentifolia*[73] and is also found in a number of bacteria. Mycobacteria and corynebacteria produce a range of trehalose-containing molecules including methylated, pyruvylated, acylated, sulfated and phosphorylated metabolites (see Chapter 10). In addition they produce a compendium of trehalose-containing oligosaccharides including D-galactosylated and D-glucosylated species.[74]

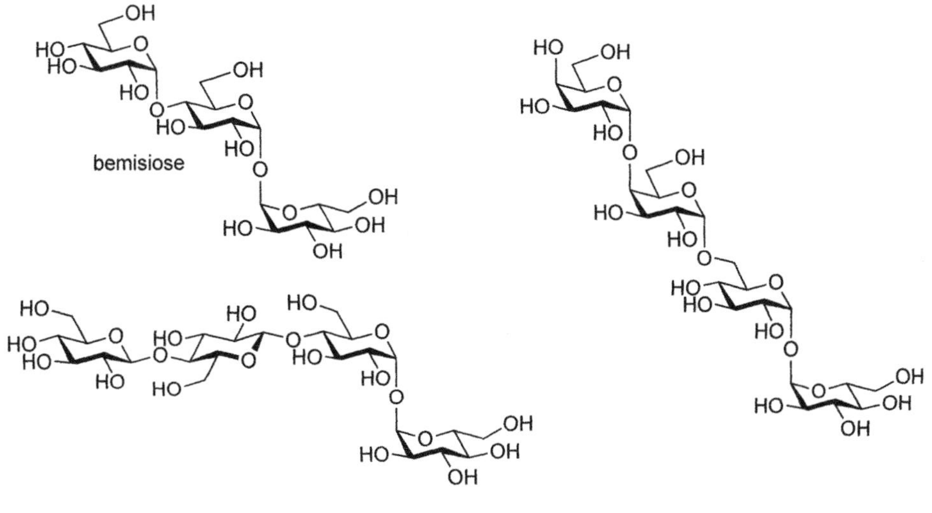

bemisiose

1,3-β-Glucans[75]

No trivial name for 1,3-β-glucans is in common use. Materials isolated from different sources and bearing subtle structural nuances frequently bear different names (e.g. callose from plant cells, paramylon from *Euglena* spp., laminarin from *Laminaria* sp., lichenin from lichens, scleroglucan from *Sclerotium glucanicum*, lentinan from *Lentinus edodes*, laricinan from *Larix laricina* and chrysolaminarin from plankton and diatoms). In plants 1,3-β-glucans are found in relatively small amounts but are widespread throughout the tissues. Callose is a 1,3-β-glucan that is found in cell walls of plants and in pollen grains and tubes. It is also produced at the plasma membrane/wall interface in response to wounding and may provide a temporary scaffold while the wound repair process proceeds. Phytoplankton and diatoms produce 1,3-β-glucans that may bear various branching and capping

References start on page 339

groups.[76] The ubiquity of these organisms in the oceans and their high productivity make their carbohydrate production capability a major contributor to the global carbon cycle. Mixed linkage β-1,3;β-1,4-glucans are found in the seeds of barley and oats, where they are the major components of endosperm walls; these materials are important components of the human diet and are related to beneficial colonic and cardiovascular function.[77]

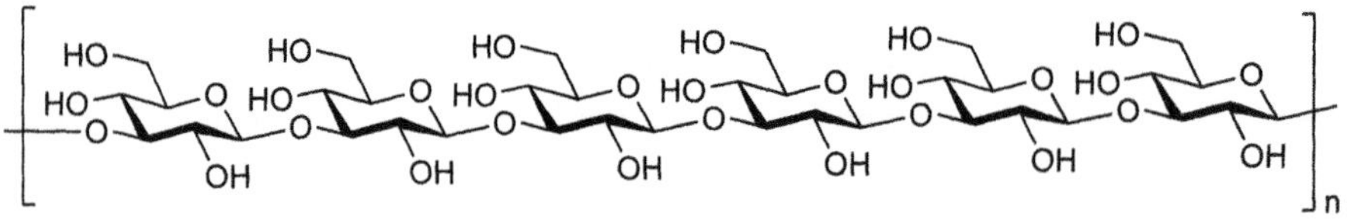

Mannans

Linear mannose polymers are found in plants, fungi and protists. Plants produce a linear β-1,4-mannose polymer that is typically modified with varying amounts of α-linked galactose and which is frequently called galactomannan. The seed of the ivory nut tree *Phytelephas macrocarpa* contains an essentially unmodified β-1,4-mannan, which was the original source of D-mannose (see Chapter 1).[78] *Leishmania* sp. produces an intracellular β-1,2-mannan; this carbohydrate appears to function as an intracellular reserve material and confers protection against stress.[79] Fungi such as *Candida albicans* produce a range of mannans in their cell wall including α-1,2-, α-1,3-, α-1,6- and β-1,2-linked structures.[80]

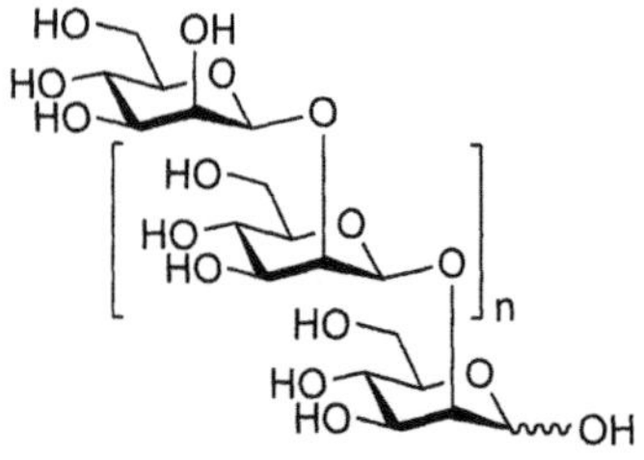

β-1,2-mannan

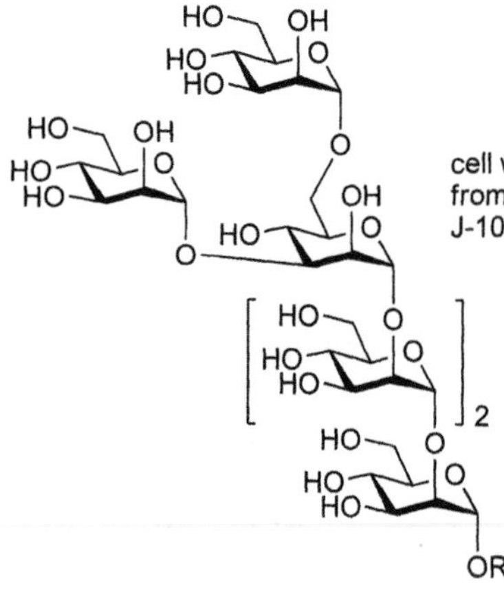

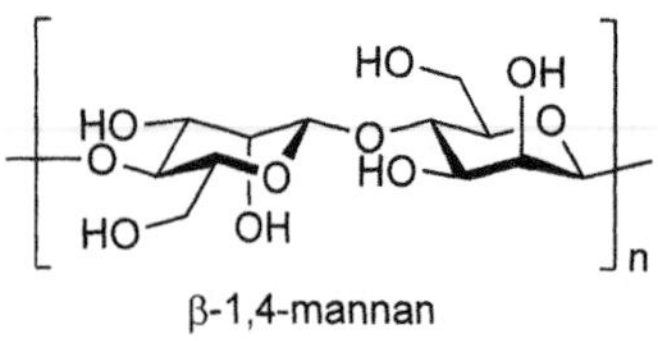

β-1,4-mannan

cell wall mannan fragment from *Candida albicans* J-1012

References

1. Marchessault, R.H. and Sundararajan, P.R. (1983). In *The polysaccharides* (G.O. Aspinall, ed.) Vol. 2, p. 11. Academic Press: New York.
2. Teeri, T.T., Brumer, H., 3rd, Daniel, G. and Gatenholm, P. (2007). *Trends Biotechnol.*, **25**, 299.
3. Somerville, C. (2006). *Annu. Rev. Cell Dev. Biol.*, **22**, 53.
4. Koyama, M., Helbert, W., Imai, T., Sugiyama, J. and Henrissat, B. (1997). *Proc. Natl. Acad. Sci. USA*, **94**, 9091.
5. Williamson, R.E., Burn, J.E. and Hocart, C.H. (2002). *Trends Plant Sci.*, **7**, 461.
6. Woodward, J. (1984). In *Topics in Enzyme and Fermentation Biotechnology*. Vol. 8, p. 9. John Wiley & Sons: New York.
7. Gilbert, H.J. and Hazlewood, G.P. (1993). *J. Gen. Microbiol.*, **139**, 187.
8. Bayer, E.A., Chanzy, H., Lamed, R. and Shoham, Y. (1998). *Curr. Opin. Struct. Biol.*, **8**, 548.
9. Aanen, D.K., Eggleton, P., Rouland-Lefevre, C., Guldberg-Froslev, T., Rosendahl, S. and Boomsma, J.J. (2002). *Proc. Natl. Acad. Sci. USA*, **99**, 14887.
10. Ito, S. (1997). *Extremophiles*, **1**, 61.
11. Balser, K., Hoppe, L., Eicher, T., Wandel, M. and Astheimer, H.-J. (1986). In *Ullmann's Encyclopedia of Industrial Chemistry* (W. Gerhartz, ed.) Vol. A5, p. 419. Wiley-VCH: Weinheim.
12. Brandt, L. (1986). In *Ullmann's Encyclopedia of Industrial Chemistry* (W. Gerhartz, ed.) Vol. A5, p. 461. Wiley-VCH: Weinheim.
13. Braun, G. (1943). *Org. Syn. Coll. Vol.* **2**, 124.
14. Daniel, J.R. (1986). In *Ullmann's Encyclopedia of Industrial Chemistry* (W. Gerhartz, ed.) Vol. A5, p. 1. Wiley-VCH: Weinheim.
15. Saenger, W. (1984). *Naturwissenschaften*, **71**, 31.
16. James, M.G., Denyer, K. and Myers, A.M. (2003). *Curr. Opin. Plant Biol.*, **6**, 215.
17. Chronakis, I.S. (1998). *Crit. Rev. Food Sci. Nutr.*, **38**, 599.
18. Guilnot, A. and Mercier, C. (1983). In *The Polysaccharides* (G.O. Aspinall, ed.) Vol. 3, p. 209. Academic Press: New York.
19. Ragauskas, A.J., Williams, C.K., Davison, B.H., Britovsek, G., Cairney, J., Eckert, C.A., Frederick, W.J., Jr., Hallett, J.P., Leak, D.J., Liotta, C.L., Mielenz, J.R., Murphy, R., Templer, R. and Tschaplinski, T. *Science*, **311**, 484.
20. Kim, S. and Dale, B.E. (2005). *Biomass Bioenergy*, **29**, 426.
21. Gibbons, B.J., Roach, P.J. and Hurley, T.D. (2002). *J. Mol. Biol.*, **319**, 463.
22. Roach, P.J. (2002). *Curr. Mol. Med.*, **2**, 101.
23. Lin, A., Mu, J., Yang, J. and Roach, P.J. (1999). *Arch. Biochem. Biophys.*, **363**, 163.
24. Thon, V.J., Khalil, M. and Cannon, J.F. (1993). *J. Biol. Chem.*, **268**, 7509.
25. Verhue, W. and Hers, H.G. (1966). *Biochem. J.*, **99**, 222.
26. Liu, W., Madsen, N.B., Braun, C. and Withers, S.G. (1991). *Biochemistry*, **30**, 1419.
27. Biwer, A., Antranikian, G. and Heinzle, E. (2002). *Appl. Microbiol. Biotechnol.*, **59**, 609.
28. Hough, L. and Phadnis, S.P. (1976). *Nature*, **263**, 800.
29. Bilger, B. (2006). *New Yorker*, **82**, 40.
30. Torres, L.L. and Salerno, G.L. (2007). *Proc. Natl. Acad. Sci. USA*, **104**, 14318.
31. Salerno, G.L. and Curatti, L. (2003). *Trends Plant Sci.*, **8**, 63.
32. Sale, J.W. and Skinner, W.W. (1922). *J. Ind. Eng. Chem.*, **14**, 522.
33. Vinjamoori, D.V., Byrum, J.R., Hayes, T. and Das, P.K. (2004). *J. Anim. Sci.*, **82**, 319.
34. Nakakuki, T. (2002). *Pure Appl. Chem.*, **74**, 1245.
35. Kanauchi, O., Mitsuyama, K., Araki, Y. and Andoh, A. (2003). *Curr. Pharm. Des.*, **9**, 333.
36. Rastall, R.A. (2004). *J. Nutr.*, **134**, 2022S.
37. Peterbauer, T. and Richter, A. (2001). *Seed Sci. Res.*, **11**, 185.

38. Ravaud, S., Robert, X., Watzlawick, H., Haser, R., Mattes, R. and Aghajari, N. *J. Biol. Chem*, **282**, 28126.
39. Sträter, P. (1986). In *Alternative sweeteners* (L.O. Nabors and R.C. Gelardi, eds) p. 217. Marcel Dekker: New York.
40. Kunz, C., Rudloff, S., Baier, W., Klein, N. and Strobel, S. (2000). *Annu. Rev. Nutr.*, **20**, 699.
41. Brodbeck, U., Denton, W.L., Tanahashi, N. and Ebner, K.E. (1967). *J. Biol. Chem.*, **242**, 1391.
42. Ramakrishnan, B. and Qasba, P.K. (2001). *J. Mol. Biol.*, **310**, 205.
43. Swallow, D.M. (2003). *Annu. Rev. Genet.*, **37**, 197.
44. Vesa, T.H., Marteau, P. and Korpela, R. (2000). *J. Am. Coll. Nutr.*, **19**, 165S.
45. Jacob, F. and Monod, J. (1961). *J. Mol. Biol.*, **3**, 318.
46. Lewis, M. (2005). *C. R. Biol.*, **328**, 521.
47. Ritsema, T. and Smeekens, S. (2003). *Curr. Opin. Plant Biol.*, **6**, 223.
48. Ritsema, T. and Smeekens, S.C. (2003). *J. Plant. Physiol.*, **160**, 811.
49. Tokunaga, T., Oku, T. and Hosoya, N. (1989). *J. Nutr.*, **119**, 553.
50. Vijn, I., van Dijken, A., Sprenger, N., van Dun, K., Weisbeek, P., Wiemken, A. and Smeekens, S. (1997). *Plant J.*, **11**, 387.
51. Hirano, S. (1986). In *Ullmann's Encyclopedia of Industrial Chemistry* (W. Gerhartz, ed.) Vol. A6, p. 231. Wiley-VCH: Weinheim.
52. Muzzarelli, R.A.A. (1999). In *Chitin and Chitinases* (P. Jollès and R.A.A. Muzzarelli, eds) p. 1. Basel: Birkhäuser Verlag.
53. Synowiecki, J. and Al-Khateeb, N.A. (2003). *Crit. Rev. Food Sci. Nutr.*, **43**, 145.
54. Merzendorfer, H. (2006). *J. Comp. Physiol. B*, **176**, 1.
55. Merzendorfer, H. and Zimoch, L. (2003). *J. Exp. Biol.*, **206**, 4393.
56. Bhattacharya, D., Nagpure, A. and Gupta, R.K. (2007). *Crit. Rev. Biotechnol.*, **27**, 21.
57. Tsigos, I., Martinou, A., Kafetzopoulos, D. and Bouriotis, V. (2000). *Trends Biotechnol.*, **18**, 305.
58. Cuero, R.G. (1999). In *Chitin and Chitinases* (P. Jollès, and R.A.A. Muzzarelli, eds) p. 315. Birkhäuser Verlag: Basel.
59. Elbein, A.D., Pan, Y.T., Pastuszak, I. and Carroll, D. (2003). *Glycobiology*, **13**, 17R.
60. Albertorio, F., Chapa, V.A., Chen, X., Diaz, A.J. and Cremer, P.S. (2007). *J. Am. Chem. Soc.*, **129**, 10567.
61. Crowe, J.H., Hoekstra, F.A. and Crowe, L.M. (1992). *Annu. Rev. Physiol.*, **54**, 579.
62. Paul, M. (2007). *Curr. Opin. Plant Biol.*, **10**, 303.
63. Cabib, E. and Leloir, L.F. (1958). *J. Biol. Chem.*, **231**, 259.
64. Gibson, R.P., Turkenburg, J.P., Charnock, S.J., Lloyd, R. and Davies, G.J. (2002). *Chem. Biol.*, **9**, 1337.
65. Tsusaki, K., Nishimoto, T., Nakada, T., Kubota, M., Chaen, H., Sugimoto, T. and Kurimoto, M. (1996). *Biochim. Biophys. Acta*, **1290**, 1.
66. Maruta, K., Hattori, K., Nakada, T., Kubota, M., Sugimoto, T. and Kurimoto, M. (1996). *Biosci. Biotechnol. Biochem.*, **60**, 717.
67. Maruta, K., Hattori, K., Nakada, T., Kubota, M., Sugimoto, T. and Kurimoto, M. (1996). *Biochim. Biophys. Acta*, **1289**, 10.
68. Maruta, K., Mitsuzumi, H., Nakada, T., Kubota, M., Chaen, H., Fukuda, S., Sugimoto, T. and Kurimoto, M. (1996). *Biochim. Biophys. Acta*, **1291**, 177.
69. De Smet, K.A., Weston, A., Brown, I.N., Young, D.B. and Robertson, B.D. (2000). *Microbiology*, **146**, 199.
70. Woodruff, P.J., Carlson, B.L., Siridechadilok, B., Pratt, M.R., Senaratne, R.H., Mougous, J.D., Riley, L.W., Williams, S.J. and Bertozzi, C.R. (2004). *J. Biol. Chem.*, **279**, 28835.
71. Tzvetkov, M., Klopprogge, C., Zelder, O. and Liebl, W. (2003). *Microbiology*, **149**, 1659.
72. Ruf, J., Wacker, H., James, P., Maffia, M., Seiler, P., Galand, G., von Kieckebusch, A., Semenza, G. and Matei, N. (1990). *J. Biol. Chem.*, **265**, 15034.

73. Hendrix, D.L. and Wei, Y.A. (1994). *Carbohydr. Res.*, **253**, 329.
74. Ohta, M., Pan, Y.T., Laine, R.A. and Elbein, A.D. (2002). *Eur. J. Biochem.*, **269**, 3142.
75. Stone, B.A. and Clarke, A.E. (1992). *Chemistry and Biology of (1,3)-β-Glucans.* Victoria: La Trobe University Press.
76. Alderkamp, A.-C., Buma, A.G.J. and van Rijssel, M. (2007). *Biogeochem.*, **83**, 99.
77. Burton, R.A., Wilson, S.M., Hrmova, M., Harvey, A.J., Shirley, N.J., Medhurst, A., Stone, B.A., Newbigin, E.J., Bacic, A. and Fincher, G.B. (2006). *Science*, **311**, 1940.
78. Stephen, A.M. (1983). In *The polysaccharides* (G.O. Aspinall, ed.) Vol. 2, p. 97. Academic Press: New York.
79. Ralton, J.E., Naderer, T., Piraino, H.L., Bashtannyk, T.A., Callaghan, J.M. and McConville, M.J. (2003). *J. Biol. Chem.*, **278**, 40757.
80. Shibata, N., Suzuki, A., Kobayashi, H. and Okawa, Y. (2007). *Biochem. J.*, **404**, 365.

Chapter 10

Modifications of Glycans and Glycoconjugates

Glycans and glycoconjugates possess diversity over and above that of polymers of amino acids and of nucleic acids through their capacity for stereochemical (α and β), linkage (1,2-, 1,3-, 1,4-, etc.) and ring (furanose, pyranose) isomerism, as well as the possibility of branching of carbohydrate chains. Additional diversity is possible through the installation of various organic and inorganic esters onto the carbohydrate units, or as linking groups between them. Examples of such groups include simple acetic acid esters found on bacterial metabolites, sulfate esters found on various glycosaminoglycans (GAGs) and, less commonly, phosphate esters located on, or within, glycan chains. Alkylation of hydroxyl groups and epimerization reactions are also common modifications of mature carbohydrate structures. Modified glycans and glycoconjugates have three possible origins: pre-glycosylational, where the modification is performed prior to the construction of a glycosidic linkage; co-glycosylational, where the modifying group is installed during the construction of, and forms part of, a linkage between neighbouring sugars; and post-glycosylational, where the modifying group or epimerization occurs on a mature glycan or glycoconjugate.[1] Examples of each of these are known, with the post-glycosylational modification being the most widely encountered. This chapter will give an overview of the occurrence, function (where known) and biosynthesis of co- and post-glycosylationally modified carbohydrates that are modified by carboxylate, sulfate and phosphate esters, by methylation and by epimerization reactions. Pre-glycosylational modifications of simple monosaccharide derivatives, or modifications involved in central metabolism, will not be discussed here as they were discussed in Chapter 6.

Before we begin, it is worth commenting that many of the problems that plague advances in studying glycoconjugates in general, such as structure elucidation and supply of homogeneous materials, are compounded in the case of modified glycoconjugates. As a result, while a great many modified glycoconjugates are known, a detailed knowledge of their roles and functions is clear only for a few.

References start on page 364

Epimerization

One of the simplest carbohydrate modifications is the epimerization of a single stereocentre in a sugar residue. Epimerizations may occur pre-glycosylationally at the level of the sugar nucleotide, which have been discussed in Chapter 6. Alternatively, epimerizations can occur after the synthesis of the glycan or glycoconjugate, as seen in the biosynthesis of certain polysaccharides that contain hexuronic acid residues.[2] Epimerizations are involved in the biosynthesis of three major glycans: β-D-mannuronic acid (ManA) to α-L-guluronic acid (GulA) in alginate biosynthesis, and β-D-glucuronic acid to α-L-iduronic acid in heparin/heparan sulfate and dermatan sulfate biosynthesis. The biosynthesis of alginates will be discussed here, and the biosynthesis of heparin/ heparan sulfate is briefly described in the next section and in Chapter 11.

Alginates are linear polysaccharides produced by brown seaweeds and certain bacteria. Bacterial alginates have attracted the greatest attention as they are produced by infectious organisms including *Pseudomonas aeruginosa* and *Burkholderia cepacia*, and allow these bacteria to colonize the airways of patients with cystic fibrosis.[3] *P. aeruginosa* alginates consist of random polymers of 1,4-linked β-D-mannuronic acid and α-L-guluronic acid, which may bear acetyl groups at O2 and O3 of the β-ManA residues. Biosynthesis commences by polymerization of GDP-D-mannuronic acid to give a linear β-D-mannuronic acid homopolymer. Next, the linear polymer is modified by the addition of acetic acid esters to some D-mannuronic acid residues. Finally, epimerization of non-acetylated β-D-mannuronic acid residues to α-L-guluronic acid occurs through the action of a C5 epimerase.[4] The mechanism of C5 epimerase is proposed to involve a general base, which removes the proton α to the carboxyl group, followed by reprotonation from the opposite face, leading to epimerization.[2] The mechanism of such epimerases is thus similar to that of lyases, which instead lead to elimination across the C4–C5 bond of uronates (see Chapter 7).[2]

Sulfation

Carbohydrate sulfation occurs only as a post-glycosylational modification. Examples of carbohydrate sulfation have been found in lower and, more commonly, higher organisms. Such modifications occur mostly on secondary metabolites, particularly those found outside the cell either on the surface or as secreted species destined for the extracellular matrix or the environment.

Sulfotransferases: Biological sulfation is mediated by enzymes termed sulfotransferases, which catalyse the transfer of a sulfuryl group from the reactive donor molecule 3′-phosphoadenosine-5′-phosphosulfate (PAPS) to a nucleophile such as a carbohydrate alcohol or amine.[5,6] In higher organisms, most carbohydrate sulfotransferases are membrane associated and are found within the Golgi apparatus. Many bacteria possess carbohydrate sulfotransferases, some of which are soluble cytosolic enzymes, whereas others are membrane associated. Diseases associated with sulfotransferase dysfunction include macular corneal dystrophy, caused by defects in a GlcNAc 6-sulfotransferase involved in keratan sulfate biosynthesis,[7] and spondyloepiphyseal dysplasia, a hereditary skeletal dysplasia resulting from defects in a chondroitin 6-O-transferase involved in chondroitin sulfate biosynthesis.[8] X-ray crystal structures of a variety of sulfotransferases from both bacteria and eukaryotic sources support a mechanism involving direct in-line transfer of the sulfuryl moiety from PAPS to the acceptor, to release 3′-phosphoadenosine-5′-phosphate (PAP).[5]

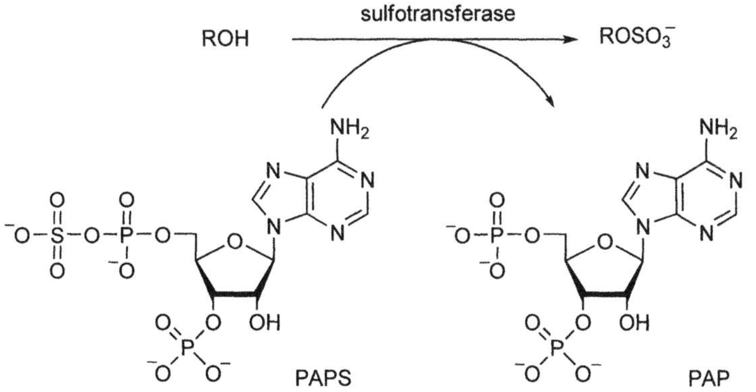

Sulfatases: Desulfation of sulfated molecules is catalysed by hydrolytic enzymes, termed sulfatases.[9]

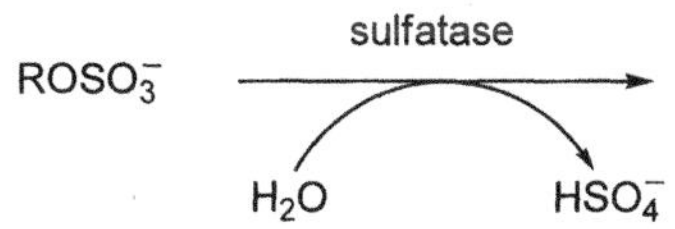

In higher organisms, most carbohydrate sulfatases are lysosomal residents and are responsible for the carefully orchestrated degradation of sulfated glycoproteins, GAGs and glycolipids (see Chapter 11).[10] Several carbohydrate sulfatases have also been found in the extracellular matrix and are used in the fine tuning of sulfated carbohydrate structures after their biosynthesis.[11] In contrast, many bacteria produce sulfatases that are secreted from the cell.[12] Diseases associated with dysfunction of carbohydrate sulfatases include several lysosomal storage disorders that lead to defects in development with a wide spectrum of severity.[13] For example, defects in iduronate-2-sulfate sulfatase lead to Hunter syndrome (MPS III), defects in heparan N-sulfatase lead to Sanfilippo A syndrome (MPS III A), defects in N-acetylglucosamine-6-sulfate sulfatase lead to Sanfilippo D syndrome (MPS III D), defects in galactose-6-sulfate sulfatase lead to Morquio A syndrome (MPS IV A) and defects in N-acetylgalactosamine-4-sulfate sulfatase lead to Maroteaux-Lamy syndrome (MPS VI).[10] Sulfatases have been proposed to operate through a variety of mechanisms, none of which have definitive evidence in their favour – common to all of the mechanisms is the involvement of an unusual formylglycine residue found in the active site and formed from a post-translational modification of the sulfatase polypeptide.[14–16] Defects in this modification machinery lead to a partial or complete loss of all sulfatase activity in a hereditary condition, termed multiple sulfatase deficiency.[17,18]

Sulfated glycosaminoglycans: GAGs are a family of linear, negatively charged oligo- and polysaccharides that are found in animal connective tissue, in the extracellular matrix, on cell surfaces and in some bacteria. In higher organisms, GAGs are found in sulfated forms including heparin, heparan sulfate, chondroitin 4-sulfate, chondroitin-6-sulfate, keratan sulfate and dermatan sulfate.

heparin/heparan sulfate, R = H or SO_3^-,
R^1 = H, Ac or SO_3^-

chondroitin sulfate, R = H or SO_3^-

dermatan sulfate, R = H or SO_3^-

keratan sulfate

Heparin: Perhaps the most widely known sulfated carbohydrate is heparin, which has been used as an antithrombotic anticoagulant for more than 70 years (see Chapter 12). More than 30 tonnes are manufactured annually, representing more than half a billion doses.[19] In its action as a blood anticoagulant, heparin binds to antithrombin III to stabilize inhibitory complexes of antithrombin III with thrombin and with factor Xa, preventing the activation of the blood coagulation cascade. Seminal studies by the groups of Sinaÿ and van Boeckel in the 1980s revealed highly sulfated pentasaccharide sequences to be the minimum required for inhibition of factor Xa,[20–22] and in 2002 a related synthetic pentasaccharide was approved for clinical use and is marketed as fondiparinux (Arixtra).[23,24] The synthesis of fondiparinux requires over 60 steps from natural carbohydrate starting materials and thus constitutes one of the most involved synthetic routes to a commercial drug.[25]

heparin pentasaccharide
R = Ac or SO$_3^-$

fondaparinux
(Arixtra)

The biosynthesis of heparin in eukaryotes resembles that of the other GAGs and begins with the synthesis of the polysaccharide chain, comprised of repeating β-GlcNAc-(1,4)-β-GlcA-(1,4)-units, in the Golgi apparatus through the action of the two corresponding glycosyltransferases (see Chapter 11). Modification of this chain then proceeds in a tightly orchestrated manner to remove some N-acetyl groups and install O- and N-sulfates, and to epimerize D-glucuronic acid units to L-iduronic acid. The bifunctional enzyme glucosamine N-deacetylase/N-sulfotransferase (NDST) initiates the process by deacetylating random N-acetyl-D-glucosamine residues in the chain to generate D-glucosamine residues. The N-sulfotransferase activity of this enzyme then catalyses the installation of an N-sulfate from PAPS, although this process is frequently incomplete with many amino groups remaining unmodified. Following the action of NDST, a C5 epimerase converts some D-glucuronic acid residues to L-iduronic acid. Sulfation then occurs at the 2-position of the L-iduronic acid (catalysed by a 2-O-sulfotransferase), succeeded by the sequential sulfation at the 6-position and then the 3-position of the D-glucosamine residues (catalysed by a 6-O-sulfotransferase and a 3-O-sulfotransferase, respectively). After the completion

References start on page 364

of the biosynthesis and transport of the mature heparin chain from the Golgi apparatus to its eventual destination, further modification may be effected by the action of Sulfs, *endo*-acting sulfatases that hydrolyse 6-*O*-sulfate esters on heparin within the extracellular matrix.[11]

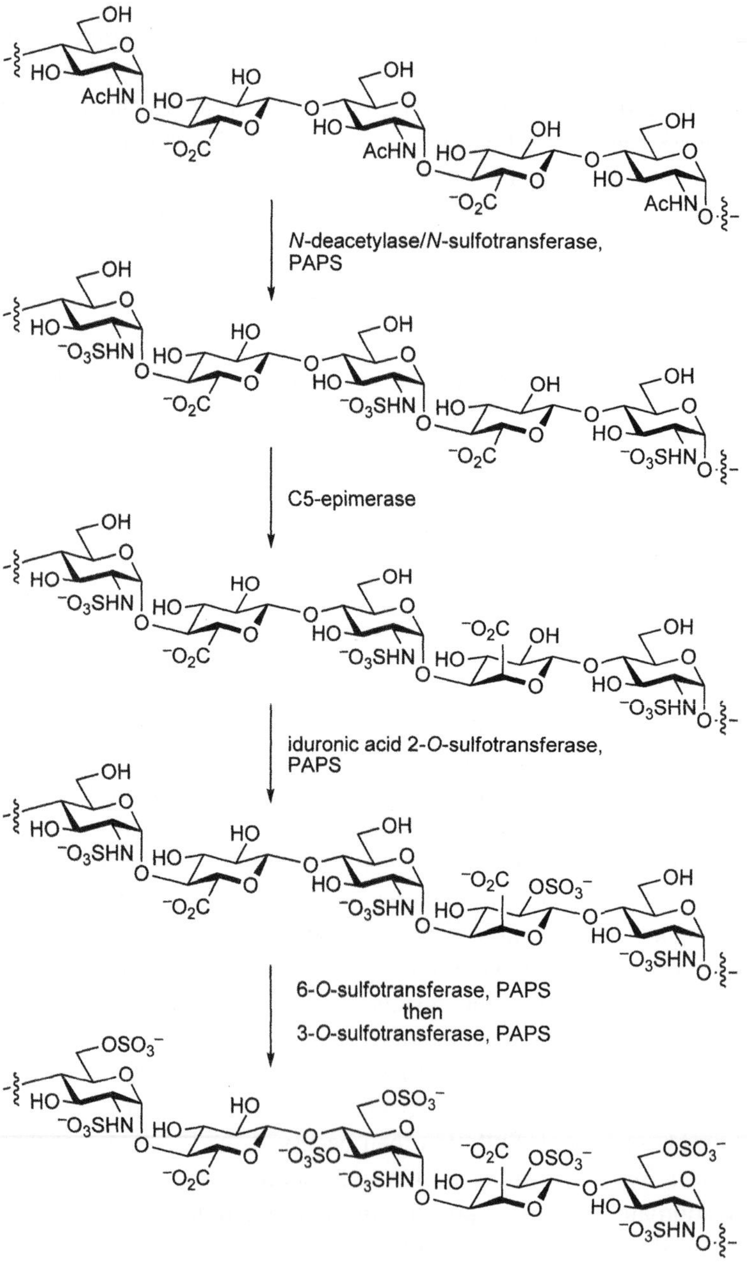

In eukaryotes, degradation of heparin (and other GAGs) occurs within the lysosome and requires the complementary action of three different types of enzymes: glycoside hydrolases, sulfatases and an acetyl transferase (see Chapter 11).[10] There are carbohydrate sulfatases specific for each substitution found in the heparin chain, namely L-iduronate-2-sulfate sulfatase, glucuronate-2-sulfate sulfatase, glucosamine-3-sulfate sulfatase, glucosamine-6-sulfate sulfatase and glucosamine-N-sulfate sulfatase (sulfamidase). These enzymes are *exo*-acting and degrade the heparin chain from the non-reducing terminus in a highly ordered fashion. Generally, removal of the exposed sulfate(s) on the non-reducing terminus is required before the terminal carbohydrate residue can be cleaved by the action of a glycosidase. Finally, an acetyltransferase is required for acetylation of free amino groups, including those exposed by the action of sulfamidase. Acetyl transfer is achieved from acetyl CoA by the action of acetyl CoA:α-glucosaminide N-acetyltransferase, and must be performed before N-acetyl-α-glucosaminidase can cleave the terminal sugar.

Nodulation factors: Nodulation factors (Nod factors) are chitooligosaccharides produced by symbiotic nitrogen-fixing bacteria as a chemical signal to elicit morphological changes in their plant host that lead to the formation of root nodules, which are then colonized by the bacteria.[26,27] Perhaps, the best characterized example of a sulfated carbohydrate is the Nod factor produced by *Sinorhizobium meliloti*.[28] In response to flavonoid signals released by the host plant alfalfa, *S. meliloti* releases a sulfated chitotetraoside, which in turn binds to a specific plant receptor, proposed to be a LysM domain-containing receptor kinase.[29–31] The presence of a single sulfate group on the *S. meliloti* Nod factor determines the host specificity of the bacterium. Mutant bacteria unable to install the sulfate group through lesions in the carbohydrate sulfotransferase NodH are unable to induce nodulation in alfalfa but gain the ability to nodulate and colonize a new plant host, vetch.[32]

R = C16:2, C16:3, C18–26(ω-1)OH fatty acids

Sulfated carbohydrates from halophilic bacteria: Halophilic archaebacteria produce a rich compendium of sulfated carbohydrates, usually, di-, tri- or tetrasaccharide glycolipids containing the diphytanylglycerol moiety.[33] Such sulfated glycolipids are membrane components and have been suggested to play a crucial role in maintaining membrane integrity in conditions of high salt concentration. The halophile *Natronococcus occultus* and various *Natronobacterium* species accumulate up to molar concentrations of trehalose-2-sulfate, which apparently acts as an osmoprotectant.[34]

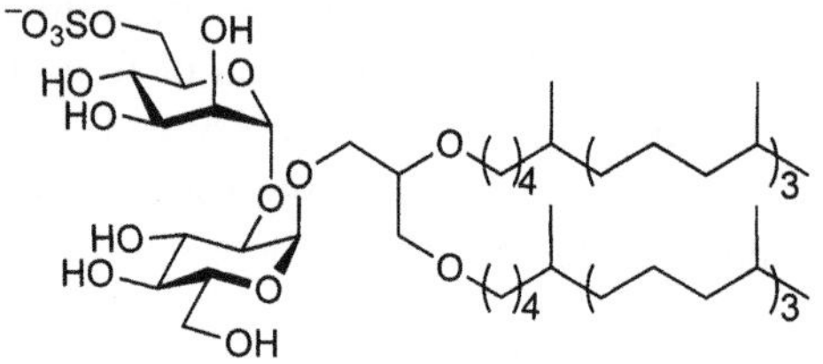

a sulfated glycolipid from
Halobacterium cutirubrum

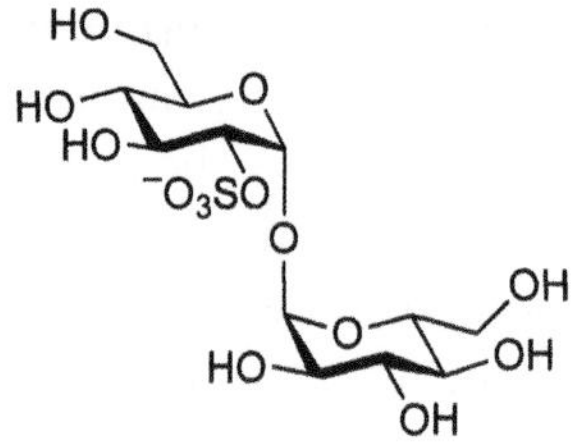

trehalose-2-sulfate

Mycobacterial sulfoglycolipids: Mycobacteria produce a range of sulfated carbohydrate metabolites.[35,36] Sulfolipid-1 is a trehalose-containing sulfated glycolipid that is found in *Mycobacterium tuberculosis* and whose presence is implicated in the virulence of this organism.[37,38] A 4-*O*-sulfated 6-deoxytalose residue was found in the glycopeptidolipid (GPL) of an ethambutol-resistant *Mycobacterium avium* strain cultured from a patient with AIDS,[39] and a sulfated GPL has also been found in *Mycobacterium fortuitum*, in what appears to be a constitutive modification by this organism.[40]

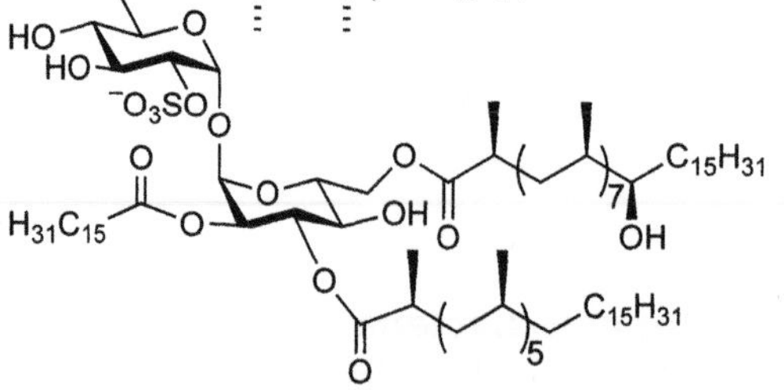

sulfolipid-1 from
Mycobacterium tuberculosis

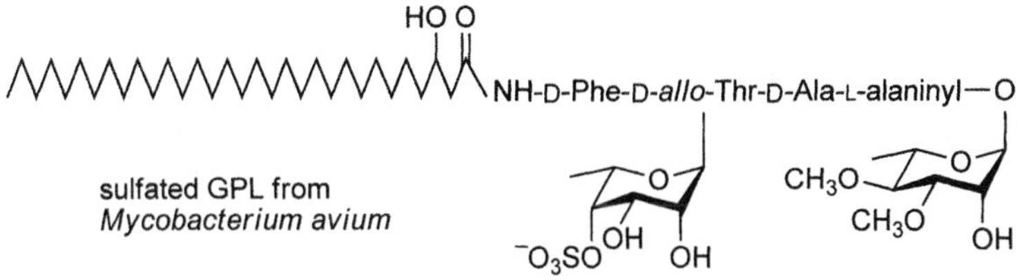

sulfated GPL from
Mycobacterium avium

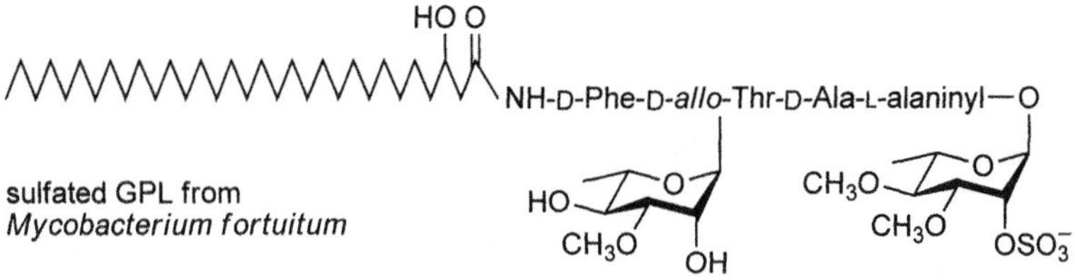

NH-D-Phe-D-*allo*-Thr-D-Ala-L-alaninyl—O

sulfated GPL from
Mycobacterium fortuitum

NH-D-Phe-D-*allo*-Thr-D-Ala-L-alaninyl—O

Sulfated nucleosides: The grass spider, *Holena curta*, produces a venom complex containing an unusual bissulfated and acetylated glycosylated nucleoside, HF-7.[41] Six other sulfated glycosylated nucleosides have been isolated from the venom of the hobo spider, *Tegenaria agrestis*, suggesting that compounds of this type may be widespread in spider venoms.[42] A related set of antibiotics are the liposidomycins, many of which contain sulfate groups.[43,44] The liposidomycins inhibit phosphoro-*N*-acetylmuramyl-pentapeptide-transferase (translocase I), an enzyme involved in the first step of the biosynthesis of lipid intermediate I in peptidoglycan biosynthesis.[45]

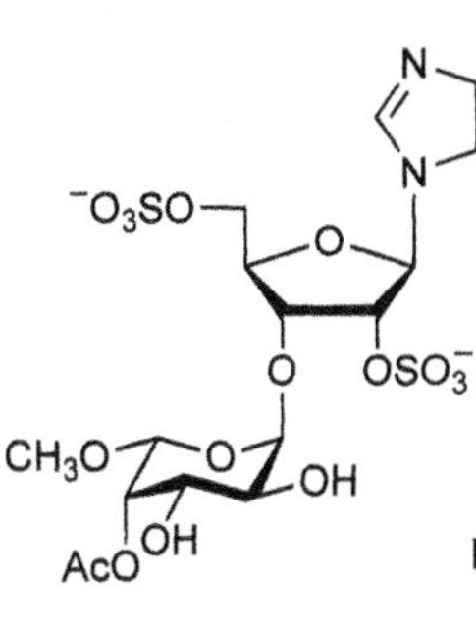

HF-7

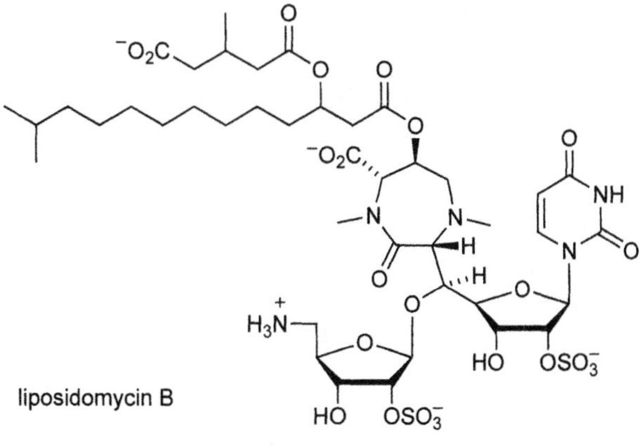

liposidomycin B

Sulfation in inflammation: In the inflammatory response, white blood cells (leukocytes) that circulate through the blood vasculature patrol for sites of damage or infection. At such sites they bind to and roll along the endothelial layer in high endothelial venules, prior to firmly adhering and migrating through the endothelial layer into the surrounding tissue. The initial rolling event is mediated by binding of carbohydrate structures to an adhesion protein, L-selectin, which is common to all leukocytes.[46] A preferred ligand for L-selectin is GlyCAM-1, found to be expressed on high endothelial venules in response to the inflammatory insult.[47] It has been shown that the presence of a family of sulfated sialyl Lewis[x] structures on GlyCAM-1 is required for L-selectin binding, including 6'-sulfated sialyl Lewis[x].[48,49]

6'-sulfated sialyl Lewis[x]

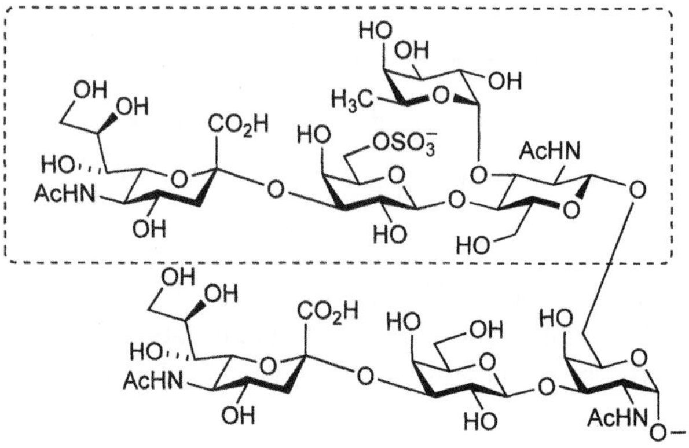

glycan structure of GlyCAM-1

Sulfatide and seminolipid: Sulfatide and seminolipid are the two major mammalian sulfoglycolipids.[50] Sulfatide is a major lipid component of the myelin sheath, a lipid-rich protective coating of nerve fibres in the central and peripheral nervous systems.[51] Defects in arylsulfatase A, a lysosomal sulfatase that desulfates sulfatide, lead to accumulation of sulfatide and result in the inherited disorder, metachromatic leukodystrophy.[52] Seminolipid is the major glycolipid in the testis and is expressed on spermatogenic cells. Genetic studies have provided evidence that seminolipid is required for spermatogenesis.[53]

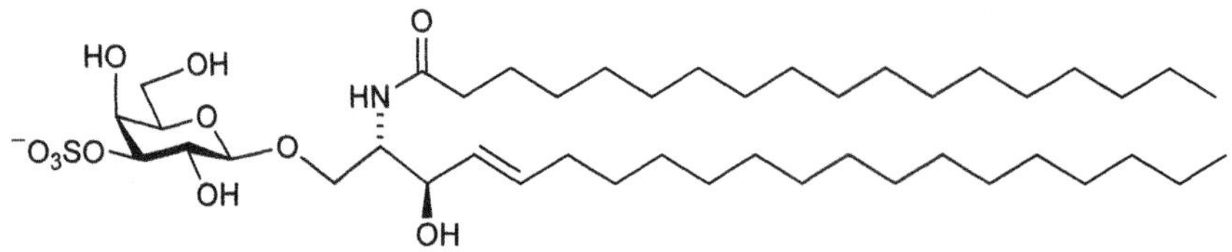

sulfatide

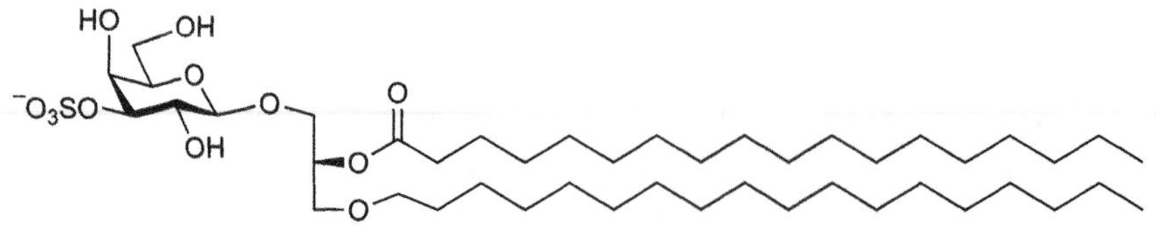

seminolipid

Phosphorylation

Phosphorylation of carbohydrates occurs widely in primary metabolism. There are significantly fewer phosphorylated carbohydrates that are secondary metabolites. Generally, such phosphorylated carbohydrates are co-glycosylationally synthesized, with post-glycosylational phosphorylation being very rare. The following subsections will give a few examples of this relatively small group of carbohydrate modifications.

Mannose-6-phosphate: After the synthesis of proteins in the rough endoplasmic reticulum (ER), they are transferred into the lumen of the ER where they undergo N-linked glycosylation and other modifications. The newly synthesized proteins migrate to the Golgi apparatus where they can be further modified and are sorted for targeting to their ultimate destination. Proteins destined for the lysosome undergo phosphorylation on terminal mannose groups found on α-1,3- and α-1,6-branches of N-linked high mannose structures, to generate mannose-6-phosphate (M6P) residues (see Chapter 11). M6P residues bind to the M6P receptor, resulting in specific targeting of such proteins to the lysosome.[54,55] The biosynthesis of M6P occurs by a two-step co-glycosylational process.[56] Within the *cis* Golgi, a phosphate diester is first formed by the *en bloc* transfer of N-acetylglucosamine 1-phosphate residues onto the 6-hydroxyls of mannosyl groups, by the action of UDP-N-acetylglucosamine:lysosomal enzyme N-acetylglucosamine-1-phosphotransferase (N-acetylglucosamine phosphotransferase).[57] In the second step, a glycoside hydrolase, N-acetylglucosamine-1-phosphodiester α-N-acetylglucosamidase (uncovering enzyme, UCE),[58,59] catalyses the hydrolysis of the glycosidic bond joining the N-acetylglucosamine group to the phosphate, leaving behind a phosphate monoester.[59] With the M6P moiety in place, the protein is transported to the lysosome through interaction with the M6P receptor. Once within the lysosome, the phosphate ester is removed by phosphatases.[60]

Phosphoglycosylation in Leishmania and other protists: Protozoa of the genus *Leishmania* are sandfly-transmitted parasites that are the causative agents of Leishmaniasis. Leishmania parasites synthesize a cell-surface lipophosphoglycan (LPG), which functions to enable the survival of the parasite against the defences of the host organisms. LPG also plays an important role in the colonization of the sandfly vector by the parasite. Upon ingestion, the parasite attaches to the gut wall of the sandfly through binding to a β-galactoside binding lectin, preventing excretion along with the rest of the digested bloodmeal, whilst offering some protection from the digestive enzymes of the insect.[61]

Leishmania LPG can be divided into four subunits: a phosphatidylinositol (PI) lipid anchor, a glycan core, a repeating phosphorylated disaccharide unit and a small oligosaccharide cap.[62] The cap consists of short neutral oligosaccharides composed of mannose and/or galactose. The repeating phosphorylated disaccharide unit of LPG is well conserved, consisting of β-Gal-(1,4)-α-Man-(1,6)-phosphate repeats and may be substituted with various branching oligosaccharide chains. Synthesis of the unique disaccharide repeating units of LPG occurs by the stepwise addition of galactose and mannose-1-phosphate residues from UDP-galactose and GDP-mannose, respectively. Two enzymes, α-mannosylphosphate transferase (MPT)[63] and β-D-galactosyltransferase (β-GalT),[64] are responsible for the synthesis of the disaccharide.

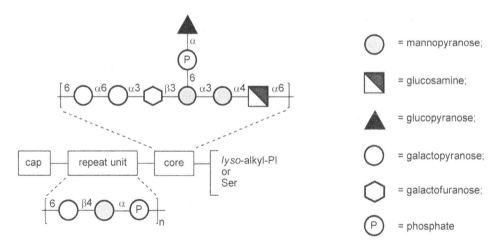

In addition to LPG, phosphoglycosylation is also widespread as a protein modification in most protists. For example, *Entamoeba histolytica*, the aeteological agent of amoebic dysentery, possesses glycosylphosphatidylinositol-anchored proteophosphoglycans with mucin-like serine-rich sequences heavily modified as phosphoserine-bearing glycan chains of general structure $[\alpha\text{-Glc-}(1,6)]_n\text{-}\beta\text{-Glc-}(1,6)\text{-}\alpha\text{-Gal-P-serine}$ (where $n = 2$–23).[65] Similarly, protein phosphoglycosylation is the most abundant form of protein glycosylation in *Leishmania* spp. with phosphoserine residues bearing repeating β-Gal-(1,4)-α-Man-P units and other sugars.[66]

Teichoic acids: The teichoic acids[a] are phosphate-rich polymers found in Gram-positive bacteria.[67] They consist of various polymers possessing phosphodiester groups linking polyols and/or sugar residues, and commonly contain D-alanine residues bound through an ester linkage and, occasionally, phosphorylcholine groups.[68,69] Lipophilic, non-covalently linked teichoic acids are found in most Gram-positive bacteria and are located in the cytoplasmic membrane and are termed membrane teichoic acids or lipoteichoic acids. Wall teichoic acids are covalently linked to peptidoglycan and are more narrowly distributed in Gram-positive bacteria.[68] Lipoteichoic acids and wall teichoic acids from different bacteria exhibit considerable structural diversity; however, all can be broken into two parts: a main chain composed of repeating monomer units joined through phosphodiester linkages and a linkage or capping unit that attaches the main chain to peptidoglycan (for wall teichoic acids) or to a glycolipid moiety (for lipoteichoic acids). Lipoteichoic acids contain main chains that are polymers of glycerol phosphate or various mono- or diglycosylglycerol phosphates that may also bear D-alanine ester or various glycosyl residues.[70] The capping group is frequently a glycosyl diacylglycerol and is joined to the main chain through a phosphodiester link at the 6-position. The main chain of wall teichoic acids consists of phosphate diester polymers of glycerol, ribitol or glycosylglycerol on which assorted sugars or D-alanine esters are attached at various intervals. The linkage group usually contains a mono- or disaccharide that is linked through the reducing end to peptidoglycan by a phosphate diester to the 6-position of *N*-acetylmuramic acid.

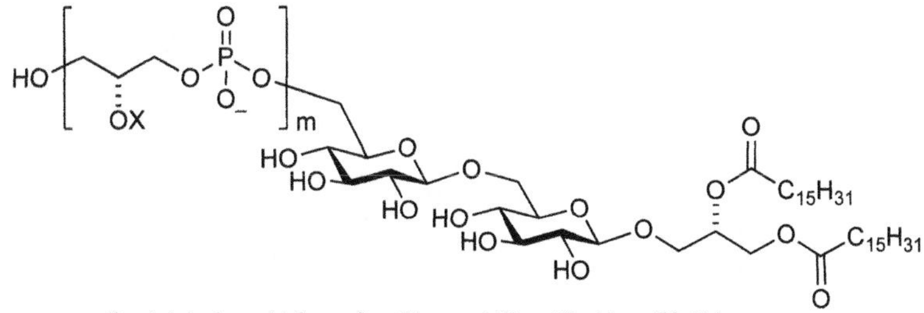

lipoteichoic acid from *Bacillus subtilus*, X = H, α-GlcNAc

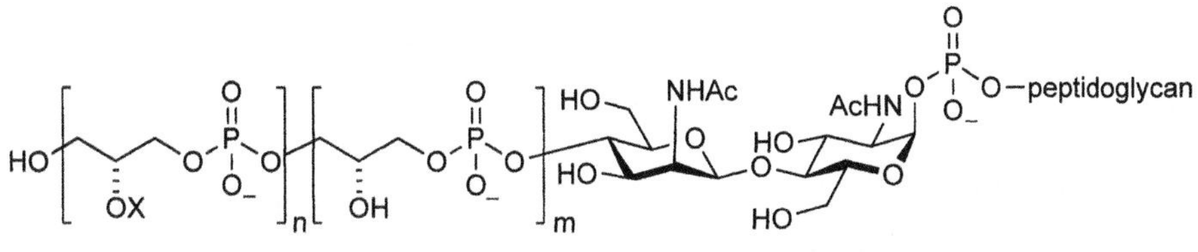

wall teichoic acid from *Bacillus subtilis* 168: X = H, D-alanine, α-Glc

---

[a] Teichos. *noun* (Greek), wall.

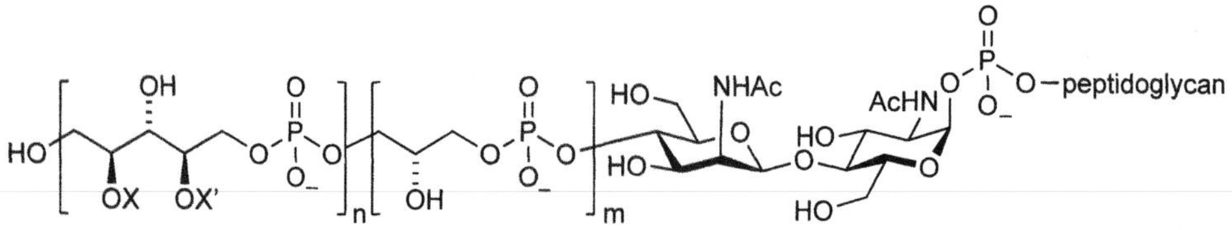

wall teichoic acid from *Staphylococcus aureus* H: X = H, D-alanine, X' = H, α- or β-GlcNAc

The donor units of the main-chain glycerol or ribitol moieties of teichoic acids are the unusual acyclic sugar nucleotides, CMP-glycerol and CMP-ribitol, respectively. Wall teichoic acids and lipoteichoic acids comprise, along with peptidoglycan, part of the polyanionic network of the cell wall that contributes to its elasticity, porosity, tensile strength and electrostatic potential.

Other phosphoglycans: A wide range of phosphoglycans has been found in various microorganisms, and only a representative few are mentioned here. Walls of *Staphylococcus lactis I3* contain a polymer of 1,6-linked N-acetyl-α-glucosamine-1-phosphate.[71] Various yeasts including *Candida albicans* contain phosphomannans that are high mannose structures containing small percentages of α-mannose-1-phospho-6'-mannose groups.[72] *Sporobolomyces* yeasts produce phosphogalactan when grown on D-glucose that contains an α-galactose-1-phospho-6'-galactose.[73] An unusual cyclic-phosphate-containing capsular polysaccharide has been isolated from *Vibrio cholerae* O139.[74,75]

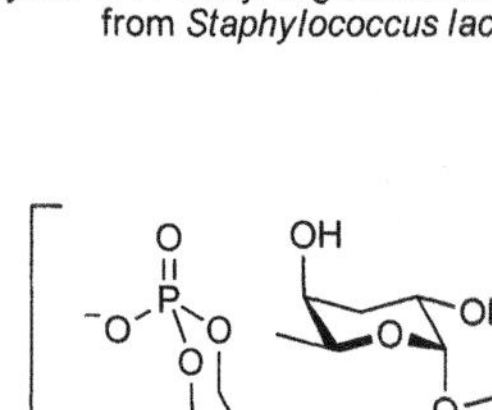

polymer of N-acetyl-α-glucosamine-1-phosphate
from *Staphylococcus lactis I3*

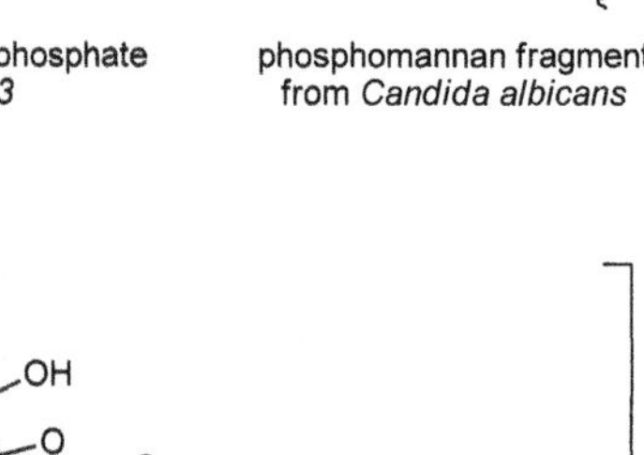

phosphomannan fragment
from *Candida albicans*

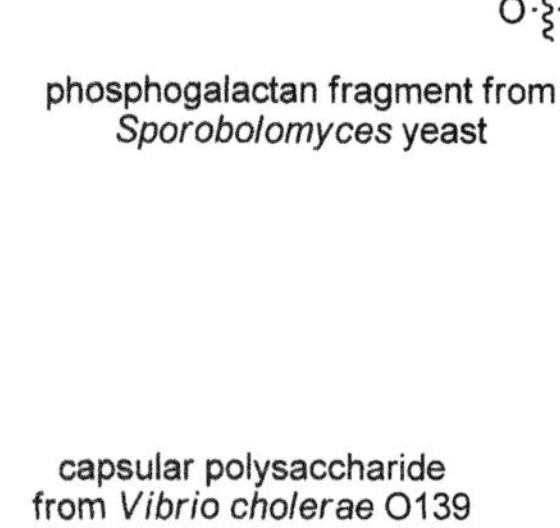

phosphogalactan fragment from
Sporobolomyces yeast

capsular polysaccharide
from *Vibrio cholerae* O139

Carboxylic Acid Esters

We have already seen several examples of organic esters of glycans that serve to illustrate that this modification is widespread. However, few of the biological imperatives for these modifications are well understood. Two modes of modifications are seen: where the sugar plays the role of an alcohol and is esterified with simple acids, or where the sugar acts as the acid, and is esterified with simple alcohols. *O*-Acylation of sugars can consist of relatively simple esterifications with acetic acid or longer chain fatty acids, which are common in many bacteria and which appear to be important for interspecies cross-talk, or with more complex cross-linked structures, such as those found in the diferulic acid ester dimers of heteroxylans. On the contrary, esterification of sugar acids occurs almost exclusively with methanol to afford methyl esters. This section will give an overview of the occurrence and structures of a few representative examples of each mode of esterification and the roles that such esterifications play.

HO⟋ + RCO$_2$H →(ester formation)→ esters of sugar alcohols

HO$_2$C⟋ + CH$_3$OH →(ester formation)→ esters of sugar acids

Acylated bacterial antigens: Lipopolysaccharide (LPS) is a major component of the outer leaflet of Gram-negative bacteria. While LPS comprises only 10–15% of molecules in the outer membrane, it is estimated that they occupy about 75% of the bacterial surface area.[28] As such, they provide one of the first lines of defence against antimicrobial molecules and stresses. Additionally, for many infectious bacteria, the LPS structure determines the nature of interactions with the host, leading to the ability to establish disease or provide a beneficial outcome. As LPS is responsible for interspecies interactions, its structure is critical and, in many cases, is rich with post-glycosylational modifications and, in particular, esterification with carboxylic acids. LPS is comprised of three regions: lipid A (also called endotoxin), core oligosaccharide and O-polysaccharide (also termed O-antigen).

References start on page 364

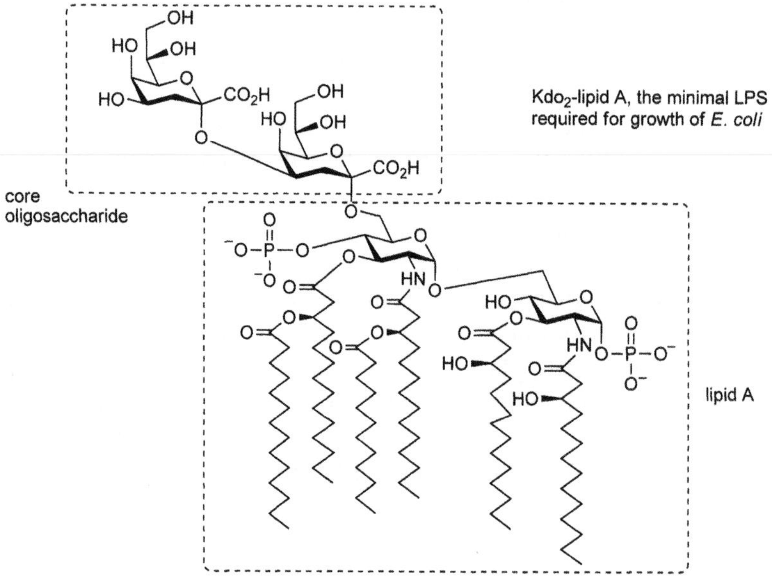

Kdo$_2$-lipid A, the minimal LPS required for growth of *E. coli*

core oligosaccharide

lipid A

Lipid A acts as an anchor into the plasma membrane of the cell and contains 4-phospho-β-GlcN-(1,6)-α-GlcN-1-phosphate, which is heavily acylated with long-chain fatty acids and may be modified to bear additional sugars. Lipid A is potently immunogenic and interacts with receptor proteins of the innate immune system such as Toll-like receptor 4; such interactions are the body's first line of defence against bacterial infection.[76]

The core oligosaccharide of LPS is joined to the sole primary alcohol of lipid A and usually consists of several L-*glycero*-D-*manno*-heptose and 2-keto-3-deoxyoctonoic acid (Kdo) sugars, plus additional neutral hexoses, which may be modified with phosphate or sulfate groups. Finally, the O-polysaccharide is attached to the core and is comprised of repeating units of three to five sugars, each of which may bear a range of post-glycosylational modifications. For example, the O-specific polysaccharide from *Burholderia gladioli* pv. *agaricicola* is a linear trisaccharide that contains D-mannose residues acetylated at O2.[77]

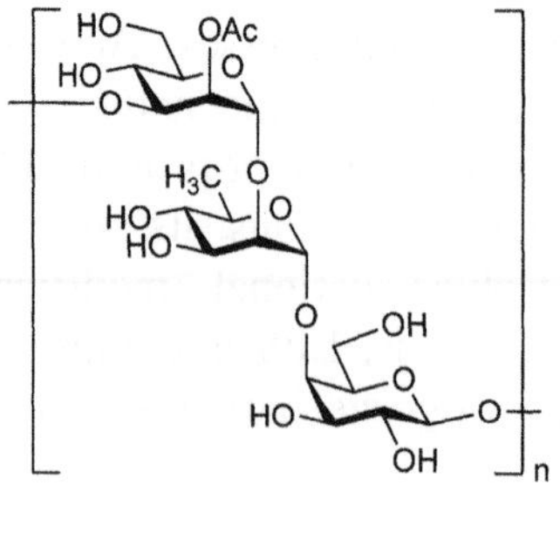

Mycobacterial fatty acid esters: Mycobacteria are a widespread group of largely pathogenic bacteria that produce prodigious amounts of lipid materials, both covalently and non-covalently attached to the cell wall macromolecule.[78] Mycobacterial peptidoglycan is covalently modified by the attachment of an arabinogalactan structure through a single phosphate diester linkage to the 6-position of *N*-acetylmuramic acid (or occasionally *N*-glycolylmuramic acid in *Mycobacterium tuberculosis*).[79] The arabinogalactan consists of a core structure of alternating β-Gal*f*-(1,6)-β-Gal*f*-(1,5)- units, which are modified at the 5-position with branched structures comprised of D-arabinofuranosyl residues; the chemical synthesis of these branched structures was discussed in Chapter 5. The terminal D-arabinofuranosyl residues are themselves modified at the 5-position with a range of complex, long-chain fatty acid esters termed mycolic acids.[80] Mycolic acids are β-hydroxy fatty acids with a long α-alkyl side chain and can contain functional groups such as cyclopropanes, ketones and methoxy groups.[81] Together the mycolylarabinogalactan–peptidoglycan (mAGP) complex comprises part of the distinctive cell wall structure of mycobacteria, is believed to be responsible for the resistance of these organisms to many antibiotics and provides protection against the antibiotic responses of the immune system to infection.[82]

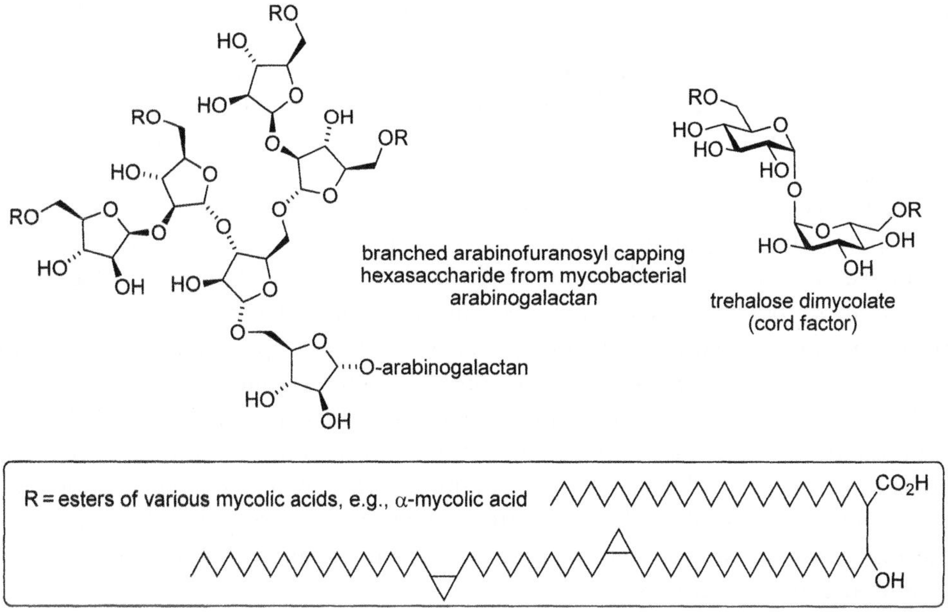

branched arabinofuranosyl capping hexasaccharide from mycobacterial arabinogalactan

trehalose dimycolate (cord factor)

O-arabinogalactan

R = esters of various mycolic acids, e.g., α-mycolic acid

Good evidence has accumulated that movement of mycolic acids from their point of synthesis within the cytoplasm to their destination in the cell wall is facilitated by a carrier molecule, the disaccharide trehalose. Mycolic acids are transferred onto the primary hydroxyls of trehalose to generate trehalose dimycolate (TDM, also termed cord factor).[83] TDM is transported out of the cell where it forms a significant non-covalently-associated portion of the cell wall.[84] Transfer of mycolic acid to the arabinan of the mAGP complex is catalysed by a set of transesterases, the antigen 85 proteins (85A, 85B and 85C).[85] The role of the antigen 85 transesterases is complex; aside from catalysing the transfer of mycolic acid from TDM to the mAGP complex, these enzymes can also catalyse the transfer of mycolic acids between TDM and trehalose to form two molecules of trehalose monomycolate, a reaction that presumably provides a ready flux of mycolyl groups for delivery to the cell wall. The antigen 85 proteins are major secreted proteins of *M. tuberculosis*, consistent with their role in cell wall manufacture. They share a common α/β hydrolase fold and possess a catalytic triad, Ser-Glu-His, which they share with lipases.[86,87]

Carboxylic acid esters in hemicelluloses: Plant cell wall polysaccharides are the most abundant organic materials on earth. Hemicelluloses are second only to cellulose as the most common type of polysaccharide found within plants.[88] Hemicelluloses include a range of non-cellulosic components (e.g. arabinoxylans, pectins, galactomannans, xyloglucans and β-glucans) and can again be modified to form esters in two possible ways: where the sugar acts as an alcohol and is esterified with various carboxylic acids (e.g. heteroxylans) or where the sugar is the acid and is esterified with the simple alcohol, methanol (e.g. pectins).[89]

Heteroxylans include various heterogeneous branched polymers consisting of a β-1,4-linked xylan core appended with L-arabinose, hexoses (D-mannose, D-glucose, D-galactose) and uronic acids (D-glucuronic, 4-O-methyl-D-glucuronic acids); these polymer chains may be further O-methylated, or esterified with acetic, ferulic (4-hydroxy-3-methoxycinnamic acid) or p-coumaric acids. The major roles of heteroxylans are to confer structural rigidity upon plant cell walls and provide protection against microbial and herbivore attack.[90] Ferulic acid esters enhance the strength of the plant cell wall by forming diferulyl cross-links between glycan polymer chains or with lignin, a complex polyphenolic compound found in the extracellular matrix. In vivo, peroxidase catalyses the oxidative coupling of ferulic acid side chains, resulting in the formation of ferulic acid dehydrodimers.[90]

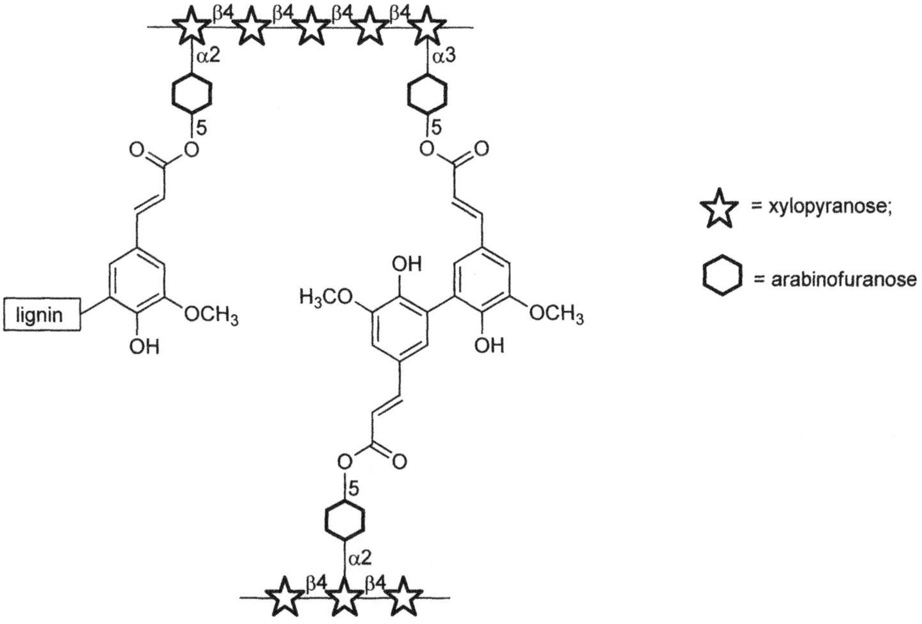

Pectins are heterogeneous, branched and highly hydrated polysaccharides that are rich in D-galacturonic acid residues. Homogalacturonans have a core structure composed of repeating α-1,4-linked galacturonic acid residues, and rhamnogalacturonan I pectins have a core of α-L-Rha-(1,4)-α-GalA-(1,2)-residues.[88] The pectin cores are modified with other carbohydrates and are heavily modified with methyl esters. They are synthesized in the Golgi of the plant cell and are de-esterified in the cell wall by enzymes termed pectin methylesterases.[91]

Biological degradation of hemicelluloses occurs through the synergistic action of main-chain degrading glycosidases and a series of carbohydrate esterases, including acetylesterases, methylesterases, p-coumaryl esterases, ferulyl esterases and pectin methylesterases.[92,93] Carbohydrate esterases are found within bacteria and fungi that catabolize hemicelluloses and have also been identified through sequencing efforts in plant genomes, where they presumably function to allow remodelling of cell wall hemicellulose during development.[94] X-ray crystal structures of most ferulyl, cinnamyl and acetylxylan esterases, and pectin methylesterases, show an α/β hydrolase fold and a conserved catalytic triad, Ser-His-Asp, which is common to lipases, suggesting that these enzymes are a subclass of this group.[95,96] A smaller subset of carbohydrate esterases is metal-ion dependent.[93,97]

Modifications of Sialic Acids

Sialic acids include the sugars *N*-acetylneuraminic acid (Neu5Ac), *N*-glycolylneuraminic acid (Neu5Gc), 2-keto-3-deoxy-D-*glycero*-D-*galacto*-nonulosonic acid (Kdn), and the rarer legionaminic and pseudaminic acids. These nine-carbon chain sugars are ubiquitous among vertebrates and are also found among certain lower orders of life. The sialic acids possess possibly the greatest post-glycosylational diversity of any of the sugars and have been found in over 50 different forms, bearing modifications including sulfation, phosphorylation, methylation, acetylation and lactylation, thereby possessing representatives of all the major classes.[98] In the case of neuraminic acid, additional diversity has been seen in mature glycoconjugates by inter- or intramolecular lactonization with the acid at C1, or intramolecular lactamization of the de-*N*-acetylated amino moiety. Such modifications possess profound implications for biological recognition processes. For example, lactamization of neuraminic acid (itself derived by de-*N*-acetylation of Neu5Ac) to Neu-1,5-lactam occurs in sialyl 6-sulfo Lewis[x] structures found on human leukocytes.[99] While de-*N*-acetylsialyl 6-sulfo Lewis[x] and the parental sialyl 6-sulfo Lewis[x] retain the ability to bind L-selectin, the lactam is devoid of selectin-binding activity. In a similar vein, the preferred ligand for influenza A and B binding is Neu5Ac, and it does not bind to 9-*O*-acetyl-Neu5Ac residues, whereas the reverse is true for the influenza C virus.[100]

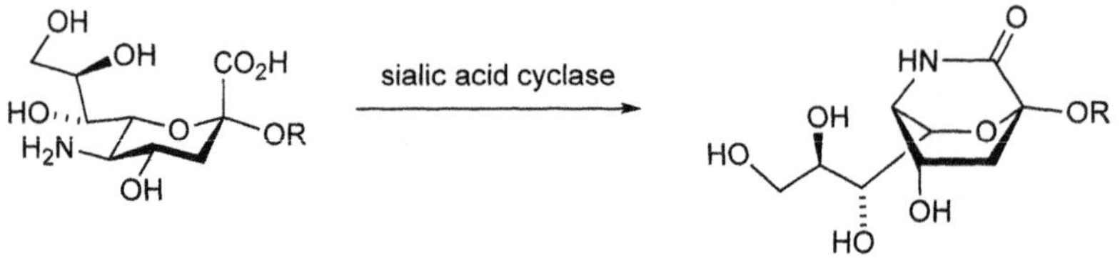

sialic acid cyclase

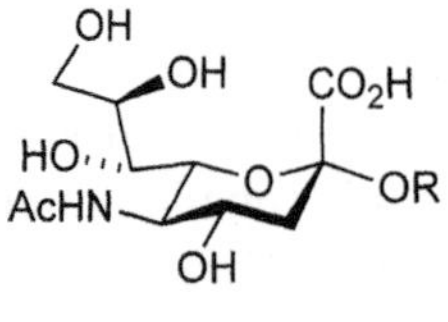

preferred ligand for
influenza A and B

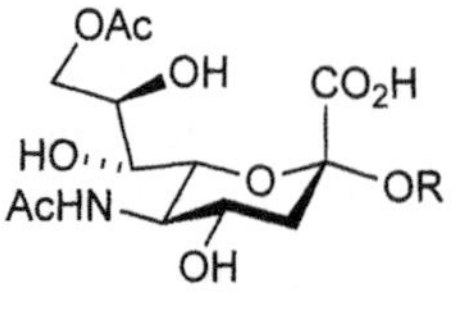

preferred ligand for
influenza C

Finally, an intriguing difference between humans and the African great apes (the chimpanzee, the bonobo and the gorilla) occurs through a mutation in the gene encoding CMP-Neu5Ac hydroxylase, an enzyme that converts CMP-Neu5Ac to CMP-Neu5Gc.[101] As a result, the African great apes possess appreciable quantities of Neu5Gc residues in tissues, whereas no Neu5Gc is detectable in human tissues, with Neu5Ac being the abundant sialic acid.[102] Such differences in a key carbohydrate involved in intercellular cross-talk may help explain the differences in susceptibility and response of humans and the great apes to diseases involving sialic acid recognition.[103]

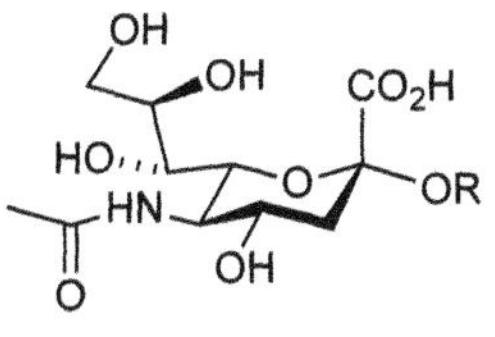

N-acetylneuraminic acid 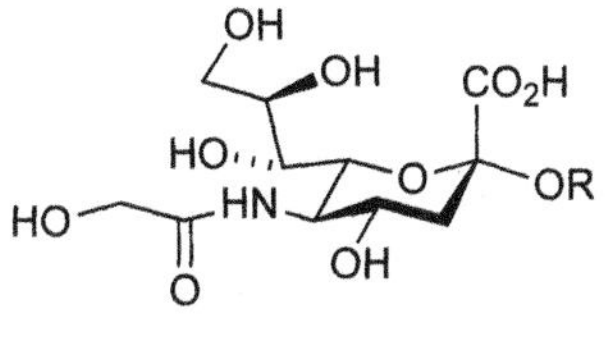   N-glycolylneuraminic acid (Neu5Gc)

Other Carbohydrate Modifications

The above sections have endeavoured to provide a brief overview of the major classes of carbohydrate modifications seen in nature. In addition to those discussed above, there are other carbohydrate modifications that endow glycoconjugates with intriguing structural accents. These are either seen more rarely, or less is understood regarding their biological roles. For example, O-methylation is widespread and occurs through various carbohydrate O-methyltransferases that utilize S-methyladenosine as the methyl donor. O-Methylation typically occurs post-glycosylationally but has been observed pre-glycosylationally, at the level of the sugar nucleotide.[104] Aside from antigenic variation, O-methylation results in large changes in hydrophobicity of the resultant methylated glycoconjugate. For example, some mycobacteria produce various 3-O-methylmannose-containing poly-saccharides (MMP) and 6-O-methylglucose-containing LPSs that form stoichiometric complexes with C_{16}- or longer acyl-CoA and that influence fatty acid biosynthesis in *Mycobacterium smegmatis*.[105]

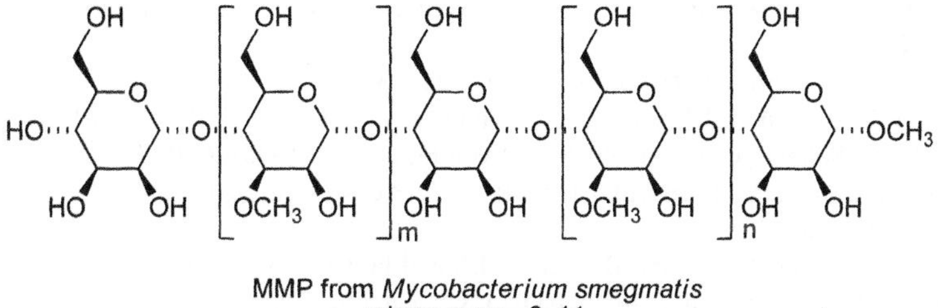

MMP from *Mycobacterium smegmatis*
where m + n = 8–11

References start on page 364

The lactyl ether in the muramic acid residue found in bacterial peptidoglycan is, of course, widespread and is a pre-glycosylational modification resulting from conversion of UDP-N-acetylglucosamine to UDP-N-acetylmuramic acid; its biosynthesis from phosphoenolpyruvate is covered in Chapter 6. A not uncommon relative of the lactyl ether modification is the pyruvyl acetal, which is found in yeast and a variety of bacterial cell surface carbohydrates,[106] particularly those from Rhizobia,[107] and also in the extractable glycolipid A from *M. smegmatis*.[108,109]

glycolipid A from *Mycobacterium smegmatis*

References

1. Yu, H. and Chen, X. (2007). *Org. Biomol. Chem.*, **5**, 865.
2. Valla, S., Li, J., Ertesvåg, H., Barbeyron, T. and Lindahl, U. (2001). *Biochimie*, **83**, 819.
3. Govan, J.R. and Deretic, V. (1996). *Microbiol. Rev.*, **60**, 539.
4. Douthit, S.A., Dlakic, M., Ohman, D.E. and Franklin, M.J. (2005). *J. Bacteriol.*, **187**, 4573.
5. Chapman, E., Best, M.D., Hanson, S.R. and Wong, C.-H. (2004). *Angew. Chem. Int. Ed.*, **43**, 3526.
6. Grunwell, J.R. and Bertozzi, C.R. (2002). *Biochemistry*, **41**, 13117.
7. Honke, K. and Taniguchi, N. (2002). *Med. Res. Rev.*, **22**, 637.
8. Thiele, H., Sakano, M., Kitagawa, H., Sugahara, K., Rajab, A., Hohne, W., Ritter, H., Leschik, G., Nurnberg, P. and Mundlos, S. (2004). *Proc. Natl. Acad. Sci. USA*, **101**, 10155.
9. Hanson, S.R., Best, M.D. and Wong, C.-H. (2004). *Angew. Chem. Int. Ed.* **43**, 5736.
10. Neufeld, E.F. and Muenzer, J. (1995). In *The Metabolic and Molecular Basis of Inherited Disease,* 7th ed. (C.R. Scriver, A.L. Beaudet, W.S. Sly and D. Valle, eds) p. 2465. McGraw-Hill: New York.
11. Morimoto-Tomita, M., Uchimura, K., Werb, Z., Hemmerich, S. and Rosen, S.D. (2002). *J. Biol. Chem.*, **277**, 49175.
12. Kertesz, M.A. (2000). *FEMS Microbiol. Rev.*, **24**, 135.
13. Meikle, P.J., Hopwood, J.J., Clague, A.E. and Carey, W.F. (1999). *JAMA*, **281**, 249.
14. Schmidt, B., Selmer, T., Ingendoh, A. and von Figura, K. (1995). *Cell*, **82**, 271.
15. Dierks, T., Schmidt, B., Borissenko, L.V., Peng, J., Preusser, A., Mariappan, M. and von Figura, K. (2003). *Cell*, **113**, 435.
16. Woo, L.L., Purohit, A., Malini, B., Reed, M.J. and Potter, B.V. (2000). *Chem. Biol.*, **7**, 773.
17. Settembre, C., Annunziata, I., Spampanato, C., Zarcone, D., Cobellis, G., Nusco, E., Zito, E., Tacchetti, C., Cosma, M.P. and Ballabio, A. (2007). *Proc. Natl. Acad. Sci. USA*, **104**, 4506.

18. Sardinello, M., Annunziata, I., Roma, G. and Ballabio, A. (2005). *Hum. Mol. Genet.*, **14**, 3203.
19. Witczak, Z.J. and Nieforth, K.A., eds (1997). *Carbohydrates in Drug Design.* Marcel Dekker, Inc: New York, USA.
20. Sinaÿ, P., Jacquinet, J.C., Petitou, M., Duchaussoy, P., Lederman, I., Choay, J. and Torri, G. (1984). *Carbohydr. Res.*, **132**, C5.
21. van Boeckel, C.A., Beetz, T., Vos, J.N., de Jong, A.J.M., van Aelst, S.F., van den Bosch, R.H., Mertens, J.M.R. and van der Vlugt, F.A. (1985). *J. Carbohydr. Chem.*, **4**, 293.
22. Petitou, M., Duchaussoy, P., Lederman, I., Choay, J., Sinaÿ, P., Jacquinet, J.C. and Torri, G. (1986). *Carbohydr. Res.*, 147, 221.
23. Petitou, M. and van Boeckel, C.A.A. (2004). *Angew. Chem. Int. Ed.*, **43**, 3118.
24. van Boeckel, C.A.A. and Petitou, M. (1993). *Angew. Chem. Int. Ed.*, **32**, 1671.
25. Codée, J.D.C., Overkleeft, H.S., van der Marel, G.A. and van Boeckel, C.A.A. (2004). *Drug Discov. Today: Technologies*, **1**, 317.
26. Cullimore, J.V., Ranjeva, R. and Bono, J.J. (2001). *Trends Plant Sci.*, **6**, 24.
27. D'Haeze, W. and Holsters, M. (2002). *Glycobiology*, **12**, 79R.
28. Lerouge, P., Roche, P., Faucher, C., Maillet, F., Truchet, G., Prome, J.C. and Denarie, J. (1990). *Nature*, **344**, 781.
29. Radutoiu, S., Madsen, L.H., Madsen, E.B., Felle, H.H., Umehara, Y., Gronlund, M., Sato, S., Nakamura, Y., Tabata, S., Sandal, N. and Stougaard, J. (2003). *Nature*, **425**, 585.
30. Limpens, E., Franken, C., Smit, P., Willemse, J., Bisseling, T. and Geurts, R. (2003). *Science*, **302**, 630.
31. Madsen, E.B., Madsen, L.H., Radutoiu, S., Olbryt, M., Rakwalska, M., Szczyglowski, K., Sato, S., Kaneko, T., Tabata, S., Sandal, N. and Stougaard, J. (2003). *Nature*, **425**, 637.
32. Roche, P., Debelle, F., Maillet, F., Lerouge, P., Faucher, C., Truchet, G., Denarie, J. and Prome, J.C. (1991). *Cell*, **67**, 1131.
33. Kates, M. (1993). *Experientia*, **49**, 1027.
34. Desmarais, D., Jablonski, P.E., Fedarko, N.S. and Roberts, M.F. (1997). *J. Bacteriol.*, **179**, 3146.
35. Mougous, J.D., Green, R.E., Williams, S.J., Brenner, S.E. and Bertozzi, C.R. (2002). *Chem. Biol.*, **9**, 767.
36. Schelle, M.W. and Bertozzi, C.R. (2006). *ChemBioChem*, **7**, 1516.
37. Mougous, J.D., Petzold, C.J., Senaratne, R.H., Lee, D.H., Akey, D.L., Lin, F.L., Munchel, S.E., Pratt, M.R., Riley, L.W., Leary, J.A., Berger, J.M. and Bertozzi, C.R. (2004). *Nat. Struct. Mol. Biol.*, **11**, 721.
38. Leigh, C.D. and Bertozzi, C.R. (2008). *J. Org. Chem.*, **73**, 1008.
39. Khoo, K.H., Jarboe, E., Barker, A., Torrelles, J., Kuo, C.W. and Chatterjee, D. (1999). *J. Biol. Chem.*, **274**, 9778.
40. López Marín, L.M., Lanéelle, M.A., Promé, D., Lanéelle, G., Promé, J.C. and Daffé, M. (1992). *Biochemistry*, **31**, 11106.
41. McCormick, J., Li, Y.B., McCormick, K., Duynstee, H.I., van Engen, A.K., van der Marel, G.A., Ganem, B., van Boom, J.H. and Meinwald, J. (1999). *J. Am. Chem. Soc.*, **121**, 5661.
42. Taggi, A.E., Meinwald, J. and Schroeder, F.C. (2004). *J. Am. Chem. Soc.*, **126**, 10364.
43. Ubukata, M., Isono, K., Kimura, K.-i., Nelson, C. and McCloskey, J.A. (1988). *J. Am. Chem. Soc.*, **110**, 4416.
44. Knapp, S., Morriello, G.J. and Doss, G.A. (2002). *Org. Lett.*, **4**, 603.
45. Kimura, K. and Bugg, T.D. (2003). *Nat. Prod. Rep.*, **20**, 252.
46. Bertozzi, C.R. (1995). *Chem. Biol.*, **2**, 703.
47. Hemmerich, S. and Rosen, S.D. (2000). *Glycobiology*, **10**, 849.
48. Hemmerich, S. and Rosen, S.D. (1994). *Biochemistry*, **33**, 4830.
49. Hemmerich, S., Bertozzi, C.R., Leffler, H. and Rosen, S.D. (1994). *Biochemistry*, **33**, 4820.

50. Ishizuka, I. (1997). *Prog. Lipid Res.*, **36**, 245.
51. Coetzee, T., Suzuki, K. and Popko, B. (1998). *Trends Neurosci.*, **21**, 126.
52. Gieselmann, V., Matzner, U., Hess, B., Lullmann-Rauch, R., Coenen, R., Hartmann, D., D'Hooge, R., DeDeyn, P. and Nagels, G. (1998). *J. Inherit. Metab. Dis.*, **21**, 564.
53. Honke, K., Zhang, Y., Cheng, X., Kotani, N. and Taniguchi, N. (2004). *Glycoconj. J.*, **21**, 59.
54. Kornfeld, S. (1987). *FASEB J.*, **1**, 462.
55. von Figura, K. and Hasilik, A. (1986). *Annu. Rev. Biochem.*, **55**, 167.
56. Hasilik, A., Klein, U., Waheed, A., Strecker, G. and von Figura, K. (1980). *Proc. Natl. Acad. Sci. USA*, **77**, 7074.
57. Reitman, M.L. and Kornfeld, S. (1981). *J. Biol. Chem.*, **256**, 11977.
58. Rohrer, J. and Kornfeld, R. (2001). *Mol. Biol. Cell.*, **12**, 1623.
59. Varki, A., Sherman, W. and Kornfeld, S. (1983). *Arch. Biochem. Biophys.*, **222**, 145.
60. Tabas, I. and Kornfeld, S. (1980). *J. Biol. Chem.*, **255**, 6633.
61. Descoteaux, A. and Turco, S.J. (1999). *Biochim. Biophys. Acta*, **1455**, 341.
62. Turco, S.J. and Descoteaux, A. (1992). *Annu. Rev. Microbiol.*, **46**, 65.
63. Carver, M.A. and Turco, S.J. (1992). *Arch. Biochem. Biophys.*, **295**, 309.
64. Carver, M.A. and Turco, S.J. (1991). *J. Biol. Chem.*, **266**, 10974.
65. Moody-Haupt, S., Patterson, J.H., Mirelman, D. and McConville, M.J. (2000). *J. Mol. Biol.*, **297**, 409.
66. McConville, M.J., Mullin, K.A., Ilgoutz, S.C. and Teasdale, R.D. (2002). *Microbiol. Mol. Biol. Rev.*, **66**, 122.
67. Archibald, A.R. and Baddiley, J. (1966). *Adv. Carbohydr. Chem.*, **21**, 323.
68. Neuhaus, F.C. and Baddiley, J. (2003). *Microbiol. Mol. Biol. Rev.*, **67**, 686.
69. Ward, J.B. (1981). *Microbiol. Rev.*, **45**, 211.
70. Stadelmaier, A., Morath, S., Hartung, T. and Schmidt, R.R. (2003). *Angew. Chem. Int. Ed.*, **42**, 916.
71. Archibald, A.R., Baddiley, J., Button, D., Heptinstall, S. and Stafford, G.H. (1968). *Nature*, **219**, 855.
72. Cutler, J.E. (2001). *Med. Mycol.*, **39**, 75.
73. Slodki, M.E. (1966). *J. Biol. Chem.*, **241**, 2700.
74. Knirel, Y.A., Paredes, L., Jansson, P.-E., Weintraub, A., Widmalm, G. and Albert, M.J. (1995). *Eur. J. Biochem.*, **232**, 391.
75. Ruttens, B. and Kovác, P. (2006). *Carbohydr. Res.*, **341**, 1077.
76. Miller, S.I., Ernst, R.K. and Bader, M.W. (2005). *Nat. Rev. Microbiol.*, **3**, 36.
77. Karapetyan, G., Kaczynski, Z., Iacobellis, N.S., Evidente, A. and Holst, O. (2006). *Carbohydr. Res.*, **341**, 930.
78. Takayama, K., Wang, C. and Besra, G.S. (2005). *Clin. Microbiol. Rev.*, **18**, 81.
79. Crick, D.C., Mahapatra, S. and Brennan, P.J. (2001). *Glycobiology*, **11**, 107R.
80. D'Souza, F.W. and Lowary, T.L. (2000). *Org. Lett.*, **2**, 1493.
81. Barry, C.E., 3rd, Lee, R.E., Mdluli, K., Sampson, A.E., Schroeder, B.G., Slayden, R.A. and Yuan, Y. (1998). *Prog. Lipid Res.*, **37**, 143.
82. Brennan, P.J. and Nikaido, H. (1995). *Annu. Rev. Biochem.*, **64**, 29.
83. Kremer, L., Baulard, A.R., and Besra, G.S. (2000). In *Molecular Genetics of Mycobacteria* (G.H. Hatful and W.R. Jacobs, Jr., eds) p. 173. ASM Press: Washington, DC.
84. Ryll, R., Kumazawa, Y. and Yano, I. (2001). *Microbiol. Immunol.*, **45**, 801.
85. Belisle, J.T., Vissa, V.D., Sievert, T., Takayama, K., Brennan, P.J. and Besra, G.S. (1997). *Science*, **276**, 1420.
86. Ronning, D.R., Klabunde, T., Besra, G.S., Vissa, V.D., Belisle, J.T. and Sacchettini, J.C. (2000). *Nat. Struct. Biol.*, **7**, 141.

87. Ronning, D.R., Vissa, V.D., Besra, G.S., Belisle, J.T. and Sacchettini, J.C. (2004). *J. Biol. Chem.*, **279**, 36771.
88. Carpita, N. and McCann, M. (2002). In *Biochemistry and Molecular Biology of Plants* (B.B. Buchanan, W. Gruissem and R.L. Jones, eds) p. 52. American Society of Plant Physiologists: Rockville, MD.
89. Henrissat, B., Coutinho, P.M. and Davies, G.J. (2001). *Plant Mol. Biol.*, **47**, 55.
90. Mathew, S. and Abeaham, T.E. (2004). *Crit. Rev. Biotechnol.*, **24**, 59.
91. Micheli, F. (2001). *Trends Plant Sci.*, **6**, 414.
92. Saha, B.C. (2003). *J. Ind. Microbiol. Biotechnol.*, **30**, 279.
93. Davies, G.J., Gloster, T.M. and Henrissat, B. (2005). *Curr. Opin. Struct. Biol.*, **15**, 637.
94. Markovic, O. and Janecek, S. (2004). *Carbohydr. Res.*, **339**, 2281.
95. Kroon, P.A., Williamson, G., Fish, N.M., Archer, D.B. and Belshaw, N.J. (2000). *Eur. J. Biochem.*, **267**, 6740.
96. Faulds, C.B., Molina, R., Gonzalez, R., Husband, F., Juge, N., Sanz-Aparicio, J. and Hermoso, J.A. (2005). *FEBS J.*, **272**, 4362.
97. Taylor, E.J., Gloster, T.M., Turkenburg, J.P., Vincent, F., Brzozowski, A.M., Dupont, C., Shareck, F., Centeno, M.S., Prates, J.A., Puchart, V., Ferreira, L.M., Fontes, C.M., Biely, P. and Davies, G.J. (2006). *J. Biol. Chem.*, **281**, 10968.
98. Angata, T. and Varki, A. (2002). *Chem. Rev.*, **102**, 439.
99. Mitsuoka, C., Ohmori, K., Kimura, N., Kanamori, A., Komba, S., Ishida, H., Kiso, M. and Kannagi, R. (1999). *Proc. Natl. Acad. Sci. USA*, **96**, 1597.
100. Rogers, G.N., Herrler, G., Paulson, J.C. and Klenk, H.D. (1986). *J. Biol. Chem.*, **261**, 5947.
101. Chou, H.H., Takematsu, H., Diaz, S., Iber, J., Nickerson, E., Wright, K.L., Muchmore, E.A., Nelson, D.L., Warren, S.T. and Varki, A. (1998). *Proc. Natl. Acad. Sci. USA*, **95**, 11751.
102. Muchmore, E.A., Diaz, S. and Varki, A. (1998). *Am. J. Phys. Anthropol.*, **107**, 187.
103. Angata, T., Varki, N.M. and Varki, A. (2001). *J. Biol. Chem.*, **276**, 40282.
104. Weisman, L.S. and Ballou, C.E. (1984). *J. Biol. Chem.*, **259**, 3464.
105. Cheon, H.S., Wang, Y., Ma, J. and Kishi, Y. (2007). *ChemBioChem*, **8**, 353.
106. Fontaine, T., Talmont, F., Dutton, G.G. and Fournet, B. (1991). *Anal. Biochem.*, **199**, 154.
107. Eckhardt, E. and Ziegler, T. (1994). *Carbohydr. Res.*, **264**, 253.
108. Kamisango, K., Saadat, S., Dell, A. and Ballou, C.E. (1985). *J. Biol. Chem.*, **260**, 4117.
109. Ziegler, T., Eckhardt, E. and Birault, V. (1993). *J. Org. Chem.*, **58**, 1090.

Chapter 11

Glycoproteins and Proteoglycans

Proteins produced by the cell may be modified with a range of carbohydrate structures. The modifications may occur prior to delivery to intracellular organelles, secretion from the cell or incorporation into membranes, or they may occur on a mature protein present in the cytosol or nucleus of the cell. There are three main types of protein modification: N-linked glycosylation, O-linked glycosylation and the attachment of glycosylphosphatidylinositol (GPI) anchors. A relatively rare modification is C-linked glycosylation. The reasons for protein glycosylation are at best only partly understood but include roles in ensuring proper folding of proteins and their degradation when incompetently folded, ensuring correct targeting of glycoproteins to intra- and extracellular destinations, altering susceptibility to degradation, modulation of immunological properties and the engagement of lectins. This chapter will review the major types of protein glycosylation found in higher organisms, with a particular emphasis on mammalian biology. While an immense number of structures is possible through the connection of the 10 common mammalian monosaccharides (Glc, Gal, Man, Sia, GlcNAc, GalNAc, Fuc, Xyl, GlcA and IdoA), systematic structural analysis reveals that only a small number of monosaccharide connections accounts for the majority of the linkages. Werz et al. have reported that of a set of 3299 oligosaccharides obtained from 38 mammalian species, 11 monosaccharide connections accounted for >75% of all glycosidic linkages.[1]

Glycoproteins are proteins that bear one or more O-, N- or C-linked glycan chains. These chains comprise typically up to 20% but, occasionally, more than 90% of the mass of the resultant conjugate. A *proteoglycan* is a peptide or protein that bears one or more covalently attached glycosaminoglycan chains. As has been noted, the distinction between a glycoprotein and a proteoglycan is somewhat arbitrary, as the same polypeptide may bear both glycosaminoglycan chains and O-, N- or C-linked glycans.[2,3] A *mucin* is a glycoprotein that carries many O-linked glycan chains that are clustered together. A *glycosphingolipid* is a mono- or oligosaccharide that is (usually) attached via glucose or galactose to ceramide. A *ganglioside* is an anionic glycolipid that contains one or more residues of sialic acid. A *GPI anchor* is a glycan structure that links a phosphatidylinositol (PI) to a phosphoethanolamine, attached as an amide, to the C-terminus of a protein.

References start on page 410

Eukaryotic cells contain an interconnected network of membrane-bound cisternae, vesicles and vacuoles that are connected to the plasma membrane and comprise the secretory and endocytic pathways.[4] Newly synthesized proteins destined for the cell surface, secretion or other compartments in the secretory pathway are initially translocated into the endoplasmic reticulum (ER). The ER comprises the start of the secretory pathway and is a relatively large, highly dynamic organelle, which is adjoined to the nucleus of the cell and can be divided into three domains: the ribosome-studded rough ER, the ribosome-free smooth ER and the nuclear envelope. Proteins are assembled by translation in the rough ER by ribosomes, multicomponent systems comprised of proteins and RNA. Protein folding occurs in the ER, and various carbohydrates may also be added here (e.g. N-linked glycan core structures and GPI anchors). Proteins may leave the ER directly or by passage through the Golgi apparatus,[a] which contains a complex post-translational modification machinery. Alternatively, folding-incompetent proteins are detected and degraded by an ER-associated degradation (ERAD) pathway. The Golgi apparatus is a polar organelle comprised of a stack of membrane-bound sacs termed cisternae. The cis face accepts vesicular traffic from the ER, and the trans face acts as the exit site for vesicular traffic. In the Golgi apparatus a complex array of protein modifications may occur including glycosylation, deglycosylation, O- and N-sulfation, and O-acetylation, resulting in the formation of O-linked glycans (e.g. mucin-type O-linked glycans and proteoglycans) and the elaboration of N-linked glycans. Additionally, specific carbohydrate structures (mannose-6-phosphate groups) may be added to ensure targeting of various, generally catabolic, enzymes to an intracellular organelle termed the lysosome. The lysosome is a digestive organelle rich in hydrolases that receives (glyco)proteins from the cytosol, the plasma membrane and the extracellular milieu that are destined for degradation, thereby allowing their constituents to re-enter cellular metabolic processes.

N-Linked Glycosylation[5]

N-Linked glycans are added to the majority of proteins that enter the secretory pathway and are important for their folding, stability and function. N-Linked glycosylation involves *en bloc* transfer of preformed glycans (see below) through a β-linkage to the side-chain nitrogen of asparagine. The asparagine residue is found in the highly conserved sequence Asn-Xxx-Thr, where Xxx is any residue but proline. While N-linked glycans possess complex and diverse structures dependent on their origin, when they are initially added to growing peptides, these glycans share a

---

[a] Camillo Golgi (1843–1926), M.D. (1865). Professor at Pavia University, Italy. Nobel Prize in Physiology or Medicine (1906).

common 14 sugar oligosaccharide. This oligosaccharide is biosynthesized as a glycophospholipid, dolichol-PP-GlcNAc$_2$Man$_9$Glc$_3$.

dolichol-PP-GlcNAc$_2$Man$_9$Glc$_3$

● = glucopyranose; ○ = mannopyranose; ■ = N-acetylglucosamine

One particularly well-understood pathway to N-linked glycoprotein synthesis is one that occurs upon infection of mammalian cells with vesicular stomatis virus (VSV). Infection with this virus results in cooption of essentially all of the cells protein synthesis apparatus to produce a single homogeneous viral coat glycoprotein, the VSV G-protein, thereby providing a useful model system for the study of glycoprotein biosynthesis, and much of the discussion that follows refers to data obtained from these studies.[5]

Biosynthesis of the lipid-linked oligosaccharide:[6] The biosynthesis of the 14 residue dolichol-PP-GlcNAc$_2$Man$_9$Glc$_3$ oligosaccharide occurs on a lipid-linked acceptor, the isoprenoid dolichol, in the ER. Biosynthesis commences on the cytoplasmic face of the ER membrane with the addition of N-acetylglucosamine phosphate from UDP-N-acetylglucosamine to dolichol phosphate by the action of N-acetylglucosaminyl phosphate transferase. The second GlcNAc residue is added by a UDP-N-acetylglucosamine-dependent glycosyltransferase. A series of GDP-mannose-dependent mannosyltransferases (ManTs) then add five mannose residues to the dolichol-PP-GlcNAc$_2$ core yielding a heptasaccharide, dolichol-PP-GlcNAc$_2$Man$_5$. At this stage the heptasaccharide must be transferred from the cytoplasmic face of the ER to the luminal face. It remains to be determined whether the translocation of this large glycolipid is mediated by a membrane-embedded flippase protein, or whether the ER membrane is sufficiently fluid to allow its rapid and spontaneous movement across the bilayer. The final steps in the construction of the dolichol-PP-GlcNAc$_2$Man$_9$Glc$_3$ tetradecasaccharide are accomplished by the sequential action of a series of ManTs and glucosyltransferases on the luminal face of the ER. Each of these sugar residues is provided by a lipid-linked glycosyl donor, dolichol-P-mannose or dolichol-P-glucose.

References start on page 410

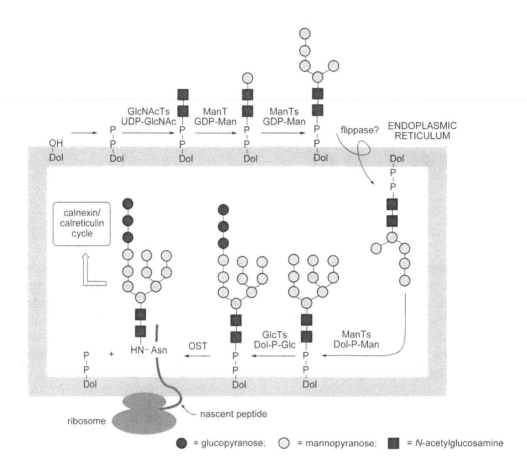

Evidence for the occurrence of steps on the exterior and interior faces of the ER was provided by the ability of membrane-impermeable reagents to disrupt early steps in the process, and the inability of the mannose-binding lectin concanavalin A to bind to the products of these reactions until the ER membrane was permeabilized.[7]

Transfer of the lipid-linked oligosaccharide: N-Linked glycosylation is a co-translational modification that occurs after synthesis of dolichol-PP-GlcNAc$_2$Man$_9$Glc$_3$. A glycosyltransferase called oligosaccharyltransferase (OST), comprising a complex of up to 10 proteins (depending on the cell type), transfers the entire glycan structure *en bloc* to the side-chain nitrogen of asparagine, found within a nascent peptide chain, during the process of translation. OST recognizes a specific conformation of the target protein, which cannot be achieved with a proline residing adjacent to the Asn acceptor, explaining the inability of the sequence Asn-Pro-Ser/Thr to be *N*-glycosylated.

N-Glycan trimming and the calnexin/calreticulin cycle: Immediately after the transfer of the pre-assembled core glycan to the Asn of an acceptor protein, two of the three terminal glucose residues are removed by α-glucosidase I. Concomitantly, synthesis of the protein is completed, and the unfolded N-glycosylated peptide chain is released from the ribosome into the lumen of the ER. The unfolded monoglucosylated N-glycan interacts with two carbohydrate-binding lectins, calnexin (CNX, a membrane-associated protein) and calreticulin (CRT, a soluble protein). CNX and CRT are chaperones that assist in protein folding. Some data suggest that the nascent N-glycoprotein can interact with CNX and CRT co-translationally, before release from the ribosome.[8] The resulting lectin/glycan complex interacts with ERp57, a thiol oxidoreductase that assists in the formation of disulfides in the protein. The CNX/CRT glycoprotein complex with ERp57 and other luminal factors are believed to form a 'folding cage', an environment that protects the unfolded peptide from interacting with other unstructured nascent chains and prevents the formation of aberrant protein aggregates. Also, formation of the lectin/glycan complex increases the residence time of the protein within the ER, allowing time for correct folding.

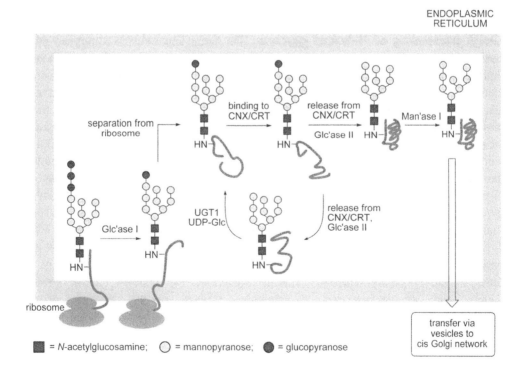

Upon release of the glycoprotein from CNX/CRT, the last glucose residue is trimmed from the N-glycan by α-glucosidase II. Deglucosylated but unfolded

N-linked glycoproteins are substrates for a luminal UDP-glucose-dependent glycoprotein glucosyltransferase (UGT1, also termed UGGT), which recognizes unfolded proteins and adds a single glucose residue to uncapped high mannose N-glycans. This allows rebinding to CNX and CRT and another attempt at correct folding. Thus, UGT1 acts as a sensor for correct folding of a protein, with addition of a glucose residue occurring only if folding is incomplete and thus allowing re-entry into the CNX-CRT cycle. Deletion of UGT1 from cells does not result in a reduction in ER quality control, suggesting that UGT1 is not the gatekeeper that prevents non-native polypetides from exiting the ER. Instead it appears that UGT1 delays the exit of unfolded proteins from cells by allowing them to re-enter the CNX/CRT cycle, in what would be a futile cycle for folding-incompetent proteins unless folding defective proteins could exit the cycle, a process achieved by ERAD. The final step of N-linked oligosaccharide processing, which occurs only for properly folded proteins in the ER prior to exit to the Golgi, is the trimming of a single mannose residue by ER α-1,2-mannosidase I to yield a $GlcNAc_2Man_8$ decasaccharide. Up until this point, the biosynthesis of all N-linked oligosaccharides is identical, but from here the biosynthetic pathways diverge dramatically.

Golgi processing of N-linked glycans: The Golgi apparatus is comprised of a series of membranous sacs termed cisternae and is separate from the ER. Proteins and lipids are transferred between these two organelles by vesicular transport. The Golgi contains functionally distinct cisternae, termed the *cis*, *medial* and *trans* cisternae. These cisternae retain an approximately constant composition while allowing most proteins to pass through the structure by a process of anterograde and retrograde vesicle traffic and cisternal maturation. Finally, vesicles bud off the trans Golgi complex, allowing for the delivery of product glycoproteins to their (extra)cellular destinations. The Golgi apparatus plays a major role in the late-stage modification of N-linked glycoproteins and contains a large number of luminally oriented glycosyltransferases, trimming glycoside hydrolases and enzymes that install functional groups such as *O*-acetyl and *O*- and *N*-sulfate residues. In the case of VSV G-protein biosynthesis, following transport of the $GlcNAc_2Man_8$ decasaccharide via vesicles to the cis Golgi, Golgi α-mannosidase I trims three mannose residues to form a $GlcNAc_2Man_5$ structure. This is transferred by vesicles to the medial Golgi, and an orchestrated modification process occurs through which N-acetylglucosaminyltransferase I adds one N-acetylglucosamine residue, Golgi α-mannosidase II trims two mannose residues and N-acetylglucosaminyltransferase II adds two additional N-acetylglucosamine residues. Finally, a fucosyltransferase (FucT) fucosylates the core N-acetylglucosamine residue. The resulting $GlcNAc_3Man_3(Fuc)GlcNAc_2$-modified protein is transferred by vesicular transport to the trans Golgi, then first galactosylated by a galactosyltransferase and then sialylated by a sialyltransferase before exit from the trans Golgi network in vesicles.

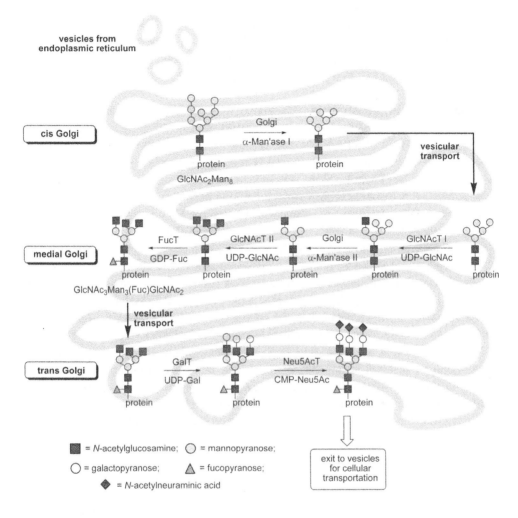

vesicles from
endoplasmic reticulum

cis Golgi

Golgi
α-Man'ase I

protein

GlcNAc₂Man₈

protein

vesicular
transport

medial Golgi

FucT
GDP-Fuc

GlcNAcT II
UDP-GlcNAc

Golgi
α-Man'ase II

GlcNAcT I
UDP-GlcNAc

protein

protein

protein

protein

protein

GlcNAc₃Man₃(Fuc)GlcNAc₂

vesicular
transport

trans Golgi

GalT
UDP-Gal

Neu5AcT
CMP-Neu5Ac

protein

protein

protein

■ = *N*-acetylglucosamine;   ○ = mannopyranose;

○ = galactopyranose;   ▲ = fucopyranose;

◆ = *N*-acetylneuraminic acid

exit to vesicles
for cellular
transportation

ER-associated protein degradation:[9–11] Misfolded proteins occur during normal cellular processes and are produced in greater abundance in conditions of stress. Proteins that fail to fold to their correct conformation in the ER are prevented from entering the Golgi apparatus and secretory pathways and are eliminated and degraded in a still incompletely understood pathway termed ERAD. ERAD plays a crucial clearance role, extracting folding-incompetent proteins from the CNX/CRT cycle and retrotranslocating unfolded glycoproteins to the cytosol, where they are ubiquitinylated and then proteasomally degraded. While it has been shown that N-glycans on folding-defective proteins are extensively demannosylated, the most important step appears to be the α-mannosidase-catalysed removal of the α-1,2-linked mannose residue that is reglucosylated by UGT1, enabling rebinding to CNX/CRT. The α-mannosidase(s) that performs this task are newly discovered proteins, EDEM1, EDEM2 and/or

EDEM3.[8] Removal of the acceptor mannose residue, which can be reglucosylated by UGT1, results in irreversible interruption of futile folding attempts in the CNX/CRT cycle. The steps leading to further degradation of the N-glycan, and retrotranslocation of the incompetently folded protein to the cytosol for degradation are poorly understood.

Diversity of N-linked glycans: The above discussion describes the biosynthesis of only one N-linked glycan. In nature, N-linked glycans display an extraordinary diversity, and even N-linked glycans on the same protein may have different structures. However, the processing of all N-linked oligosaccharides is identical up to the formation of the $GlcNAc_2Man_8$ structure formed immediately after the CNX/CRT cycle, so that all N-linked glycans share a common $GlcNAc_2Man_3$ pentasaccharide core (five non-core Man residues are cleaved in the case of VSV G-protein glycans). The oligosaccharides that result from this common pentasaccharide can be classified into three groups. *High mannose oligosaccharides* have two to six mannose residues added to the core pentasaccharide; *complex oligosaccharides* contain varying numbers of sialylated *N*-acetyllactosamine units as well as a fucose residue attached to the core GlcNAc, and a bisecting GlcNAc residue linked to the β-linked mannose residue; *hybrid oligosaccharides* contain features of both the complex and the high mannose structures and usually contain a bisecting GlcNAc. The complex oligosaccharides may be further modified by the addition of extra branches on the α-linked mannose residues or the addition of extra sugars that elongate and introduce branching into the outer chains. Further post-glycosylational modifications occur through the action of Golgi-resident sulfotransferases and *O*-acetyltransferases that add *N*- and *O*-sulfate and *O*-acetyl groups, respectively, to maturing N-linked oligosaccharides.

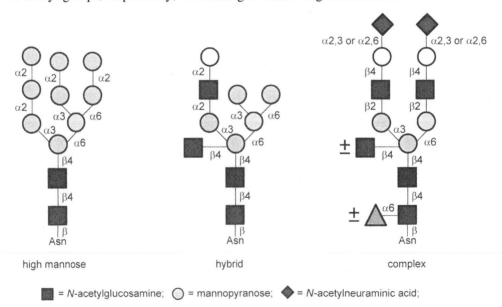

high mannose   hybrid   complex

■ = *N*-acetylglucosamine; ◯ = mannopyranose; ◆ = *N*-acetylneuraminic acid;

◯ = galactopyranose; ▲ = fucopyranose

Inhibitors of N-linked glycoprotein biosynthesis: Inhibitors of N-linked glycoprotein biosynthesis and degradation have proven important tools in the study of these pathways. Tunicamycin inhibits the synthesis of dolichol-PP-GlcNAc, the first lipid-linked carbohydrate in the pathway for the synthesis of dolichol-PP-GlcNAc$_2$Man$_9$, through inhibition of GlcNAc phosphorotransferase. Tunicamycin therefore decreases the amount of dolichol-PP-GlcNAc available for the synthesis of N-linked glycoproteins and is a widely used reagent for the study of the effect of the complete loss of N-linked glycosylation. Bacitracin, a cyclic polypeptide, forms a complex with dolichol-PP that prevents its dephosphorylation, thereby preventing the biosynthesis of lipid-linked glycoprotein precursors. Amphomycin, also a peptide antibiotic, forms a complex with dolichol-P, thereby preventing its utilization in N-linked glycoprotein biosynthesis.[12] As mentioned in Chapter 8, Imperiali and co-workers have reported the development of a membrane-permeable cyclic peptide inhibitor of OST. This compound was designed with knowledge of the preference of this enzyme for a specific conformation of the Asn-Xxx-Ser/Thr consensus sequence and has been modified by the inclusion of a membrane-permeable peptide to ensure that the inhibitor crosses the intracellular membrane of the ER.[13]

oligosaccharyltransferase inhibitor

tunicamycins

NH-Val-Thr-Ala-Ala-Val-Leu-Leu-Pro-Val-Leu-Leu-Ala-Ala-Pro-CO$_2^-$

Various sugar-shaped nitrogen heterocycles are especially useful in blocking processing glycosidases involved in N-linked glycan biosynthesis and degradation.[14,15] These inhibitors do not interfere with protein folding but lead to glycoproteins bearing truncated N-glycans. Examples include glucosidase inhibitors: australine, an inhibitor of α-glucosidase I, and castanospermine and deoxynojirimycins (see Chapter 7), inhibitors of α-glucosidase I and II. These inhibitors lead to accumulation of N-glycans bearing varying numbers of glucose residues. Swainsonine and mannostatin

A are inhibitors of Golgi α-mannosidase II, which cause accumulation of high-mannose oligosaccharides.[14,16] Deoxymannojirimycin and kifunensine are inhibitors of α-mannosidase I and lead to accumulation of $GlcNAc_2Man_{7-9}$ oligosaccharides. Kifunensine is also an inhibitor of the demannosylation step that leads to ERAD.[11]

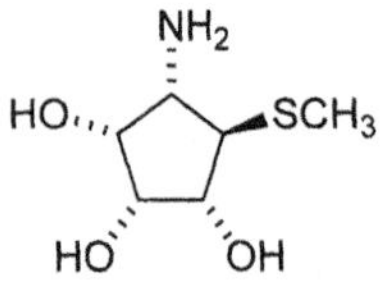

mannostatin A 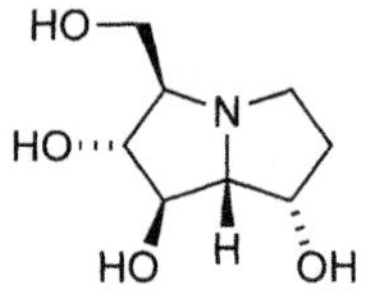   australine 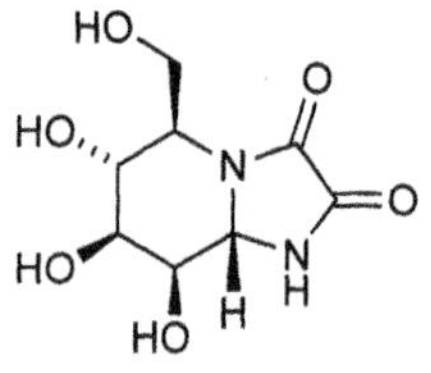   kifunensine

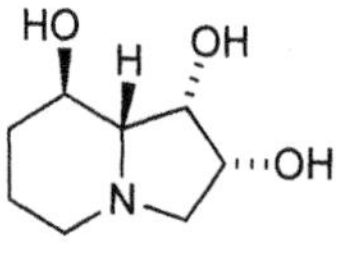

swainsonine 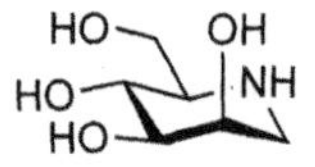   1-deoxymannojirimycin

Enveloped viruses such as HIV can evade immune recognition by using host N-glycosylation machinery to protect potentially immunogenic coat proteins. The major protein component of the HIV-1 coat is gp160, a heterodimer of gp120 and gp41. When gp120 is expressed in the presence of N-butyldeoxynojirimycin, an inhibitor of glucosidases I and II, part of this protein is misfolded. This misfolding does not prevent formation of the viral coat but prevents fusion of the virus with the host cell and prevents entry into the cell.

Modification of N-Linked Glycans for Lysosomal Targeting

Proteins destined for lysosomes undergo modification to install mannose-6-phosphate residues, which interact with the mannose-6-phosphate receptor resulting in translocation to the lysosome. Two enzymes in the cis Golgi generate this residue; first UDP-N-acetylglucosamine:lysosomal-enzyme N-acetylglucosamine-1-phosphotransferase (N-acetylglucosamine-phosphotransferase) transfers N-acetyl-glucosamine 1-phosphate onto the 6-hydroxyl of a mannose residue and, secondly, a glycoside hydrolase, N-acetylglucosamine-1-phosphodiester α-N-acetylgluco-samidase (uncovering enzyme, UCE),[17,18] cleaves the glycosidic bond joining the N-acetylglucosamine group to the phosphate, leaving behind a mannose-6-phosphate group. Upon arrival in the lysosomes, the phosphate ester is cleaved by lysosomal phosphatases. Other details of these processes have already been discussed in Chapter 10.

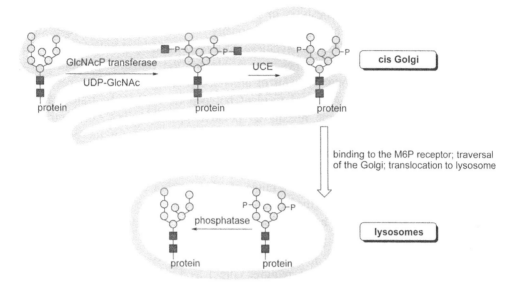

O-Linked Mucins/Proteoglycans, Blood Group Antigens and Xenorejection

Linkage of a glycan to oxygen of a serine, threonine or tyrosine side chain on a protein results in O-linked glycosylation, which occurs in lower and higher organisms. While some use the term *O*-glycosylation to refer solely to modification of serine or threonine with GalNAc, necessitating the activity of a polypeptide GalNAc transferase, here, the term will be used in its broadest sense and different structures will be qualified with other, more specific terms.

'Mucin-type' O-linked glycosylation:[3] The most abundant form of O-linked glycosylation in higher eukaryotes is the 'mucin-type', which is characterized by the presence of an *N*-acetyl-α-D-galactosamine attached to the hydroxyl group of serine or threonine. Mucin-type O-linked glycans are typically found clustered together in 'mucin domains' on membrane-bound and secreted proteins. Dense clusters of 10–100 O-linked glycans can occur in a single mucin and can comprise the major proportion of the molecular weight of the resultant conjugate. For example, the mucin polypeptides are proteins that are highly modified by O-linked glycans and form epithelial mucins. *O*-Glycosylation in mucins occurs on tandem repeat sequences rich in proline, serine and threonine that lack conserved structures and upon *O*-glycosylation likely adopt extended solution conformations. Other proteins possess relatively few scattered O-glycans that can coexist with N-linked glycans in proteins with well-defined folded structures.

References start on page 410

O-Linked glycan biosynthesis is in many ways simpler than N-linked glycan biosynthesis, in that it occurs post-translationally, by the addition of single sugar residues, rather than through the biosynthesis of a preformed lipid-linked glycan that is transferred *en bloc*. The enzymes responsible for the initiation of mucin-type O-linked glycosylation through transfer of an *N*-acetyl-α-D-galactosamine residue to the polypeptide substrate are Golgi resident polypeptide α-GalNAc transferases;[19] analysis of the human genome reveals 24 putative ppGalNAcTs.[20] In contrast to N-linked glycosylation, there is no simple sequence consensus for the addition of GalNAc residues, although predictive methods exist.[21] Following formation of the α-GalNAc-Ser/Thr linkage, many different reactions occur to form a mature O-linked glycan. These mature glycans possess a range of different core structures, which can be used to classify their subtype. The most common subtypes are those that possess a 'core 1' structure (also known as the Thomsen-Friedenreich [TF] or T antigen), resulting from the addition of a β-1,3-linked galactose residue by core 1 β-1,3-galactosyltransferase. Core 2 O-glycans possess a β-1,6-GlcNAc residue added to core 1 structures; the corresponding core 2 GlcNAcT requires core 1 as a substrate. Core 3 O-glycans possess a β-1,3-GlcNAc linked to α-GalNAc-Ser/Thr, and core 4 O-glycans possess a β-1,6-GlcNAc linked to core 3. Core 1–4 O-glycans comprise the majority of O-glycan structures found in nature; however, many other modifications are possible.

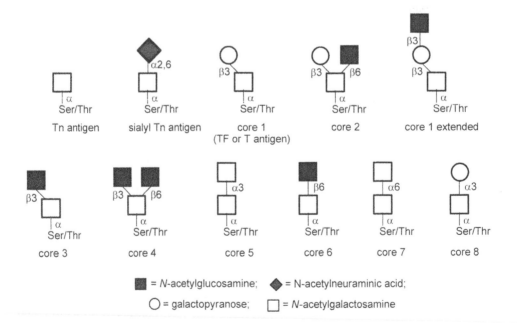

All mucin-type O-linked glycans are formed from α-GalNAc-Ser/Thr through the action of Golgi resident glycosyltransferases.[20] The core structures are built by individual core-specific glycosyltransferases. It is likely that the production of individual O-linked glycoforms depends on competition between individual glycosyltransferases that are expressed in different regions within the Golgi, and in different amounts depending on developmental and pathological cues. Despite considerable recent progress in the understanding of the biosynthesis of mucin-type O-linked glycosylation, only a few functions for this abundant type of post-translational modification have been reported, some of which are discussed below. Chapter 10 has already detailed the function of another example of O-linked glycans, as ligands for carbohydrate-binding proteins, termed selectins, that are involved in inflammation.

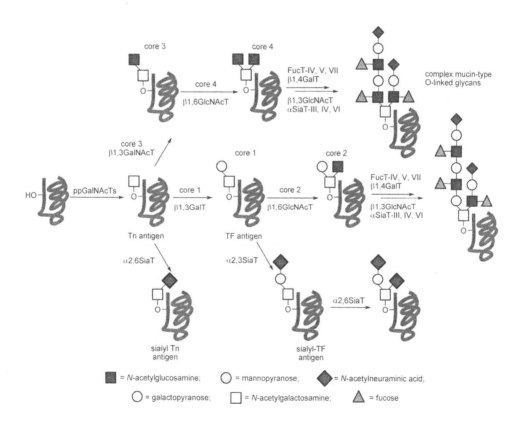

The blood group antigens: The first human blood group division, the ABO system (also known as the ABO histo-blood group system),[22] was discovered by Landsteiner and colleagues.[23,b] These workers were able to show that human blood could be separated into groups on the basis of whether the serum from one individual would cause the red blood cells of another to clump together, or agglutinate. At the time the underlying molecular reason(s) for this behaviour was unknown, but the result proved invaluable for the safe practice of blood transfusion and led to the discovery of other blood systems (e.g. rhesus). The cause of agglutination seen by Landsteiner is now known: the serum constituents are antibodies, and the antigens that react with these antibodies are mainly three oligosaccharides displayed on the surface of red blood cells. The differences between individuals arise from differences in a series of genetically encoded glycosyltransferases.

H antigen   A antigen   B antigen

The A, B and O (or H, for heterogenic, being present on most red blood cells, independent of blood type) blood group antigens are oligosaccharides displayed on various types of N-linked, O-linked or lipid-linked precursor glycans.[22,24] Type 1 chains correspond to N-linked, O-linked or lipid-linked glycan cores where the terminal GlcNAc residues are modified by β-1,3-linked galactose to generate a 'neo-lactosamine' unit; type 2 chains refer to N-linked, O-linked and lipid-linked glycan cores where the terminal GlcNAc is modified by β-1,4-linked galactose to generate an N-acetyllactosamine unit; type 3 chains refer to O-linked glycans where the Tn antigen is modified with a β-1,3-linked galactose (core 1); and type 4 chains refer to lipid-linked glycans bound to ceramide bearing a terminal β-1,3-linked galactose residue.

---

[b] Karl Landsteiner (1868–1943), professorships at University of Vienna, Rockefeller Institute. Nobel Prize in Physiology or Medicine (1930).

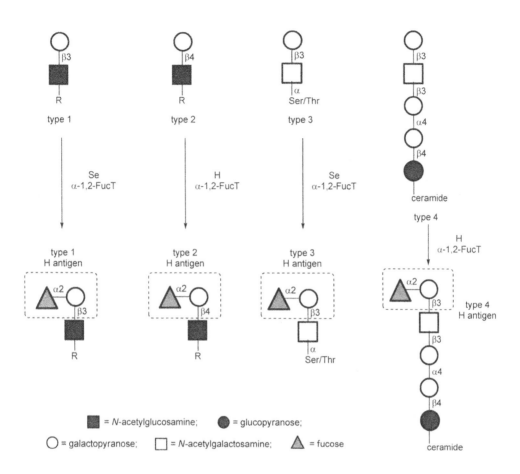

The A, B and H antigens are formed on these precursor glycans through the action of distinct glycosyltransferases found at three locations within the genome: the *ABO*, *H* and *Secretor* (*Se*) loci. Each individual will have two copies of each gene, corresponding to alleles. The synthesis of the antigens commences with the action of two α-1,2-FucTs found at the *H* and *Se* loci. The H α-1,2-FucT is expressed on red blood (erythrocyte) progenitor cells and modifies type 2 and 4 precursors to generate type 2 and 4 H antigens.[3] The Se α-1,2-FucT is expressed in epithelial cells and uses type 1 and 3 precursors to generate type 1 and 3 H antigens, which are found in epithelia within the gastrointestinal tract and on secreted proteins.[3] The A or B blood group antigens are formed from the type 1, 2, 3 or 4 H antigens by glycosyltransferases encoded at the ABO locus. Blood group A glycan is formed by the A α-1,3-GalNAcT encoded by the *A* allele at this locus. Blood group B glycan is formed by the B α-1,3-GalT encoded by the *B* allele at this

References start on page 410

locus. The *O* allele encodes a non-functional truncated glycosyltransferase that is unable to modify the preformed H antigen and thus is a null allele at the *ABO* locus. Alternatively, blood group O can arise from the O^2 enzyme, which is a full-length form of the A transferase with three substitutions (Pro74Ser, Arg176Gly and Gly268Arg) that possesses no detectable transferase activity.[25] As every individual has two alleles, a range of possible gene combinations exist. *AA* or *AO* individuals will produce A antigen along with some H antigen (arising from incomplete utilization of precursors); *BB* or *BO* individuals will produce B and H antigens; *AB* individuals will produce A, B and H antigens; and *OO* individuals will produce only H antigen.

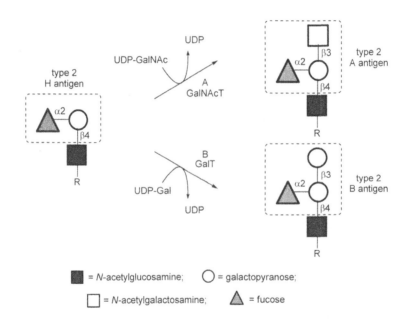

The ABO blood system arises because of the formation, early in life, of two antibodies towards the A and B antigens found on red blood cells: the anti-A and anti-B antibodies, respectively. These antibodies always occur in the serum when the corresponding antigen is absent from the red blood cell, likely because of exposure to ubiquitous environmental bacteria and fungi that display these antigens on their surface. If an individual displays a particular antigen on their cell surface, they perceive this structure as self and cannot generate an antibody response; thus, most individuals will not generate anti-H antibodies because a substantial fraction of their H structures are not converted to A or B structures even in the presence of functional

A or B genes.[c] Incorrectly matched blood transfusions result in a strong immune response by IgM-class antibodies against ABO oligosaccharide structures that are absent from the recipient's red blood cells. These IgM antibodies result in complement-dependent lysis of the red blood cells, leading to hypotension, shock, renal failure or death from circulatory collapse. Such complications can be avoided by the simple practice of ensuring blood types are a match between donor and recipient, and thus that red blood cells of the donor are deficient in ABO antigens that are lacking in the recipient. In practice this is done by blood typing and cross-matching. In typing, blood from the donor is chosen to match the recipients ABO blood type. In cross-matching, a sample of the recipient's serum is mixed with red blood cell from the prospective donor and examined under a microscope to ensure that agglutination does not occur. By such simple means, essentially all ABO blood incompatibility transfusion events have been eliminated.

Blood group (phenotype)	Genotype	Antigen on red cells	Antibodies in serum	Compatible recipient	Compatible donor
A	*AA, AO*	A	anti-B	A,O	A,AB
B	*BB, BO*	B	anti-A	B,O	B,AB
AB	*AB*	AB		A,B,O,AB	AB
O	*OO*		anti-A and anti-B	O	A,B,O,AB

The genes for the ABO locus were cloned in 1990.[24] Remarkably, the A and B glycosyltransferases differ by only four amino acid substitutions: Arg176Gly, Gly235-Ser, Leu266Met and Gly268Ala. These two glycosyltransferases utilize a common acceptor, the H antigen, and the differences in their donor specificity (UDP-galactose versus UDP-*N*-acetylgalactosamine) must arise from these mutations. X-ray structures of the A and B glycosyltransferases in complex with an H antigen acceptor and UDP have revealed that two of the four amino acids are in proximity to the donor or the acceptor.[26] Despite this conceptually appealing structural basis for the differences in specificity of these enzymes, mutagenesis studies reveal that all four amino acids are required for the observed activity and specificity of the glycosyltransferases.[27]

Enzymatic removal of antigen structures from the surface of red blood cells was originally proposed as a means to generate universal blood cells that should not result in

[c] A rare ABO blood group called the Bombay phenotype has been reported, where individuals lack functional H and Se α-1,2-fucosyltransferases. This leads to an absence of H, A and B antigens and the production of antibodies that react with blood from all standard donors. The first individual with this blood group was from the city of Bombay. para-Bombay refers to individuals who lack only functional H α-1,2-fucosyltransferases.

References start on page 410

ABO blood group incompatibility reactions. Clinical trials of B red blood cells treated with coffee bean α-galactosidase to remove α-galactose residues yielded positive results;[28] however, the low activity of the enzyme towards the B antigen required as much as 1–2 g of enzyme per unit of blood for adequate conversion.[29] Through the screening of 2500 fungal and bacterial isolates, two glycoside hydrolases with high specificity for the A and B antigens were isolated. *Elizabethkingia meningosepticum* *N*-acetyl-α-galactosidase and *Streptomyces griseoplanus* α-galactosidase were identified as optimal enzymes for conversion of A or B red blood cells to group O, requiring 60 mg and 2 mg of each enzyme, respectively, for conversion of a single unit of red blood cells in 60 min.[30] Such 'enzyme-converted universal O group' cells hold great promise for overcoming limited supplies of donor blood for a variety of applications.[31]

Xenotransplantation and the α-1,3-Gal epitope: Xenotransplantation refers to the transplantation of organs between species and has the potential to overcome the critical shortage of organs such as kidney, lung and heart, and also to provide a cure for type 2 diabetes through supply of pancreatic islet cells.[32] The pig is considered to be the most likely species to act as a transplant donor for humans owing to its ability to breed rapidly and have large litters and owing to its similar organ size and physiology to humans. However, transplantation of organs from most mammals into humans results in hyperacute rejection within minutes to hours, and later rejection events. Initial rejection of the xenograft occurs because of antibody responses to carbohydrate antigens present on the graft tissue surface, including the ABO blood group antigens described earlier, and the xenograft antigen, α-Gal-(1,3)-β-Gal-(1,4)-GlcNAc (termed 'α-1,3-Gal'). Hyperacute rejection is not unique to xenotransplantation and also occurs when organs are transplanted across blood group compatibility groups.

Biosynthesis of the α-1,3-Gal epitope occurs by the action of α-1,3-galactosyltransferase and UDP-galactose on *N*-acetyllactosamine of a type 2 precursor. This galactosyltransferase is present in most mammals but is not functional in humans and Old World monkeys. As α-1,3-Gal is expressed by many environmental microorganisms to which we are exposed, humans possess natural antibodies against this antigen; these specific antibodies comprise approximately 1% of all circulating antibodies.

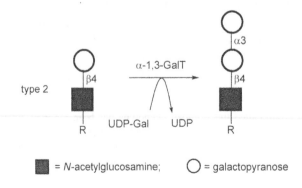

Hyperacute rejection can be prevented by absorbing α-1,3-Gal antibodies in vivo through competitive binding to externally administered soluble α-Gal epitope or ex vivo by immunoabsorption.[32] However, a secondary rejection process, acute humoral xenograft rejection, develops days or weeks after transplantation. A more promising route to reducing xenograft rejection has been to genetically modify the donor animal. The greatest attention has focussed on deleting or modifying the α-1,3-Gal epitope. Pigs with homozygous deletion of α-1,3-galactosyltransferase have been developed, and these pigs are negative for α-1,3-Gal as determined using a lectin.[33] These pigs have been used as xenograft donors in transplantation experiments into baboons; transplanted hearts survived up to 6 months, and kidney transplants survived for up to 83 days.[34] However, these experiments used powerful immunosuppression techniques that are not feasible for humans. In the future, α-1,3-Gal-deficient pigs should provide an experimental model for studying the finer details of xenograft destruction and a genetic platform for engineering additional genetic modifications to overcome other barriers to xenograft survival.

O-Linked *N*-Acetyl-β-D-glucosamine[35–37]

O-Linked *N*-acetyl-β-D-glucosamine (O-GlcNAc) on Ser or Thr is found on a wide variety of nuclear and cytoplasmic proteins within plant and animal cells. O-GlcNAc has several differences from classical N-linked glycosylation and O-linked glycosylation discussed earlier. First, O-GlcNAc is not formed within the luminal compartments of the ER or Golgi apparatus but occurs within the cytoplasm or nucleoplasm. Secondly, O-GlcNAc is not elongated through the addition of further sugars. Finally, O-GlcNAc is a dynamic modification and can be added and removed on a timescale similar to that of protein phosphorylation. Cycling of O-GlcNAc occurs through the action of two proteins, the UDP-*N*-acetylglucosamine-dependent O-GlcNAc transferase, and O-GlcNAcase, a retaining glycoside hydrolase that uses neighbouring group participation by the 2-acetamido group of the GlcNAc residue.[38]

A clear and demonstrable role for O-GlcNAc has remained elusive; however, many interesting features of this dynamic post-translational modification have been noted. O-GlcNAc has been observed to exhibit several types of competitive interplay with *O*-phosphate.[35] Reciprocal same-site occupancy, where the two post-translational modifications compete at the same site within a protein, occurs in several proteins including the transcription factor c-Myc and endothelial nitric oxide synthase. Adjacent-site occupancy can occur where O-GlcNAc and *O*-phosphate are found at adjacent sites, and occupancy of one site by one modification precludes the other occurring at the adjacent site. This situation has been observed in the tumour suppressor p53.

References start on page 410

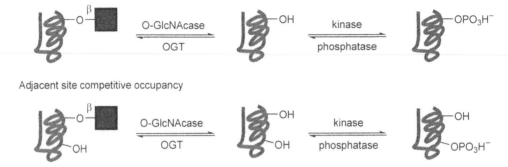

Humans possess three genes encoding N-acetylhexosaminidases capable of degrading terminal N-acetyl-β-glucosaminides: HexA and HexB, lysosomal resident proteins that form dimeric isozymes, and O-GlcNAcase. Highly selective inhibitors of O-GlcNAcase have been developed that take advantage of differences in the tolerance of these GlcNAcases for N-substitution. Lysosomal hexosaminidases were found to be intolerant of increasing steric bulk at nitrogen, whereas O-GlcNAcase can tolerate modestly sized substituents.[38] The known broad-spectrum GlcNAcase inhibitors NAG-thiazoline[38] (also discussed in Chapter 7) and PUGNAc[39,40] were modified to install a bulky substituent on nitrogen, in the former case providing inhibitors with 1500-fold selectivity against O-GlcNAcase over the lysosomal enzymes.

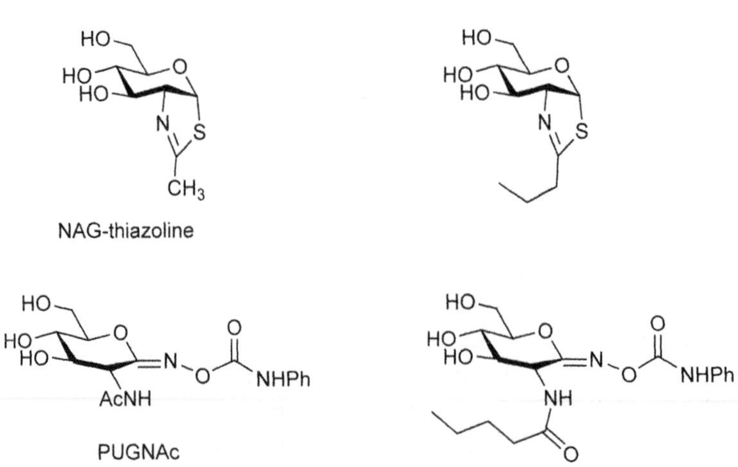

NAG-thiazoline

PUGNAc

Glycosylphosphatidylinositol Membrane Anchors

Many cell surface proteins in eukaryotes are anchored to the membrane through a glycophospholipid termed GPI that is covalently attached to the C-terminus of the protein through an amide bond. GPI anchors are found in many different organisms and are particularly abundant in single-celled eukaryotes, including fungi and protists. A wide range of proteins can be found attached to a GPI anchor including hydrolytic enzymes, cell adhesion molecules, prion proteins and various antigenic proteins.[41] GPI anchors provide a way to anchor proteins to a membrane that is complementary to peptide transmembrane domains; examples of the same gene product attached to a membrane through either a GPI anchor or a peptide transmembrane domain are known, the latter resulting from differential RNA splicing. GPIs in all organisms share a common core structure but possess wide differences in the adorning groups.

R_1, R_2 = fatty acid, alkyl or ceramide
R_3 = H or fatty acid
R_4 = H or α-Gal-(1,2)-α-Gal-(1,6)-[α-Gal-(1,2)]-α-Gal-(1,3)-
R_5 = H or β-GalNAc-(1,4)-
R_6, R_7 = H or phosphoethanolamine
R_8 = H, α-Man-(1,2)-, α-Man-(1,2)-[α-Man-(1,3)]-α-Man-(1,2)-
   or α-Glc-(1,2)-

The majority of research into GPI structure and biosynthesis has been performed in mammals, yeast and protozoan pathogens including the African trypanosomes. The African trypanosome has proved a particularly useful object for study, as each parasite is covered with a dense monolayer of 10^7 molecules of variant surface glycoprotein (VSG), which is anchored to the cell surface with a GPI anchor.[42] A single trypanosome expresses only a single type of VSG at a time but has hundreds of genes encoding immunologically distinct VSG variants. The expressed variant changes regularly, allowing the organism to evade the host immune response; despite significant differences in primary sequences, all VSG molecules are believed

References start on page 410

to share similar three-dimensional structures allowing them to pack into functionally identical protective coats. The ease of cultivation of the parasite and the abundance and homogeneity of its coat protein allow purification of substantial quantities of the VSG and led to the first complete structural characterization of a GPI anchor,[43] followed shortly thereafter by the determination of the structure of a mammalian GPI anchor, from Thy-1 glycoprotein.[44,45]

representative GPI structure from
rat Thy-1 glycoprotein

major GPI structure from *Trypanosoma brucei*
variant surface glycoprotein variant 117

GPI is synthesized and installed onto proteins post-translationally in a process that bears considerable similarity to the initial stages of N-linked glycoprotein biosynthesis.[46,47] The carbohydrate backbone of the GPI anchor is prepared on the exterior and interior membrane of the ER and, once complete, is transferred to the protein. The first step of GPI synthesis is the transfer of *N*-acetylglucosamine from UDP-*N*-acetylglucosamine to PI. This step occurs on the cytoplasmic face of the ER, and in mammals it is catalysed by a complex of a UDP-*N*-acetylglucosamine-dependent glycosyltransferase, PIG-A, and several auxiliary proteins. The second step in GPI synthesis is catalysed by the deacetylase PIG-L, which results in the deacetylation of GlcNAc-PI to form GlcN-PI. Next, GlcN-PI is acylated on inositol to form GlcN-(acyl)PI. Acyl-CoA is the acyl donor and, in humans, this step is catalysed by PIG-W. While a range of acyl groups may be transferred, palmitylation is the major form observed. At this stage it is hypothesized that the GlcN-(acyl)PI intermediate undergoes a shift across the ER membrane to the luminal face in a process catalysed by a hypothetical flippase. Next, the internalized GlcN-(acyl)PI is mannosylated twice by two dolichol-P-mannose-dependent ManTs to generate Man$_2$GlcN(acyl)PI, and then ethanolaminephosphate (EtN-P) is transferred to the 3-position of the first mannose residue. One more mannose residue is added by a dolichol-P-mannose-dependent mannosyltransferase to generate Man$_2$(EtN-P)ManGlcN(acyl)PI. A second unit of EtN-P is added to the 6-position of the reducing end mannose, generating (EtN-P)ManMan(EtN-P)ManGlcN(acyl)PI. At this stage a third EtN-P may be added to the second mannose to generate a GPI bearing three phosphoethanolamine units. The outline provided above gives a general description of GPI biosynthesis in mammals, but differences exist both in the order of biosynthetic steps and in the attachment of other groups onto the core structure in yeast, trypanosomes and other protozoa.[42,48] GPI anchors have been found to be essential for the virulence or survival of various pathogenic protozoa, and the differences in the GPI biosynthesis pathways have been proposed as potential drug targets.

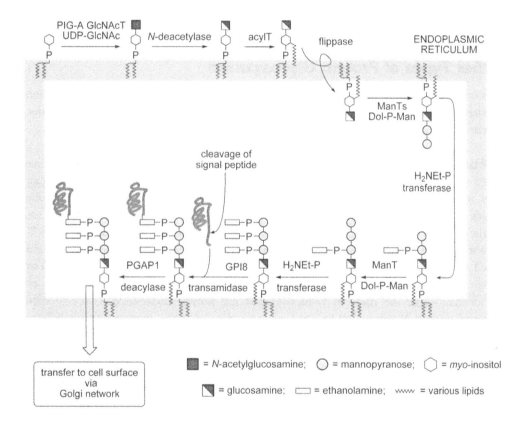

Transfer of the mature GPI to a protein generates an amide bond to the C-terminus of the protein. Proteins to be GPI anchored possess two signal sequences.[49] One signal sequence encodes for translocation across the ER membrane and is common to secreted and cell surface membrane proteins. The other is a C-terminal sequence that encodes for its own cleavage and replacement with the GPI anchor in what is believed to be a transamidation reaction, catalysed by a transamidation complex containing the transamidase enzyme GPI8.[49] After attachment of the GPI to a protein, inositol deacylation can occur resulting in the removal of the palmityl group on the inositol ring. This step is catalysed by the

References start on page 410

ER-resident deacylase PGAP1 in mammalian cells. GPI-anchored proteins are then transported to the cell surface through the Golgi apparatus. A variety of post-ER modifications can occur on mature GPI-anchored proteins to install side-chain carbohydrates and are believed to occur in the Golgi apparatus.[42]

A mutation in the glycosyltransferase PIG-A, required for the first step of GPI biosynthesis in humans, leads to the haemolytic disease paroxysmal nocturnal haemoglobinuria (PNH).[50] PNH is an acquired haematopoietic stem cell disorder that results from cells being deficient in the expression of GPI-anchored proteins. A clinical symptom is intravascular haemolysis caused by own complement.

Other Types of Protein Glycosylation

Above, the major types of O- and N-linked glycosylation have been discussed. However, there are many other types of glycosylation found in higher organisms. Some of the more abundant forms found in mammals are discussed below.

O-Fucose: *O*-Fucosylation is relatively widespread in higher organisms and has been found on proteins involved in blood coagulation, cell signalling and metastasis.[51–53] In *O*-fucosylation, L-fucose is α-linked to serine or threonine resides in the consensus sequence Cys-X-Ser/Thr-X-Pro-Cys. Examples of *O*-fucosylated proteins include urokinase, tissue plasminogen activator, factor VII, factor XII and Notch-1. O-Fucose is installed by the GDP-fucose-dependent protein FucT OFut1, a soluble enzyme found in the ER.[51] O-Fucose is frequently extended by other sugars; α-Neu5Ac-(2,6)-β-Gal-(1,4)-β-GlcNAc-(1,3)-α-Fuc-Ser is found on factor XI. While the roles of *O*-fucosylation are poorly understood, *O*-fucosylated proteins termed Notch receptors are known to play key roles in determining cellular fate and in the control of growth during development.[51] Notch receptors are membrane associated and their ligands, Delta and Jagged/Serrate, are also *O*-fucosylated. The ability of Notch to bind its ligands is modulated by the O-fucose-specific β-1,3-*N*-acetylglucosaminyltransferase Fringe, which is found in the Golgi apparatus.[54]

C-Mannose: An unusual form of protein glycosylation is *C*-mannosylation wherein C2 of tryptophan is mannosylated through an α-*C*-glycosidic linkage. *C*-Mannosylation was first reported in the mid-1990s, and relatively little is known of its function. *C*-Mannosylation occurs through a microsomal mannosyltransferase that utilizes dolichol-P-Man as donor;[55] this suggests that *C*-mannosylation occurs within the ER.[56] The consensus sequence for *C*-mannosylation is Trp-Xxx-Xxx-Trp, with *C*-mannosylation occurring on the first tryptophan residue. *C*-Mannosylation may occur in closely positioned repeat sequences in one protein and can coexist with other forms of protein glycosylation. The protein thrombospondin-1 is a platelet protein that possesses four *C*-mannosyl groups and two O-linked fucose residues.[57]

O-Mannose glycans: *O*-Mannosylation of serine and threonine residues is a modification common to fungi such as *Saccharomyces cerevisiae* and *Pichia pastoris*.[58] These structures include high mannose mannans that possess linear α-Man-(1,3)-α-Man-(1,3)-α-Man-(1,2)-α-Man-(1,2)-α-Man-Ser/Thr and that may also be modified by mannose-1-phosphate residues. The majority of mammalian O-mannose glycans are variations of the tetrasaccharide α-Neu5Ac-(2,3)-β-Gal-(1,4)-β-GlcNAc-(1,2)-α-Man-Ser/Thr. O-Mannose glycans have been found in brain glycopeptides and in α-dystroglycan, a highly glycosylated transmembrane protein that is part of the dystrophin-associated glycoprotein complex and that is involved in the pathogenesis of many forms of muscular dystrophy. The biosynthesis of O-mannose glycans is best understood in *S. cerevisiae* where the initial step is the formation of an α-mannosyl serine or threonine catalysed by a dolichol-P-Man-dependent protein, *O*-mannosyltransferase, in the ER.[59] The *S. cerevisiae* protein *O*-mannosyltransferase PMT4 will only glycosylate proteins that are membrane associated with a Ser/Thr-rich domain facing the lumen of the ER.[60] Further mannosylation events occur in the Golgi apparatus. Defects in the synthesis of O-mannose chains lead to several forms of congenital muscular dystrophies including Walker–Warburg syndrome (WWS), Fukuyama congenital muscular dystrophy and muscle-eye-brain disease.[58] These diseases lead to abnormal glycosylation of α-dystroglycan, and mutations in the human protein *O*-mannosyltransferase gene have been found in some sufferers of WWS.

References start on page 410

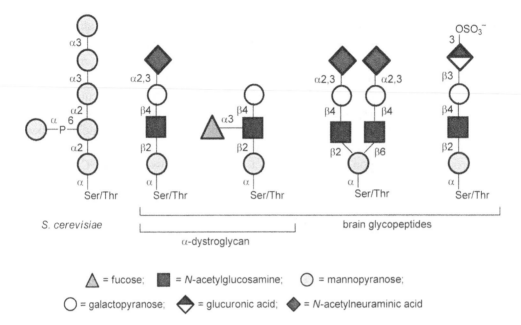

= fucose;   ■ = N-acetylglucosamine;   ◯ = mannopyranose;

◯ = galactopyranose;   ◈ = glucuronic acid;   ◆ = N-acetylneuraminic acid

Rare protein modifications: A range of other protein glycosylations have been observed.[56] These include α-Glc-(1,2)-β-Gal-O-hydroxylysine as found in collagen; O-β-glucose to serine or threonine, which may be further elaborated with xylose residues as found in epidermal growth-factor-like domains for factors VII and IX and protein Z;[61] and glucose α-linked to tyrosine as found in the core of glycogen. Other types of glycosylation different from those discussed above have been observed in plants and bacteria, but are outside the scope of this discussion. The apparent rarity of such modifications may be a result of limitations in analytical techniques for their detection rather than being a true reflection of their abundance.

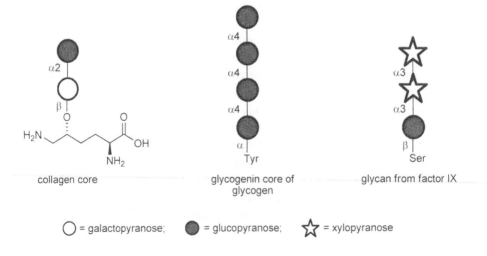

◯ = galactopyranose;   ● = glucopyranose;   ☆ = xylopyranose

Proteoglycans and Glycosaminoglycans

Glycosaminoglycans (GAGs) are linear polysaccharides comprised of disaccharide units, each of which is composed of an acetamido sugar (*N*-acetyl-D-glucosamine or *N*-acetyl-D-galactosamine) and a uronic acid (D-glucuronic or L-iduronic acid) or D-galactose units. GAGs are commonly *O*-sulfated to varying degrees and may also be *N*-deacetylated and sulfated on nitrogen.

Proteoglycans are proteins to which one or more glycosaminoglycan chain is attached.[2] Proteoglycans may contain only a single GAG chain (e.g. decorin) or more than 100 chains (e.g. aggrecan). Proteoglycans are found secreted from the cell in the extracellular matrix (e.g. aggrecan) or in intracellular secretory granules (e.g. serglycin) or may remain attached to the cell through a GPI anchor (e.g. glypican) or through a membrane-spanning domain (e.g. syndecan). Most proteoglycans also possess O- and N-linked glycans that are typical of those found on other glycoproteins.

GAGs include hyaluronan, chondroitin and dermatan sulfate, keratan sulfate, heparin and heparan sulfate. With the exception of hyaluronan, all GAGs are synthesized as proteoglycans in the Golgi apparatus and possess a covalent linkage to protein. In the case of chondroitin sulfate/dermatan sulfate and heparin/heparan sulfate, the core linkage is through a xylose residue that is β-linked to serine. This core xylose residue is elaborated to β-GlcA-(1,3)-β-Gal-(1,3)-β-Gal-(1,4)-β-Xyl-Ser, which forms a bridge to the GAG chain. The biosynthesis of the linkage tetrasaccharide commences through the action of UDP-xylose-dependent proteoglycan core protein β-xylosyltransferase, an enzyme localized to the Golgi apparatus.[62,63] The second step of the biosynthesis occurs through the action of proteoglycan UDP-galactose-dependent β-xylose β-1,4-galactosyltransferase I.[64] Defects in this enzyme have been detected in individuals with progeroidal Ehlers–Danlos syndrome, a condition that is associated with a reduction in the number of GAG chains attached to certain proteoglycans. The third step of the core tetrasaccharide biosynthesis is catalysed by β-1,3-galactosyltransferase II.[65] The final step is catalysed by a UDP-glucuronic acid–dependent β-1,3-glucuronosyltransferase.[66]

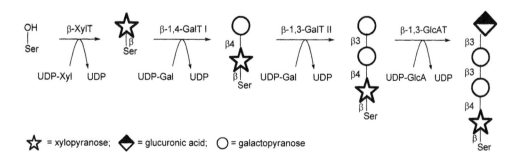

☆ = xylopyranose;   ◆ = glucuronic acid;   ○ = galactopyranose

References start on page 410

The linkage tetrasaccharide may be phosphorylated on O2 of xylose and sulfated on the 4- and 6-positions of either of the galactose residues. It is unclear whether these modifications precede or follow installation of the GAG chain onto the core tetrasaccharide; however, it is clear that phosphorylation of the core xylose residue follows elaboration by β-1,4-galactosyltransferase I, and sulfation of the core galactosides promotes transfer by β-1,3-glucuronosyltransferase.[67] The core tetrasaccharide may be elaborated into different GAG chains dependent on the next processing step. Addition of α-GlcNAc leads to heparin and heparan sulfate, and addition of β-GalNAc leads to chondroitin sulfate and dermatan sulfate.

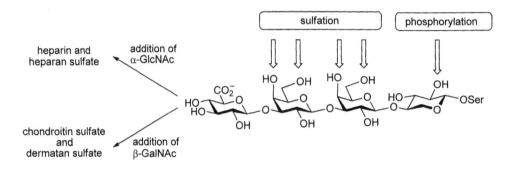

Hyaluronan: Hyaluronan is the simplest glycosaminoglycan and one of the longest linear carbohydrates found in vertebrates. It consists of β-GlcNAc-(1,3)-β-GlcA-(1,4)- repeats. Hyaluronan is an important component of skin, cartilage, connective tissues, umbilical cord, synovial fluid and the vitreous of the eye. Chains range from 10^5 to 10^7 Daltons and a polymer of 10^6 Daltons is approximately 2 μm in length.[3] At a concentration of 10 mg ml^{-1}, the viscosity is 5000 times that of water, and this confers rigidity to its solution in tissues. Hyaluronan is the only GAG not found attached to a protein core and is synthesized at the inner face of the plasma membrane by hyaluronan synthases using UDP-glucuronic acid and UDP-N-acetylglucosamine.[68] Hyaluronan synthases are transmembrane proteins that encode two distinct glycosyltransferase activities in a single polypeptide chain.

hyaluronan

Chondroitin sulfate/dermatan sulfate: Chondroitin sulfate and dermatan sulfate proteoglycans are produced by most vertebrate cells and are major components of connective tissue matrix and the cell surface and basement membranes. They are involved in the control of basic cellular processes including cell division, scaffolding functions in connective tissue and promotion of neurite outgrowth in the brain.[69] Chondroitin sulfate contains repeating chains of β-GlcA-(1,3)-β-GalNAc-(1,4)- units. Dermatan sulfate is closely related to chondroitin sulfate and contains some β-L-IdoA-(1,3)-β-GalNAc-(1,4)- units.

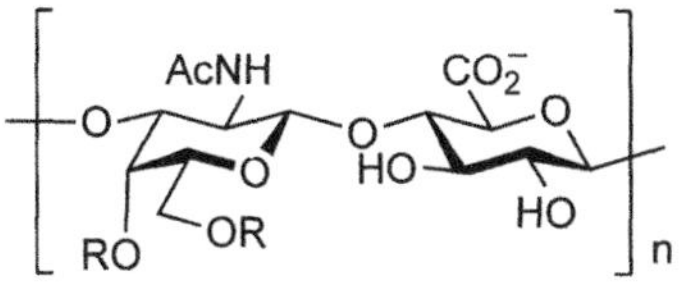

chondroitin sulfate, R = H or SO_3^-

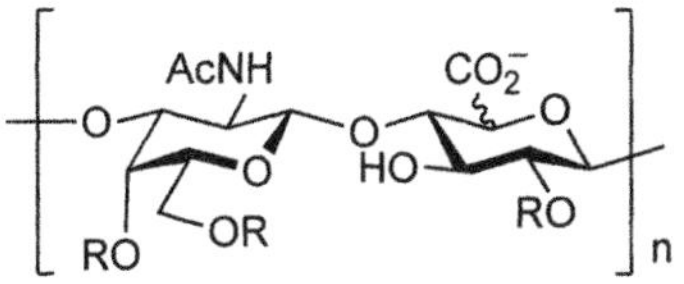

dermatan sulfate, R = H or SO_3^-

The carbohydrate backbone of chondroitin sulfate is formed from the core tetrasaccharide β-GlcA-(1,3)-β-Gal-(1,3)-β-Gal-(1,4)-β-Xyl-Ser discussed above by the action of a bifunctional glycosyltransferase, chondroitin synthase, which possesses both N-acetyl-β-galactosaminyltransferase and β-glucuronosyltransferase activities.[69,70] This enzyme is located in the medial and trans Golgi. Chondroitin sulfate is variably sulfated, with sulfate esters found as GalNAc-4-sulfate, GalNAc-6-sulfate, GalNAc-4,6-disulfate and small amounts of GlcA-2-sulfate. These sulfation steps appear to occur in the Golgi apparatus concomitant with the formation of the carbohydrate backbone of the GAG chain. Dermatan sulfate is formed from the polymer backbone of chondroitin sulfate by the action of chondroitin-glucuronate C5 epimerase, which epimerizes individual D-glucuronic acid residues to L-iduronic acid.[71] Epimerase activity appears to be closely associated with O-sulfation steps that may preclude re-epimerization back to D-glucuronic acid. The sulfation pattern of dermatan sulfate is similar to that of chondroitin sulfate, but while GlcA-2-sulfate is rare in chondroitin sulfate, IdoA-2-sulfate is common in dermatan sulfate.

References start on page 410

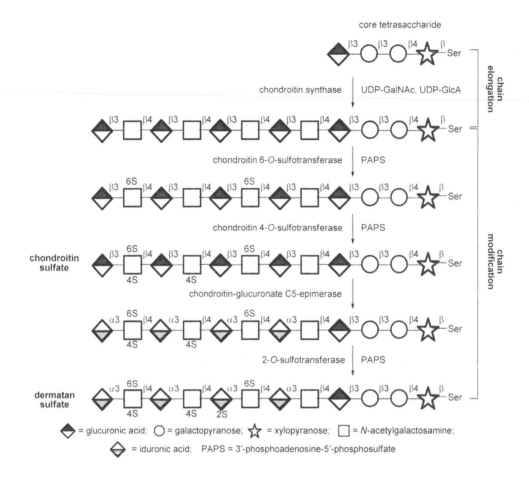

Keratan sulfate: Keratan sulfate is a β-1,3-linked poly-*N*-acetyllactosamine, with sulfate residues found on the 6-positions of both galactose and *N*-acetylglucosamine, and is found in cartilage, cornea and brain. In the eye, keratan sulfate proteoglycans maintain the regular spacing of collagen fibrils, allowing passage of light; defects in sulfation cause distortions of fibril organization and corneal opacity, resulting in macular corneal dystrophy. The GAG chain of keratan sulfate may be attached in three different ways to the core protein.[72] Keratan sulfate I refers to proteoglycans where the GAG chain is attached to Asn in core proteins by a complex-type N-linked branched oligosaccharide. Keratan sulfate II refers to proteoglycans where the GAG chain is attached to a core protein through an O-glycan linked through α-Gal-Ser/Thr. Keratan sulfate III refers to attachment of the GAG chain to the core protein through a mannose–Ser linkage.[72] Most keratan sulfate chains are capped with *N*-acetylneuraminic acid, *N*-acetyl-β-D-galactosamine, or α-D-galactose.[72]

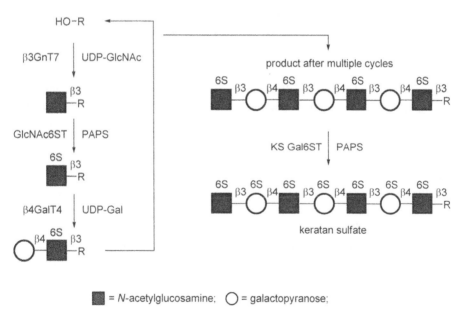

keratan sulfate

The keratan sulfate carbohydrate chain is formed by the action of the β-1,4-galactosyltransferase β4GalT4 and β-1,3-N-acetylglucosaminyltransferase β3GnT7. Sulfation of N-acetylglucosamine residues occurs by the action of N-acetylglucosaminyl-6-sulfotransferase (GlcNAc6ST), and sulfation of galactose residues is performed by keratan sulfate galactose-6-sulfotransferase (KS Gal6ST). The former enzyme transfers sulfate only to terminal N-acetylglucosamine residues, whereas the latter enzyme can sulfate internal galactose residues.[73] This suggests that sulfation of keratan sulfate on N-acetylglucosamine must occur during elongation of the GAG chain, whereas sulfation on galactose can occur after polymerization.

■ = N-acetylglucosamine;   ◯ = galactopyranose;

PAPS = 3′-phosphoadenosine-5′-phosphosulfate

Heparin and heparan sulfate: Heparin and heparan sulfate are two closely related GAGs. Heparin is produced exclusively by mast cells, which are specialized cells found in the skin, the mucosa of the lungs and the digestive tract, as well as in the mouth, parts of the eye and the nose. In these cells, heparin is found attached to the protein serglycin and resultant proteoglycan is localized in granules within mast cells.[74]

References start on page 410

Pharmaceutical heparin, a protein-free degradation product of heparin proteoglycans, is used widely as an anticoagulant and is prepared from various tissues obtained from slaughterhouses (see Chapters 10 and 12).[74] Heparan sulfate is found as a component of proteoglycans produced by a range of cells in essentially all animal tissues. Considerable confusion surrounds the precise definition of heparin and heparan sulfate.[3] Heparin may be defined as having more than 1.8 sulfate groups per hexosamine, an L-iduronic acid content ≥70% and a high proportion of glucosamine-N-sulfates. Heparan sulfate has fewer than 1.8 sulfate groups per hexosamine, a lower L-iduronic acid content and a lower proportion of glucosamine N-sulfation.

heparin/heparan sulfate, R = H or SO₃⁻,
R¹ = H, Ac or SO₃⁻

Heparin and heparan sulfate are synthesized as proteoglycans in the Golgi apparatus on proteins that are modified with the core tetrasaccharide β-GlcA-(1,3)-β-Gal-(1,3)-β-Gal-(1,4)-β-Xyl-Ser by the action of the bifunctional co-polymerases of the EXT family.[75,76] These proteins contain both α-1,4-N-acetylglucosaminyltransferase and β-glucuronosyltransferase activities in a single polypeptide chain, and use the Leloir donors UDP-N-acetylglucosamine and UDP-glucuronic acid. Whilst the chain undergoes polymerization, several modification steps occur. The first step is performed by a bifunctional N-deacetylase/N-sulfotransferase, which acts to N-deacetylate some GlcNAc residues in the growing carbohydrate, and install N-sulfate groups. Usually, the two activities are tightly coupled, but some free amino groups escape N-sulfation, resulting in a mixture of N-acetyl, N-sulfate and free amino groups in the final polymer. Second, C5 epimerization of D-glucuronic acid residues to L-iduronic acid occurs through the action of glucuronic acid C5 epimerase. This enzyme prefers to act on glucuronic acid residues that are located on the reducing side of GlcNSO₃ but does not act on 2-O-sulfated uronic acid residues or those that are adjacent to O-sulfated glucosamine residues, suggesting that it acts after N-deacetylation/N-sulfation but prior to any O-sulfation reactions. The third step is the 2-O-sulfation of uronic acid residues by uronosyl-2-O-sulfotransferase. This enzyme acts on both D-glucuronic acid and L-iduronic acid residues but has greater activity on the latter. There is some evidence that GlcNAc deacetylation, *N*-sulfation, uronic acid C5 epimerization and uronic acid 2-*O*-sulfation occur simultaneously during polymer biosynthesis. The final modification to the GAG chain leading to the mature heparin or heparan sulfate occurs through the action of GlcNAc 3-*O*-sulfotransferase and 6-*O*-sulfotransferase.

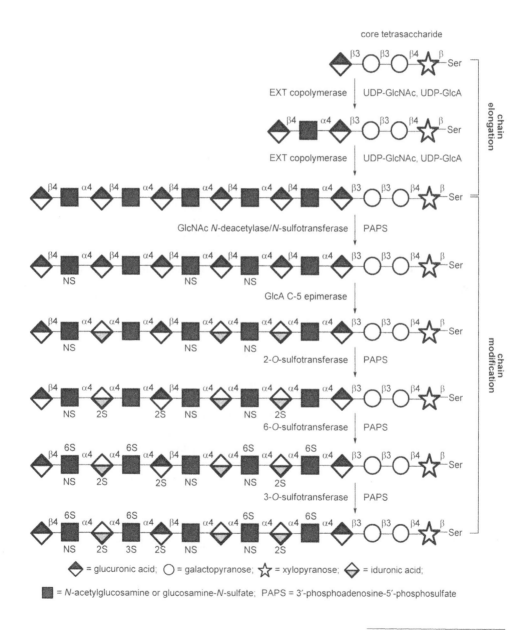

References start on page 410

Lysosomal Degradation of Glycoconjugates[3,77]

Lysosomes are acidic membranous organelles populated by a range of degradative enzymes. Lysosomes represent the final destination for most glycoproteins, proteoglycans and glycolipids. The lumen of the lysosome has pH approximately 5, and most enzymes that act within the lysosome have their pH optima between 4 and 5.5. Most carbohydrate structures destined for degradation in the lysosome undergo highly sequenced deconstruction. Substitutions such as *O*-acetyl, *O*- and *N*-sulfate, and *O*-phosphate groups must be removed before *exo*-acting glycosidases cleave the sugar chain. The enzymes associated with these steps have been unravelled with the help of a series of rare human genetic disorders termed lysosomal storage disorders (Tables 1, 2 and 3). Lysosomal storage disorders represent a group of greater than

Table 1 Disorders of glycoprotein degradation: glycoproteinoses

Diagnosis	Deficiency
α-Mannosidosis types I and II	α-Mannosidase
β-Mannosidosis	β-Mannosidase
Aspartylglucosaminuria	Aspartyl-*N*-acetyl-β-glucosaminidase
Sialidosis (mucolipidosis I)	Sialidase
Schindler (Kawasaki) types I and II	*N*-Acetyl-β-galactosaminidase
Galactosialidosis	Protective protein/cathepsin A
Fucosidosis	α-Fucosidase

Table 2 Disorders of glycosaminoglycan degradation: mucopolysaccharidoses

Diagnosis	Deficiency
MPS I; Hurler-Scheie	α-Iduronidase
MPS II; Hunter	Iduronate-2-sulfatase
MPS III A; Sanfilippo A	Heparan *N*-sulfatase
MPS III B; Sanfilippo B	*N*-Acetyl-α-glucosaminidase
MPS III C; Sanfilippo C	Acetyl CoA:α-glucosaminide acetyltransferase
MPS III D; Sanfilippo D	*N*-Acetylglucosamine-6-sulfatase
MPS IV A; Morquio A	Galactose-6-sulfatase
MPS IV B; Morquio B	β-Galactosidase
MPS VI; Maroteaux-Lamy	*N*-Acetylgalactosamine-4-sulfatase
MPS VII; Sly	β-Glucuronidase

Table 3 Disorders of glycolipid, O-linked mucin-type glycosylation degradation, glycogen storage or lysosomal targeting

Diagnosis	Deficiency
Tay-Sachs	β-Hexosaminidase A
Sandhoff	β-Hexosaminidase A and B
GM1 gangliosidosis	β-Galactosidase
GM2 gangliosidosis	β-Hexosaminidase
Sialidase	Sialidase
Fabry	α-Galactosidase
Gaucher	β-Glucosylceramidase
Krabbe	β-Galactoceramidase
Metachromatic leukodystrophy	Arylsulfatase A (cerebroside sulfatase)
Multiple sulfatase deficiency	Formylglycine-generating enzyme
I-cell disease	UDP-N-acetylglucosamine-1-phosphotransferase
Pompe disease (glycogen storage disease type II)	α-Glucosidase

40 distinct genetic diseases that result from deficiencies of specific lysosomal proteins or proteins involved in lysosome biogenesis. Individually, lysosomal storage disorders are rare but as a group have a prevalence of 1 per 7700 live births in Australia.[78] Lysosomal storage disorders lead to defective processing of specific lysosomal substrates, resulting in their accumulation within lysosomes, tissues or excretion in urine. The lysosomal disorders result in a range of developmental and neurological diseases with a wide spectrum of severity.

N-Linked glycoprotein degradation:[79] Proteases degrade the protein of an N-linked glycoprotein before N-glycan catabolism commences, resulting in the glycan being linked only to the amino acid asparagine. α-Fucosidase cleaves core and peripheral fucose residues, and then aspartyl-N-acetyl-β-glucosaminidase cleaves the N-acetylglucosamine-Asn bond. $endo$-N-Acetyl-β-glucosaminidase cleaves the reducing-end N-acetylglucosamine residue, and then sialidase, β-galactosidase, N-acetyl-β-hexosaminidase, α-mannosidase and β-mannosidase sequentially break down the remaining glycan chain. Excess N-glycan precursors and N-glycans released by cytosolic peptide N-glycanase are also targeted to the lysosome and degraded through a similar process.

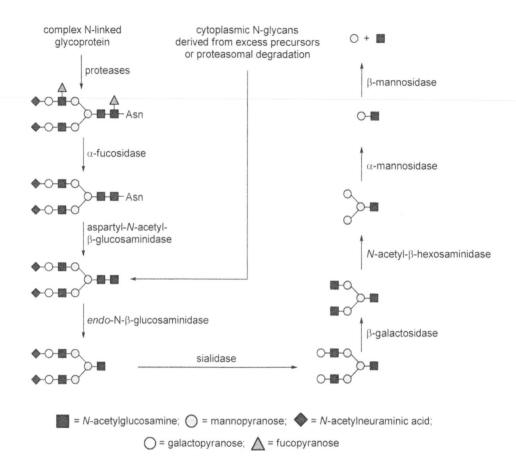

= N-acetylglucosamine; ○ = mannopyranose; ◆ = N-acetylneuraminic acid;
○ = galactopyranose; △ = fucopyranose

Glycosaminoglycan degradation: The various GAGs are degraded in highly ordered processes. Hyaluronan is found unattached to protein and is degraded through the action of hyaluronidase, which breaks the long chain into tetrasaccharides and longer fragments, through the cleavage of the β-GlcNAc linkage. The smaller fragments are then broken down to monosaccharides by β-glucuronidase and N-acetyl-β-hexosaminidase.

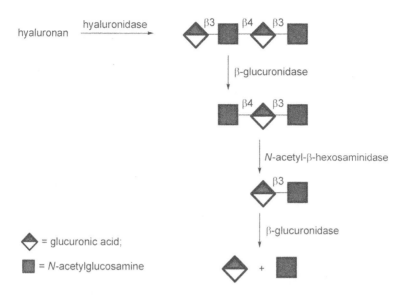

Heparan sulfate chains are cleaved by an *endo*-glucuronidase, termed heparanase, to shorter chains. These shorter chains are cleaved in a highly sequenced fashion from the non-reducing terminus. IdoA-2-sulfatase or GlcA-2-sulfatase cleaves 2-*O*-sulfate residues from a terminal uronic acid residue, enabling cleavage of the uronic acid by α-iduronidase or β-glucuronidase. The glucosamine residue is next targeted for cleavage. However, prior to cleavage *O*-sulfate residues must be removed by 3-, 4- and 6-*O*-sulfatases. If the glucosamine residue is *N*-sulfated, this residue must also be removed by the action of *N*-sulfatase. The resulting glucosamine residue is not a substrate for a glycoside hydrolase and must first be *N*-acetylated by a membrane-bound *N*-acetyltransferase. This enzyme utilizes acetyl-CoA from the cytosol and transfers the acetyl moiety to the terminal glucosamine, generating *N*-acetylglucosamine. This terminal residue can be cleaved by *N*-acetyl-α-glucosaminidase.

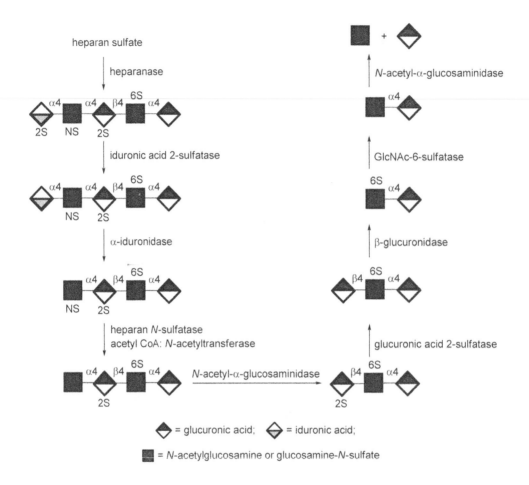

A process similar to that of heparan sulfate degradation applies to chondroitin sulfate and dermatan sulfate degradation. Thus, parent chains are first cleaved into smaller fragments by *endo*-hexosamindases. Next, sulfatases cleave sulfate residues from terminal sugars before *exo*-acting α-iduronidase, β-glucuronidase or *N*-acetyl-β-hexosaminidases A or B, act to remove single sugar residues. An alternative pathway exists for degradation of dermatan sulfate/chondroitin sulfate where *N*-acetyl-β-hexosaminidase A removes *N*-acetylgalactosamine-4-sulfate, followed by sulfatase cleavage of the sulfate ester on the monosaccharide.

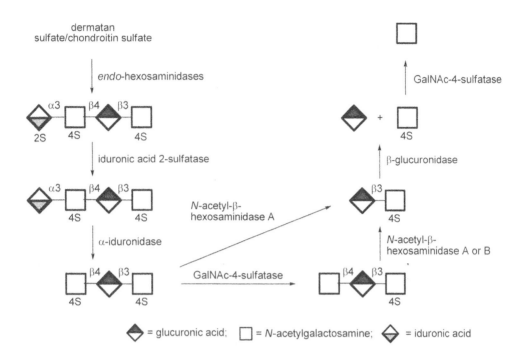

= glucuronic acid;   □ = N-acetylgalactosamine;   = iduronic acid

Keratan sulfate is degraded through the sequential action of sulfatases and *exo*-glycosidases without the involvement of *endo*-glycosidases. Desulfation of galactose residues precedes β-galactosidase cleavage of terminal galactose residues, leaving exposed GlcNAc-6-sulfate. This may be cleaved by desulfation followed by *N*-acetyl-β-hexosaminidase A or B cleavage or, alternatively, *N*-acetyl-β-hexosaminidase A can remove GlcNAc-6-sulfate prior to sulfatase cleavage of the sulfated monosaccharide.

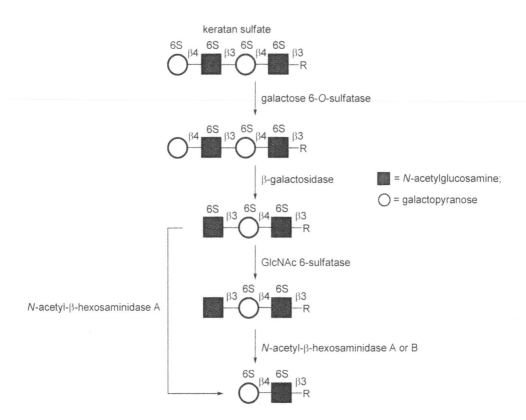

Treatment of lysosomal storage disorders with imino sugar inhibitors:[80] Cellular homeostasis requires that a balanced cycle of biosynthesis and degradation occurs, preserving cellular function. The lysosomal storage disorders generally arise because of disruption to the catabolic processes mediated by the lysosome, leading to various pathologies. Several approaches have been proposed for the treatment of lysosomal storage disorders. Enzyme replacement therapy has been utilized where the aberrant enzyme is supplied exogenously, through bone marrow transplantation, or by gene therapy. However, the high cost of this approach is a significant burden on the health-care system. Two alternative approaches have emerged in recent years that use small molecules to treat lysosomal glycolipid storage disorders: substrate reduction therapy and chaperone-mediated therapy.

Substrate reduction therapy aims to use drugs to partially inhibit biosynthetic enzymes, thereby reducing accumulation of undegraded substrates.[80] N-Butyldeoxynojirimycin (miglustat, Zavesca), previously discussed in Chapters 7 and 8, is the first drug that acts by substrate reduction to be approved and is used for the treatment of type I Gaucher disease.[81] Gaucher disease occurs through defects in the action of β-glucosylceramidase, causing accumulation of glycosphingolipids. *N*-Butyldeoxynojirimycin is an inhibitor of ceramide glucosyltransferase and acts to reduce the formation of glycosphingolipids, preventing their accumulation.

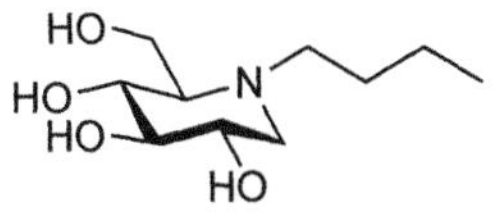

N-butyldeoxynojirimycin

Chaperone-mediated therapy aims to restore function to defective lysosomal enzymes by assisting in protein folding and trafficking, thereby preventing clearance and degradation of folding-defective proteins by normal cellular proteolysis pathways.[80] Many imino sugars are reversible, tight-binding inhibitors of lysosomal enzymes and confer sufficient stability on their target enzymes to protect against misfolding or inactivation. For this to occur, these inhibitors must bind to their target enzyme in the ER or later stages of the secretory pathway *en route* to the lysosome. Several alkylated deoxynojirimycin-type imino sugars such as deoxygalactonojirimycin, *N*-butyldeoxygalactonojirimycin, and *N*-adamantylmethyloxypentyldeoxynojirimycin[82] are currently under investigation in clinical and pre-clinical settings for the treatment of heritable glycosphingolipidoses. A series of hexosaminidase inhibitors has been assessed as pharmacological chaperones for the treatment of Tay-Sachs and Sandhoff diseases, which result from mutations that prevent the obtention and retention of the native fold of *N*-acetyl-β-hexosaminidase A. The most promising compound, NAG-thiazoline, could restore intralysosomal hexosaminidase levels well above the critical 10% level believed to be the threshold for disease.[83]

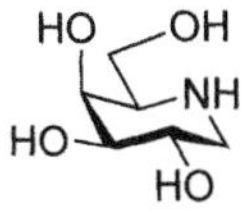

deoxygalactonojirimycin 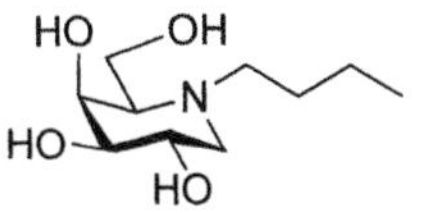   *N*-butyldeoxygalactonojirimycin 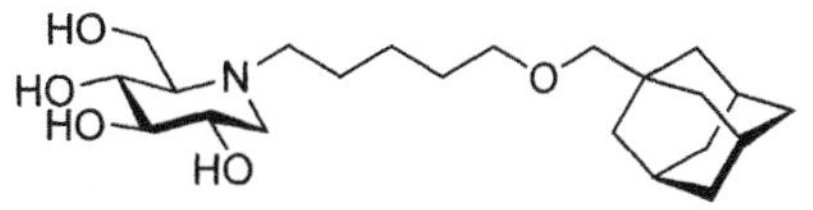   *N*-adamantylmethyloxypentyldeoxynojirimycin

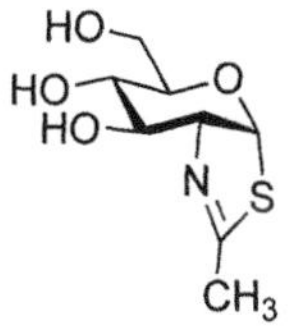

NAG-thiazoline

References start on page 410

References

1. Werz, D.B., Ranzinger, R., Herget, S., Adibekian, A., von der Lieth, C.-W. and Seeberger, P.H. (2007). *ACS Chem. Biol.*, **2**, 685.
2. Kjellén, L. and Lindahl, U. (1991). *Annu. Rev. Biochem.*, **60**, 443.
3. Varki, A., Cummings, R., Esko, J., Hart, G. and Marth, J., eds (1999) *Essentials of Glycobiology.* Cold Spring Harbor: New York.
4. Roth, J. (2002). *Chem. Rev.*, **102**, 285.
5. Kornfeld, R. and Kornfeld, S. (1985). *Annu. Rev. Biochem.*, **54**, 631.
6. Burda, P. and Aebi, M. (1999). *Biochim. Biophys. Acta*, **1426**, 239.
7. Snider, M.D. and Robbins, P.W. (1982). *J. Biol. Chem.*, **257**, 6796.
8. Moreman, K.W. and Molinari, M. (2006). *Curr. Opin. Chem. Biol.*, **16**, 592.
9. Helenius, A. and Aebi, M. (2001). *Science*, **291**, 2364.
10. Rudd, P.M., Elliott, T., Cresswell, P., Wilson, I.A. and Dwek, R.A. (2001). *Science*, **291**, 2370.
11. Ruddock, L.W. and Molinari, M. (2006). *J. Cell Sci.*, **119**, 4373.
12. Banerjee, D.K. (1989). *J. Biol. Chem.*, **264**, 2024.
13. Eason, P.D. and Imperiali, B. (1999). *Biochemistry*, **38**, 5430.
14. Elbein, A.D. (1991). *FASEB J.*, **5**, 3055.
15. Winchester, B. and Fleet, G.W. (1992). *Glycobiology*, **2**, 199.
16. Kawatkar, S.P., Kuntz, D.A., Woods, R.J., Rose, D.R. and Boons, G.J. (2006). *J. Am. Chem. Soc.*, **128**, 8310.
17. Rohrer, J. and Kornfeld, R. (2001). *Mol. Biol. Cell*, **12**, 1623.
18. Varki, A., Sherman, W. and Kornfeld, S. (1983). *Arch. Biochem. Biophys.*, **222**, 145.
19. Abeijon, C. and Hirschberg, C.B. (1987). *J. Biol. Chem.*, **262**, 4153.
20. Hang, H.C. and Bertozzi, C.R. (2005). *Bioorg. Med. Chem.*, **13**, 5021.
21. Julenius, K., Mølgaard, A., Gupta, R. and Brunak, S. (2005). *Glycobiology*, **15**, 153.
22. Clausen, H. and Hakomori, S.-i. (1989). *Vox Sang.*, **56**, 1.
23. Watkins, W.M. (1966). *Science*, **152**, 172.
24. Hakomori, S. (1999). *Biochim. Biophys. Acta*, **1473**, 247.
25. Lee, H.J., Barry, C.H., Borisova, S.N., Seto, N.O., Zheng, R.B., Blancher, A., Evans, S.V. and Palcic, M.M. (2005). *J. Biol. Chem.*, **280**, 525.
26. Patenaude, S.I., Seto, N.O., Borisova, S.N., Szpacenko, A., Marcus, S.L., Palcic, M.M. and Evans, S.V. (2002). *Nat. Struct. Biol.*, **9**, 685.
27. Rose, N.L., Palcic, M.M. and Evans, S.V. (2005). *J. Chem. Ed.*, **82**, 1846.
28. Kruskall, M.S., AuBuchon, J.P., Anthony, K.Y., Herschel, L., Pickard, C., Biehl, R., Horowitz, M., Brambilla, D.J. and Popovsky, M.A. (2000). *Transfusion*, **40**, 1290.
29. Goldstein, J., Siviglia, G., Hurst, R., Lenny, L. and Reich, L. (1982). *Science*, **215**, 168.
30. Liu, Q.P., Sulzenbacher, G., Yuan, H., Bennett, E.P., Pietz, G., Saunders, K., Spence, J., Nudelman, E., Levery, S.B., White, T., Neveu, J.M., Lane, W.S., Bourne, Y., Olsson, M.L., Henrissat, B. and Clausen, H. (2007). *Nat. Biotechnol.*, **25**, 454.
31. Daniels, G. and Withers, S.G. (2007). *Nat. Biotechnol.*, **25**, 427.
32. Yang, Y.-G. and Sykes, M. (2007). *Nat. Rev. Immunol.*, **7**, 519.
33. Phelps, C.J., Koike, C., Vaught, T.D., Boone, J., Wells, K.D., Chen, S.H., Ball, S., Specht, S.M., Polejaeva, I.A., Monahan, J.A., Jobst, P.M., Sharma, S.B., Lamborn, A.E., Garst, A.S., Moore, M., Demetris, A.J., Rudert, W.A., Bottino, R., Bertera, S., Trucco, M., Starzl, T.E., Dai, Y. and Ayares, D.L. (2003). *Science*, **299**, 411.
34. Milland, J., Christiansen, D. and Sandrin, M.S. (2005). *Immunol. Cell Biol.*, **83**, 687.
35. Wells, L., Vosseller, K. and Hart, G.W. (2001). *Science*, **291**, 2376.
36. Hart, G.W., Housley, M.P. and Slawson, C. (2007). *Nature*, **446**, 1017.

37. Zachara, N.E. and Hart, G.W. (2002). *Chem. Rev.*, **102**, 431.
38. Macauley, M.S., Whitworth, G.E., Debowski, A.W., Chin, D. and Vocadlo, D.J. (2005). *J. Biol. Chem.*, **280**, 25313.
39. Kim, E.J., Perreira, M., Thomas, C.J. and Hanover, J.A. (2006). *J. Am. Chem. Soc.*, **128**, 4234.
40. Stubbs, K.A., Zhang, N. and Vocadlo, D.J. (2006). *Org. Biomol. Chem.*, **4**, 839.
41. Ferguson, M.A. and Williams, A.F. (1988). *Annu. Rev. Biochem.*, **57**, 285.
42. Ferguson, M.A. (1999). *J. Cell Sci.*, **112**, 2799.
43. Ferguson, M.A., Homans, S.W., Dwek, R.A. and Rademacher, T.W. (1988). *Science*, **239**, 753.
44. Homans, S.W., Ferguson, M.A., Dwek, R.A., Rademacher, T.W., Anand, R. and Williams, A.F. (1988). *Nature*, **333**, 269.
45. Tse, A.G., Barclay, A.N., Watts, A. and Williams, A.F. (1985). *Science*, **230**, 1003.
46. Kinoshita, T. and Inoue, N. (2000). *Curr. Opin. Chem. Biol.*, **4**, 632.
47. Maeda, Y., Ashida, H. and Kinoshita, T. (2006). *Methods Enzymol.*, **416**, 182.
48. McConville, M.J. and Ferguson, M.A. (1993). *Biochem. J.*, **294**, 305.
49. Zacks, M.A. and Garg, N. (2006). *Mol. Membr. Biol.*, **23**, 209.
50. Kinoshita, T., Ohishi, K. and Takeda, J. (1997). *J. Biochem.*, **122**, 251.
51. Okajima, T. and Matsuda, T. (2006). *Methods Enzymol.*, **417**, 111.
52. Harris, R.J. and Spellman, M.W. (1993). *Glycobiology*, **3**, 219.
53. Peter-Katalinic, J. (2005). *Methods Enzymol.*, **405**, 139.
54. Haltiwanger, R.S. and Stanley, P. (2002). *Biochim. Biophys. Acta*, **1573**, 328.
55. Doucey, M.A., Hess, D., Cacan, R. and Hofsteenge, J. (1998). *Mol. Biol. Cell*, **9**, 291.
56. Spiro, R.G. (2002). *Glycobiology*, **12**, 43R.
57. Hofsteenge, J., Huwiler, K.G., Macek, B., Hess, D., Lawler, J., Mosher, D.F. and Peter-Katalinic, J. (2001). *J. Biol. Chem.*, **276**, 6485.
58. Willer, T., Valero, M.C., Tanner, W., Cruces, J. and Strahl, S. (2003). *Curr. Opin. Struct. Biol.*, **13**, 621.
59. Lehle, L., Strahl, S. and Tanner, W. (2006). *Angew. Chem. Int. Ed.*, **45**, 6802.
60. Hutzler, J., Schmid, M., Bernard, T., Henrissat, B. and Strahl, S. (2007). *Proc. Natl. Acad. Sci. USA*, **104**, 7827.
61. Nishimura, H., Kawabata, S., Kisiel, W., Hase, S., Ikenaka, T., Takao, T., Shimonishi, Y. and Iwanaga, S. (1989). *J. Biol. Chem.*, **264**, 20320.
62. Gotting, C., Kuhn, J., Zahn, R., Brinkmann, T. and Kleesiek, K. (2000). *J. Mol. Biol.*, **304**, 517.
63. Schon, S., Prante, C., Bahr, C., Kuhn, J., Kleesiek, K. and Gotting, C. (2006). *J. Biol. Chem.*, **281**, 14224.
64. Almeida, R., Levery, S.B., Mandel, U., Kresse, H., Schwientek, T., Bennett, E.P. and Clausen, H. (1999). *J. Biol. Chem.*, **274**, 26165.
65. Bai, X., Zhou, D., Brown, J.R., Crawford, B.E., Hennet, T. and Esko, J.D. (2001). *J. Biol. Chem.*, **276**, 48189.
66. Wei, G., Bai, X., Sarkar, A.K. and Esko, J.D. (1999). *J. Biol. Chem.*, **274**, 7857.
67. Gulberti, S., Lattard, V., Fondeur, M., Jacquinet, J.C., Mulliert, G., Netter, P., Magdalou, J., Ouzzine, M. and Fournel-Gigleux, S. (2005). *J. Biol. Chem.*, **280**, 1417.
68. Itano, N. and Kimata, K. (2002). *IUBMB Life*, **54**, 195.
69. Sugahara, K., Mikami, T., Uyama, T., Mizuguchi, S., Nomura, K. and Kitagawa, H. (2003). *Curr. Opin. Struct. Biol.*, **13**, 612.
70. Silbert, J.E. and Sugumaran, G. (2002). *IUBMB Life*, **54**, 177.
71. Maccarana, M., Olander, B., Malmstrom, J., Tiedemann, K., Aebersold, R., Lindahl, U., Li, J.P. and Malmstrom, A. (2006). *J. Biol. Chem.*, **281**, 11560.
72. Funderburgh, J.L. (2002). *IUBMB Life*, **54**, 187.
73. Kitayama, K., Hayashida, Y., Nishida, K. and Akama, T.O. (2007). *J. Biol. Chem.*, **282**, 30085.

74. Linhardt, R.J. (2003). *J. Med. Chem.*, **46**, 2551.
75. Esko, J.D. and Selleck, S.B. (2002). *Annu. Rev. Biochem.*, **71**, 435.
76. Sugahara, K. and Kitagawa, H. (2002). *IUBMB Life*, **54**, 163.
77. Scriver, C.R., Beaudet, A.L., Valle, D. and Sly, W.S., eds (2006). *Metabolic and Molecular Bases of Inherited Diseases Online*, 8th Ed., acccessed 22/11/07. McGraw-Hill.
78. Meikle, P.J., Hopwood, J.J., Clague, A.E. and Carey, W.F. (1999). *JAMA*, **281**, 249.
79. Winchester, B. (2005). *Glycobiology*, **15**, 1R.
80. Butters, T.D., Dwek, R.A. and Platt, F.M. (2005). *Glycobiology*, **15**, 43R.
81. Elstein, D., Hollak, C., Aerts, J.M.F.G., van Weely, S., Maas, M., Cox, T.M., Lachmann, R.H., Hrebicek, M., Platt, F.M., Butters, T.D., Dwek, R.A. and Zimran, A. (2004). *J. Inherit. Metab. Dis.*, **27**, 757.
82. Wennekes, T., Berg, R.J., Donker, W., Marel, G.A., Strijland, A., Aerts, J.M., Overkleeft, H.S. (2007). *J. Org. Chem.*, **72**, 1088.
83. Tropak, M.B., Reid, S.P., Guiral, M., Withers, S.G. and Mahuran, D. (2004). *J. Biol. Chem.*, **279**, 13478.

Chapter 12

Classics in Carbohydrate Chemistry and Glycobiology

Increasingly, the greatest advances in science are being made at the interfaces of disciplines. In the areas of carbohydrate chemistry and glycobiology, synergistic interactions among synthetic organic chemists, biochemists and biologists are leading to new therapies that promise to improve human health through combating disease. The ultimate goal is the design of new therapeutics with exquisite selectivity and without deleterious side-effects. These studies are being enabled by the development of detailed molecular-level insights into the role of carbohydrates in various disease states and the power of chemical synthesis to deliver molecules of remarkable complexity on demand. This last chapter presents four accounts of interdisciplinary research aimed at the development of carbohydrate-based therapeutics where synthetic chemistry efforts have played a vital role.

The Immucillins: Transition-state Analogue Inhibitors of Enzymic *N*-Ribosyl Transfer Reactions

Enzymes achieve their remarkable catalytic capabilities largely through preferential binding of the transition-state structures over ground-state structures. Transition-state theory demands that the ratio of the affinity of an enzyme for the transition state over the ground state is equal to its catalytic prowess: $K_{ts}/K_d = k_{cat}/k_{uncat}$.[1,2] For effective enzyme catalysts, particularly those that catalyse difficult, intrinsically slow reactions, the rate accelerations achieved can approach 10^{19}. As enzyme affinities for substrates are typically in the range of micromolar to millimolar, molecules that are able to perfectly mimic the structure of the transition state are expected to bind especially strongly, with dissociation constants $(K_i) > 10^{-20}\,M$ being possible.[3]

References start on page 442

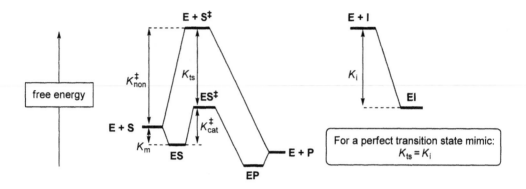

In practice a transition-state structure has partial bonds, unnatural bond angles and non-integer charge distributions, meaning that it is essentially impossible to create a perfect chemically stable mimic. Nonetheless, transition-state mimicry remains a highly sought ideal. One exceptionally successful application of this principle has arisen from a 15-year collaboration between the research groups of Vernon Schramm at the Albert Einstein College of Medicine, USA, and Richard Furneaux at Industrial Research Limited, New Zealand, leading to some of the most effective inhibitors known for *any* enzyme, and a series of promising new drugs that are in clinical trials.[4] The approach has been to (1) conduct basic studies to determine the structure of the transition state for the enzyme of interest, (2) design and synthesize molecules that mimic the transition state and, finally, (3) integrate medicinal chemistry principles to mature rationally designed lead compounds into clinical candidates.

N-Ribosyl transfer reactions occur widely in a range of processes essential for cellular function. *N*-Ribosyl hydrolases catalyse the hydrolytic cleavage of purine, pyrimidine and related bases from nucleosides and nucleotides. These enzymes are involved in DNA repair, RNA depurination by plant toxins (e.g. ricin) and salvage pathways that allow cellular recycling. Nucleoside phosphorylases are related enzymes that cleave nucleosides by phosphorolysis and have roles in nucleoside salvage, with the resulting ribosyl phosphates re-entering cellular biosynthetic processes. These two classes of enzymes can act by inversion or retention of stereochemistry, the latter being achieved by a two-step mechanism involving stepwise inversions by an enzymic nucleophile. The transition states for these processes can differ substantially, even for enzymes that share a common mechanism. Design of an inhibitor requires detailed information about the structure of the transition state for the enzyme of interest – a blueprint for inhibitor design. The two enzymes to be discussed here are purine nucleoside phosphorylase (PNP), and 5′-methylthioadenosine-*S*-adenosylhomocysteine nucleosidase (MTAN).

PNP is a human enzyme that catalyses the phosphorolysis of *N*-ribosidic bonds of 6-oxypurine nucleosides and 6-oxypurine-2′-deoxynucleosides and is involved in nucleoside recycling. A rare genetic deficiency of PNP results in accumulation of deoxynucleoside substrates, causing a disease termed T-cell deficiency. Inhibition of this pathway has the potential to treat T-cell proliferative diseases such as T-cell lymphomas.

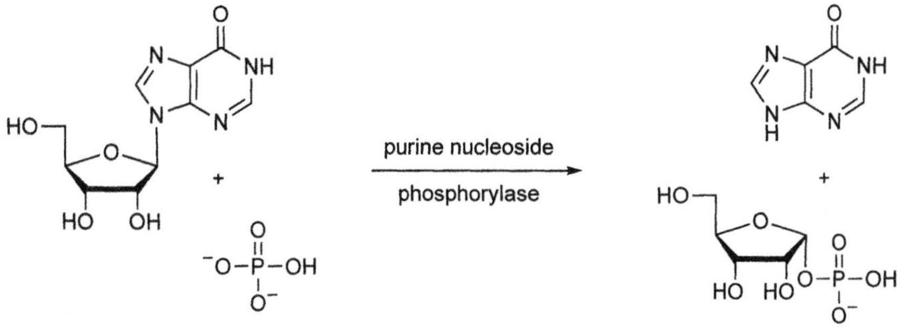

MTAN is found in pathogenic bacteria and catalyses the hydrolysis of 5′-methylthioadenosine (MTA) to adenine and 5-methylthioribose. Adenine is recycled into the nucleotide pool. MTA is also a by-product of polyamine synthesis, and its accumulation leads to feedback inhibition of this pathway. Inhibition of MTAN has therefore been proposed as an antibiotic target.

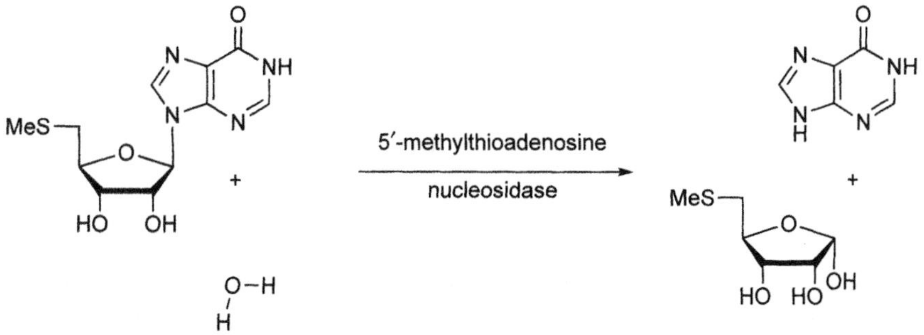

PNP transition-state structure and inhibitor design:[5] Kinetic isotope studies using isotopically labelled substrates, in combination with computational chemistry, can deliver profound insight into transition-state structure. Kinetic isotope effects reveal the effect upon reaction rate of substrates modified by the introduction of isotopes at specified positions. Synthesis of a range of purine nucleosides with isotopes at specific positions allows the investigation of kinetic isotope effects for the enzyme-catalysed reaction. However, as the enzyme initially chosen for study, bovine PNP, has a large forward commitment factor for phosphorolysis, the arsenolysis reaction was studied instead.[6] These kinetic isotope studies revealed that, at the transition state, bond formation to the nucleophile was only nascent (bond order 0.01) and the leaving group still possessed a significant partial bond to the anomeric centre (bond order 0.37). In addition, partial charge development occurs on the anomeric carbon and endocyclic oxygen. Given these characteristics of the transition-state structure, the design of an inhibitor incorporating these features was undertaken, namely immucillin H. This molecule incorporates a single bond (approximately 1.5 Å) between the anomeric centre and the nucleobase (now a linkage to carbon, not nitrogen), mimicking the partial bond to the base in the transition state, and the endocyclic nitrogen can be protonated, providing mimicry of the charge development on the anomeric centre and endocyclic oxygen in the transition-state structure.

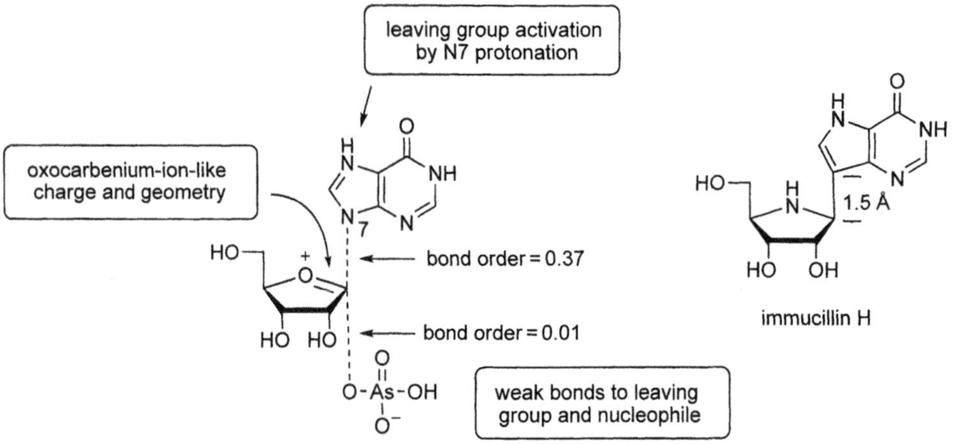

Immucillin H was synthesized from 'D-gulonolactone'.[8] Formation of the diisopropylidene derivative was achieved by transacetalation with 2,2-dimethoxypropane, followed by reduction with LiAlH$_4$ to a diol. The diol was converted to the dimesylate, and then cyclized to the pyrrolidine with benzylamine. Selective hydrolysis of the 5,6-isopropylidene group gave a vicinal diol, which was cleaved with periodate followed by borohydride reduction of the intermediate aldehyde to afford an alcohol. Protection of the alcohol as a silyl ether and hydrogenolytic debenzylation afforded a new pyrrolidine.

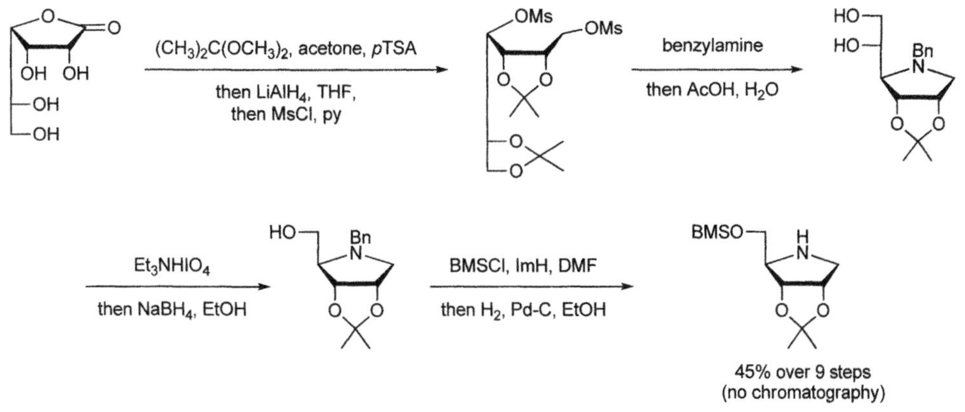

45% over 9 steps
(no chromatography)

With the silylated pyrrolidine in hand, synthesis of the imino-*C*-glycoside needed to be achieved. The imine was generated by oxidation of the pyrrolidine to the *N*-chloro derivative with *N*-chlorosuccinimide, followed by elimination with LiTMP. Addition of the lithiated-protected heterocycle to the imine was achieved in diethyl ether containing anisole.[9] Finally, deprotection afforded the hydrochloride of immucillin H.

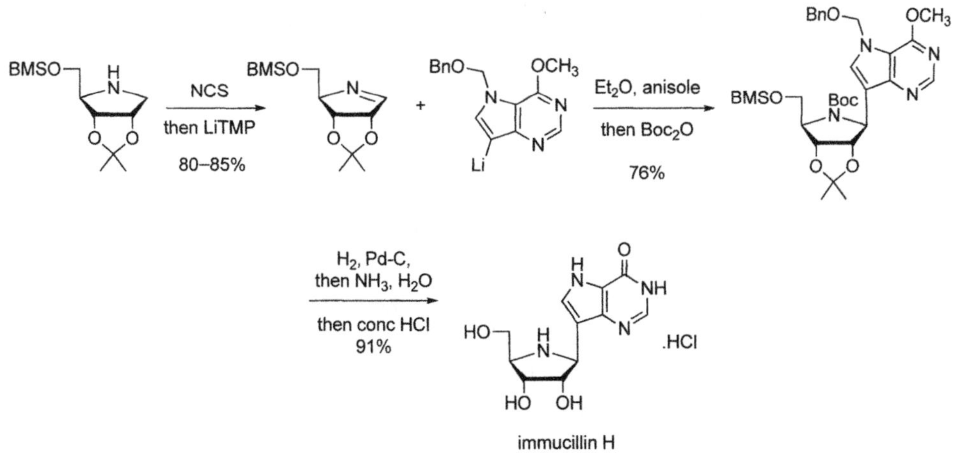

immucillin H

MTAN transition-state structure and inhibitor design: Kinetic isotope effect studies for the MTAN-catalysed hydrolysis of MTA revealed that at the transition state, almost complete bond cleavage to the nucleobase has occurred while at the same time there is essentially no participation by the nucleophilic water.[7] This leads to significant charge development on the anomeric carbon and endocyclic oxygen. These features were proposed to be mimicked by a 5'-MeT-DADMe-immucillin A. This molecule possesses a methylene spacer between the nucleobase and the ribose moiety, resulting in lengthening of the spacing between the anomeric centre and the nucleobase (approximately 2.5 Å). Charge development at the transition state should be replicated by protonation of the nitrogen incorporated in place of the anomeric carbon.

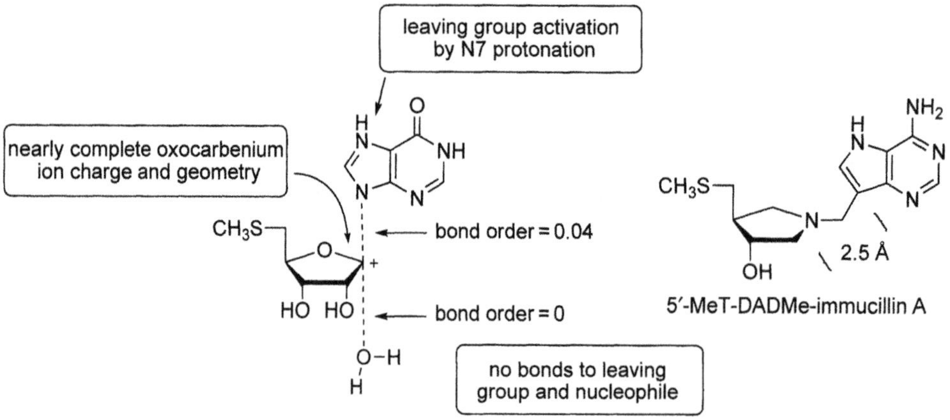

5'-MeT-DADMe-immucillin A was prepared from D-xylose. Conversion of D-xylose to the 1,2-isopropylidene acetal, followed by silylation of the primary hydroxyl of the diol, gave a secondary alcohol.[8] This alcohol was converted to the exo-cyclic methylene derivative via the ketone. Hydroboration/oxidation, activation as a mesylate, and substitution with azide gave the primary azide, possessing a branched-carbon backbone. Treatment of this azide with acid afforded an intermediate aldehyde, which was converted to the pyrrolidine by intramolecular reductive amination. Protection of the amine with a Boc group and periodate cleavage/reduction afforded an N-Boc-protected diol. Introduction of the 5-methylthio substituent was achieved by way of an intermediate mesylate and subsequent treatment with sodium methanethiolate.[9] Removal of the Boc group was achieved by treatment with methanolic HCl. Finally, introduction of the heterocycle was achieved by a Mannich reaction with the nucleobase and formaldehyde.[10]

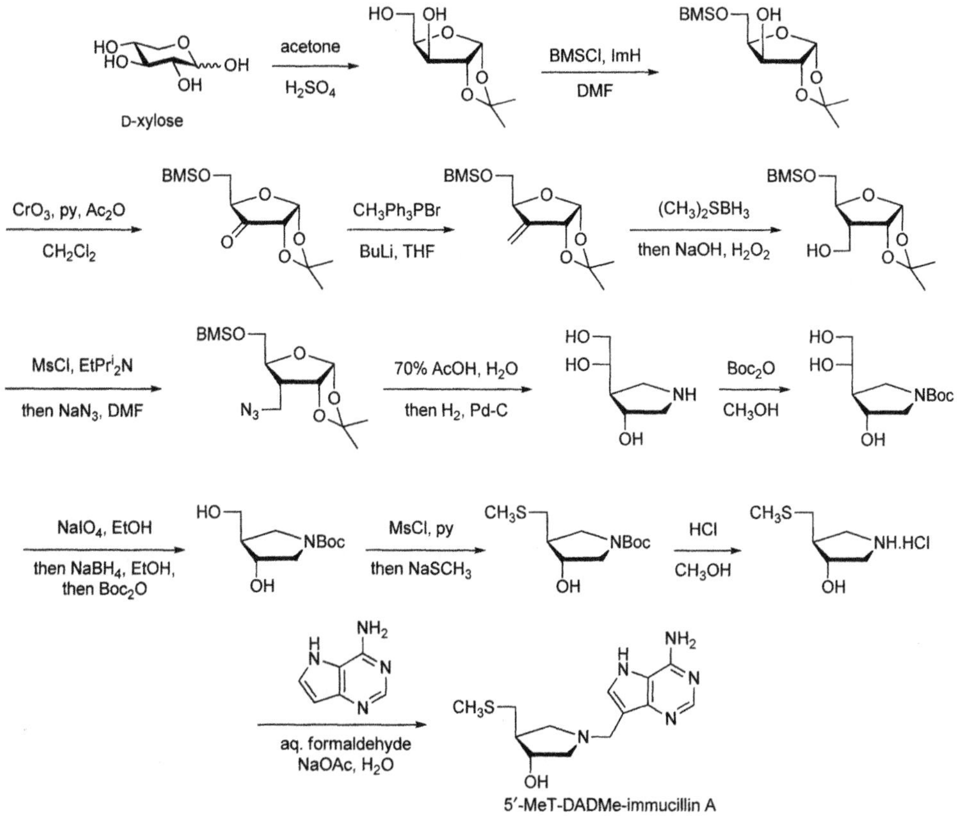

5′-MeT-DADMe-immucillin A

Immucillin H was assessed as an inhibitor of calf spleen PNP and erythrocyte PNP. The inhibitor exhibited slow-onset, tight binding inhibition, where addition of the inhibitor to reaction mixtures of the enzyme and substrate first results in modest inhibition, which increases as a function of time. These results are consistent with a two-step mechanism in which the inhibitor binds to form a reversible enzyme–inhibitor complex $(E+I \rightarrow E.I)$, which converts into a more tightly bound enzyme–inhibitor complex $(E.I \rightarrow E.I^*)$. Two inhibition constants can be extracted: K_i, representing the equilibrium constant between $E+I$ and $E.I$; and K_i^*, representing the equilibrium between $E+I$ and $E.I^*$. Accordingly, immucillin H was shown to be an extremely potent inhibitor, with a K_i value of 41 nM and a K_i^* value of 23 pM for bovine PNP, and a K_i^* value of 56 pM for human PNP.[11] More recently, kinetic isotope effect studies of the arsenolysis reaction of human PNP revealed it to possess a dissociative transition-state structure more akin to that for MTAN described above, with almost complete formation of an oxocarbenium ion and complete cleavage of the bond to leaving group, whilst at the same time the bond to nucleophile is essentially non-existent. Accordingly, the 'second generation' inhibitor, DADMe-immucillin H, was investigated as an inhibitor of human PNP and was found to be an 8.5 pM inhibitor of this enzyme.[12]

DADMe-immucillin H

X-ray crystallographic analysis of the enzyme–inhibitor complex of bovine PNP and immucillin H reveals that essentially every hydrogen-bond donor/acceptor site is fully engaged in favourable interactions, providing a structural rationale for the effective inhibition of the enzyme by this compound.[13]

The exceptional potency of immucillin H and DADMe-immucillin H suggested their potential for the inhibition of T-lymphocytes. Previous studies with nanomolar inhibitors of PNP had failed to inhibit T-lymphocytes because of lack of potency – it is known that >95% inhibition of PNP is required for significant reduction in T-cell function. Gratifyingly, immucillin H was an inhibitor of human T-lymphocytes, suggesting that it may have utility in the treatment of diseases characterized by abnormal T-cell growth or activation.[14] Oral dosing of immucillin H in mice resulted in effective inhibition of PNP activity within minutes, and it required 4 days for recovery to 50% PNP activity from a single dose (0.8 mg kg^{-1}).[15] Oral administration of DADMe-immucillin H at the same dosage leads to similarly effective inhibition of PNP activity. However, the rate of recovery of PNP activity after a single dose of DADMe-immucillin H was lower, with 11.5 days being required for recovery to 50% activity, a rate that approximately matches the rate of protein resynthesis, suggesting that the off-rate for inhibitor dissociation is not higher than resynthesis of the target enzyme! At the time of writing, immucillin H (Fodosine) is in Phase II clinical trials for relapsed/resistant T-cell leukaemia and cutaneous T-cell leukaemia.[16] Further investigations have shown that related acyclic derivatives are effective inhibitors of human PNP.[17] These 'third generation' PNP inhibitors include a triol that has been found to be as effective as DADMe-immucillin H for inhibition of human PNP, with a K_i^* value of 8.6 pM, and an achiral derivative of 'tris base' that is a sub-nanomolar inhibitor of malarial PNP. More recent studies have reported the synthesis of an achiral azetidine that is a sub-nanomolar inhibitor of PNPs.[18]

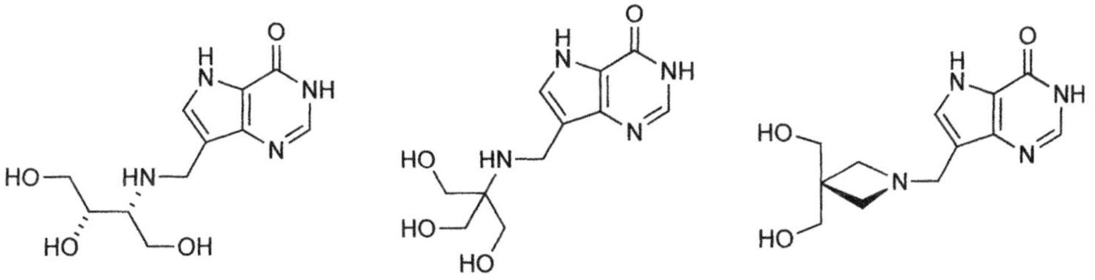

5′-MeT-DADMe-immucillin A was initially assessed as an inhibitor of *E. coli* MTAN.[9] Analogous to immucillin H, this compound was found to exhibit slow-onset, tight binding inhibition, with a dissociation constant (K_i^*) value of 2 pM. Also studied was 5′-MeT-immucillin A, which was also an effective inhibitor with K_i^* value of 77 pM. A structural rationale for inhibitor binding was provided for these two inhibitors and the *E. coli* enzyme by X-ray crystallography of enzyme–inhibitor complexes.[19] A key feature of both structures was the formation of favourable interactions between the protonated nitrogens of the inhibitors and a water molecule in the active site, which was proposed to take the role of the nucleophilic water involved in substrate hydrolysis. Distance analysis of the 'nucleophilic' water, the 'leaving group' base and the anomeric centre showed that 5′-MeT-immucillin A was a better mimic of an early transition state than 5′-MeT-DADMe-immucillin A, whereas the latter was a better mimic of the highly dissociated transition state of the *E. coli* MTAN. These features have been exploited to allow the use of these two inhibitors as probes for transition-state structure. Thus, the ratio of the inhibition constants for 5′-MeT-DADMe-immucillin A and 5′-MeT-immucillin A can be used to distinguish between early and late transition states for a range of bacterial MTAN enzymes. MTANs with early transition states exhibit a ratio of K_i^*(ImmA)/ K_i^*(DADMe) of approximately 2, whereas for MTANs with late transition states the ratio is >10.[20]

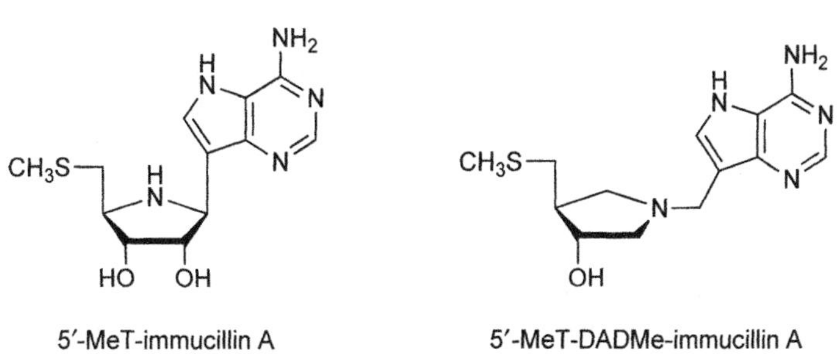

5′-MeT-immucillin A   5′-MeT-DADMe-immucillin A

Further improvements to the potency of 5′-MeT-DADMe-immucillin A were achieved by systematic structural variation.[9] It was known that the pocket in MTAN that binds the 5'-methylthio substitituent of the substrate is hydrophobic and can bind groups larger than a simple MeS moiety. Consequently, substituents were introduced into 5'-MeT-DADMe-immucillin A, on sulfur, to investigate the effect upon inhibition. 5'-Me, 5'-Et, 5'-Pr and 5'-Bu groups led to an increase in affinity, reflected in K_i^* values of 2000, 950, 580 and 296 fM, respectively. A range of substituents was introduced into the 5'-position, ultimately leading to 5'-pClC$_6$H$_4$T-DADMe-immucillin A, with a K_i^* value of 47 fM. This inhibitor binds 43-fold tighter than 5'-MeT-DADMe-immucillin A and is one of the most powerful non-covalent inhibitors known for any enzyme.

5'-pClC$_6$H$_4$T-DADMe-immucillin A

Development of a Candidate Anti-toxic Malarial Vaccine

Vaccination is undoubtedly one of the most elegant methods for combating disease. By marshalling the formidable resources of the immune system, vaccines prompt the body to provide its own defences against infection. The process of immunization was probably first achieved by Edward Jenner[a] who used cowpox to vaccinate against smallpox.[b] Ultimately, a more sophisticated but related approach has led to the eradication of smallpox from the world, as declared by the World Health Organization in 1980.

Malaria is caused by the parasite *Plasmodium falciparum*, which infects 5–10% of the world's population and kills 2 million people annually. The malarial infection typically results in fever, seizures, coma and cerebral edema, symptoms that have striking similarity to bacterial infection and that can be mimicked by chronic salicylate poisoning.[21] Over a century ago, Camillo Golgi and other malaria

---

[a] Edward Jenner (1749–1823), M.D. University of St Andrews (1792). Elected Fellow of the Royal Society for observations on the nesting behaviour of the cuckoo (1788). In 1796 he carried out the first vaccination experiment on 8-year-old James Phipps, by inserting pus from a cowpox pustule into the boy's arm. Jenner subsequently proved that having been inoculated with cowpox, Phipps was immune to smallpox.

[b] Coined by Jenner from the Latin *vaccinus*, from *vacca* = cow.

scientists demonstrated that malarial fever is synchronous with the developmental cycle of the blood-stage parasite. Together, these observations support the hypothesis that the pathological reactions that occur upon infection by the parasite are initiated by a malarial toxin.[22] Glycosylphosphatidylinositol (GPI) of the parasite, which is expressed in free and protein conjugated forms, has the properties predicted for a toxin, can induce cytokine and adhesion expression in macrophages and vascular endothelium, and can be lethal in vivo.[23,24] Anti-toxin vaccines are effective public health measures that are used for protection against tetanus and diphtheria and prevent subjects from becoming ill, rather than preventing parasite multiplication. These vaccines use a 'toxoid', an inactivated form of the toxin, to induce host immunity and protect against subsequent toxin. This section will describe work by the groups of Louis Schofield at the Walter and Eliza Hall Medical Institute, Australia and Peter Seeberger, at the Massachusetts Institute of Technology, USA, and the ETH, Switzerland, to develop a candidate anti-toxin vaccine for malaria.

Determination of the structure and biological activity of the malarial GPI: Glycolipids were extracted from *P. falciparum* cultured in the presence of radiolabelled [^{3}H]mannose, [^{3}H]glucosamine and [^{3}H]ethanolamine precursors. Determination of the results of cleavage by a range of enzymes allowed definition of the major structures of the GPI-anchor precursors as containing three and four mannose residues.[25] Different parasite lines were subsequently investigated, revealing that all investigated strains showed a very similar series of GPIs.[26] These studies demonstrate that the GPI structure is likely to be conserved among all *P. falciparum* subtypes, even those isolated from geographically diverse locations, and emphasize the potential for a general anti-toxin vaccine. Isolated *P. falciparum* GPI elicits a range of inflammatory effects, and deacylation by enzymatic or chemical means renders the resulting carbohydrate non-toxic.[27]

A delipidated glycan can be generated by sequential treatment of the GPI with methanolic ammonia and phosphatidylinositol phospholipase C.[25] This non-toxic carbohydrate structure was chosen as the antigen from which to develop a vaccine candidate.

A widely used approach to the development of synthetic vaccines is to conjugate the antigen (the hapten) to an immunogenic carrier protein. Conjugation to generate a large molecular species ensures long-term residence of the resultant conjugate in the blood system, and the immunogenicity of the carrier protein ensures that the hapten is recognized by the immune system and results in an immune response. However, acquisition of the glycan in homogeneous form and sufficient quantities requires chemical synthesis.

The initial route described to the GPI glycan was performed candidate anti-toxin vaccine was performed entirely in the solution phase.[22] The synthetically challenging construction of the α-linkage between glucosamine and inositol was the first glycosylation performed. Thus, a 2-azido thioglycoside was condensed with a selectively protected inositol derivative to provide the masked α-GlcNH$_2$-myo-inositol core. Elaboration of the core with a series of mannose-derived trichloroacetimidate donors generated the complete glycan structure, in fully protected form.

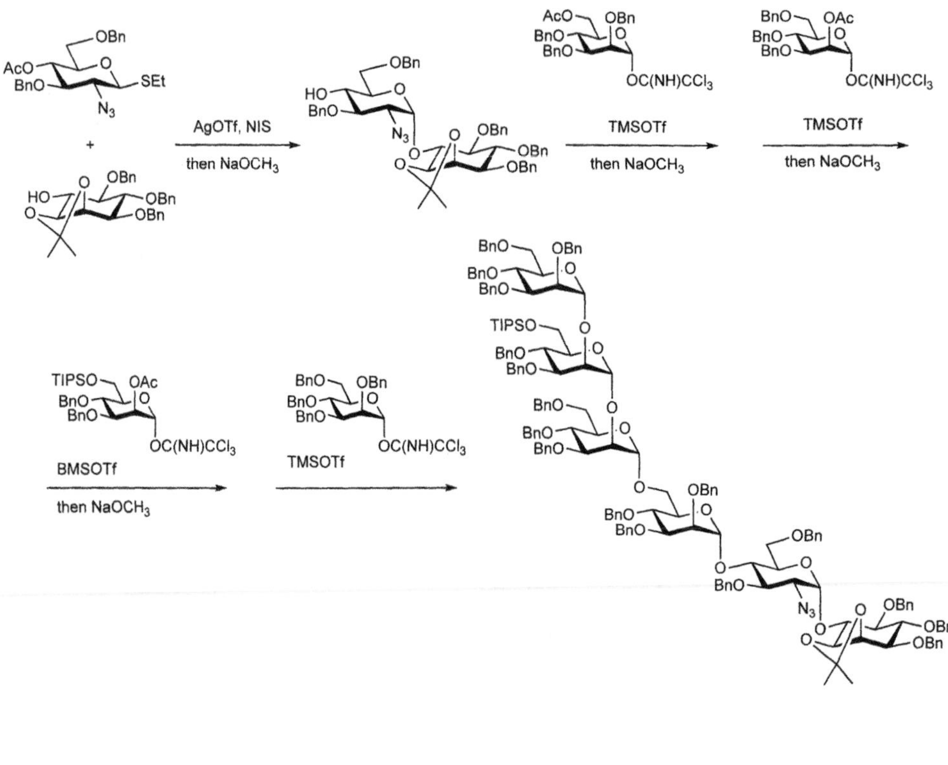

Installation of the cyclic phosphate and the phosphoethanolamine side chain was achieved prior to cleavage of the cyanoethyl group (DBU); subsequent reductive debenzylation under Birch conditions then gave the GPI glycan.

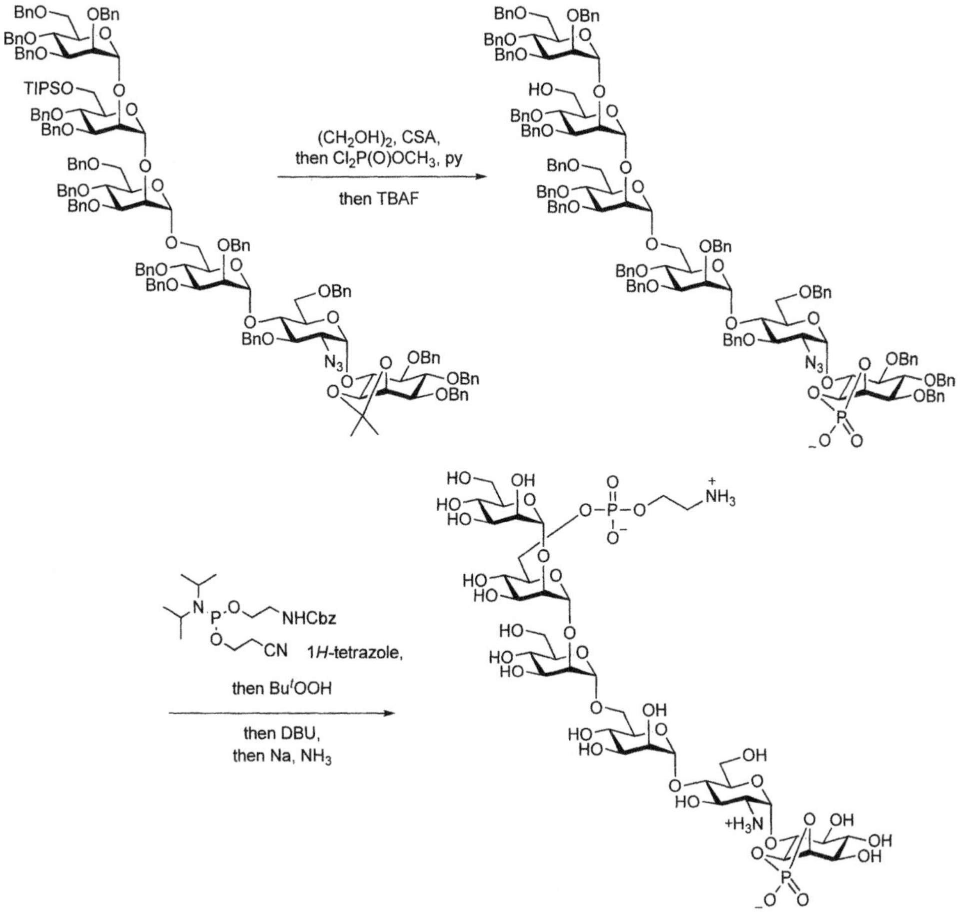

The antigen next needed to be presented in a form that allowed generation of an antigen-targeted immune response. A widely used carrier protein is keyhole limpet haemocyanin (KLH), which is potently immunogenic and displays free lysine-derived amino groups that are available for conjugation. The immunogen was prepared by treatment of the synthetic GPI glycan with 2-iminothiolane, generating a thiol off the phosphoethanolamine group. Finally, the resultant thiol was conjugated to KLH pre-activated with maleimide.

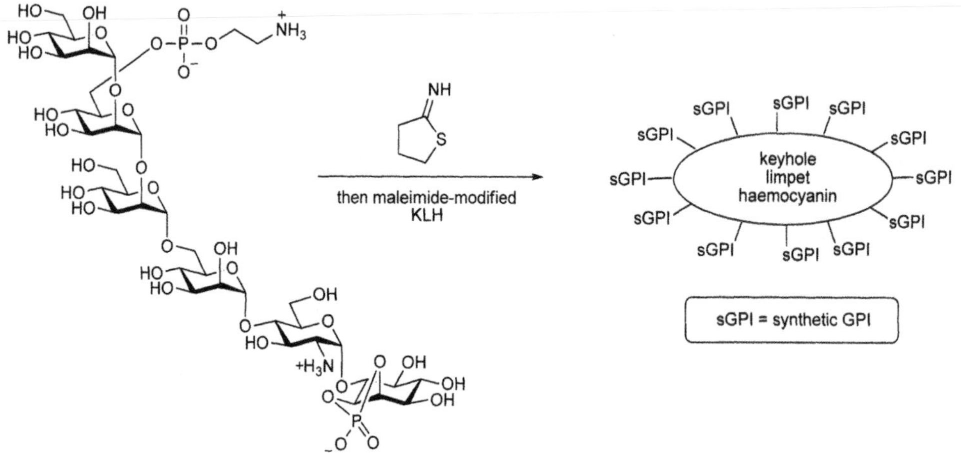

Solid-phase oligosaccharide synthesis of glycans promises to simplify significantly the preparation of complex carbohydrate structures and has been discussed in Chapter 5.[28,29] However, owing to the complexity of many glycans and the intricacies of carbohydrate chemistry, a generalized approach to solid-phase oligosaccharide synthesis has proved elusive and may never be entirely achievable. However, many remarkable inroads into the area of solid-phase synthesis have been made. Seeberger and co-workers have applied automated solid-phase oligosaccharide synthesis technology to develop an improved synthesis of the mannan portion of the vaccine candidate glycan.[30] Thus, the tetramannoside was rapidly assembled using an octenediol-functionalized Merrifield resin and glycosylations with five equivalents of donor, with each glycosylation being performed twice to ensure high coupling efficiencies. Removal of the acetyl protecting group at each step was achieved with sodium methoxide, providing a glycosyl acceptor ready for the next glycosylation. The tetramannoside was cleaved from the resin by olefin cross-metathesis with ethylene using Grubbs' catalyst, affording a 4-pentenyl tetrasaccharide. The overall yield for the 4-pentenyl tetrasaccharide over the entire sequence of glycosylations was 44% (HPLC peak area) and was achieved in just 9 hrs from the pre-assembled monosaccharide glycosyl donors.[29]

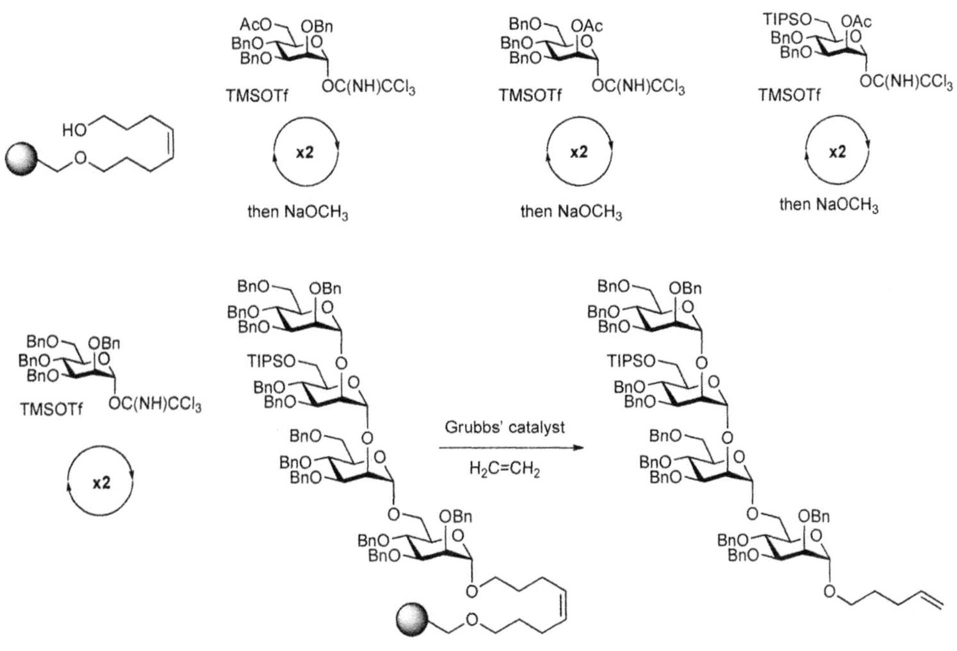

Whilst the 4-pentenyl tetrasaccharide might conceivably act as a glycosyl donor to a suitably protected GlcNH$_2$-*myo*-inositol acceptor, model studies revealed that 4-pentenyl donors failed to yield the expected products. Thus, the tetrasaccharide was converted to a trichloroacetimidate, which functioned as a moderate glycosyl donor, yielding the complete glycan structure.

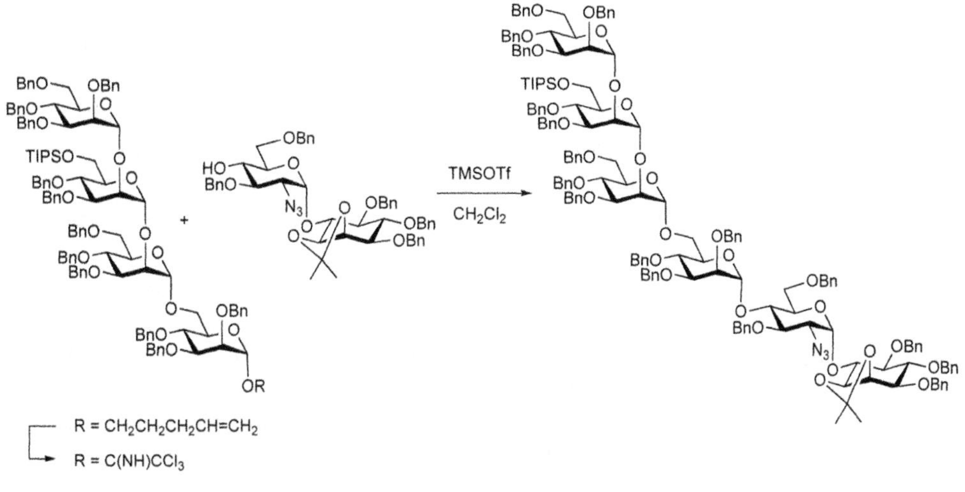

R = CH$_2$CH$_2$CH$_2$CH=CH$_2$
R = C(NH)CCl$_3$

The synthetic GPI–KLH conjugate was used to vaccinate mice, and gave rise to antibodies with activity against glycan.[22] Notably, the anti-GPI antibodies did not cross-react with mammalian GPIs, possibly owing to the structural differences between malarial and mammalian GPI structures. Mice immunized with the synthetic GPI–KLH conjugate were significantly protected against severe malaria, as indicated by improved rates of survival, when challenged with *Plasmodium berghei*, a murine model for human malaria. Consistent with the mode of action of the vaccine being against the malaria toxin, parasite levels in immunized animals were not significantly different from non-immunized animals. Current work is focused on defining the optimal structure and presentation of the glycan antigen,[31,32] developing an effective immunogenic carrier protein (of parasite or bacterial origin), and establishing optimum dosing regimes in pre-clinical studies.[24] Ultimately, clinical trials will be needed to evaluate whether a synthetic GPI anti-toxin vaccine will reduce malarial mortality and morbidity in the field with a suitable safety profile.

Synthetic Carbohydrate Anti-tumour Vaccines

Aberrant glycosylation is a hallmark of the cancer phenotype. Thus, a carbohydrate vaccine against tumour cells has long been a keenly sought goal.[33,34] Unlike anti-infection vaccines, which protect against future infections, anti-cancer vaccines aim to assist in the eradication of malignant cells post-diagnosis and face a long list of potential problems including tumour cell–induced immunosuppression even in the presence of immunogenic antigens, the intrinsically poor immunogenicity of most carbohydrates and the fact that many tumour antigens are self-antigens that are present in lower amounts on healthy tissues. The following section details research aimed at exploiting the identification of specific glycolipid and glycoprotein glycans that are over-expressed on the surfaces of malignant cells. The work has aimed to develop a cell-free, fully synthetic anti-tumour vaccine that could be used to elicit highly specific and robust immune responses against malignant cells that bear the requisite glycans. The goal is to induce the body to produce antibodies that could act within the blood system to clear circulating tumour cells and micrometastases, through complement-mediated cell lysis, or other mechanisms.

Globo-H is a hexasaccharide that was originally isolated as a ceramide-linked glycolipid from the human breast cancer cell line MCF-7.[34] The glycolipid is found at the cell surface and has been characterized with the aid of a monoclonal antibody, MBr1. Immunohistological analysis using MBr1 demonstrated that Globo-H is found in a range of other cancers including pancreas, stomach, uterine endometrium and prostate. While there is some evidence that Globo-H is found on normal tissues, it is believed to be localized on these tissues in areas where the immune system has restricted access.

Globo-H was therefore chosen as a candidate glycan for the development of an anti-tumour vaccine.

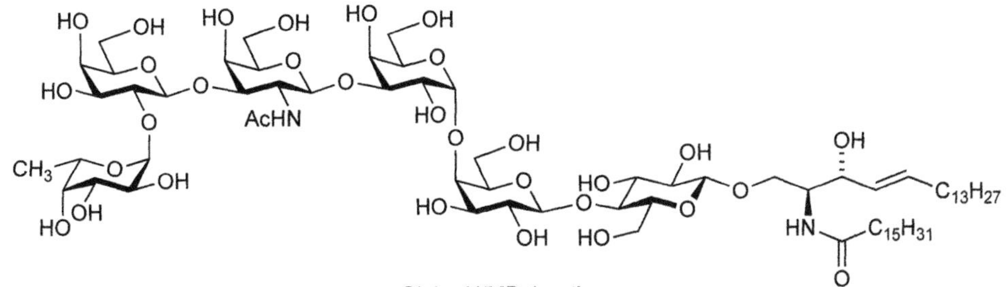

Globo-H/MBr1 antigen

Several total syntheses of Globo-H have been achieved. The first, by Danishefsky and co-workers and outlined here, utilized the glycal assembly method for the construction of three of the six glycosidic linkages and was partially discussed in Chapters 4 and 5.[35,36] The synthetic approach used by these workers favoured the disconnection of the hexasaccharide into two trisaccharides, which were joined in a [3 + 3] coupling. Synthesis of the non-reducing-end trisaccharide was achieved using two sequential glycosylations. Thus, regioselective glycosylation of the allylic 3-hydroxy group of a D-galactal acceptor with a 1,2-anhydrosugar, itself obtained from dimethyldioxirane oxidation of a 3,4-carbonate-protected D-galactal donor, gave a disaccharide glycal.

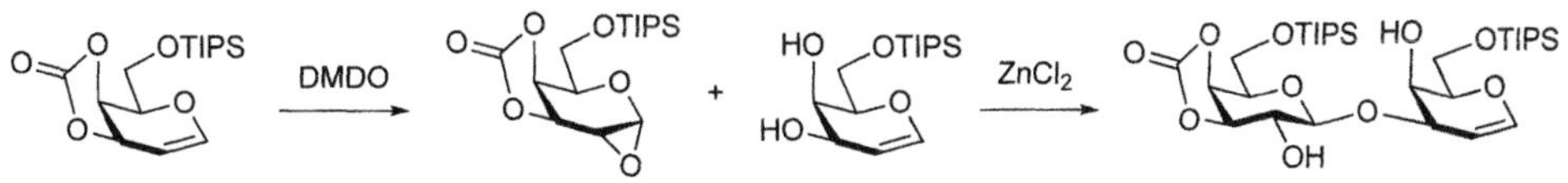

The disaccharide glycal was regioselectively glycosylated on the 2'-hydroxy group, formed in the preceding glycosylation, by an L-fucosyl fluoride donor. While the resulting trisaccharide glycal could conceivably be used in a sulfonamidoglycosylation with the reducing-end trisaccharide, it was found to lack the required reactivity and was therefore converted to a thioglycoside donor. Thus, treatment of the glycal with iodonium dicollidine perchlorate and phenylsulfonamide generated a 2-deoxy-2-iodo glycosyl sulfonamide. Treatment of this species with lithium ethanethiolate resulted in migration of the sulfonamido group from C1 to C2, via an intermediate aziridine, yielding the thioglycoside.

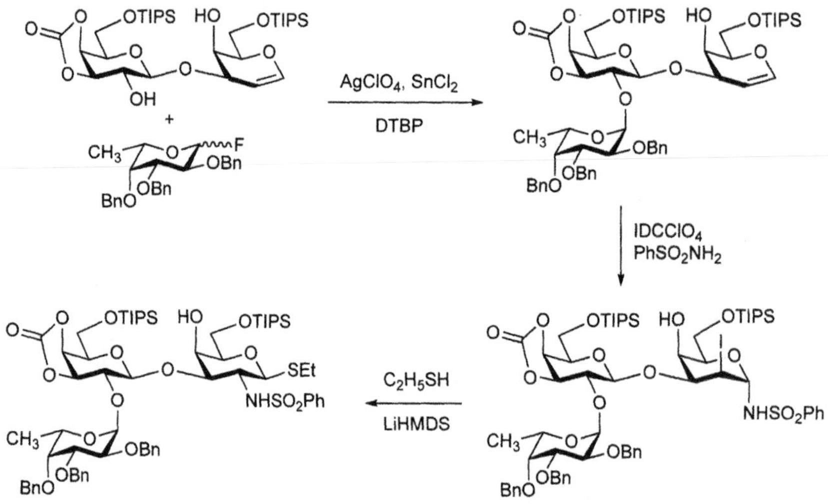

The reducing-end trisaccharide was assembled by the condensation of a D-galactosyl fluoride and a lactal acceptor (itself assembled from D-galactal and D-glucal). Removal of the sole 4-methoxybenzyl protecting group afforded a new glycosyl acceptor, and methyl triflate–mediated glycosylation by the non-reducing-end thioglycoside afforded a fully protected hexasaccharide in a 10:1 anomeric ratio in favour of the desired β-anomer. A remarkable feature of this glycosylation is the presence of an unprotected hydroxyl in the glycosyl donor. In a related work, it was shown that protection of this hydroxyl as an acetate led to a reversal in stereoselectivity of the glycosylation to 1:5 in favour of the unwanted α-anomer.

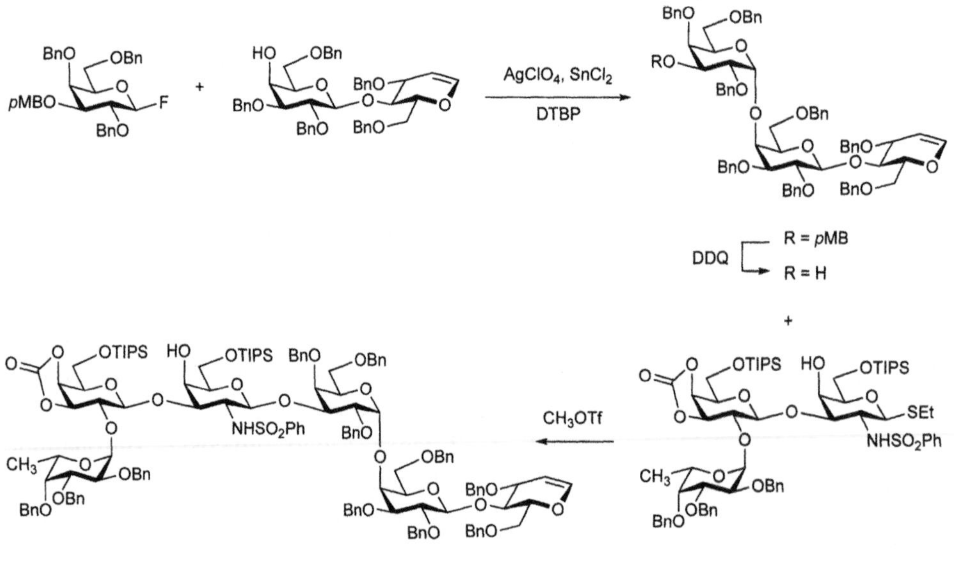

Most tumour antigens, including carbohydrates, are generally poor immunogens and require conjugation with an immunogenic carrier in order to elicit a significant immune response. Global deprotection of the protected hexasaccharide glycal prepared above, and reacetylation, afforded an acetylated hexasaccharide. This was epoxidized with dimethyldioxirane and the resultant 1,2-anhydrosugar used to glycosylate allyl alcohol. Deacetylation then provided the deprotected allyl glycoside. The alkene of the allyl glycoside was converted to an aldehyde by ozonolysis, and this was reductively coupled to KLH, resulting in presentation of approximately 350 Globo-H epitopes per molecule of KLH.

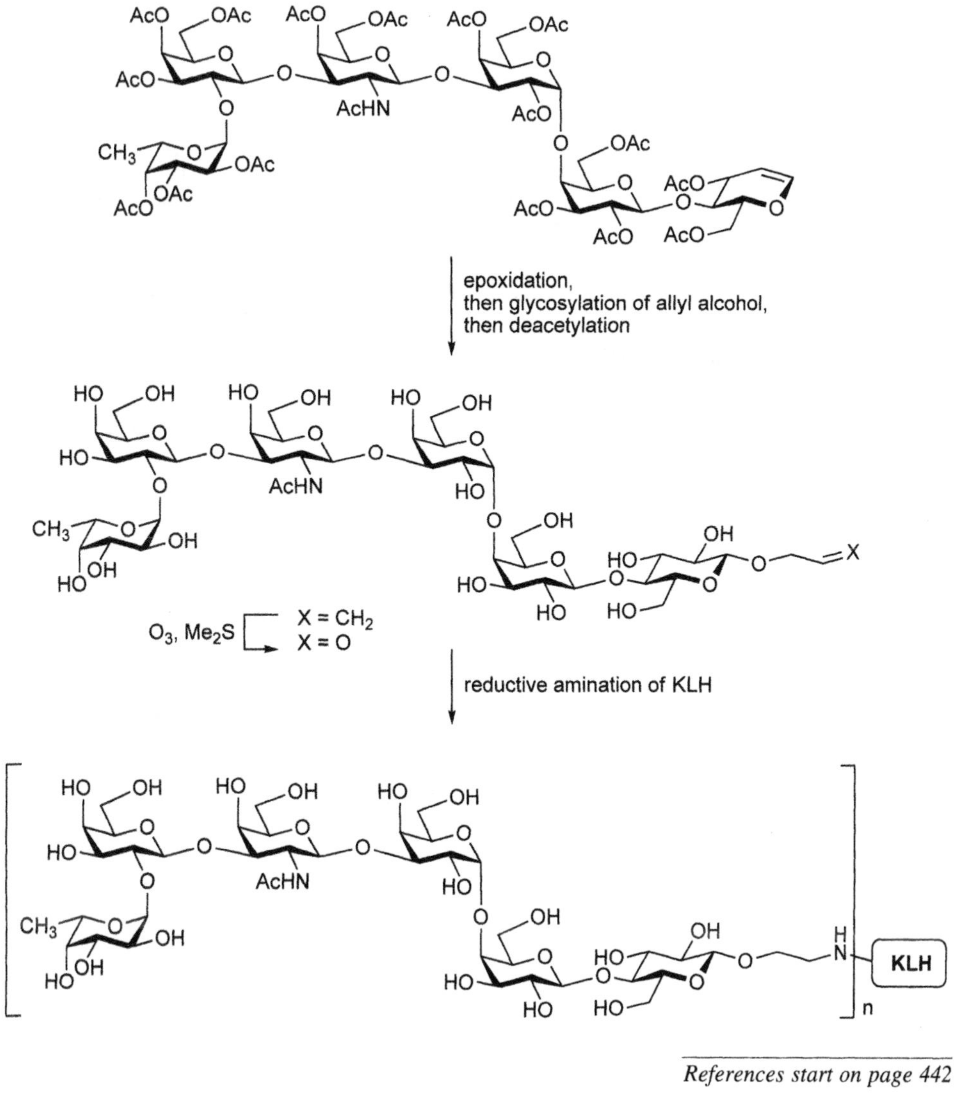

As mentioned above, the carbohydrate antigen of Globo-H was conjugated to a carrier protein with the expectation that this would enhance its immunogenicity. Even so, such subunit vaccines are inherently less immunogenic than those employing attenuated microorganisms. Many antigens are therefore administrated with adjuvants, immunostimulatory molecules that are themselves not immunogenic but serve to enhance or prolong the immune response. One of the most promising adjuvants in terms of potency and favourable toxicity is a semi-purified extract from the South American tree *Quillaja saponaria Molina*, termed QS-21.[37] This extract consists of a range of complex triterpenoid saponins, including QS-21A$_{api/xyl}$.[38,c] QS-21 is currently in use in a range of clinical trials for cancer, HIV-1 and malaria, and has been converted to a semi-synthetic analogue, GPI-0100, which shows improved tolerance, lower toxicity and greater aqueous stability.[39]

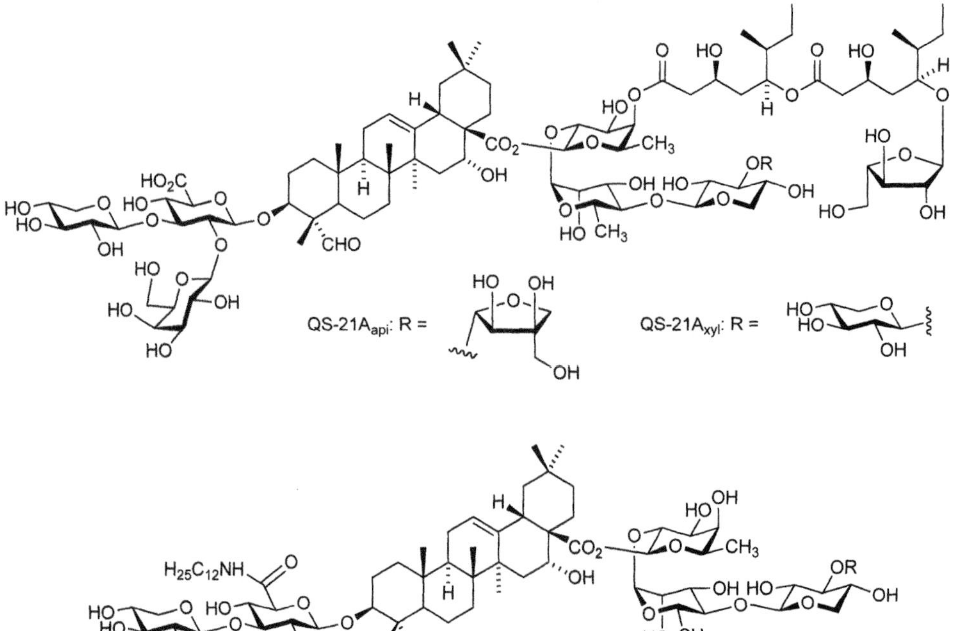

QS-21Aapi: R =
QS-21Axyl: R =

GPI-0100; R = various sugar residues

[c] '21' refers to it being isolated in the 21st fraction by HPLC.

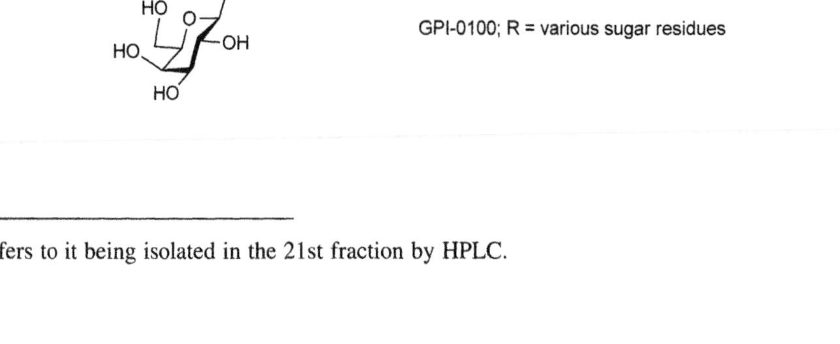

Immunization of mice with the Globo-H-KLH conjugate in the presence of QS-21 resulted in high-titre IgM and IgG responses against the Globo-H antigen.[40] The vaccine also elicits high-titre IgM antibodies against Globo-H in humans and was shown to be safe. At the time of writing, this conjugate is scheduled for Phase II/III clinical trial evaluation for the treatment of breast cancer.

A potential limitation to vaccine constructs such as those detailed above, which bear a single tumour–associated carbohydrate antigen, is a consequence of the heterogeneity of carbohydrates displayed on the cell surface of tumour cells. Even within a particular cancer type in a single patient, tumour cells display a range of cell-surface antigens, and a vaccine targeted towards a single tumour–associated antigen would likely lead to only limited benefit. For this reason, Danishefsky and co-workers have focused on the construction of so-called unimolecular, multivalent vaccines. In these vaccines, several carbohydrate antigens are displayed on a single polypeptide backbone, which is itself conjugated to a carrier protein. Among one of the most advanced such constructs is a unimolecular pentavalent vaccine construct that displays Lewis^y, Globo-H, Sialyl-Tn, Tn and TF antigens, all of which are associated with breast and prostate cancer.[33,41]

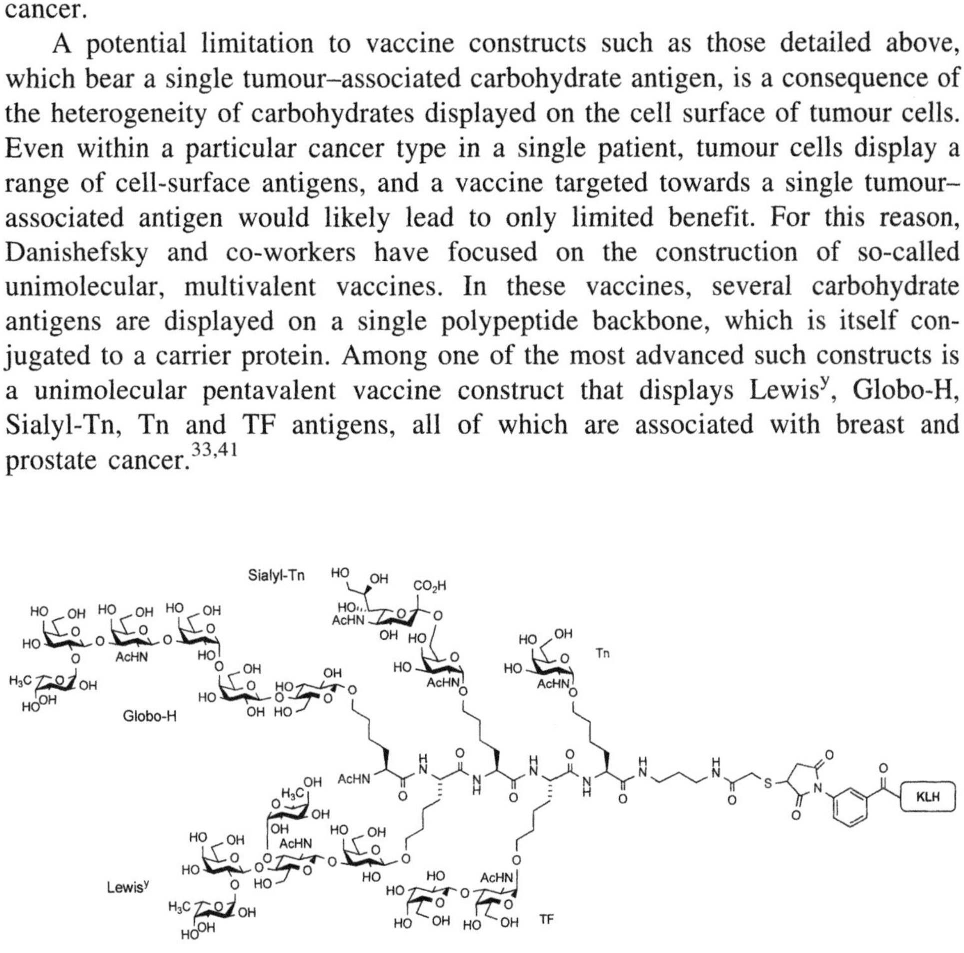

Carbohydrate antigens coupled to foreign carrier proteins such as KLH may cause strong B-cell responses against the carrier protein itself, resulting in suppression of antibody responses against the carbohydrate antigen. Additionally, carbohydrate antigens have typically resulted in the formation of relatively weak and short-lived IgM antibodies, without involvement of helper T-cells that would lead to more prolonged and higher-affinity IgG antibodies. As more detail has emerged on the function of the immune system, alternative

References start on page 442

strategies have been put forward to overcome these problems. Typically, upon administration of a carbohydrate-carrier protein conjugate, the entire construct must be taken up by B-cells, leading to the production of relatively low avidity IgM antibodies. Ideally, maturation of the antibody recognition of the carbohydrate antigen requires a class switch of IgM antibodies to IgG antibodies, leading to more efficient soluble antibodies and a stronger immune response. Such a class switch requires that antigen-presenting cells, which process the construct and present smaller peptide fragments bearing the carbohydrate entity on the major histocompatibility complex class II molecule (MHC II), recruit helper T-cells. In practice, with vaccines derived from conjugation of a carbohydrate epitope and a carrier protein, after processing of the conjugate, unmodified peptides derived from the carrier protein are typically presented by MHC II molecules to helper T-cells, resulting in the observed immune suppression.

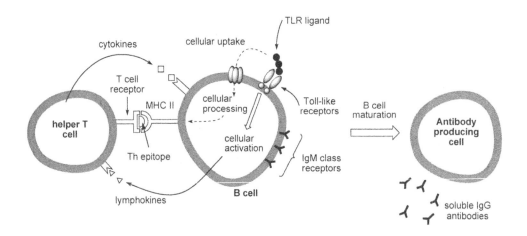

Boons and co-workers have recently demonstrated an alternative approach to generating a strong IgG response to a synthetic carbohydrate-based vaccine, through the synthesis of a three-component vaccine.[42,43] Presentation of the construct by an antigen-presenting cell was ensured by incorporation of tripalmityl-cysteine-tetralysine (PAM$_3$CysK$_4$). PAM$_3$CysK$_4$ is a known ligand for a class of cell-surface receptors called the Toll-like receptors, which are components of the innate immune system and ensure cellular activation by the conjugate. The PAM$_3$CysK$_4$ moiety was attached to a polio-derived peptide that is documented to be presented by mouse MHC II, thereby ensuring efficient presentation of the carbohydrate antigen. The carbohydrate antigen (B epitope) was a peptide-borne Tn antigen, which was conjugated to generate a tripartite construct. Upon presentation of this antigen by MHC II molecules on the surface of an antigen-presenting cell, interaction with a helper T-cell results in the release of cytokines and antibody maturation to IgG class. Notably, this construct contains only those parts needed for presentation by the immune system, ensuring that antibody responses against unwanted parts of the construct are limited and the potential for immune suppression is reduced. The tripartite construct was administered with the immunoadjuvant QS-21 and led to exceptionally high titres of IgG-class antibodies that could recognize cancer cells expressing the tumour-associated carbohydrate.

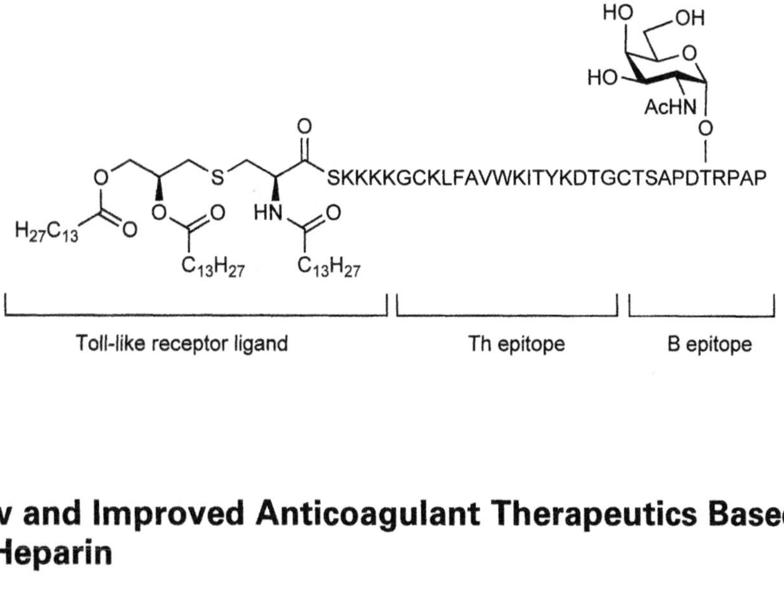

Toll-like receptor ligand   Th epitope   B epitope

New and Improved Anticoagulant Therapeutics Based on Heparin

The clotting of blood is a natural response to injury to the vascular system, without which the body would be unable to stem the flow of blood from a cut. However, the normal propensity for blood to clot must be overcome in many surgical procedures and in extra-corporeal procedures such as heart-lung oxygenation and kidney dialysis.[44] Moreover, the clotting of blood must be controlled in diseases such as coronary heart disease, where atherosclerosis leads to disruption of blood flow and platelet activation resulting in the formation of a blood clot, which can lead to a heart attack or stroke. One of the major anticoagulant therapeutics is the glycosaminoglycan heparin, and related derivatives. Heparin is one of the most effective and widely used drugs but suffers from a range of problems. This section will discuss the discovery of heparin and the development of new and improved heparin analogues that overcome the many shortcomings and problems associated with heparin.

References start on page 442

The discovery and development of heparin: In a search for substances from mammalian tissues that cause blood to clot, in 1916 McLean and Howell found a substance, first termed 'heparophosphatide', and later heparin, which prevented blood coagulation.[44,45] More than 50 years passed before even the most rudimentary structural characterization was finally completed, revealing that heparin was a sulfated polymer of D-glucosamine, L-iduronic acid and D-glucuronic acid.[44] Nonetheless, the therapeutic potential of heparin was rapidly recognized, and a number of groups attempted to develop a commercial process for its production, first from dog liver and then from more abundant beef liver. However, the concurrent development of a pet food industry in Canada and the USA drove up the cost of this raw material and required the development of an alternative source. A collaboration between the University of Toronto and Connaught Laboratories led to a commercial process for heparin production from beef lung and intestines (and later porcine intestines), allowing preparation of sufficient material for human trials.[44,46] Heparin was approved for use in humans in 1937, 1 year before the Federal Food, Drug, and Cosmetic Act of 1938 was passed by the US Congress, containing new provisions including the requirement that new drugs must be shown safe before marketing. Possibly, as a consequence of the lax rules in place during the time of its approval, heparin use has many serious side-effects including a risk of osteoporosis, haemorrhagic complications and heparin-induced thrombocytopoenia (HIT), which in severe cases can lead to death. Heparin has been cited as the drug responsible for a majority of drug-related deaths in patients who are otherwise 'reasonably healthy'.[47] Owing to its complex dose/activity profile and narrow safety window, heparin is typically only used in in-patients, in which case, careful monitoring of the patient can be performed.

The coagulation cascade:[44,48] Blood coagulation is a complex process that involves signal amplification of a blood-clotting stimulus by a series of enzymes where enzyme precursors (denoted by Roman numerals) are converted to active enzymes (denoted by Roman numerals, followed by lowercase 'a'), which then act on the next set of enzyme precursors, converting them in turn into active enzymes. The intrinsic (or contact activation) pathway is activated by the contact of blood with foreign surfaces and to sub-endothelial surfaces exposed during blood vessel damage. The extrinsic (or tissue factor) pathway is activated by the excretion of tissue factor by injured cells. Both of these pathways converge to form factor Xa, an enzyme that catalyses the formation of a burst of thrombin. Thrombin catalyses the conversion of soluble fibrinogen to insoluble strands of fibrin, which are cross-linked by a transglutaminase, factor XIIIa, to form a cross-linked blood clot.

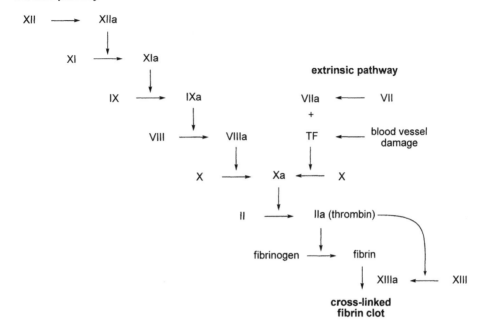

An imbalance in the blood-clotting system leads to either a pro-thrombotic or a pro-haemorrhagic state. The body contains several natural inhibitors of the coagulation cascade that prevent pro-thrombotic states arising and assist in maintaining hemostasis. One of the most important of these inhibitors is the protein antithrombin III (ATIII), which upon binding to heparin causes inactivation of factors VIIa, IXa, Xa, XIa and XIIa, and thrombin. All of these interactions may potentially contribute to hemostasis; however, clinicians have found that low doses of heparin, which do not result in significant blood coagulation when assessed in simple coagulation assays, nonetheless effectively prevent thrombosis.[48] This has led to the hypothesis that inhibition of factor Xa may be more clinically relevant for an anti-thrombotic therapeutic than inhibition of thrombin. As a consequence, significant efforts have been directed at developing modified heparins with improved factor Xa inhibition activity and reduced anti-thrombin activity so as to exert a fine control over the blood clotting process.

The simplest approach to enhance ATIII-mediated factor Xa inhibition over ATIII-mediated thrombin inhibition is to reduce the molecular weight of the heparin preparation.[48] Low molecular weight heparins (LMWHs) are produced from heparin by various enzymatic or chemical depolymerization processes followed by fractionation.[44] Studies of the biochemical, pharmacological and

References start on page 442

chemical structures of LMWHs from different sources demonstrate that they are not chemically or biochemically equivalent and have different pharmacological properties.[49] Despite the potential complications that these variations may cause, LMWHs have largely displaced heparin from the drug market, and their improved safety profile, more predictable dose/activity response and prolonged anti-thrombotic activity allow their use in an outpatient setting. However, LMWHs still suffer from some side-effects, in particular HIT.[50]

In the 1970s several crucial observations were made that helped cast light on the mechanism of action of heparin.[51] It was found that only a small fraction of heparin molecules bind with high affinity to the blood-clotting inhibitor ATIII. Moreover, the high affinity fraction accounted for essentially all of the anti-coagulant activity of unfractionated heparin. This suggested that there was a specific chemical structure within some heparin chains that was the preferred ligand for ATIII. A Herculean effort by a number of research groups led to the identification of several related pentasaccharides, which are the minimal structure required for high-affinity binding to ATIII.[51] The biological activity of these pentasaccharides was confirmed by their synthesis,[52–54] which showed that they were not only able to bind to ATIII (with an affinity of 50 nM) but also were able to enhance ATIII-mediated inhibition of factor Xa. The ATIII-binding penta-saccharides contain a 3-O-sulfate group on the central glucosamine, which is especially rare in heparin and, in the absence of this group, the affinities of the pentasaccharides for ATIII are >2 mM.[55]

heparin pentasaccharide
R = Ac or SO₃⁻

A model for the role of the pentasaccharide in catalysing the inhibition of factor Xa by ATIII has been proposed.[56] In this model, binding of the pentasaccharide to ATIII results in a conformational change of ATIII, leading to exposure of a loop region. This loop region is recognized by factor Xa and allows the formation of a ternary pentasaccharide/ATIII/factor Xa complex. Finally, the pentasaccharide dissociates from the complex, where it is able to catalyse the formation of a new ATIII/factor Xa complex. Evidence for this model has been obtained through X-ray crystallographic analysis of ATIII, ATIII in complex with a synthetic pentasaccharide, and an ATIII-binding pentasaccharide/ATIII/factor Xa complex.[57,58]

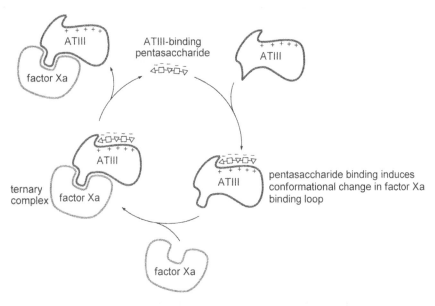

Ultimately, the methyl glycoside of the ATIII-binding pentasaccharide (termed fondaparinux) was selected as a drug development candidate and, in 2002, following clinical trials, was registered as a drug (Arixtra) in the USA and Europe for the prevention of venous thromboembolic events following knee- and hip-replacement surgery and after hip fractures.[56] The synthesis of fondaparinux is one of the most complex of any small-molecule drug synthesis and takes over 60 steps.[59] Nonetheless, multi-kilogram synthesis of the drug is performed industrially. Fondaparinux possesses a half-life of 17 hrs in healthy humans, which allows for once-daily administration, rather than two to three times daily for LMWH. Fondaparinux does not result in HIT and has been used in patients who have previously exhibited this complication with LMWHs.[60,61]

fondaparinux
(Arixtra)

As mentioned above, the synthesis of fondaparinux is complex, which has driven efforts to simplify its structure. Detailed structure-activity studies of the ATIII-binding pentasaccharide showed that all of the $NHSO_3^-$ groups of the pentasaccharide can be replaced with more synthetically tractable OSO_3^- groups and that all of the hydroxy groups can be methylated, with little effect on the activity of the resultant derivatives.[48] Additionally, the 2-*O*-sulfate on the L-iduronic acid moiety can be replaced by a 2-*O*-methyl group. Finally, sulfation of the 3-hydroxy group on the reducing-end sugar provides an enhancement of ATIII-binding ability. When all of these modifications were incorporated into a single molecule, the resulting pentasaccharide, termed idraparinux, is not only much easier to synthesize than fondaparinux (approximately 25 steps) but also possesses higher activity and a longer duration of action (half-life in humans of 120 hrs), potentially allowing once-weekly administration. Idraparinux is currently in Phase III clinical trials.

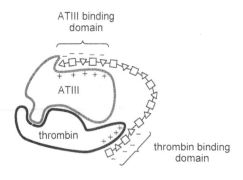

idraparinux

While the ATIII-binding pentasaccharide is able to catalyse the binding of ATIII to factor Xa, it does not mediate the binding of ATIII to thrombin. While in some cases this is desirable, in other cases anti-thrombin activity might be beneficial. For example, blood clots exhibit pro-coagulant activity that appears to arise from clot-bound factor Xa and thrombin, which leads to activation of the coagulation system and propagation of a clot.[62] Inhibition of thrombin and factor Xa may therefore provide improved anti-thrombotic activity over inhibition of factor Xa alone. A model for heparin-assisted ATIII-inhibition of thrombin was suggested on the basis of biochemical data and structures of ATIII, thrombin and ATIII bound to the ATIII-binding pentasaccharide.[56] In this model, ATIII-mediated anti-thrombin activity arises from the formation of a ternary complex between ATIII, heparin and thrombin, with the ATIII-binding pentasaccharide domain binding to ATIII and facilitating the interaction of ATIII with thrombin through a thrombin-binding domain in the heparin chain. According to this model, the thrombin-binding domain is found towards the non-reducing end of the ATIII-binding pentasaccharide, and the linker between the ATIII-binding domain and the thrombin-binding domain does not interact with either protein.

ATIII binding
domain

ATIII

thrombin

thrombin binding
domain

Consideration of the heparin/ATIII/thrombin model led to the proposal of an ingenious approach to overcome one of the most serious problems with heparin, HIT.[63] HIT is caused by an immunoallergic reaction to the immunogenic platelet factor 4–heparin complex.[50] Platelet factor 4 (PF4) appears to bind to any anionic heparin sequence, with the size for optimal PF4 binding being a 16-mer, and with at least eight consecutive negatively charged sugars being required for binding. The model described above for the heparin/ATIII/thrombin ternary complex suggests that at least 14 sugar residues are required for successful formation of the ternary complex, and biochemical studies of model compounds demonstrated that in fact at least a 15-mer was required. Thus, at first glance it might appear that it would be impossible to find an oligosaccharide that exhibits thrombin inhibition without interacting with PF4. However, the model for the pentasaccharide/ATIII/thrombin complex described above suggests that the sugars linking the ATIII-binding domain and the thrombin-binding domain do not interact with either protein. Taken together, these hypotheses led to the synthesis of a 16-mer (termed SR123781), possessing a thrombin-binding domain at the non-reducing end, a neutral linker region and an ATIII-binding pentasaccharide (derived from idraparinux) at the reducing end. Notably, this 16-mer does not possess any more than five consecutive negatively charged sugars and so does not interact with PF4. This molecule displays good anti-factor Xa activity, superior anti-thrombin activity to heparin and no interaction with PF4. X-ray crystallographic analyses of two SR123781/ATIII/thrombin ternary complexes are in remarkable agreement with the proposed model for their interaction.[64,65]

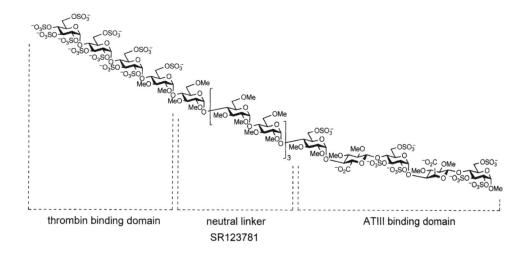

thrombin binding domain   neutral linker   ATIII binding domain

SR123781

References

1. Mader, M.M. and Bartlett, P.A. (1997). *Chem. Rev.*, **97**, 1281.
2. Wolfenden, R. and Snider, M.J. (2001). *Acc. Chem. Res.*, **34**, 938.
3. Withers, S.G., Namchuk, M. and Mosi, R. (1999). In *Iminosugars as Glycosidase Inhibitors: Nojirimycin and Beyond* (A.E. Stütz, ed.) p. 188. Wiley-VCH: Weinheim.
4. Evans, G.B. (2005). *Aust. J. Chem.*, **57**, 837.
5. Taylor, E.A. and Schramm, V.L. (2005). *Curr. Top. Med. Chem.*, **5**, 1237.
6. Schramm, V.L. (2003). *Acc. Chem. Res.*, **36**, 588.
7. Singh, V., Lee, J.E., Nunez, S., Howell, P.L. and Schramm, V.L. (2005). *Biochemistry*, **44**, 11647
8. Filichev, V.V., Brandt, M. and Pedersen, E.B. (2001). *Carbohydr. Res.*, **333**, 115.
9. Singh, V., Evans, G.B., Lenz, D.H., Mason, J.M., Clinch, K., Mee, S., Painter, G.F., Tyler, P.C., Furneaux, R.H., Lee, J.E., Howell, P.L. and Schramm, V.L. (2005). *J. Biol. Chem.*, **280**, 18265.
10. Evans, G.B., Furneaux, R.H., Tyler, P.C. and Schramm, V.L. (2003). *Org. Lett.*, **5**, 3639.
11. Miles, R.W., Tyler, P.C., Furneaux, R.H., Bagdassarian, C.K., Schramm, V.L. (1998). *Biochemistry*, **37**, 8615.
12. Evans, G.B.; Furneaux, R. H.; Lewandowicz, A., Schramm, V.L., Tyler, P.C. (2003). *J. Med. Chem.*, **46**, 5271.
13. Kicska, G.A., Tyler, P.C., Evans, G.B.,Furneaux, R.H., Shi, W., Fedorov, A., Lewandowicz, A., Cahill, S.M., Almo, S.C. and Schramm, V.L. (2002). *Biochemistry*, **41**, 14489.
14. Kicska, G.A., Long, L., Horig, H. Fairchild, C., Tyler, P.C., Furneaux, R.H., Schramm, V.L. and Kaufman, H.L. (2001). *Proc. Natl. Acad. Sci. USA*, **98**, 4593.
15. Lewandowicz, A., Tyler, P.C., Evans, G.B., Furneaux, R.H., Schramm, V.L. (2003). *J. Biol. Chem.*, **278**, 31465.
16. Korycka, A., Blonski, J.Z. and Robak, T. (2007). *Mini-Rev. Med. Chem.*, **7**, 976.
17. Taylor, E.A., Clinch, K., Kelly, P.M. Li, L., Evans, G.B., Tyler, P.C. and Schramm, V.L. (2007). *J. Am. Chem. Soc.*, **129**, 6984.
18. Evans, G.B., Furneaux, R.H., Greatrex, B., Murkin, A.S., Schramm, V.L., Tyler, P.C. (2008). *J. Med. Chem.*, **51**, 948.
19. Lee, J.E., Singh, V., Evans, G.B., Tyler, P.C., Furneaux, R.H., Cornell, K.A., Riscoe, M.K., Schramm, V.L. and Howell, P.L. (2005). *J. Biol. Chem.*, **280**, 18274.
20. Gutierrez, J.A., Luo, M., Singh, V, Li, L., Brown, R.L., Norris, G.E., Evans, G.B., Furneaux, R.H., Tyler, P.C., Painter, G.F., Lenz, D. H. and Schramm, V.L. (2007). *ACS Chem. Biol.*, **2**, 725.
21. Clark, I.A. and Schofield, L. (2000). *Parasitol. Today*, **16**, 451.
22. Schofield, L., Hewitt, M.C., Evans, K., Siomos, M.A. and Seeberger, P.H. (2002). *Nature*, **418**, 785.
23. Schofield, L., McConville, M.J., Hansen, D., Campbell, A.S., Fraser-Reid, B., Grusby, M.J. and Tachado, S.D. (1999). *Science*, **283**, 225.
24. Schofield, L. (2007). *Microbes Infect.*, **9**, 784.
25. Gerold, P., Dieckmann-Schuppert, A. and Schwarz, R.T. (1994). *J. Biol. Chem.*, **269**, 2597.
26. Berhe, S., Schofield, L., Schwarz, R.T. and Gerold, P. (1999). *Mol. Biochem. Parasitol.*, **103**, 273.
27. Schofield, L. and Hackett, F. (1993). *J. Exp. Med.*, **177**, 145.
28. Plante, O.J., Palmacci, E.R. and Seeberger, P.H. (2001). *Science*, **291**, 1523.
29. Seeberger, P.H. (2003). *Chem. Commun.*, 1115.
30. Hewitt, M.C., Snyder, D.A. and Seeberger, P.H. (2002). *J. Am. Chem. Soc.*, **124**, 13434.
31. Seeberger, P.H., Soucy, R.L., Kwon, Y.U., Snyder, D.A. and Kanemitsu, T. (2004). *Chem. Commun.*, 1706.
32. Kwon, Y.U., Soucy, R.L., Snyder, D.A. and Seeberger, P.H. (2005). *Chem. Eur. J.*, **11**, 2493.
33. Ouerfelli, O., Warren, J.D., Wilson, R.M. and Danishefsky, S.J. (2005). *Expert Rev. Vaccines*, **4**, 677.

34. Danishefsky, S.J. and Allen, J.R. (2000). *Angew. Chem. Int. Ed.*, **39**, 836.
35. Bilodeau, M.T., Park, T.K., Hu, S., Randolph, J.T., Danishefsky, S.J., Livingston, P.O. and Zhang, S. (1995). *J. Am. Chem. Soc.*, **117**, 7840.
36. Park, T.K., Kim, I.J., Hu, S., Bilodeau, M.T., Randolph, J.T., Kwon, O. and Danishefsky, S.J. (1996). *J. Am. Chem. Soc.*, **118**, 11488.
37. Galonic, D.P. and Gin, D.Y. (2007). *Nature*, **446**, 1000.
38. Kim, Y.J., Wang, P., Navarro-Villalobos, M., Rohde, B.D., Derryberry, J., Gin, D.Y. (2006). *J. Am. Chem. Soc.*, **128**, 11906.
39. Marciani, D.J., Press, J.B., Reynolds, R.C., Pathak, A.K., Pathak, V., Gundy, L.E., Farmer, J.T., Koratich, M.S. and May, R.D. (2000). *Vaccine*, **18**, 3141.
40. Slovin, S.F., Ragupathi, G., Adluri, S., Ungers, G., Terry, K., Kim, S., Spassova, M., Bornmann, W.G., Fazzari, M., Dantis, L., Olkiewicz, K., Lloyd, K.O., Livingston, P.O., Danishefsky, S.J. and Scher, H.I. (1999). *Proc. Natl. Acad. Sci. USA*, **96**, 5710.
41. Ragupathi, G., Koide, F., Livingston, P.O., Cho, Y.S., Endo, A., Wan, Q., Spassova, M.K., Keding, S.J., Allen, J., Ouerfelli, O., Wilson, R.M. and Danishefsky, S.J. (2006). *J. Am. Chem. Soc.*, **128**, 2715.
42. Ingale, S., Wolfert, M.A., Gaekwad, J., Buskas, T. and Boons, G.J. (2007). *Nat. Chem. Biol.*, **3**, 663.
43. Bundle, D.R. (2007). *Nat. Chem. Biol.*, **3**, 605.
44. Linhardt, R.J. (2003). *J. Med. Chem.*, **46**, 2551.
45. Linhardt, R.J. (1991). *Chem. Ind.*, **2**, 45.
46. Rutty, C.J. (1996). *CONNTACT*, **9**.
47. Porter, J. and Jick, H. (1977). *JAMA*, **237**, 879.
48. van Boeckel, C.A.A. and Petitou, M. (1993). *Angew. Chem. Int. Ed. Engl.*, **32**, 1671.
49. Linhardt, R.J., Loganathan, D., al-Hakim, A., Wang, H.M., Walenga, J.M., Hoppensteadt, D. and Fareed, J. (1990). *J. Med. Chem.*, **33**, 1639.
50. Baglin, T.P. (2001). *J. Clin. Pathol.*, **54**, 272.
51. Petitou, M., Casu, B. and Lindahl, U. (2003). *Biochimie*, **85**, 83.
52. Petitou, M., Duchaussoy, P., Lederman, I., Choay, J., Sinaÿ, P., Jacquinet, J.C. and Torri, G. (1986). *Carbohydr. Res.*, **147**, 221.
53. Sinaÿ, P., Jacquinet, J.C., Petitou, M., Duchaussoy, P., Lederman, I., Choay, J. and Torri, G. (1984). *Carbohydr. Res.*, **132**, C5.
54. van Boeckel, C.A., Beetz, T., Vos, J.N., de Jong, A.J.M., van Aelst, S.F., van den Bosch, R.H., Mertens, J.M.R. and van der Vlugt, F.A. (1985). *J. Carbohydr. Chem.*, **4**, 293.
55. Petitou, M., Duchaussoy, P., Lederman, I., Choay, J. and Sinaÿ, P. (1988). *Carbohydr. Res.*, **179**, 163.
56. Petitou, M. and van Boeckel, C.A.A. (2004). *Angew. Chem. Int. Ed.*, **43**, 3118.
57. Imberty, A., Lortat-Jacob, H. and Pérez, S. (2007). *Carbohydr. Res.*, **342**, 430.
58. Johnson, D.J.D., Li, W., Adams, T.E. and Huntington, J.A. (2006). *EMBO J.*, **25**, 2029.
59. Codée, J.D.C., Overkleeft, H.S., van der Marel, G.A. and van Boeckel, C.A.A. (2004). *Drug Discov. Today: Technologies*, **1**, 317.
60. Kuo, K.H. and Kovacs, M.J. (2005). *Hematology*, **10**, 271.
61. Giangrande, P.L.F. (2002). *Int. J. Clin. Prac.*, **56**, 615.
62. Hérault, J.-P., Cappelle, M., Bernat, A., Millet, L., Bono, F., Schaeffer, P. and Herbert, J.-M. (2003). *J. Thromb. Haemost.*, **1**, 1959.
63. Petitou, M., Hérault, J.-P., Bernat, A., Driguez, P.-A., Duchaussoy, P., Lormeau, J.-C. and Herbert, J.-M. (1999). *Nature*, **398**, 417.
64. Li, W., Johnson, D.J., Esmon, C.T. and Huntington, J.A. (2004). *Nat. Struct. Mol. Biol.*, **11**, 857.
65. Dementiev, A., Petitou, M., Herbert, J.-M. and Gettins, P.G.W. (2004). *Nat. Struct. Mol. Biol.*, **11**, 863.

Appendix I

Reagents for O-Protecting Group Removal

Functional Group	NaOCH₃	H⁺	H₂NCSNH₂	N₂H₄	H₂[b]	DDQ	Bu^tOK	F⁻	Zn	NBS	NH₃[c]	Ph₃P	Pd(0) or Pd(II)	Bu₃SnH	Na/NH₃	NaOH
OAc	√[d]	√		√			√				√					√
OBz	√			√			√									√
OCOCH₂Cl	√		√	√												
OPiv							√				√					
OLev	√			√			√									√
OBn					√					√					√	
OpMB		√			√	√				√					√	
OAll					√		√			√			√		√	
OTr		√			√											
OBMS		√						√		√						
OBPS		√						√								
OSiEt₃		√				√		√					√			
OSiPr^i₃		√						√								
OCH₂CCl₃		√							√							
OCH₂CH₂Si(CH₃)₃		√[e]						√								
O(CH₂)₃CHCH₂										√						
PhCHO,O^f		√			√											
4MeOC₆H₄CHO,O		√			√	√				√					√	
Me₂CO,O		√													√	
Dispiroacetals		√														

Reagent(s) for Deprotection[a]

---

[a] A blank entry indicates the protecting group is (probably) stable to the reagent(s).

[b] In the presence of a catalyst such as Pd-C.

[c] As a solution in methanol or ethanol.

[d] "√" indicates a reaction under normal conditions to generate the alcohol; the choice of solvent may be an important consideration.

[e] Trifluoroacetic acid has been reported to cleave trimethylsilylethyl glycosides (see Chapter 2).

[f] A benzylidene acetal.

Reagents for *N*-Protecting Group Removal

Functional Group	Reagent(s) for Deprotection															
	NaOCH₃	H⁺	H₂NCSNH₂	N₂H₄	H₂	DDQ	Bu^tOK	F⁻	Zn	NBS	NH₃	Ph₃P	Pd(0) or Pd(II)	Bu₃SnH	Na/NH₃	NaOH
NHCbz					√					√					√	
NHBoc		√														
NHCOCCl₃	√			√	√ᵍ						√			√ᵍ		
NHCOCF₃	√			√							√					√
NHCO₂CH₂CCl₃									√							
NHCO₂All							√						√		√	
NPhth				√												√
NTCP	√			√												√
NAc₂	√ᵍ										√					√ᵍ
NHDTPM				√								√				
N₃					√					√					√ʰ	
NBn₂					√										√	

ᵍ The product is an amide.

ʰ The product is on occasion the alkane!

Appendix II

Carbohydrate Nomenclature

By now, the reader will have gained some idea of the basic rules of carbohydrate nomenclature. Fortunately, these rules have been reformulated and are readily available in several places as the 'Nomenclature of carbohydrates':

Pure Appl. Chem., 1996, **68**, 1919.
Adv. Carbohydr. Chem. Biochem., 1997, **52**, 43.
Carbohydr. Res., 1997, **297**, 1.
J. Carbohydr. Chem., 1997, **16**, 1191.
www.chem.qmw.ac.uk/iupac/2carb/index.html

These new rules are not cast in stone, and the use of older nomenclature may have advantages in certain cases. The International Union of Pure and Applied Chemistry (IUPAC) has very recently introduced an International Chemical Identifier (InChI), which is a computer-readable chemical identifier defined by an IUPAC standard algorithm (www.iupac.org/inchi/). Such identifiers will ultimately make the job of naming complex molecules obsolete.

The Literature of Carbohydrates

Reference literature: *Chemical Abstracts* still remains the most important reference source for literature on carbohydrates. For those who have access to the hard copy, the 14th Collective Index enables searching up to 2001, utilizing the Author, General Subject, Chemical Substance or Formula Indexes.

CA Selects (Plus) (http://www.cas.org/PRINTED/caselects.html) is a fortnightly publication that provides all the relevant abstracts in a certain subject area, e.g. Carbohydrates.

SciFinder (http://www.cas.org/SCIFINDER/) *Scholar* (http://www.cas.org/SCIFINDER/SCHOLAR/) uses the CAS database to bring the chemical and scientific literature onto one's desktop. This extraordinary search engine has changed the nature of chemical research forever.

Beilstein's *Handbuch der Organischen Chemie* is an ingenious system that details individual classes of chemical compounds in a particular volume (Hauptwerk) and then updates the information with supplements; carbohydrates are to be found in volumes 17, 18, 19 and 31. *CrossFire Beilstein*, an online version in English, has essentially replaced the hard copy of Beilstein. Notably, these two resources provide physical data on the abstracted compounds.

Rodd's Chemistry of Carbon Compounds provides another source of literature on carbohydrates, located in volumes I^F (1967) and I^G (1976) and their supplements (1983, to I^{FG} and 1993, to $I^E/I^F/I^G$).

Methoden der Organischen Chemie (Houben-Weyl) is another multi-volume work that describes, in volume E14a/3, various aspects of the chemistry of carbohydrates and their derivatives. An online searchable version is available as *Science of Synthesis*.

Angewandte Chemie International Edition includes an 'Index of reviews' covering the past 45 years, and this contains some excellent articles on carbohydrates.

Primary literature: General papers on all aspects of the chemistry, biochemistry and biology of carbohydrates now appear in all primary journals, and this is simply a reflection of the increased interest in carbohydrates shown by mainstream chemists and biochemists. There are various specialist journals devoted to carbohydrates, namely *Carbohydrate Research* (1965–), the *Journal of Carbohydrate Chemistry* (1982–), *Carbohydrate Polymers* (1981–), *Glycobiology* (1990–) and *Glycoconjugate Journal* (1984–). *Carbohydrate Polymers*, in fact, regularly reports a 'Bibliography of carbohydrate polymers' that lists all the papers published on each of the major biopolymers.

Monographs and related works: *Methods in Carbohydrate Chemistry* (1962–) is an excellent series that provides discussion, references and experimental procedures for a host of transformations in carbohydrates; volume II is probably one of the most valuable works ever published for carbohydrate chemists.

Advances in Carbohydrate Chemistry (1945–1968) and *Advances in Carbohydrate Chemistry and Biochemistry* (1969–) provide a set of excellent reviews on many aspects of carbohydrate chemistry, biochemistry and biology.

Specialist Periodical Reports, Carbohydrate Chemistry (1968–2000) was an annual review of the carbohydrate literature. Although the series has been discontinued (there are rumours that it will return), the individual volumes still provide one of the most rapid ways of accessing information on a particular topic.

The Monosaccharides (1963), by Jaroslav Staněk and co-authors, is a bible for carbohydrate chemists. The four-volume treatise, *The Carbohydrates. Chemistry and Biochemistry*, edited by Ward Pigman and Derek Horton, is again a 'must' for all researchers in carbohydrates. The current edition (volume IA, 1972; volume IB, 1980; volumes IIA and IIB, 1970) is really showing its age, but all efforts by Horton to organize a new edition have met with frustration (and failure).

Carbohydrates – a Source Book (1987), edited by Peter M. Collins, is of use if one wishes to consult a 'dictionary' of common compounds that lists the relevant physical data and gives references for methods of preparation. A new (second) edition is now available, under the banner of *Dictionary of Carbohydrates* (2005), and is accompanied by a CD-ROM.

Handbook of Reagents for Organic Synthesis. Reagents for Glycoside, Nucleotide and Peptide Synthesis (2005) obviously contains sections of use to the synthetic carbohydrate chemist. This work is a follow-up to the more general but older *Encyclopedia of Reagents in Organic Synthesis* (1995).

Comprehensive Natural Products Chemistry (1999) has published a whole volume (volume 3, *Carbohydrates and their derivatives including tannins, cellulose, and related lignins*) devoted to various aspects of carbohydrates.

Ullman's Encyclopedia of Technology has useful articles on applications of carbohydrates.

Recent edited works: A popular trend in many areas of science has been the publication of works that emanate either from a conference or from the desire of one person (the editor) to present recent progress in a certain field. As such, these works contain articles by many different authors, and there are the obvious problems associated with differing styles and presentation, and the inevitable overlap of some material.

Modern Methods in Carbohydrate Synthesis (eds Khan, S.H. and O'Neill, R.A.; Harwood Academic, 1996) was the first of these modern edited works and contains many useful articles on all aspects of carbohydrate chemistry.

Preparative Carbohydrate Chemistry (ed. Hanessian, S.; Marcel Dekker Inc., 1997) again contains many chapters, some more up to date than others, and a multi-chapter section on unpublished aspects of the 'remote activation' concept. A novel aspect of the book is the inclusion of experimental details for the theme reaction(s) of each chapter.

Carbohydrate Chemistry (ed. Boons, G.-J.; Blackie, 1998) contains a wealth of information about carbohydrates and has powerful chapters dealing with the synthesis of glycosides and the chemistry of neoglycoconjugates.

Bioorganic Chemistry: Carbohydrates (ed. Hecht, S.M.; Oxford University Press, 1998) is the final of three volumes on bioorganic chemistry and is unique in that the 13 chapters are designed to fit into a one-semester course.

Carbohydrate Mimics: Concepts and Methods (ed. Chapleur, Y.; Wiley-VCH, 1998) is a compilation of works from laboratories around the world that describes the synthesis of carbohydrate mimics such as azasugars, C-linked sugars, carbasugars, aminocyclopentitols and carbocycles.

Essentials of Glycobiology (eds Varki, A., Cummings, R., Esko, J., Freeze, H., Hart, G. and Marth, J.; CSHL Press, 1999) focuses on the role of carbohydrates in biology and is probably the best text available in the area of glycobiology; a new edition is imminent.

Carbohydrates in Chemistry and Biology (eds Ernst, B., Hart, G.W. and Sinaÿ, P.) is a four-volume work published by Wiley-VCH (2000), which is in direct competition with the three-volume *Glycoscience: Chemistry and Chemical Biology* (eds Fraser-Reid, B., Tatsuta, K. and Thiem, J.; Springer, 2001). Both works cover virtually every aspect of glycochemistry and glycobiology; an update of the latter has appeared (2008).

Glycochemistry: Principles, Synthesis and Applications (eds Wang, P.G. and Bertozzi, C.R.; Marcel Dekker, 2001) followed hard on the heels of the two multi-volume works above, covering some of the same topics but offering the 'basics' in a much smaller work.

Carbohydrates (ed. Osborn, H.M.I.; Academic Press, 2003) is the fourth in the series, 'Best synthetic methods', and is very strong on the construction of the glycosidic linkage and the subsequent oligosaccharides.

Carbohydrate-Based Drug Discovery (ed. Wong, C.-H.; Wiley-VCH, 2003) is a two-volume work with obviously a slightly different emphasis, namely biological aspects and therapeutic applications of carbohydrates and some derivatives.

The Organic Chemistry of Sugars (eds Levy, D.E. and Fügedi, P.; CRC Press, 2006) covers, in one volume of almost 1000 pages, most aspects of carbohydrate chemistry and biology.

Comprehensive Glycoscience (ed. Kamerling, J.P.; Elsevier, 2007) is the latest in these edited works and covers, in four volumes, all aspects of the subject.

Recent textbooks: As well as supplying the scientific community with the latest literature and review articles, it is also necessary to provide textbooks for use by undergraduates, postgraduates and young researchers. A textbook demands a certain style of the author(s), to present a goodly part of a subject in an understandable, friendly and readable manner.

Carbohydrate Chemistry: Monosaccharides and Their Oligomers by Hassan S. El Khadem (Academic Press, 1988) was just about the first of a rush of new-wave textbooks on carbohydrates. The emphasis is on monosaccharides, with only a handful of literature references.

Modern Carbohydrate Chemistry by Roger W. Binkley (Marcel Dekker, 1988) gives a reasonable overview of monosaccharide chemistry with a more generous offering of literature references.

Monosaccharides: Their Chemistry and Their Roles in Natural Products by Peter M. Collins and Robert (Robin) J. Ferrier (Wiley, 1995) is essentially the second edition of a small book (*Monosaccharide Chemistry*, Penguin, 1972) that took the carbohydrate community by storm. The present book, written by two wise, wistful and knowledgeable carbohydrate chemists, is a mine of information, presumably gleaned from the years associated with 'Specialist periodical reports, carbohydrate chemistry'.

Carbohydrates: Structure and Biology by Jochen Lehmann was originally published in German (*Kohlenhydrate. Chemie und Biologie*, Thieme, 1996) and subsequently translated by Alan H. Haines (Thieme, 1998). The English version does not contain the chapter, 'Chemical aspects', which is in the German version. Chapter 2 of the English version, entitled 'Biological aspects', is an excellent summary of the role of carbohydrates in biology (glycobiology). This book is another 'must' for any self-respecting carbohydrate chemist and represents excellent value for money.

Essentials of Carbohydrate Chemistry by John F. Robyt (Springer, 1998) is another book with a biological emphasis; there is a heavy accent on aspects of the chemistry of sucrose throughout the book.

The Molecular and Supramolecular Chemistry of Carbohydrates by Serge David (Oxford University Press, 1998) provides an overview of the physical, chemical and biological properties of carbohydrates. With 18 chapters (and 320 pages), the book is wide ranging in its coverage.

Carbohydrate Chemistry by Benjamin G. Davis and Antony J. Fairbanks (Oxford University Press, 2002) is an affordable Oxford Chemistry Primer most suited to those about to enter the field.

Essentials of Carbohydrate Chemistry and Biochemistry by Thisbe K. Lindhorst was first published in 2000 (Wiley-VCH). A second (2003) and a third (2007) edition have appeared that include some basic references, lacking in the first edition.

Miscellaneous: *Stereochemistry of Carbohydrates* by J.F. Stoddart (Wiley-Interscience, 1971) is still one of the most authoritative publications on almost all aspects of carbohydrate constitution, configuration, conformation and isomerism.

Carbohydrate Building Blocks by Mikael Bols (Wiley, 1996) spends some 60 pages discussing the chemistry of carbohydrates and then finishes with the main thrust of the book, a compendium of 'building blocks' derived from carbohydrates for the synthesis of natural products.

Carbohydrate Synthons in Natural Products Chemistry: Synthesis, Functionalization, and Applications (eds Witczak, Z.J. and Tatsuta, K.; American Chemical Society/Oxford University Press, 2003) follows on very much from the Bols book.

Monosaccharide Sugars: Chemical Synthesis by Chain Elongation, Degradation, and Epimerization by Zoltán Györgydeák and István F. Pelyvás (Academic Press, 1998) describes a myriad of reactions for the elongation, degradation and isomerization of monosaccharides. A highlight of the book is the inclusion of detailed experimental descriptions of many of the transformations discussed.

Epilogue

Life goes on at a frenetic pace, and chemistry, of particular interest carbohydrate chemistry, is swept up in it. With masses of information available at the press of a key, we forge ahead, often looking neither left nor right, achieving great things that we convey to an often despairing audience in some convention or symposium. There is very little time for reflection before one has to move onto the next grant application, to feed the hungry co-workers, and to keep the great ball of achievement rolling.

This was not always the case. Many of us experienced mentors who were aware of all of the chemical literature in their chosen field, who puzzled over a result and who questioned something that did not seem to fit in with past observations. So were discovered new phenomena, new chemical reactions or perhaps the key to unravelling a novel structure. Fortunately, some of those people are still with us today; if not, they have left us a legacy in print.

Presented here are two very different articles, one from Stephen Angyal (University of New South Wales, Australia) and the other, in fact, three related pieces, from William (Bill) Whelan (University of Miami, Florida, USA). Angyal's article is beautiful in its simplicity, describing the preparation of an unnatural monosaccharide, L-ribose, by metal-ion isomerization of L-arabinose. Whelan's three pieces (under the banner of 'The Wars of the Carbohydrates'), dealing with common public misconceptions concerning high fructose corn syrup, fructose and maltose, are beautifully written and succinct. These are a 'must read' for any budding postgraduate student.

Short Communication

Aust. J. Chem. **2005**, *58*, 58–59

CSIRO PUBLISHING

www.publish.csiro.au/journals/ajc

L-Ribose: an Easily Prepared Rare Sugar

Stephen J. Angyal[A]

[A] School of Chemistry, University of New South Wales, Sydney NSW 2052, Australia.
Email: s.angyal@unsw.edu.au

An old method has been updated to provide an easy one-step process for the synthesis of L-ribose.

Manuscript received: 6 September 2004.
Final version: 13 November 2004.

L-Ribose (Fig. 1) does not occur in nature and is very expensive to purchase. Recently, it has been much in demand as the starting material for the synthesis of nucleoside enantiomers and other biologically important compounds. In the last few years, six papers[1–6] have appeared on the synthesis of L-ribose or its derivatives (esters, acetals) from a variety of starting materials, such as D-glucose, D-ribose, D-mannonolactone, D-galactose, and L-arabinose. They all employ sequences of reactions that require some expensive reagents and several organic solvents. Enzymic conversions of other sugars have also been explored for the preparation of L-ribose.[7]

However, it should be noted that in 1973, Bílik and Čaplovič[8] published a simple synthesis of L-ribose by molybdate-catalyzed epimerization of the readily available L-arabinose—a reaction now known as the Bílik reaction. The synthesis is one step, and except for a catalyic amount of molybdic acid does not require any expensive reagents, while the solvents used are water and a small amount of methanol. The process can be carried out in 2–3 days. Although the yield of L-ribose is only 20%, this being in equilibrium with arabinose, most of the unreacted arabinose is recoverable and can be used again.

Interestingly, five of the papers on the synthesis of L-ribose do not quote Bílik's work at all. Only Pitsch[1] mentions it, but at the same time dismisses it because of the low yield and the cumbersome purification step. The sugars were originally separated by chromatography on a cation-exchange column in the barium form. Bilik himself was not satisfied with this method and offered another, longer alternative via the *N*-phenyl-L-ribosamine intermediate. However, cation-exchange chromatography has greatly improved since

then. Barium forms only weak complexes with sugars, but the lanthanide metals form much stronger complexes. We found that a neodymium column[9] is preferable to the barium column as the column can be much smaller and the flow rate faster. Bílik's chromatography lasted 100 h, while we completed a run half this size in one working day. He used a column of 120 cm length, whereas ours was only 29 cm and the solvent used was water. Bílik treated his reaction mixture with charcoal and then removed the molybdic acid. We found that neither procedure was necessary because molybdic acid and any coloured impurities come off the chromatographic column before the sugars. Moreover, ribose undergoes complexation much better than the other sugars present (arabinose, also lyxose and xylose, which are by-products of the reaction) and emerges from the column last as a pure compound. The column can then be re-used without any need for regeneration.

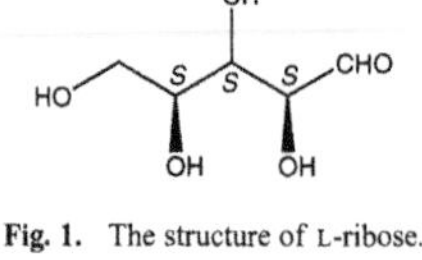

Without doubt, this is the simplest, quickest, and cheapest way to make L-ribose.

Experimental

Preparation of the Neodymium Column

Dowex AG 50W-X2(H⁺) resin (200–400 mesh, 300 mL) was gently stirred with water in a beaker. After the mixture settled, the water was decanted and the process was repeated twice in order to remove fine particles. A solution of neodymium chloride hexahydrate (5 g) in water was added to the resin, the resultant mixture was gently stirred, and was then poured into the glass column (4.3 × 29 cm) and allowed to settle. The column was washed with water until the effluent was no longer acidic.

Preparation of L-Ribose

L-Arabinose (25.0 g) and molybdenic acid ($MoO_3 \cdot H_2O$, Univar; 0.25 g) were heated in water (100 mL) at 90°C for 9 h. The clear and only slightly coloured solution was evaporated, and the residue was dissolved in methanol (50 mL) and left at 5°C overnight. As arabinose crystallized out, it adhered to the glass surface and needed to be broken up and stirred in order to increase its recovery. Filtration gave 16.0 g of arabinose (64% recovery). The mother liquor was then concentrated by evaporation (the presence of some methanol assists the chromatographic separaton), water (20 mL) was added, and the solution was poured into the chromatographic column.

Fig. 1. The structure of L-ribose.

The column was eluted at the rate of $150\,mL\,h^{-1}$ with $50\,mL$ fractions being collected. The contents of each fraction were tested by TLC on cation-exchange thin-layer plates (Polygram Ionex-25 SA, Macherey–Nagel) in the calcium form (R_F for ribose was 0.55, while for the other pentoses was 0.65–0.75). The other sugars appeared in the early fractions, while L-ribose emerged after a distinct gap at between 700 and $1500\,mL$. Evaporation of these fractions gave $4.8\,g$ of L-ribose (19%, 53% calculated on arabinose not recovered). This ribose was determined to be pure from its ^1H NMR spectrum, which was identical to that of crystalline D-ribose and did not contain signals as a result of any other substances.

Ribose is not readily crystallized and evaporation of its solution gives a syrup that tends to remain syrupy for weeks even in the presence of seed crystals.[10] The recommended method for crystallization[8,10] from absolute ethanol is not satisfactory as much of the sugar remains in the mother liquor. As a result, if the ribose is only an intermediate in the preparation of other substances, it is not worth trying to crystallize, and the syrup should be used for any subsequent reactions.

References

[1] S. Pitsch, *Helv. Chim. Acta* **1997**, *80*, 2286.

[2] G. G. Sivets, T. V. Klennitskaya, E. V. Zhernosek, I. A. Mikhailopulo, *Synthesis* **2002**, 253. doi:10.1055/S-2002-19805

[3] H. Takasashi, Y. Iwai, Y. Hitomi, S. Ikegami, *Org. Lett.* **2002**, 2401. doi:10.1021/OL026141I

[4] M. E. Jung, Y. Xu, *Tetrahedron Lett.* **1997**, *38*, 4199. doi:10.1016/S0040-4039(97)00870-8

[5] Z.-D. Shi, B.-H. Yang, Y.-L. Wu, *Tetrahedron Lett.* **2001**, *42*, 7651. doi:10.1016/S0040-4039(01)01623-9

[6] M. J. Seo, J. An, J. H. Shim, G. Kim, *Tetrahedron Lett.* **2003**, 3051. doi:10.1016/S0040-4039(03)00552-5

[7] (*a*) Z. Ahmed, K. Izumori, *Bangladesh J. Sci. Ind. Res.* **2000**, *35*, 90.
    (*b*) Z. Ahmed, T. Shimonishi, S. H. Bhuiyan, T. Saiki, M. Utamuro, M. Takada, K. Izumori, *J. Biosci. Bioeng.* **1999**, *88*, 444. doi:10.1016/S1389-1723(99)80225-4

[8] V. Bílik, J. Čaplovič, *Chem. Zvesti* **1973**, *27*, 547. *Czech Pat.* *149,472* **1973**; *Chem. Abstr.* **1974**, *81*, 487.

[9] S. J. Angyal, D. C. Craig, *Carbohydr. Res.* **1993**, *241*, 1. doi:10.1016/0008-6215(93)80089-W

[10] R. L. Whistler, M. L. Wolfrom, *Meth. Carbohydr. Chem.* **1962**, *1*, 86.

IUBMB *Life*, 56(5): 287–288, May 2004

Website Horrors

The Wars of the Carbohydrates : Part 1

William J. Whelan
e-mail: wwhelan@miami.edu

Two sugars are currently the topics of heated debate in the United States in terms of their alleged undesirability in foodstuffs. These are fructose and maltose.

HIGH FRUCTOSE CORN SYRUP (HFCS)

Xylose isomerase (EC 5.3.1.5) interconverts xylose and xylulose. In the 1960s a form of the enzyme was discovered that would interconvert glucose and fructose such that, starting from glucose, there was formed a mixture of the two sugars that was not far from the 50/50 composition of invert sugar, traditionally made by hydrolyzing sucrose and used in the food industry instead of sucrose because it does not crystallize and give a brittle texture to foodstuffs, where this would be undesirable. The US corn industry was quick to seize on this discovery and by the mid-1970s had developed an efficient enzymic conversion process whereby corn starch is hydrolyzed to glucose and then isomerized. The name 'high fructose corn syrup' is applied to this product. The percentage fructose at equilibrium is about 42, but can be increased to 55% and products corresponding to both extremes are on the market.

HFCS is competitive in price with sucrose and has been an outstanding success. By 1985 it accounted for a 150 lb per capita consumption in the US, all gained at the expense of sucrose. It has become the second most abundant product made from starch, after fuel alcohol. The Corn Annual 2003, a publication of the Corn Refiners Association (CRA), shows that in the 15 years between 1987 and 2002 the US sale of HFCS has increased from the equivalent of 358 million bushels of corn to 545 million. The increase in US per capita consumption of caloric sweeteners in that time (23.1 lbs) is mostly represented by the increase in the consumption of HFCS (16.9 lbs) which now ranks equally with sucrose in terms of the percentage of total caloric sweeteners used (42%). Undoubtedly, the breakthrough in the use of HFCS came about when it was accepted by cola manufacturers as a replacement for the sucrose traditionally used. Perhaps it has totally displaced sucrose in the US cola market. One wonders

in fact how much of the sucrose used in colas (sucrose is still used in Europe), in fact reaches the consumer as sucrose, because of the high acidity of cola, which is laced with phosphoric acid.

One of the attractive things about HFCS both to the manufacturers and, in turn, the consumers is its relative sweetness. Fructose itself, relative to sucrose (100), is 170. All this is tempting in the potential to be able to sweeten food and drinks and use the slogan – 'Same sweetness, lower calories'. In these days, that would get everybody's attention.

Latterly, with the realization of the alarming increase in obesity of the American population, HFCS has come under fire as an alleged leading cause of obesity. Certainly, graphs showing the plot of increased HFCS consumption against increase in obesity are rather persuasive. Since the word got around that I am supposed to know something on the subject, I have been called by newspaper and magazine reporters wanting more information with respect to the aforesaid correlation. I pointed out that someone once plotted the salaries of Presbyterian ministers in Philadelphia over the years against the price of rum in Jamaica and obtained an equally persuasive correlation. I followed this up by pointing out that HFCS is not on the market in Europe and yet Western Europe has been seeing the same sort of increase in obesity. My contention is that it is all a matter of calories. The real correlation with obesity is the total caloric consumption, whatever the composition of the food. The fact that HFCS is not sold in Western Europe is not because of any concern about its nutritive properties. Rather it is the knowledge that HFCS could put the European sugar beet industry out of business, in the same way that its sale in the USA has devastated the domestic sugar industry. High tariffs discourage sales of HFCS in Europe. It is also a political plus for HFCS that it has damaged the economy of Cuba because of Cuba's former heavy dependence on sucrose exports in order to obtain hard foreign currency.

The HFCS manufacturers have not been slow to defend their products and have established a website at HFCSfacts.com The general tenor of the 'facts' on the website was established under the heading:

Received 3 May 2004; accepted 3 May 2004

Address correspondence to: e-mail: wwhelan@miami.edu

ISSN 1521-6543 print/ISSN 1521-6551 online © 2004 IUBMB
DOI: 10.1080/15216540412331270481

'The Truth About High Fructose Corn Syrup', where it was stated that:

'There are no published studies regarding the effects of HFCS on insulin production'.

I invited the CRA to key 'HFCS Insulin' into PubMed. Shortly thereafter, this claim disappeared.

The website also included another since-removed article by Sandy Szwarc entitled 'Trick or Treat?' The CRA drew attention to this by writing:

'Finally a report that gets it right'.

I felt provoked to write to the CRA to comment that their use of the word 'right' could not be viewed as in the context of 'accurate' but that it was 'right' in the sense of agreeing with their opinion. The misstatements included the following:

'Fructose is in such food as Jerusalem artichokes'. I pointed out that the fructose in Jerusalem artichokes is not food, at least for humans. Fructose occurs there in the form of a polymer not digestible by humans. But then Szwarc went to write that 'fructose is found in such foods as the starch of corn kernels'. How this chemical monstrosity got onto the website of the supplier of US corn starch is perplexing. But the same inference is also to be found at www.chemistry.eku.edu/ BROCK/Causes where we read that 'a hybrid corn has been developed with high fructose in its syrup.'

In further defense of fructose we read that 'the body uses all types (of sugars) in the same way', something which is certainly not the case with fructose (*1*) and indeed is at the basis of the concerns that some people have about this sugar. And then finally:

'All carbohydrates are broken down to simple sugars before being absorbed by the body'.

This highly misleading statement should in fact read 'All *digestible* carbohydrates…'. Given the awareness of such dietary regimes as the Atkins' Diet and the South Beach Diet one would have thought that most people would know by now what is meant by 'net carbs' (digestible carbohydrate) vs. total carbs (digestible plus indigestible). There seems to be little

realization that many of the carbohydrate polymers that we consume are indigestible. Even the American Diabetes Association was putting out this same misleading statement until I protested to them.

Reading the articles on the CRA website one begins to realize that they have a defensive posture. They point out repeatedly that, from a nutritional point of view, there is essentially no difference between HFCS and sucrose, something that I would find no difficulty in agreeing with. But the defenders of HFCS are having to contest a contention that fructose in HFCS is metabolized differently than fructose in sucrose. I know of no evidence for this and can think of no reason why this should be the case, assuming that fructose enters the circulation directly from HFCS and from sucrose after having been hydrolyzed by intestinal sucrase.

What is making the HFCS defenders nervous is the abundant evidence that consumption of fructose alone can certainly give rise to problems. These are occasioned by the uncontrolled rate at which the liver seizes on dietary fructose and phosphorylates it. Hypoglycemia is one of the consequences, a seeming paradox that the level of circulating blood glucose may fall as a result of consuming another sugar.

Thus, the CRA website states that:

'It is important that experimental studies concerned with health effects of HFCS incorporate both sugars into the test regimen in equal proportions. Few reports purporting to link HFCS with health effects have done so; most seek to show an effect for (sic) exaggerated levels of fructose in isolation from glucose'. Another section goes on to state that 'humans rarely consume fructose as a sole source of sugar'.

Pure fructose is fortunately too expensive to be offered on a large scale as the ultimate in sweetness with low calories, but not too expensive for it to be included in a new brand of bottled water, a high-profit-margin commodity.

The saving grace about HFCS, if one can express it in those terms, is that it contains glucose, so that the hypoglycemic effect of fructose is automatically combated.

And the second sugar involved in the wars of the carbohydrates - maltose? Watch for the next installment.

REFERENCE

1. Van Schaftingen, E. (2002) *IUBMB Life* **53**, 319–320.

IUBMB *Life*, 56(9): 571, September 2004

New York Times Horror

The Wars of the Carbohydrates: Part 2

William J. Whelan
e-mail: wwhelan@miami.edu

I promised that my next missive on carbohydrates would focus on maltose. And it will. But I could not resist inserting this piece, as a follow-up to Part 1, on fructose (*1*).

Writing in the New York Times for 7 September, 2004, a Thomas G. Burns used the topical subject of obesity and fructose consumption to launch an attack on US tariffs. Here are two paragraphs from his letter:

'Although excess consumption of any sugar can lead to obesity, cane and beet sugar (glucose) and corn syrup (fructose) do not have equal effects on our digestive systems. Ingesting large amounts of fructose has been shown in animal studies to cause insulin resistance, impair glucose tolerance, increase insulin levels, raise triglycerides and contribute to high blood pressure.'

A large human study has confirmed the link between fructose and Type 2 diabetes.

I was quick to put pen to paper (figuratively) and wrote the following to the Times. Since they did not publish it, I take the opportunity to add it to this series.

Under the heading "Sugar is Sweet... and Unhealthy", Thomas G. Burns stated that: 'cane and beet sugar (glucose) and corn syrup (fructose) do not have equal effects on our digestive systems'.

He goes on to list several deleterious effects of "ingesting large amounts of fructose". The reader will be forgiven for concluding that corn syrup is bad for you in a way that cane and beet sugar are not.

'This is voodoo chemistry. Cane and beet sugar are not "glucose". They are sucrose, a molecule containing one part each of glucose and fructose that most of us are able to absorb from the gut after splitting sucrose into the two parts. And it is corn syrup that is glucose, not 'fructose'. Your correspondent has confused corn syrup with high fructose corn syrup (HFCS), manufactured from corn syrup. HFCS is a mixture of approximately equal parts of glucose and fructose, a kind of predigested sucrose. As articles of diet, there can be no difference between sucrose and HFCS.

Consuming excessive amounts of any of these products is, nutritionally, bad news for the body. But Mr. Burns' specious premise should not be used as a stick with which to beat the sugar quota system.

REFERENCE

1. Whelan, W. J. (2004) The Wars of the Carbohydrates: Part 1. *IUBMB Life* **56**, 287–288.

Received 27 September 2004; accepted 27 September 2004
Address correspondence to: e-mail: wwhelan@miami.edu

ISSN 1521-6543 print/ISSN 1521-6551 online © 2004 IUBMB
DOI: 10.1080/15216540400013929

IUBMB *Life*, 56(10): 641, October 2004

Taylor & Francis
healthsciences

New York Times Horror

The Wars of the Carbohydrates: Part 3: Maltose

William J. Whelan
wwhelan@miami.edu

In 'The South Beach Diet' (St. Martin's Press, 2003; 83 weeks on the New York Times best-seller list) Arthur Agatston, MD, a cardiologist from nearby Mount Sinai Medical Center, Miami Beach, alarmed the major US brewer by criticizing the consumption of beer because of its alleged content of maltose and the fact that maltose has a high glycemic index.

The glycemic index (GI) is explained on the South Beach Diet website. It is a measure of the rise in the level of glucose that occurs in the bloodstream after food is ingested. The arrival of glucose causes the release of insulin, which in turn causes the body to store excess sugar as fat, inhibit the burning of previously stored fat and signals the liver to make cholesterol.

Agatston and others advocate the consumption of foods with a low glycemic index, the basis of the criticism of maltose.

The GI figures that Agatston quotes are fructose 32, glucose 137 (relative to white bread at 100) and maltose 150.

Evidently, Agatston's message got around because on 23 April 2004 Anheuser-Busch, the aforesaid brewing company, took out full-page advertisements in 31 newspapers, including The New York Times. The message read:

'Have a Beer
With Your South Beach Diet.
There's No Maltose to Worry About.'

Agatston, when challenged, conceded readily and with good grace that Anheuser-Busch was generally right about maltose. There is maltose in beer but most of it is fermented out. He said he hired a graduate student to look into the question.

I decided to do the same with an undergraduate student and an assistant working in my laboratory for the summer, Tiffany Biason and Mari Park. Together we measured the total carbohydrate content of 8 beers and the amount of glucose released by a mixture of alpha-amylase and glucoamylase, identical to the enzymes that we use in our digestive system to bring about the conversion of starch into glucose. The measurements provide, on the one hand, 'total carbs' and on the other the 'net carbs' that give rise to the metabolizable glucose that determines the glycemic index.

We also carried out paper chromatography of the beers and confirmed what generally had been agreed both by Agatston and Anheuser-Busch, that there was essentially no maltose to be seen. But we could see very clearly larger molecules that corresponded to the net carbs we had measured. These are collectively termed maltodextrins.

In the beer-making process, starch is acted on by a mixture of amylases to form glucose, maltose, maltotriose, and higher saccharides containing both alpha-1, 4- and alpha-1 6-glucosidic linkages. Yeast readily ferments glucose and maltose to ethanol, but has a harder time with the higher saccharides, the maltodextrins.

What is the significance of all this? The significance is that the glycemic index of maltodextrins is exactly the same as that of maltose. As long as there are significant amounts of maltodextrins in the beer, the claim by Anheuser-Busch that there is no maltose in beer is irrelevant. If Agatston's claim that maltose gives you a beer belly is correct, then so should the maltodextrins. I have not seen this aspect mentioned in the debate about maltose.

It seems to me that Anheuser-Busch, by proclaiming that 'there is no maltose to worry about', while ignoring that beer contains something, in terms of what is being argued about, that may be just as bad as maltose, has succeeded in grasping the tar baby of American folklore.

And, when a Washington University School of Medicine nutritionist is quoted on Realbeer.com as saying that 'the carbohydrates in beer are not sugar. Basically all the sugar is converted to alcohol during fermentation', he is also begging the question. Whatever one may choose to call the net carbohydrates in beer, they undergo exactly the same metabolic fate as does maltose, when we digest them, and they have the same glycemic index.

For the record, the net carbs determined by Ms. Biason in Budweiser beer corresponded to 2.5% wt/vol and about half that amount in Bud Light. The lowest quantities of net carbs that she found were in Michelob Ultra Beer, at 0.14% and Miller Lite at 0.11%, which border on the insignificant.

Received 27 September 2004; accepted 30 September 2004
Address correspondence to: e-mail: wwhelan@miami.edu

ISSN 1521-6543 print/ISSN 1521-6551 online © 2004 IUBMB
DOI: 10.1080/15216540400022458

Index

Printed and bound by CPI Group (UK) Ltd, Croydon, CR0 4YY

03/10/2024

01040329-0013